# A History of Energy

## Northern Europe from Stone Age to the Present Day

***Other books by the author:***

*Life-cycle Analysis of Energy Systems: From Methodology to Applications*, 2011.

*Hydrogen and Fuel Cells, 2nd edition*, 2011 (previous edition 2005).

*Renewable Energy Reference Book Set* (ed., 4 volumes of reprints), 2010

*Renewable Energy – Physics, Engineering, Environmental Impacts, Economics & Planning*, 4th edition, 2010 (previous editions 1979, 2000 and 2004)

*Renewable Energy Focus Handbook* (with Breeze, Storvick, Yang, Rosa, Gupta, Doble, Maegaard, Pistoia and Kalogirou), 2009

*Renewable Energy Conversion, Transmission and Storage*, 2007

*Life-cycle analysis of energy systems* (with Kuemmel and Nielsen), 1997

*Blegdamsvej 17*, 1989

*Superstrenge*, 1987

*Fred og frihed*, 1985

*Fundamentals of Energy Storage* (with Jensen), 1984

*Energi for fremtiden* (with Hvelplund, Illum, Jensen, Meyer and Nørgård), 1983

*Energikriser og Udviklingsperspektiver* (with Danielsen), 1983

*Skitse til alternativ energiplan for Danmark* (with Blegaa, Hvelplund, Jensen, Josephsen, Linderoth, Meyer and Balling), 1976

*Music across times and fences*, a book project in progress

Further information on the author's work at http://energy.ruc.dk

# A History of Energy

## Northern Europe from Stone Age to the Present Day

Bent Sørensen

Roskilde University
Institute of Environmental, Social and Spatial Change
Universitetsvej 1, P.O.Box 260
DK-4000 Roskilde, Denmark

earthscan
from Routledge

First published 2012 by
Earthscan, 2 Park Square, Milton Park, Abingdon, Oxon OX14 4RN

Simultaneously published in the USA and Canada by
Earthscan, 605 Third Avenue, New York, NY 10017

*Earthscan is an imprint of the Taylor & Francis Group, an informa business*



This publication has received support from the Danish Government Agency of Research and Innovation, through funds from the Danish State Lottery Pool.



British Library Cataloguing in Publication Data
A catalogue record for this book is available from the British Library.

Library of Congress Cataloging in Publication Data
Sørensen, Bent, 1941-
A history of energy : Northern Europe from the Stone Age to the present day / Bent Sørensen.
p. cm.
Includes bibliographical references and index.
TJ163.25.E853S68 2012
333.790948--dc23
2011029882

ISBN: 978-1-84971-384-9 (hbk)
ISBN: 978-1-84971-385-6 (pbk)
ISBN: 978-0-203-13674-4 (ebk)

Typeset by the author, using spelling, hyphenation and word capitalization style specified by publisher.

## Table of contents

# Preface

This book discusses temporal variations in quality and quantity of energy provision and use, as well as the interplay between energy-consuming activities and other events in the human environment and the human societies. The time horizon of the inquiry is from the earliest knowledge of human energy-related activities and through the subsequent stages of human exploitation of our planet, to the present epoch. The book constitutes a first venture into this territory, and in particular the energy history of Denmark (with excursions to neighbouring regions) will be used as a case story. The aim is to lay the foundation for energy history as a new discipline, or at least sub-field in historical sciences, in the hope that the methodology will be subjected to further development and find uses in studies of other regions in the world.

The field is new because energy questions have until now only enjoyed a sporadic mention in historical analysis, never a comprehensive treatment as an important determinant in the development of human activities and social constructs. The closest analogy to energy history is probably the newly coined field of "environmental history", so far explored mainly by short case studies (Hughes, 2001; Hornborg *et al.*, 2007), or agricultural history (Mazoyer and Roudart, 2006). One relevant precursor is the book "Energy in world history" (Smil, 1994), which after a brief sketch of human history proceeds to dwell on technology history. The history of technology has for a long time enjoyed an important place in historical literature, primarily thanks to the encyclopaedic publication of Singer, Holmyard and Hall (1954 and following years). Other similar work of more recent publication date include König (1992). In a way, the tradition of focussing on the development of technology goes back to late-medieval/early Renaissance work on mining and related techniques by Agricola (1531), and the work on windmills by Ramelli (1588).

The energy-historical focus on technology is somewhat similar to the traditional historical focus on royalty and wars. I follow the alternative trend that has emerged in the science of historical studies, to assess the issues of basic importance to the average member of a given human society by posing questions such as: how does environmental setting influence lifestyle and conditions for survival, and how do skills and technology change the outlook for development of a given society? Data on historical life conditions can be found in the interdisciplinary crossings of archaeology, geophysics, traditional history, environmental and life sciences. Their importance for human settlements is explored by studies in anthropology. The classical history toolbox of written documents, plus finds of tools and other artefacts, can be extended into periods of insufficient primary data by estimating the constraints posed by the patterns of flora and fauna, by the elements of climate and associated requirements for energy in terms of diet, heat and other energy forms needed first to ensure basic survival (food, shelter) and then to support all the further activities developed when societies become more sophisticated.

My general writing approach has been to sketch enough of the pattern of life and its environmental prerequisites in each period to be able to estimate and discuss energy use. A more complete history of events, developments and conditions for the lives of individual citizens must be sought elsewhere, but without some overview the energy-history would not emerge in an understandable context. The inspiration has worked in both ways: conventional political and social history has provided the entrance to focussing on energy use, and on the other hand, the energy angle has often inspired me to look at other aspects of history in a new way, often taking me to analyse (and in some cases arrival at new conclusions) in areas not having any evident connection to the energy investigation that prompted the wandering of thoughts.

The coverage of more than 100 000 years entails the need to be careful in selecting issues felt to be important in the development, and leaving out other ones not believed to be decisive or interesting enough to need inclusion in a mere 500-page account. Apart from inviting a degree of subjective choice, I have tried to let the energy angle determine this requirement and sharpen the attention to what is relevant in the long range. However, history is also full of entertaining events not qualified as cornerstones in the long term, and I have purposely included a selection of such narratives, because they help make periods in history come alive. In any case, I do not claim completeness in coverage.

Denmark is interesting in an energy context: in recent times because of its unique early development of renewable energy solutions, such as wind power, on a more than marginal scale; in recent-millennia history because of its international orientation and extensive trade–sometimes with a sword in one hand; and in the distant past because of its considerable stock of data bearing witness to a continued struggle with changing climate, including the massive impact upon plant and animal life caused by temperature excursions and intermittent invasion of ice covering large parts of the country.

Finally, I would like to state clearly that an account like this one necessarily will involve interpretations and assessments of a kind that will to an extent reflect the values system of the author. Objectivity is an important goal to exploit to the fullest extent, but history writing can never escape subjective traits imbedded in the social fabric prevailing at a given moment in time and the author's position relative to the range of contemporary ideas, which is of course why history writing has been changing from period to period. I hope to have made it clear when I am making comments of a normative nature, mainly to be found in the concluding chapter. My value basis has three cornerstones: the individual human rights as expressed by the United Nations declaration; a belief that humans should be careful with major manipulations of nature, as our knowledge of consequences is incomplete and likely to remain so; and finally, the belief that the role of institutions trying to control the behaviour of individual citizens (by political, religious or other kinds of coercion) should remain limited. Solidarity in maintaining conduct based on such norms is in my opinion necessary for ensuring long-terms sustainability of a large population on our finite-sized planet.

The project has been supported by the Danish Government Agency of Research and Innovation by funds from the State Lottery Pool, and I would like to express my warm thanks to the many people and institutions helping me over the several years spent preparing and writing this book. In particular I would like to acknowledge the many large and small museums and individual photographers, who have put their artwork at my disposal for pictures in the book, and particularly the Danish National Museum, the Danish Royal Library, the Copenhagen Mail & Tele Museum and the Danish State Archives. Care has been taken not to misrepresent illustrations, but in a few cases a slight retouching has been necessary in order to meet the printing contrast requirements or avoid rasterization from low-quality printed sources. Tinting of black-and-white photos is occasionally done in the way generally common when the pictures were taken some 100 years ago. Finally, I want to express my deep appreciation of the service provided by the Roskilde University Library, both through their extensive collections and by assistance in borrowing hard-to-find books or articles on historical subjects from other libraries around the world.

*Bent Sørensen, Gilleleje, March 2011*

*References:*

Agricola, G., 1531. *De re metallica* (English translation by H. & L. Hoover, 1950), Dover Publications, New York.
Hornborg, A., McNeill, J., Martinez-Alier, J. (eds.), 2007. *Rethinking environmental history.* Altamira Press, Lanhan.
Hughes, J., 2001. *An environmental history of the world.* Routledge, London.
König, W. (ed.), 1992. *Propyläen Technikgeschichte* (5 volumes in German). Propyläen Verlag, Berlin.
Mazoyer, M., Roudart, L., 2006. *A history of World agriculture.* Earthscan, London.
Ramelli, A., 1588. *Le diverse et artifisiose machine.* Published in Paris by author; facsimile reprint 1991 by Edizione il Polifilo, Milan.
Singer, C., Holmyard, E., Hall, A. (eds.), 1954. *A history of technology* (Vol. I; subsequent volumes and updates with additional editors). Clarendon Press, Oxford.
Smil, V., 1994. *Energy in world history.* Westview Press, Boulder, Colorado.

# 1

# Introduction: Climate background and Eem visitors

## 1.1 Setting the stage

One sometimes sees history divided into the period with written sources, the period for which cultural objects have been found, and the period with no such tangible evidence. The first two periods are sometimes called "the historical period" and "the archaeological period".

I do not think these distinctions are particularly useful. The period with no written or archaeological finds still allows a number of conclusions to be reached (e.g. regarding conditions for life and diversity). Methods may involve climate reconstructions, geological models, genetic development models and simulation of resource competition and niche occupancies. Some of these models may even allow falsification by suitable observations in nature or experiments.

In the period with archaeological findings, an approach needs to be adopted that prevents the actual finds from overshadowing other (and perhaps more important) factors. For example, there could be other types of artefacts, which have not survived or physically could not have survived until present society, and these artefacts might have delivered a substantially different picture of the society in which they functioned. We are preoccupied with the stone tools of the palaeolithic period, because of their abundance, but talk little of what clothes and shelter were used during time periods too far away to have delivered evidence of such items. For high-latitude ice age human survival, these are key factors, and our research becomes enriched once we begin to study scenarios for survival in the winter of Saalian France at –20 °C with different types of footwear. Finding the boots is not a condition for insight, and the colleague who tells you that the Neanderthals could not have worn tight footwear because the sewing needle had not been invented may have to yield.

The challenges do not become smaller as one passes from the period of museum artefacts to the period of prolific writing. Written documents are of course perfect tools for attempting to manipulate history. Kings and emperors, religious institutions and political figures will do anything to improve their image or that of their cause in history, and other historical agents are keen to follow suit. Historical research is a continuous struggle to see through such attempts at history falsification and get to the bottom of the matter. The volume of written documentation rarely tells which things are important and which are not. Today, nonsense documents are sometimes produced with the sole purpose of hiding more important issues, and even if there is not any conspiracy involved, written junk information still surpasses the true information by orders of magnitude.

Thus my overall suggestion is that all of the historical periods, with written, archaeological or no tangible sources, have large margins of uncertainty and an immense need for somebody to sort true from false by a maximum of unbiased scientific analysis and synthesis.

In this work, I have used Denmark as a case study for the energy history of a region. "Denmark" should be taken in a flexible sense, as the notion of nations only exists in a relatively short and recent period. Even during this period, Denmark has, like most other European nations, not had a constant size. Most European countries have, at some point in time, been a dominant force on the continent, and have subsequently lost that position. During most of its existence, Denmark comprised the Jutland peninsula and most of the surrounding islands, the provinces of Skåne, Halland and Blekinge (conquered by Sweden a little over 300 years ago), and the provinces of Schleswig and Holstein (which have been handed back and forth between Denmark and various German states, in connection with wars and for the most recent border adjustment by popular referendum). In the context of this book, former colonies such as Greenland (now making the transition from being part of the Danish state over extensive home rule to independent status) are not included, and nor are the Norwegian remnants (such as Iceland) adopted by Denmark when the union between Denmark and Norway was broken by the Swedish appropriation of Norway proper.

The alternative of defining "Denmark" as where the Danes live would be equally difficult, because nearly all populations in Europe have been through several migration phases, sometimes moving from one end of the continent to the opposite one. At the modern origin some 4000–5000 years ago, the people living in the south of Jutland are referred to as Germanian. They presumably descended from the Stone Age people living in this area at earlier times, and these had again come from regions further south in Europe. They subsequently spread both north and south, to the rest of Denmark and Scandinavia and to the North of current Germany. Neither culture nor language allows a precise delimitation of what is Danish. Just over 2000 years ago, another wave of Danes, the Cimbrians and the Teutons, marched south and tried to settle near current Belgrade. The next waves were those of Angles, Jutes and Vikings (also coming from other parts of Scandinavia), settling between 400 and 1100 in England and France, and establishing strongholds from Ireland to Portugal, Spain and Italy (Kinder and Hilgemann, 1964). In this way, through many earlier and subsequent movements in south and central Europe, a beautifully multicultural Europe was formed and readied to serve as a basis for current cooperation. Some time after 700, a proper Danish state was formed.

In the palaeolithic era there is even less point in singling out the people living in Denmark. The European population moved north and south in response to climate change, and the interaction, not least through trade, with people in the Near East was a most important source of new ideas. In consequence, this study will simply focus on a Denmark similar to that which exists today, but will make excursions to adjacent areas whenever relevant.

The style of history-telling will be very selective: to explore living conditions to the extent required to determine energy use, and occasionally to identify impacts of energy issues on other traits of society. This means following everyday life and when society becomes stratified, then to explore the effects of this on energy provision and energy usage, but not to go into much detail on issues such as wars, fights over thrones and territories, and exploitation and revolt between the classes.

The rest of this chapter gives an overview of environmental conditions induced by changes in climate and then concentrates on the development of human (Neanderthal) societies in Europe during the warm interglacial period called the Eem period. Exploring the energy conditions during this part of the Stone Age will set the stage for several of the following chapters by introducing methods and illustrating climate development, population size and available technology across the time periods, to be subjected to further scrutiny in the subsequent chapters.

## 1.2 Climatic conditions in the European interglacial period

During the interglacial period from about 130 000 years to about 115 000 years ago (130–115 ky BP), a modest number of humans, presumably Neanderthals, ventured to exploit the hunting possibility in the northern parts of Europe.* It is likely that at the higher latitudes they were only seasonal visitors, although a few may have stayed over during the winter to avoid the long march back to where they came from (perhaps in current mid-Germany, where a fairly large amount of Neanderthal remains have been found).

What were the environmental conditions these people had to cope with, and how did conditions change during the following millennia? The conditions of interest include temperature, solar radiation, precipitation, windiness, land and sea distribution and topology, soil type, vegetation and fauna, ice and glaciers, and, for several of these quantities, their seasonal variation.

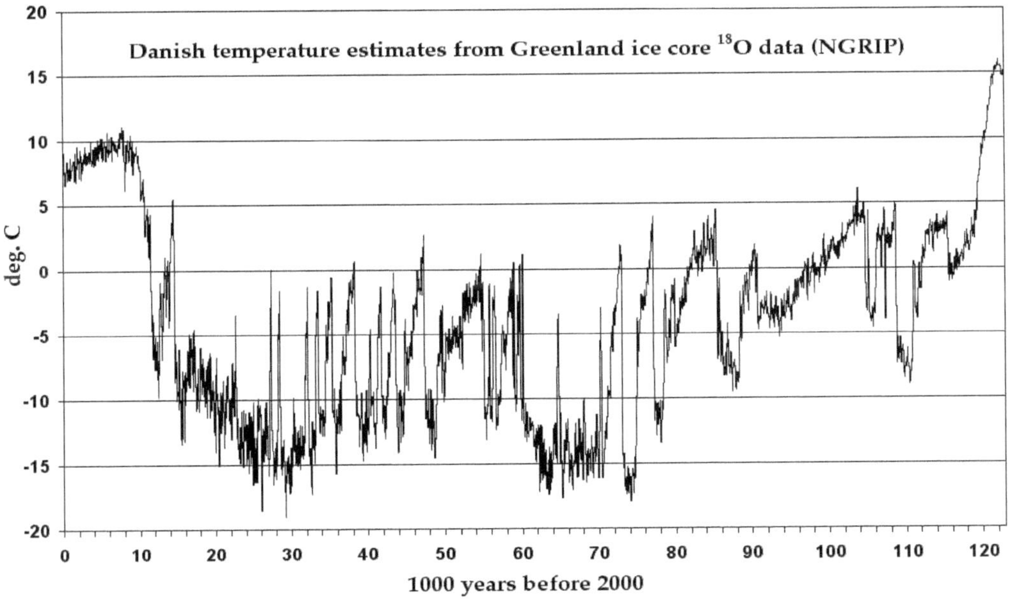


*Figure 1.1. Temperatures in Denmark deduced from 50-year averages of ice core data from NGRIP, Greenland (based on Huber et al., 2006).*

Let me begin with temperature and solar radiation. The major cyclic behaviour in solar radiation (the 105 000-year Milankovitch cycle induced by changes in eccentricity of the Earth's orbit) has caused periodic shifts between warming and cooling, with periods of extensive

---

* In the following, the abbreviation "BP" shall be used to denote a number of years before the present, with "present" being taken as year 2000 and the abbreviation "BP" used only for much earlier times. More recent dates are given using the presently widely employed zero (some 2010 years ago). Thus years not accompanied by a "BP" are years after the presently employed zero.

glaciation accompanied by changes in atmospheric carbon dioxide content and ocean salinity. Increased ice cover changes the surface albedo (to make the surface reflect more sunlight) and thereby provides a feedback to surface temperatures (Jansen *et al.*, 2007; Sørensen, 2004). Further changes in the obliquity of the Earth's orbit and the precession of its perihelion cause variations in the geographical distribution of solar radiation, with periods of 41 000 years and 23 000 to 19 000 years. The changes in temperature that may be inferred from ice core samples and ocean sediment analysis are depicted in Figs. 1.1, 1.3 and 1.4, where they have been adjusted by an overall shift to give an approximate temperature profile relevant for the location of Denmark. The variation in solar radiation due to variations in the Earth's orbital parameters is shown in Fig. 1.2, also for Denmark.

Figure 1.1 shows that mean temperatures in Denmark may have varied by over 30 °C between the most severe part of the last ice age (just under 30 000 years ago) and the warm Eem period some 125 000 years BP, and that even during the middle of the recent (Weichselian) ice age, temperatures varied up and down by up to 20 °C over fairly short intervals of time. How certain are we that the Greenland ice records reflect temperatures in Denmark? The presence of various amounts of the oxygen isotope variant $^{18}O$ in the water falling as snow in Greenland and being incorporated in the ice cores (drilled vertically down to the bottom of the ice sheet covering Greenland) constitutes the basic measured data. The $^{18}O$ is then interpreted as having been incorporated somewhere else into the clouds responsible for the Greenland snow, say over Europe, and is thus reflecting the conditions at that location. Temperature is an important determinant for the amount of $^{18}O$ in the water that forms clouds and the Greenland ice core measurements therefore tell us something of the temperature – not in Greenland, but in the regions where the water came from. A typical path with today's climate for water to arrive in Greenland is, in winter, from middle Europe northeast to the Barents Sea, then turning west to the Denmark Strait and finally turning north across Greenland. In summer, a typical path goes east from the central North American continent, then turning north towards Cape Farewell and further north into Greenland. The "$^{18}O$-temperature" seen at the NGRIP borehole in mid-Greenland will be lower in winter than in summer. Going back in time (deeper into the borehole), the ice will become increasingly compressed, and after a while seasonal variations can no longer be seen. The lowering of sea level due to ice formation at peaks in the glacial periods may alter ocean current paths as well as wind circulation patterns, and the borehole $^{18}O$-temperature may at such times be reflecting real temperatures at other locations than currently. The large temperature variations of the ice core records (e.g. during the period from 50 000 to 30 000 years BP) may therefore reflect changes in posting locations combined with modest changes in circulation, or they may reflect real abrupt temperature changes (e.g. in Europe).

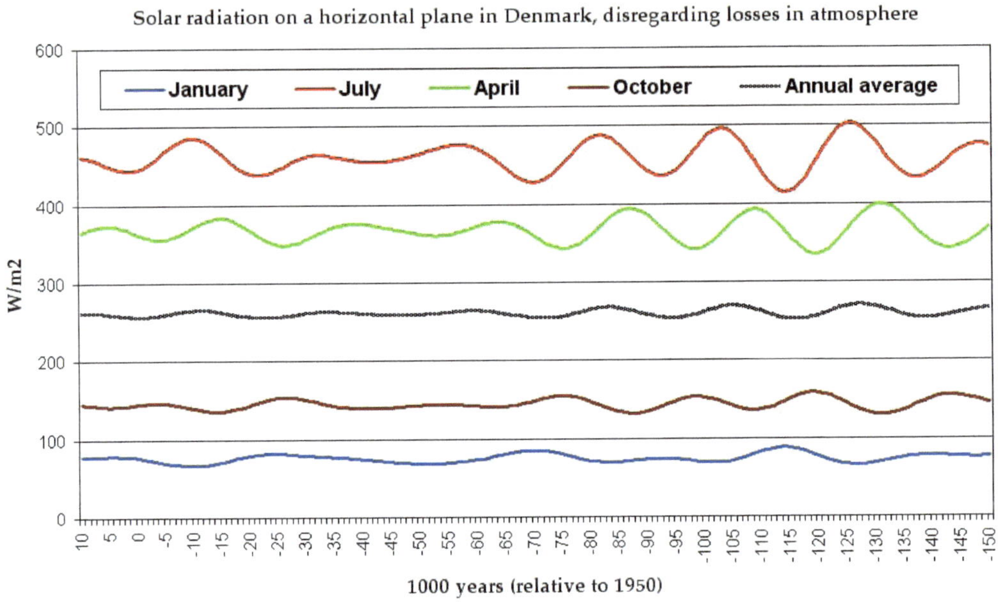


*Figure 1.2. Historical and near-future levels of solar radiation in Denmark, calculated from variations in orbital parameters and without attenuation in the atmosphere (based on data from Berger et al., 2007).*

In Antarctica, where few large land masses are adjacent, one does not see quite such large variations in ice core $^{2}$H-temperatures, and the overall temperature excursions during the last 300 000 years are also much smaller than in the Arctic records (Fig. 1.3): some 12 °C versus 35 °C in the north. On the other hand, the recent excursion around 14 000 years BP in the Greenland record (Fig. 1.1) is corroborated by pollen data from Denmark, as discussed in the following chapter on the latest deglaciation period. A reasonable conclusion is that the Antarctic data do not tell us much about detailed climate change in the north, but only about the rough extent of ice ages.

A third indicator of temperature is from the $^{18}$O-analysis of benthic cores. They are cores drilled at ocean bottoms and reflect the conditions here through foraminiferal calcite (i.e. from shells of eukaryotic species living near the bottom sediments). The data reflect not only temperature, but also salinity and sea depth, which vary substantially when ice sheets form or disappear (cf. Fig. 1.5). Figure 1.4 gives an average of data from a large number of sites, both in the Atlantic Ocean and in the Pacific, and in both the Northern and Southern hemisphere. The variations in $^{18}O$ are plotted on a pseudo-Danish temperature scale, only as an indication, because of the other factors influencing the data. They are similar to the ice core records, but not identical to the Antarctic one, and even less so to the Arctic record. This could be expected because Arctic–Antarctic records themselves differ, and the conditions at ocean floors have a complex relationship to climate elsewhere. Benthic data become interesting by going back more than 5 million years, far beyond what is depicted in Fig. 1.4. One sees some 6 major ice ages, and before them variations on a considerably smaller scale, but more rapidly varying. One also notes that the ice ages do not develop identically and do not have precisely the same length. Most of the differences can be at least partially understood from the complex variations in the Earth's orbital parameters, some of which do not change the total amount of radiation received at the top of the Earth's atmosphere, but change the distribution over the globe.

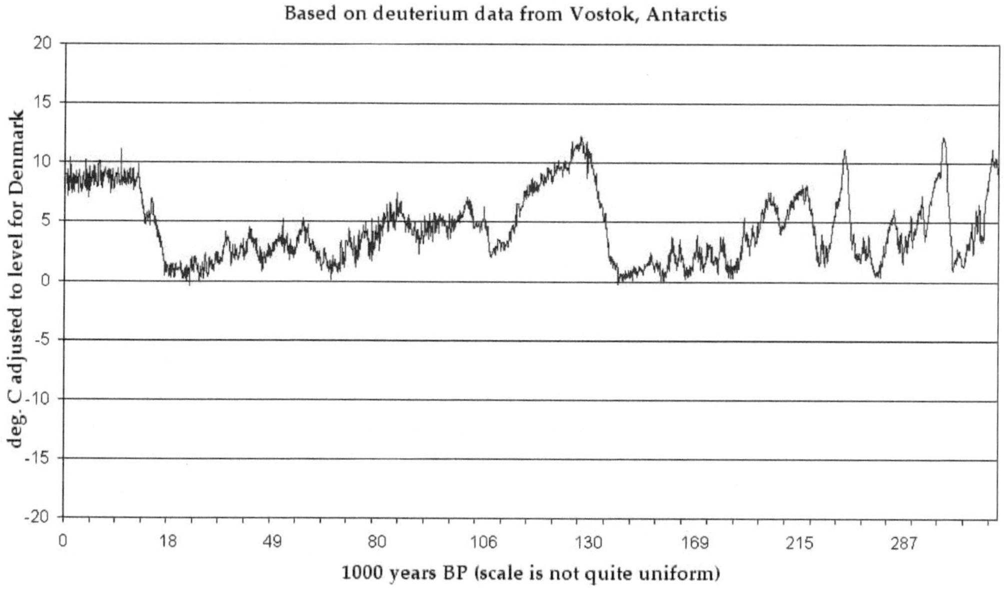


*Figure 1.3. 250 000 years of temperature estimates (shifted to Danish average temperatures for easy comparison with Fig. 1.1) based on $\delta\,^2H$ data from Vostok, Antarctica (based on Petit et al., 1999).*

Figure 1.2 shows the solar radiation variability at a particular location (Denmark) and one notes large excursions (some 18% in summer, 4% as an annual average and displacement between seasons) during the period 150 000 to 60 000 years ago, and again during the last 30 000 years (and continuing during the forthcoming ones). This corresponds well with the temperature records of Fig. 1.1, but the much smaller variation 60 000 to 30 000 years ago again places doubt on the interpretation of the rapid ice-core-based temperature variations during this period.

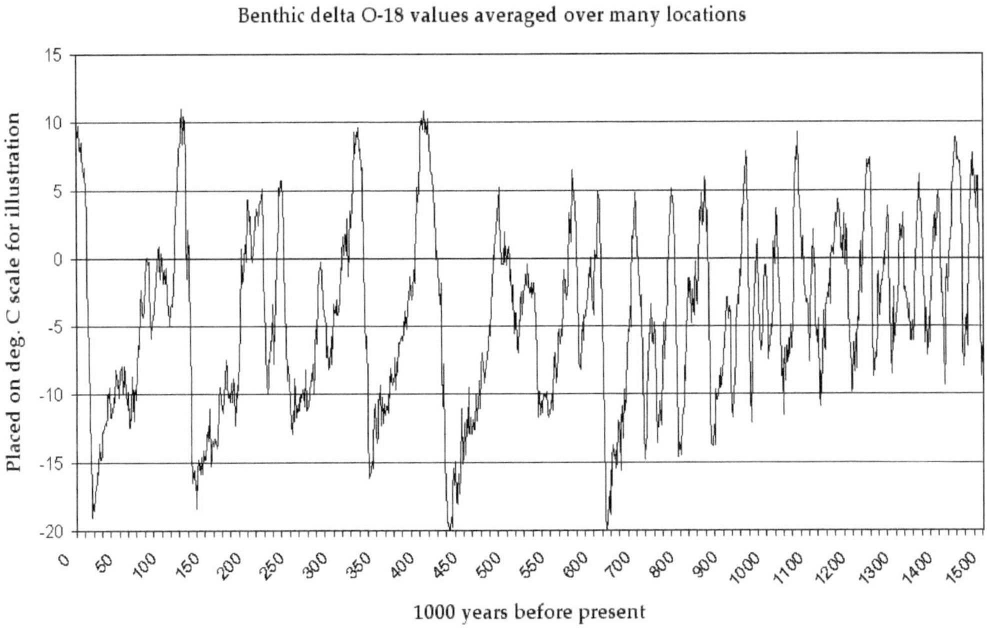


*Figure 1.4. Average benthic $\delta^{18}O$, with shift to Danish temperature scale imposed for illustration (based on Lisiecki and Raymo, 2005).*

An important determinant for the early migration of human populations between continents and regions is the existence of contiguous land masses allowing passage without having to construct seagoing vessels (Manning, 2006). Periods of substantially lower sea levels as compared with the present ones have existed as a result of glaciation, which deprives the oceans of the water masses transformed to ice sheets. Fig. 1.5 shows the variations in relative sea level during the last 140 000 years, coming from changes in ice volume. Locally, these would be superimposed with effects of land depression due to the mass of the ice (and delayed upheaval due to elasticity of the sediments) and to stemming up water in some locations as a result of oceanic circulation and the barriers set by land contours.

Ice sheet models describing the formation and melting of ice sheets (over land), ice shelves (extensions with water below) and sea ice (floating) have been formulated and are increasingly incorporated in climate models. Continental depressions and rebound upheaval make use of a model seeing the Earth's crust as floating on a fluid mantle. Large-scale motion of continental plates is taken as unimportant at the timescales considered, but would play a role if several million years were considered. Efforts to model a full cycle of glaciation onset, development and ultimately disappearance, using the varying solar radiation as input to the models, has been undertaken by Tarasov and Peltier (1997) and by Bintanja *et al.* (2002), giving a fair agreement with available data, but still missing the power to predict the precise situation facing humans trying to survive during the glaciation period, and particularly its middle period. The late deglaciation modelling has been much more successful and will be detailed in the section dealing with that time period.

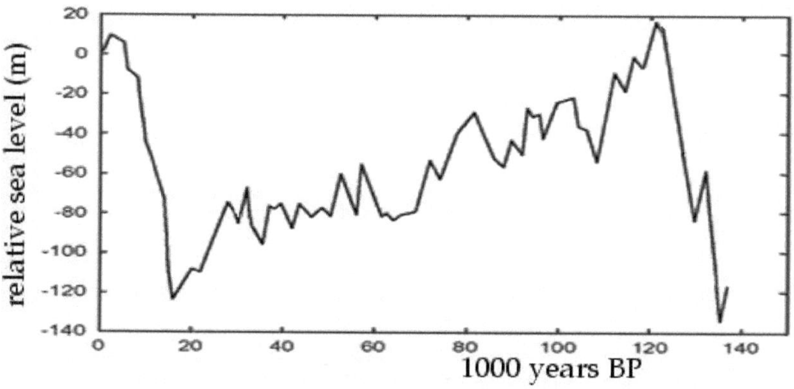


*Figure 1.5. Sea level change corresponding to variations in the water bound in ice caps, but neglecting dynamic effects from currents and shore topology, as well as local depressions from ice loads and rebounds (based on ice and sediment studies by Shackleton, 1987; Bard et al., 1990; Siddall et al., 2007; recent analyses by Skinner and Shackleton, 2006, and by Waelbroeck et al., 2008, propose to put the Eem maximum back to about 130 000 years BP).*

According to Figs. 1.3 and 1.4, the maximum glaciation position during the late Saalian ice age some 143 000 years ago was associated with temperatures as low as, or lower than, those of the recent Weichselian ice maximum about 29 000 years ago. A sketched reconstruction of the extent of the maximum Saalian ice sheet over Eurasia is shown in Fig. 1.6. Hardly any climate model calculations exist for this period, so the graph is constructed from sedimentary information from individual sites. The extent of the ice sheet underlines the possibly substantial difference between the following thawing and warm Eem period and the more recent period of Weichselian deglaciation. The Eem period has been covered by several climate calculations and also has substantial amounts of supporting data capable of shedding light on the conditions for human settlers. This is a subject of the following section. The globally extending difference between temperatures during the Eem and the present interglacial periods shown in Fig. 1.7 is a further indication of the differences in behaviour between the two hemispheres.

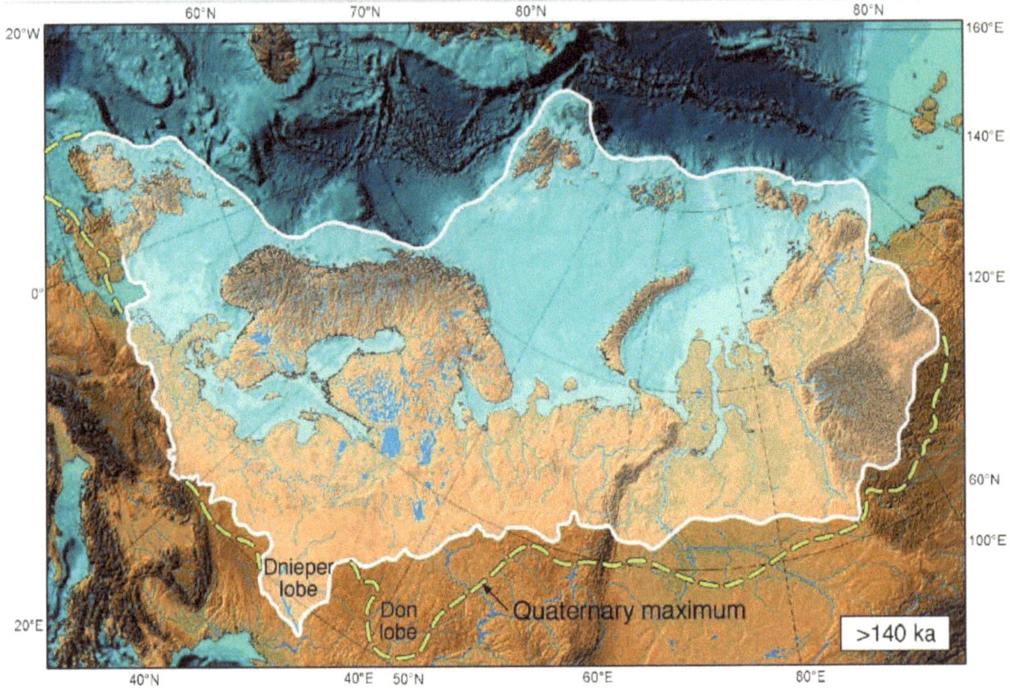


*Figure 1.6. Estimated maximum extent of the Eurasian ice sheet during the Saalian ice age (260–140 ky BP). The dashed line indicates the estimated maximum during the entire Quaternary period, occurring twice at about 450 and 650 ky BP, according to the benthic evidence shown in Fig. 1.4 (from Svendsen et al., 2004, which again is based on several earlier sources).*

The calculations behind Fig. 1.7 use solar radiation input as it was 125 000 years BP in a coupled atmosphere–ocean model (called ECHO-G) with a spatial resolution of about 3.75° in the atmosphere and an ocean model with consideration of sea-ice (Kaspar and Cubasch, 2007). The temporal (cf. Fig. 1.2) and geographical location dependence of the solar input is sufficient for the model results to exhibit marked hemispherical differences in temperature deviations from the present, which characterize the Greenland and Antarctic data (Figs. 1.1 and 1.3), i.e. much less change in the Southern hemisphere. The varying effects on the Northern hemisphere temperatures, with regions of increased and others of decreased temperatures during certain parts of the year, are consistent with the complex Eem temperature curve for the Arctic (see Fig. 1.8 below), quite different from the smooth Antarctic temperature behaviour of Fig. 1.3.

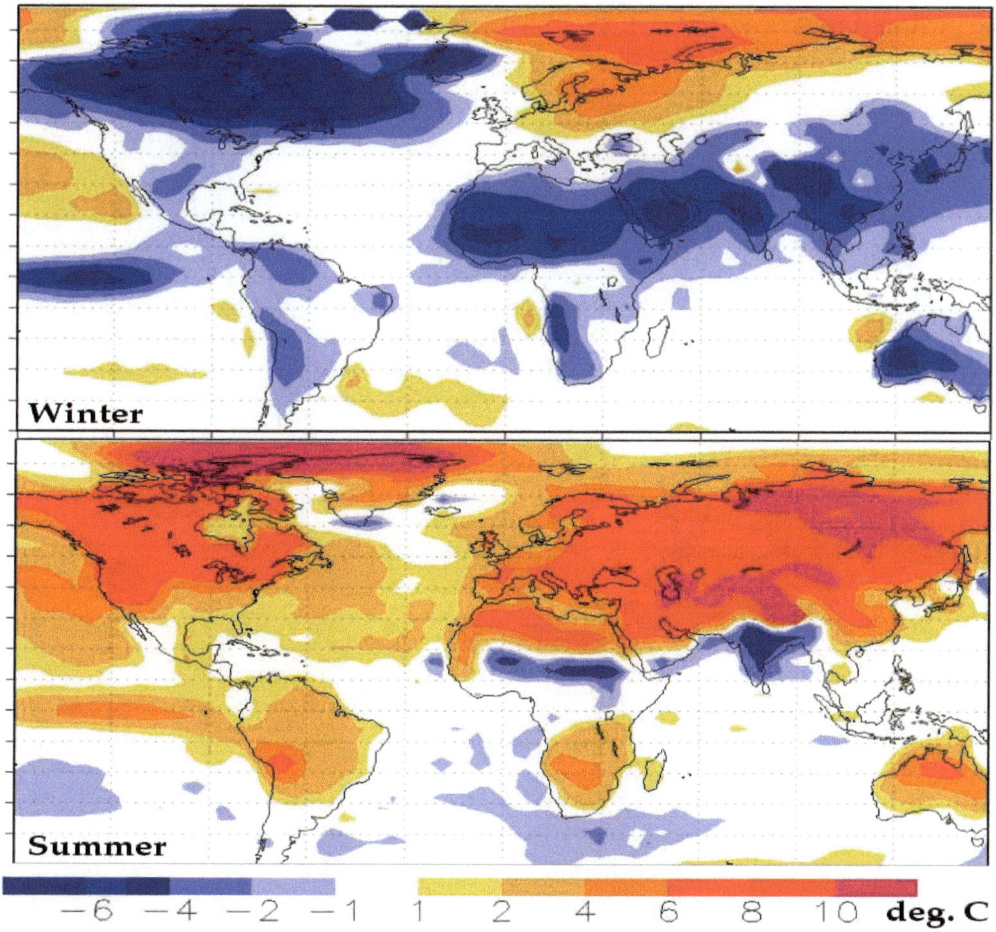


*Figure 1.7. Temperature deviations in the interglacial Eem period (125 ky BP) relative to recent, pre-industrial times (~1880), for northern winter (above) and summer (below), calculated with an atmosphere–ocean coupled climate model (Kaspar and Cubasch, 2007).*

## 1.3 Details of the Eem interglacial period

Although the Eem period around 120 000 years BP was at least as warm as than the present period, there are important differences between the Eem climate in North Europe and the present one, of relevance for the living conditions in human settlements. Because the NGRIP temperatures in Fig. 1.1 do not cover the early Eem deglaciation, the earlier GRIP data are presented in Fig. 1.8. Although they are the oldest temperatures in this data set (corresponding to the most compressed parts of the ice core), and therefore possibly less accurate, there is overall agreement with the newer data of Fig. 1.1 during periods of overlap (but differences in the shorter-term oscillations).

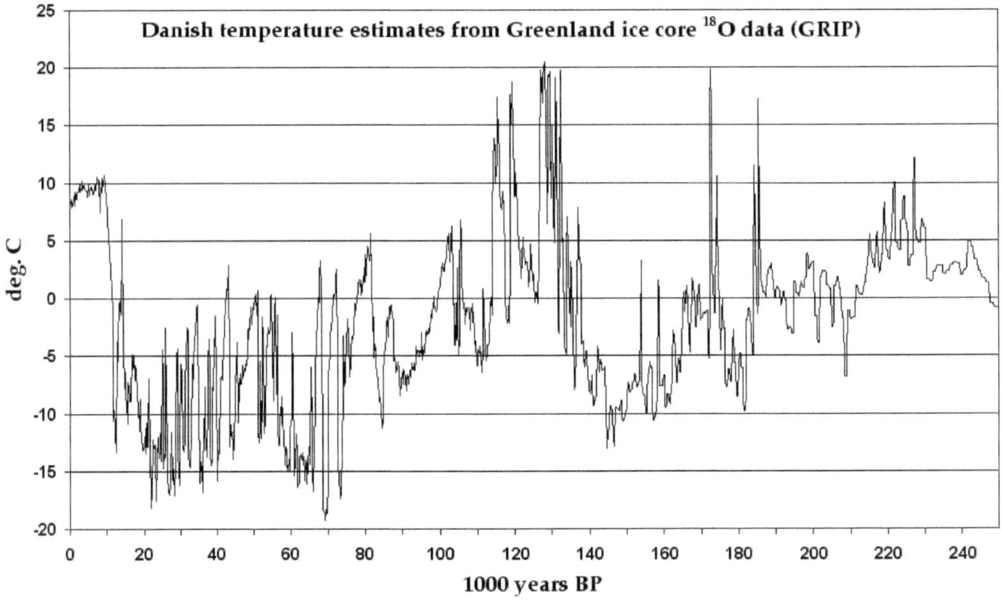


*Figure 1.8. Temperatures in Denmark based on GRIP ice core data, Greenland (Johnsen et al., 1997, adapted to Danish conditions).*

It is therefore quite possible that the complex temperature development through the period 130–115 ky BP, which has no counterpart in the Antarctic or in the benthic data of Figs. 1.3 and 1.4, is real. The stunning 20 °C temperature drop at about 129 ky BP, first for a 6000-year period and then for another one of about 1000 years length, in the middle of the Eem period (around 121 ky BP), may be connected with the complex behaviour seen in Fig. 1.7, where the winter temperature is lower than today in North America, while it is higher in North Europe and Asia. As noted, changes in oceanic circulation (and in consequence atmospheric circulation) may change the origin of Arctic precipitation water and produce rapid temperature excursions in the ice core $^{18}O$ data. Still, the calculated variation in temperatures at 60° N latitude is not nearly as large as that of the ice core data, so a closer look at both available data and calculations may be in order.

The model calculations are sensitive to what is included in the model, to geographical resolutions and to the mode of calculation. Climate models may be run as equilibrium models for parameters pertaining to a specific point in time. In this case, one calculates for many model-years, until the solutions become stable. This may take from 20 years to 50 000 years, depending on the details of the model (Donat and Kaspar, 2006; Calov *et al.*, 2005a). Models that reach stability fairly rapidly may be used in a transient mode, where input is changed at suitable time-steps during the calculation. The results shown in Figs. 1.9 to 1.10 are from a transient calculation, using the same computer code as the one used for the equilibrium calculation in Fig. 1.7. However, the transient calculation is performed for an extremely long period (-140 ky to +30 ky), and therefore the calculation is set to jump 100 time-steps at a time. This prevents many aspects of the oceanic and atmospheric circulation from being captured, and the 124 ky BP transient calculation (shown in Fig. 1.11 as difference from modern climate, rather than the incremental changes from 130 ky BP shown in Figs. 1.9 and 1.10) finds that the substantial winter warming over Eurasia found in Fig. 1.7 has turned into a net cooling in most places. This is also the case for another equilibrium and transient set of calculations with a lower resolution (5.6°), but including the effect of varying vegetation (Donat and Kaspar, 2006). The importance of vegetation in climate models has been stressed (e.g. by Calov *et al.*, 2005b). They made model calculations using an extremely simple model (latitude resolution 10° and longitude resolution 50°), but including a land ice model and, in one set of runs, the presence of mineral dust (e.g. from erosion or from volcanoes), which is known to affect ice melting. The results showed that a realistic amount of dust in the atmosphere reduced the ice formation and avoided the presence of excessively large Eurasian ice sheets otherwise found during the period from 115 to 105 ky BP. One would further like to include the shoreline change and land depression/rebound that is associated with ice sheet presence and melting as part of the modelling effort. This has been done in a deglaciation model neglecting full ocean circulation (Peltier, 1994), but not together with the other model improvements. The bottom line is that current computer power allows models with reasonably high resolution to be used for simple atmosphere–ocean calculations, but forces the investigator to use increasingly coarse resolutions, when the aim is to capture the effects of increasing amounts of the detailed time-dependent real-life climate determinants.

Approaching the issue from the other side, that of observation, the evidence is also necessarily incomplete. Recognizing the need for assessing

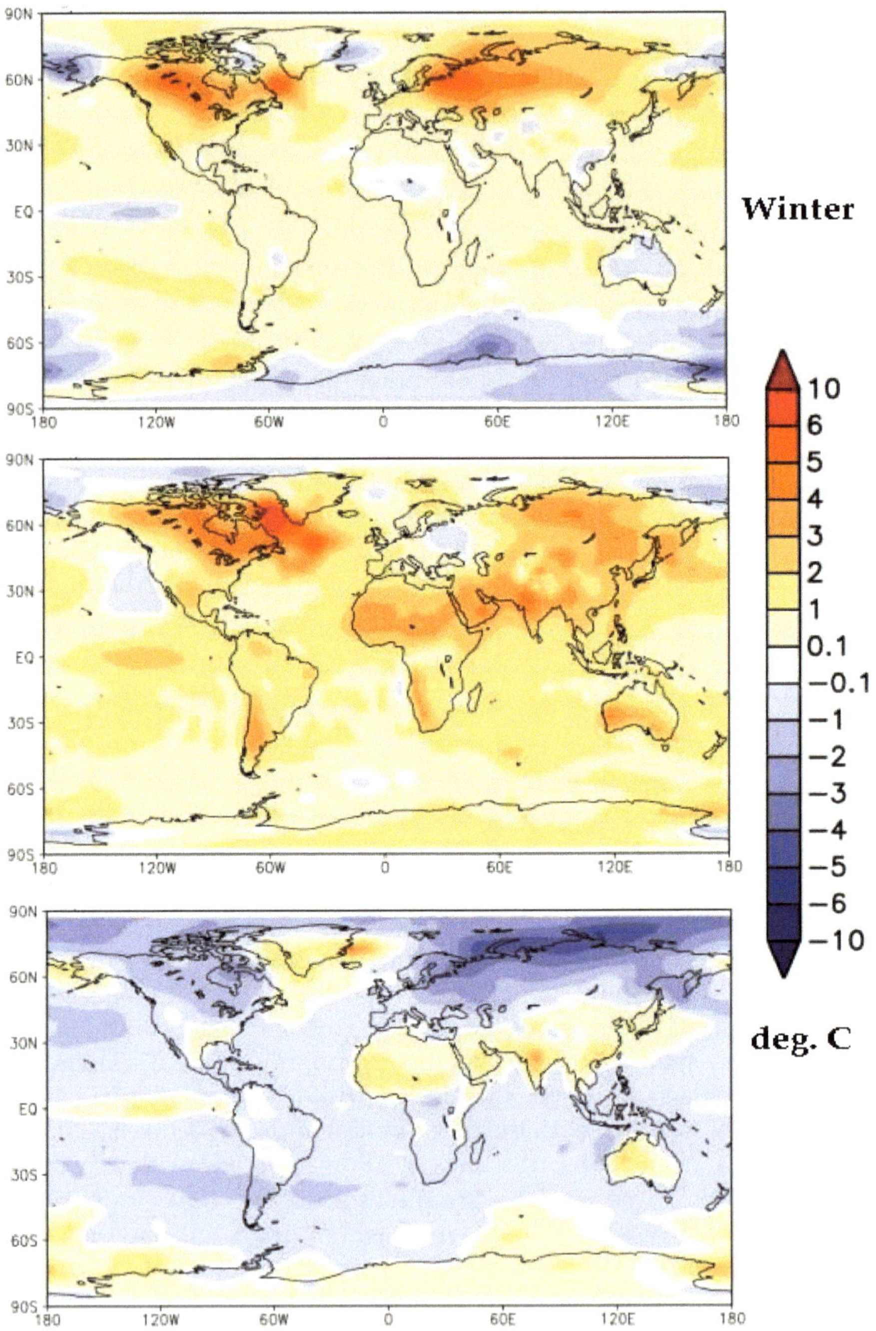


*Figure 1.9. Winter temperature change, 130 to 125 ky BP (top), 125 to 120 ky BP (middle) and 120 to 115 ky BP (bottom; Lohmann and Lorenz, 2007).*

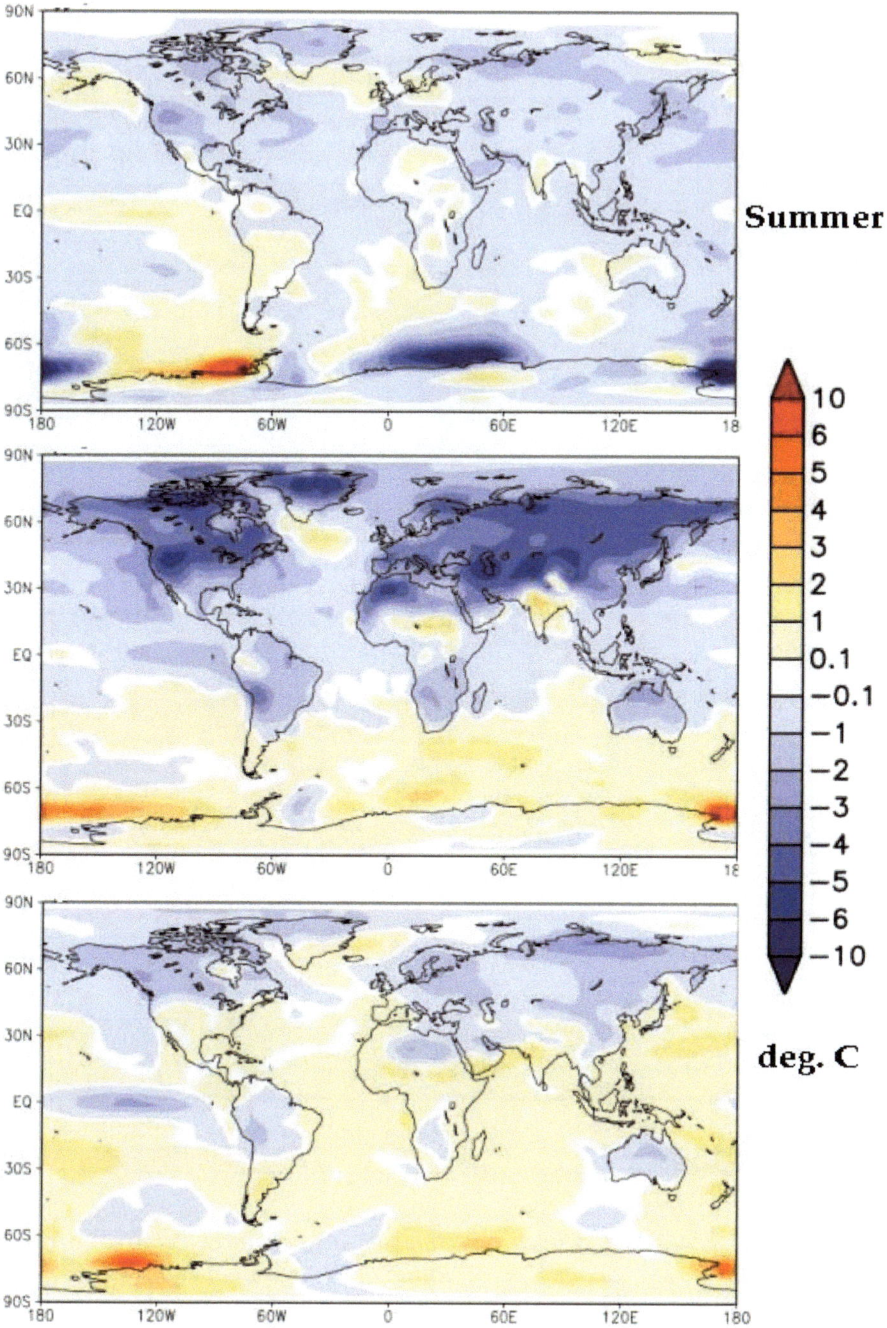


*Figure 1.10. Summer temperature change, 130 to 125 ky BP (top), 125 to 120 ky BP (middle) and 120 to 115 ky BP (bottom; Lohmann and Lorenz, 2007).*

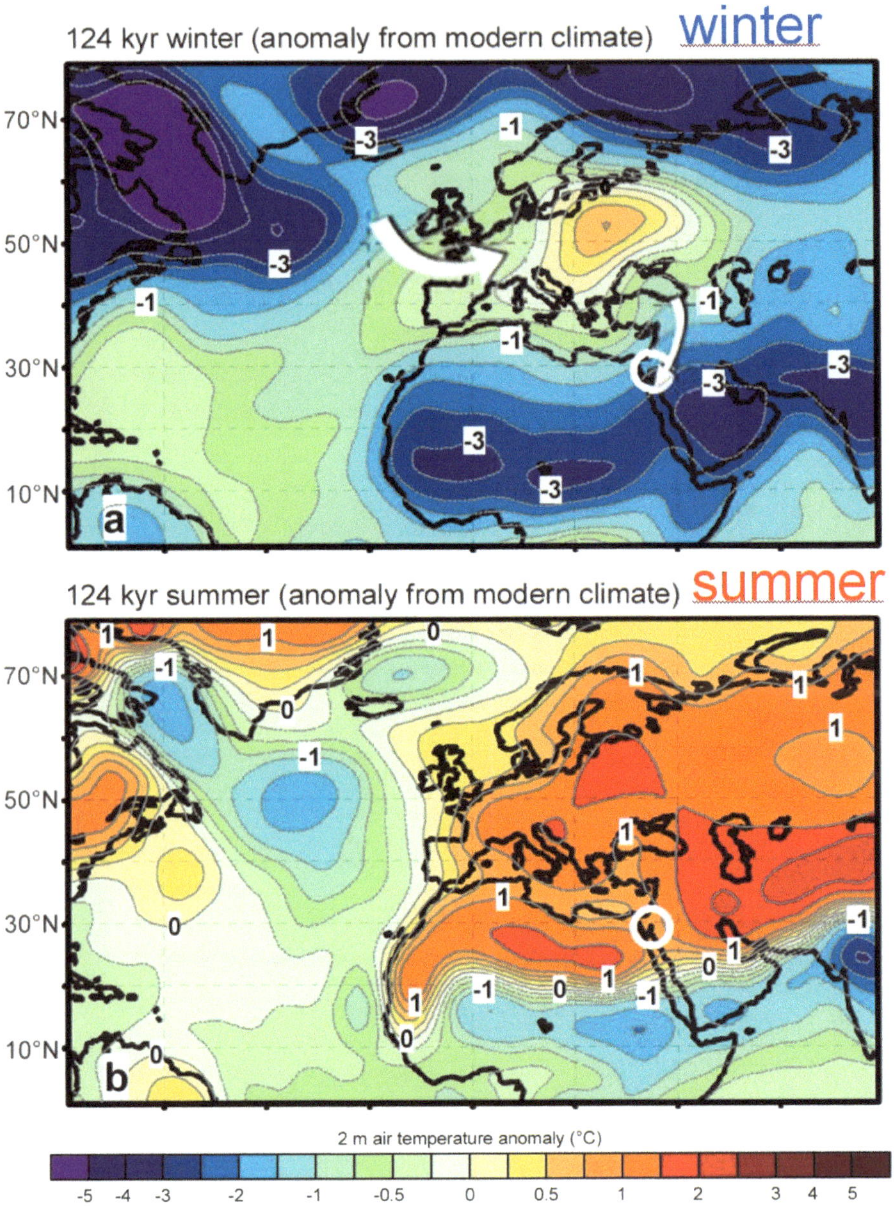


*Figure 1.11. 124 ky BP winter and summer temperatures relative to recent, pre-industrial period: another presentation of the transient calculation of Figs. 1.9–1.11 (Lohmann and Lorenz, 2007, supplementary information CD).*

seasonal variations in climate variables, studies of sediment cores with identified pollen content are relevant. One example relevant for Denmark is the core from Bispingen near Hamburg, where the sediment layers appear little disturbed and allow temperatures to be estimated from pollen of species with known temperature thresholds for growth (see Fig. 1.12; Kühl and Litt, 2007). Other sites are Gröbern (southeast Germany) and Grand Pile (Alsace; Rousseau *et al.*, 2007).

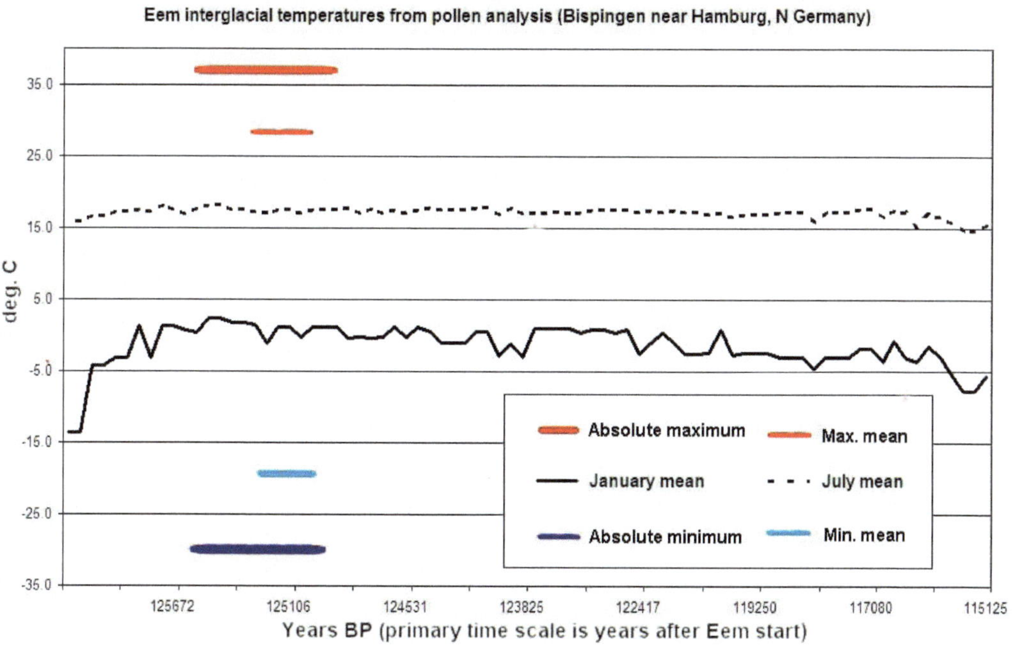


*Figure 1.12. Eem temperatures deduced from pollen records near Hamburg, just south of Denmark (based on Kühn and Litt, 2007; Kühn et al., 2007).*

The layers of the sediment pollen record shown in Fig. 1.12 are generally proportional to time. However, there is a greatly compressed section corresponding to the period from 122 to 119 ky BP, with hardly any pollen traces. This could be related to the sudden cooling indicated in the Greenland record of Fig. 1.8 (exact absolute timing is always somewhat uncertain, for data as well as for models), suggesting a 3000-year period with little plant growth, due to very low temperatures. This supports the view that the Eem warming period is possibly much more complex than the more recent interglacial one. The records from Grand Pile in Alsace are similar to those in Fig. 1.12 for the period given here, but between 115 ky BP and 100 ky BP a dramatic 20 °C drop in winter temperatures occurs for the period 110–103 ky BP, while the summer temperatures just continue their modest reduction around 115 ky BP by a further couple of degrees to 100 ky BP.

The transient climate model calculation illustrated in Figs. 1.9 and 1.10

has the winter temperature rise from 130 to 125 ky BP (by some 1–2 °C in Denmark, more in Asia and North America), then a further rise from 125 to 120 ky BP (less than 1 °C in Denmark, but much more in North America, and now also in Equatorial and Southern Hemisphere areas), and finally a drop from 120 to 115 ky BP (by about 1 °C in Denmark, much more in northern Asia and North America, but still warming in most Equatorial regions). The calculated summer temperature, on the other hand, generally cools in all three periods, except for some areas of warming in the Southern Hemisphere. In Denmark, after less than 1 °C warming from 130 to 125 ky BP, a 2–3 °C cooling follows from 125 to 120 ky BP, and a further 1–2 °C cooling from 120 to 115 ky BP. The predicted conditions for human survival, in the millennia after 130 ky BP, where Danish temperatures (according to Fig. 1.8) were very similar to the present, thus involve winter temperatures a couple of degrees higher, but declining summer temperatures from 125 ky BP to a value some 4 °C below those of 130 ky BP by 115 ky BP. This may initially have made it attractive for Neanderthal populations (known to be present in Germany) to explore the hunting possibilities in Denmark, but less so as the Eem period progressed.

At the other end of the Nordic region, in Finland, summer temperatures at the start of the Eem interglacial were similar to those in Denmark, but winter temperatures were some 15 °C lower (which is some 5 °C below the current winter temperatures). An increase by 5 °C in winter temperatures took place between 130 and 120 ky BP (reaching current winter temperature levels), while summer temperatures dropped by some 4° C during the same period, followed by a further lowering.

The reconstruction of sea levels in the Baltic regions shown in Fig. 1.13 is based on data, primarily of microfauna found in stratified sediment, with characteristics of sea provenance and layer dating to the Eem. Sea levels derived in this way support the Equatorial data behind Fig. 1.5, suggesting relative sea levels some 10 m higher than at present. Superimposed on this is the effect of land depression by the up to 3000 m ice sheets having covered the area (Fig. 1.6). This depression is much larger over Sweden and Finland than over Denmark, because of the diminishing ice thickness towards the ice rims. Most of the subsequent upheaval would have taken place in the first 3000–6000 years after the ice had gone, so the map constructed in Fig. 1.13 is relevant only for the early part of the Eem period. Assuming the ice melting to have taken place between 148 and 130 ky BP, raising the general sea level by 140 m, the depression in the area around Finland must have been of a similar magnitude (relative to today's topology) for the map to be realistic. This is quite possible, comparing with ice models such as that of Peltier (1994), but the rebound would have restored the topology by 125 ky BP to much more closely resemble the present shorelines. It is thus fair to consider the shorelines of Fig. 1.13 as pertaining only to a relatively short period at about 130 ky BP.

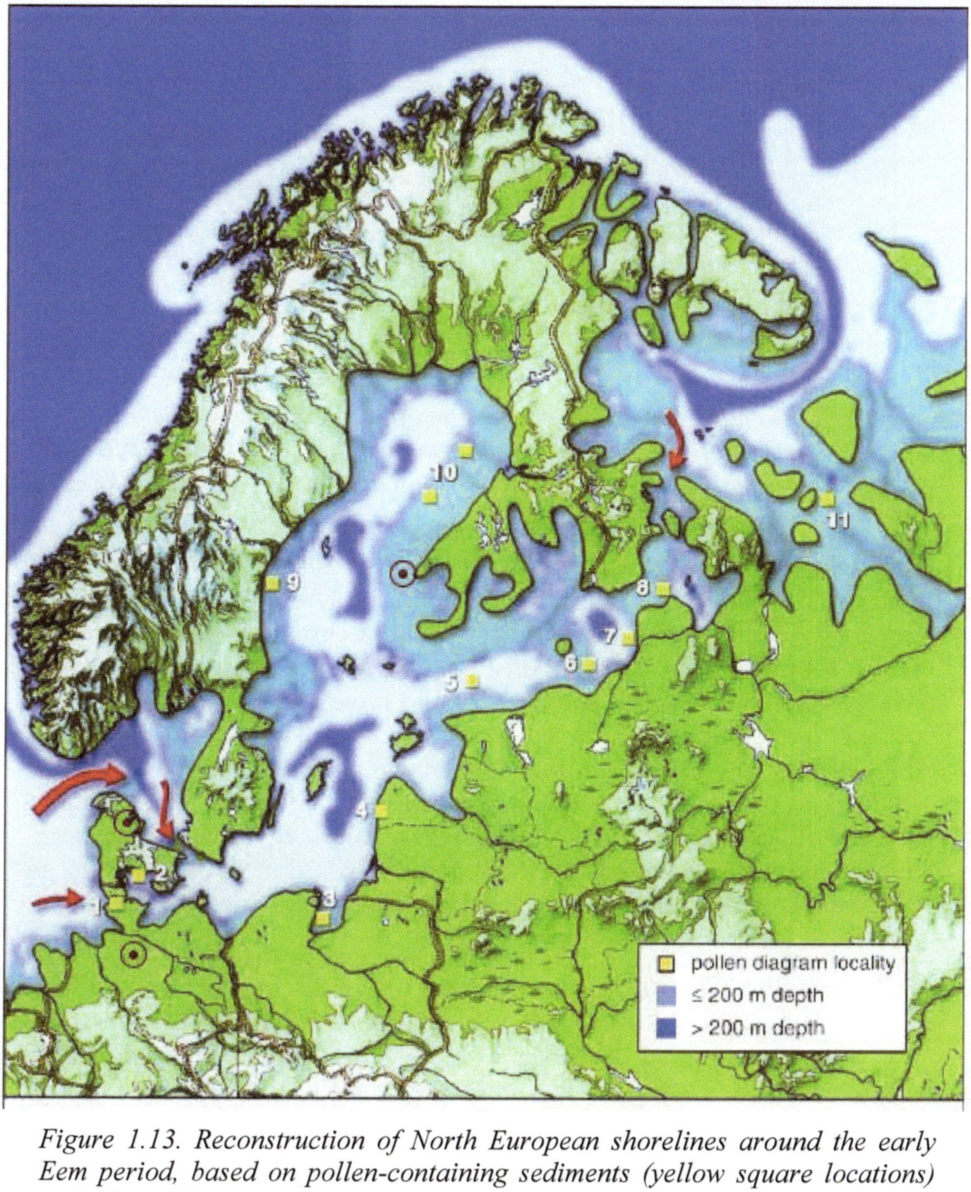


*Figure 1.13. Reconstruction of North European shorelines around the early Eem period, based on pollen-containing sediments (yellow square locations) and other indicators of water level (Funder et al., 2002). Three sites with signs of possible human presence are shown as encircled black dots.*

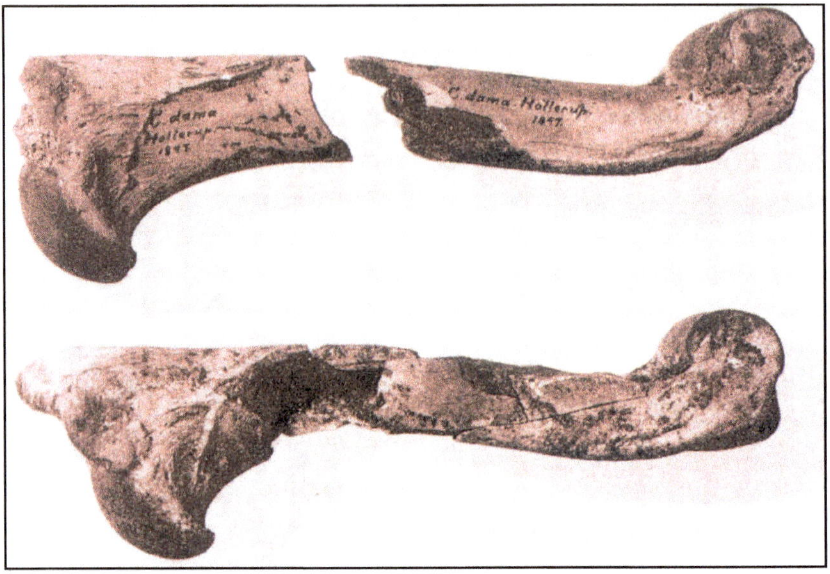

*Figure 1.14. Fallow deer (Dama dama) bones found at Hollerup, Denmark (see encircled dot in Fig. 1.13) and dated to Eem period. The bones have been split with a blunt instrument, interpreted as the work of marrow-eating humans (Møhl-Hansen, 1958).*

During this period, the reconstruction suggests that the North Sea through the Baltic Sea was directly connected to the White Sea. This feature, which is not captured in the general climate models discussed above (because they do not consider differential depression and upheaval during ice growth and melt), could have important effects on the oceanic circulation, and the same is true of the subsequent closing of this passageway. For example, one may speculate that such abrupt changes in ocean circulation and the implied changes of wind circulation are at least contributing reasons for the massive mid-Eem temperature drop in the Greenland ice core data, a drop that appears too large to reconcile with temperature variations in any one European or North American location.

Turning now to the question of human presence in the Nordic area during the Eem period, there are two indications. One consists of two fallow deer bones (Fig. 1.14) found at Hollerup in Jutland, marked by an encircled dot in Fig. 1.13. The bones are dated to the Eem period, and have been punctured/split by a blow with some instrument, in a way typical of Stone Age people wanting to eat the bone marrow. There is no sign of ash remains around, indicating that the site is not a dwelling site, but rather the site where the animal was cut into parts and the marrow delicacy eaten, but the remainder left or carried on for later consumption. In other words, the find indicates the possible passage of a Neanderthal hunting party, but not any permanent settlement. The interpretation of the bone splitting can always be questioned, but it is consistent with a general assessment of bone marks from later finds (Noe-Nygaard, 1989).

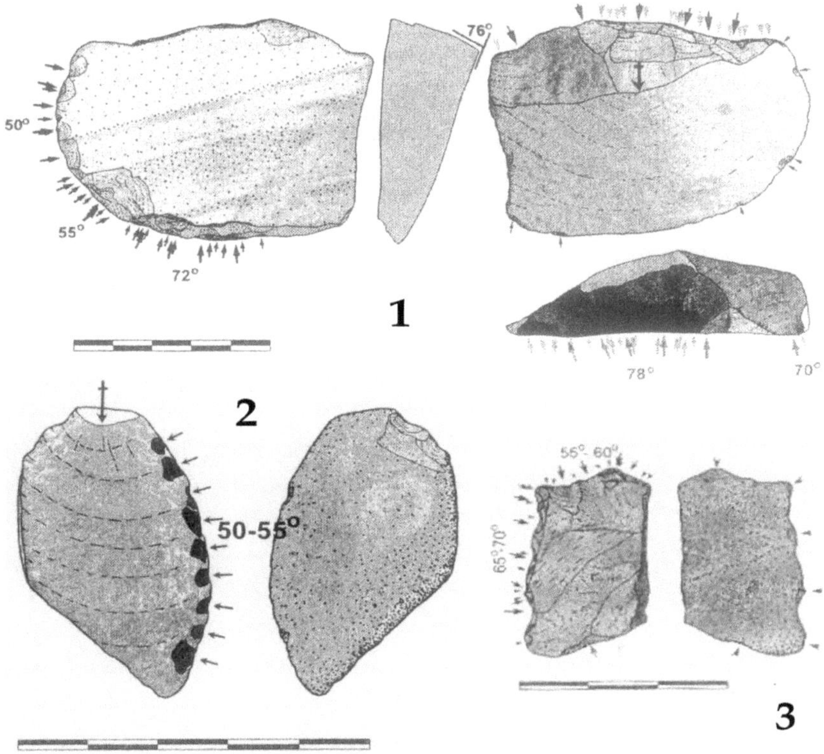


*Figure 1.15. Wolf Cave Eem artefacts made from quartzite (1) and red siltstone: scraper (2) and denticulate (3). Arrows indicate blow positions with some blow angles noted (Schulz, 2006; cf. Fig. 1.16).*

Further north, at the Baltic shore of Finland (another encircled dot on Fig. 1.13, possibly underwater at the beginning of the Eem period but on the shore around 120 ky BP; today some 100 km inland), one finds the Susiluola (Wolf) Cave. The stratification of layers in a cave having experienced subsequent ice cover and actions of ice and water streams can be problematic, but there is reasonable evidence for dating two layers to the Eem period (Schulz, 2001). One of the layers contains a "paved" floor and burnt stones, possibly indicating a fireplace. Both layers contain lithic artefacts bearing several traces of processing, such as initially coarse and then finer striking by a pointed instrument on the edge of the stones shown in Fig. 1.15. As always in archaeology, one must consider the possibility that natural processes produced these marks, but in this case such criticism has been rejected as having a very small probability (Schulz, 2006). The conclusion is that Neanderthal hunters did occupy the cave during the late Eem period, and probably for more than a brief summer passage. The argument for this is the very nature of the stone artefacts. Stone tools made by Neanderthals in other regions during this period use high-quality flint stone, which, however, is not present in this part of Finland. The artefacts found in Wolf Cave use sandstone, quartzite and red siltstone, which (particularly the last) must have been unfamiliar to the Neanderthal dwellers. This indicates ability to adapt techniques to new materials, and also that the Neanderthal stay was long enough to force them to renew their hunting and food preparation equipment (one should also note the level of intelligence involved in identifying substitute raw materials and adapting the striking technique to materials with properties different from what the Neanderthals were accustomed to).

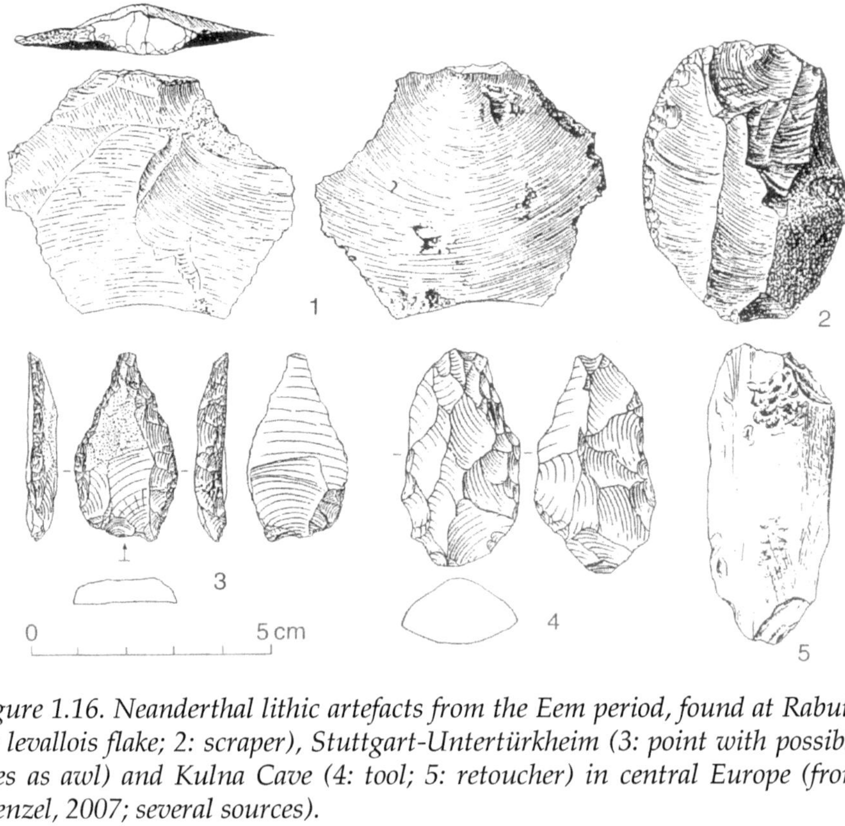


*Figure 1.16. Neanderthal lithic artefacts from the Eem period, found at Rabutz (1: levallois flake; 2: scraper), Stuttgart-Untertürkheim (3: point with possible uses as awl) and Kulna Cave (4: tool; 5: retoucher) in central Europe (from Wenzel, 2007; several sources).*

Considering the distance to known Neanderthal sites in central Europe, it appears possible that the Wolf Cave dwellers would have preferred to stay during the winter rather than walking maybe 1000 km back to where they came from. The climatic comments above also indicate that, during the latter part of the Eem, winter temperatures in Finland were similar to current ones.

How were the Neanderthals able to survive in a winter climate that suits us fine today, with insulated and heated houses, big overcoats and warm cars, but which really was quite harsh for creatures without fur and with only modest subcutaneous fat? Not to speak of the even harsher conditions characterizing not the short interglacial periods, but the two or more ice ages that humans lived through from the first arrival in Europe some 640 ky ago of *Homo heidelbergensis* (suggested to have been a common ancestor to both *H. neanderthalensis* and the latter *H. sapiens sapiens* (the "modern human"), but with *H. sapiens sapiens* forming outside Europe according to Mounier *et al.*, 2009; geographical dispersal patterns would be simpler to explain if the split occurred before *H. heidelbergensis* came to Europe). I shall attempt to outline the survival requirements and discuss how the Neanderthals may have met these requirements.

Present humans need a temperature above 20–22 °C to carry out outdoor activities without clothes, and a temperature above 28 °C to sleep outside without clothes. The Neanderthals presumably had similar requirements, although average adult body mass were higher than that of modern humans and height smaller. The more sturdy structure has been associated with higher cold tolerance (Aiello and Wheeler, 2004), but if this is so, the difference is probably fairly small.

During the warmest part of the Eem period, temperatures at the Neanderthal site at Lehringen (a few kilometres from the Bispingen site, indicated in Fig. 1.13, with the temperature records shown in Fig. 1.12 and consistent with the calculated ones of Fig. 1.7) were approximately 1 °C (like today) in the coldest month and 17–18 °C for the warmest (1–2 °C above today's), on average. The Danish Hollerup site would have had similar temperatures. Like today, there would have been periods of considerably lower and higher temperatures. This means, as discussed in more detail below, that a single layer of clothes would have sufficed in the warmest month, but that several layers of clothes, or clothes with exceptionally high insulation properties, plus snow-tight footwear and socks, would often have been needed for winter daytime activities, and some sort of shelter during sleep. The shelter requirement, even in the warmest interglacial times, explains the many cave sites occupied by Neanderthals: here they could keep warm and away from the wind, assuming a number of fireplaces in use, The 60 ky BP Kebara Cave Neanderthal site in Israel has a "large" number of preserved hearths (Speth, 2006) and the Rose Cottage cave in South Africa, occupied around 30 ky BP, has as many as 48; Wadley, 2006). Eemian winds were similar to the present ones in summer, but stronger in winter (and predominantly from west; Kaspar and Cubasch, 2007), so wind chill would have contributed to the shelter requirement at night. According to the climate model calculation used for Fig. 1.7, precipitation was similar to today during all seasons. However, the limitations of the model assumptions could affect this statement. Indications from plant growth suggest the possibility of higher precipitation.

Many of the known populated Eemian sites are not cave sites (Wenzel, 2007), and, according to the observations made above, there probably was a need for shelter in winter. This might have been wickerwork mats extended between poles and eventually covered by hides from prey animals, materials unlikely to have left much in the way of archaeological traces. From more recent times, building poles can usually be spotted by aerial photography sensitive to soil depression and looser material where objects have been protruding. However, the Eem sites are usually lake basins or travertine layers formed by melting ice and water streams, unlikely to have left imprints of pole positions.

As regards clothes, the evidence is a little better, because the tools used to prepare hides and furs for clothing are abundant. The variety of such tools indicates that they had several uses other than readying game for cooking and eating. In addition to scrapers and retouchers, a number of knife-like blades and pointed stone tools have been found at Neanderthal sites (see Figs. 1.16 and 1.17; the latter collection is from the more recent period at about 40 ky BP). That these were used to make clothes fit without openings permitting wind and frost to get at the human body seems undisputable in the light of the clothing requirements, even if there is no evidence passed on to tell us exactly which techniques were used. Use of small bones to tie two layers of hide with bored holes together has been suggested. A more likely suggestion may be threading a strip or ribbon of leather (cut from a hide using the blade tools) through a series of holes in the skins to be worn as clothes, thus effectively knitting two pieces together. The use of string or adhesive tape made of resin-producing bark to fasten handles on knife blades is known (e.g. from the Neumark-Nord-1 Eem site; Wenzel, 2007).

The quite elaborate and intelligent techniques already used by Neanderthals during the Eem period have raised questions regarding their capacity for symbolic thought. For example, at several Eem Neanderthal sites quite large collections of deer antlers have been found.

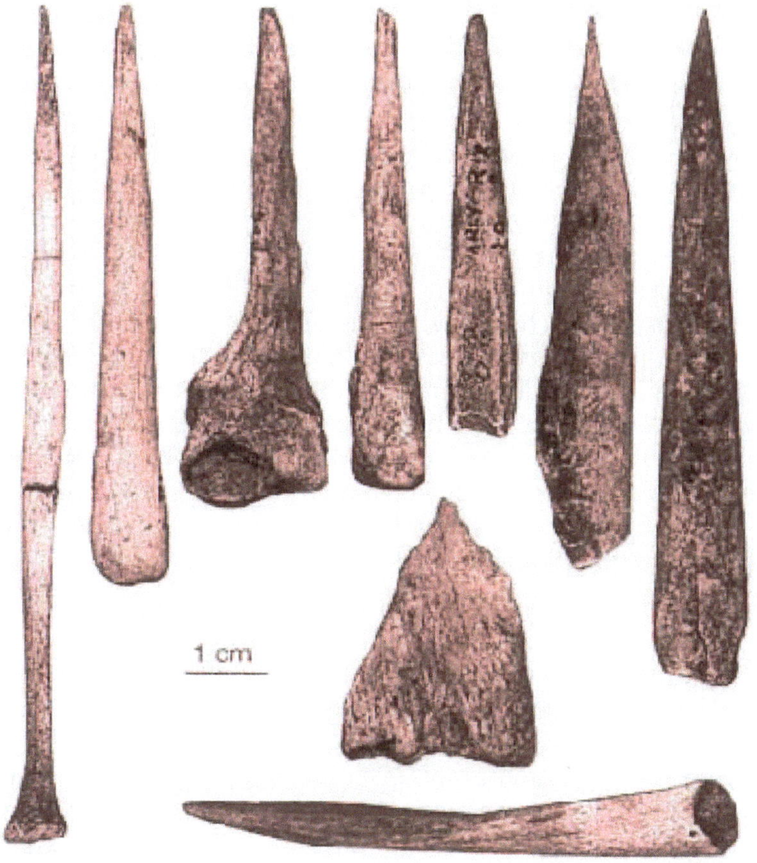


*Figure 1.17. Bone awls from Grotte de Renne, in a Chatelperronian layer, France, estimated to be clothes-making tools from a Neanderthal site active 40 ky BP (adapted from d'Errico, 2003).*

The reasons for keeping antlers could be similar to those behind hunters' displays of antlers in present times, indicating pride after a successful hunt, as well as appreciation of interesting design, both involving what would be defined as symbolic thinking, as opposed to the evidence of practical problem-solving associated with both the tools that have been found and with those (clothes, constructed shelters) that have been inferred. Other evidence is in the burial practices, where defleshing and decoration of human bones for keepsake or for a secondary burial has been found in several instances (Wenzel, 2007). Some researchers take this as evidence for cannibalism (Defleur *et al.*, 1999), but the behaviour is known from other societies (e.g., Neolithic Danes and American Indians) to have taken place without such a connection.

The diet of European Neanderthals in the Eem period consisted almost entirely of meat, with nuts and berries added when they were in season but overall constituting only a small proportion. Later (55–40 ky BP) analysis of teeth shows a similar pattern (e.g. Richards *et al.*, 2008). The meat was primarily from big game, such as horse, mammoth, elephant, rhinoceros and various species of deer (Patou-Mathis, 2000), with smaller prey added some 60–80 ky later (Stiner, 2006). A possible reason would be the quicker moves of small prey, requiring attack by spear-throwing, which from analysis of shoulder and upper arm bone asymmetries seem to have been uncommon in the Eem period (Rhodes and Churchill, 2009). The Eem Neanderthals did use spears, of which one perfect 2.4-m-long specimen, one of the few examples of surviving wooden artefacts from the Eem, was found between the ribs of a straight-tusk elephant at Lehringen (marked in Fig. 1.13); but it appears that the hunting technique usually involved close approach to the prey animal and penetration with several such spears (Wenzel, 2007).

The energy for cooking the meat can probably be inferred with help from similar activities in modern (last 200 years, now gradually disappearing) hunter societies. Based on the Neolithic sites, including those in South Africa mentioned above, one may assume that there were several fireplaces, both inside and outside the shelter used. They would function as cooking devices, as space heaters (particularly in winter), as light sources at night, and as social gathering and activity centres. An open-air fire consuming 10–50 kg of dry wood per hour would convert energy at a rate of 10 000–100 000 W, while an indoor fireplace consuming around 1 kg of wood per hour would produce 1000 W (Sørensen, 2004; W (watt) measures flow of energy per unit of time).

A Neanderthal group living in a cave or a (possibly multi-unit) built shelter would typically have comprised 15–50 individuals, with 25 being considered optimal (Hassan, 1981). Assuming an average adult Neanderthal mass of 80 kg for male and 75 kg for female (Lahr and Foley, 1998), the energy requirement at rest would on average have been 92 W for male and 77 W for female, according to the linear models of Sorensen and Leonard (2001), as worked out by Sørensen (2009). For modern adult humans, the weight difference between male and female is larger: 78 W for an average male, 58 W for females, with large variations (even excluding abnormal overweight or undernourished people in parts of the current world). The maximum sustainable energy conversion of food by a human is about 330 W, of which an average of 100 W can be used to perform work (Spitzer, 1954; Sørensen, 2004).

The rate of heat loss to the air from the body surface of a modern or Neanderthal person takes place through processes of heat radiation, convection and conductance. It depends on the temperature difference between the exposed skin surface and the ambient temperature, the exposed body surface area, and an effective conductivity that would be constant only for the purely conductive heat loss (which is tiny because air is a poor heat conductor). The radiative loss has a strong dependence on the temperatures, and the convective part depends on motion of the air and is thus dependent on wind speed at the exposed skin surface, but even with no macroscopic wind, there is always chaotic eddy motion that will promote convection. In the human heat-loss case looked at here, it will be the dominant contribution (Sørensen, 2004, 2009).

If the heat loss takes place from all parts of a naked human body, then the exposed area for modern males would be around 1.9 $m^2$ and for modern females around 1.55 $m^2$. For Neanderthals, the male surface would also be around 1.9 $m^2$, while for the Neanderthal women it would be around 1.75 $m^2$. The presence of wind to transport the heat away from the body surface will increase the heat loss. At the average Eem wind speed just above 5 m $s^{-1}$ the rate of heat loss would roughly triple relative to that of no macroscopic wind. During a 25 m $s^{-1}$ storm, more than 6 times as much heat will be lost, compared to the no-wind situation.

Heat transfer between the interior of the human body (assumed requiring a temperature of 37 °C) and the interface with the surrounding air will normally be accomplished by conduction, which is described by a thermal resistance for each layer that the heat has to go through. Considering all these processes, one can calculate the minimum ambient temperature that is endurable and consistent with a constant body temperature of 37 °C under given assumptions about any muscle work performed. Depressed body temperatures can only be survived for a limited period of time (Sørensen, 2009).

A naked male Neanderthal asleep in a cave (without wind), insulated only by a shallow subcutaneous fat layer, would be able to endure air temperatures down to 26.9 °C. This corresponds to a skin surface temperature of 36.8 °C (Sørensen, 2009). The amount of fat (or other layers not requiring full body temperature) obviously varies over the body, but is generally 40% higher for females, for whom body fat averages about a quarter of the total weight (Frisch, 2004; Ng, 2006). High levels of body fat are specific to humans and do not apply to other primates.

Continuing to consider a cave-sleeping Neanderthal male (and neglecting at first the heating of the surroundings by the human heat losses, which would not be correct if the person were sleeping in a small hut), one can now explore how an amount of further insulation can lower the acceptable temperature. Also neglected at this stage is heat from fireplaces in the cave. Suppose that our Neanderthal person is sleeping under a cover made from a reindeer, mammoth or bison skin (wrapped into it or having a straw bed underneath), with a heat conduction of 0.035 W $m^{-1}$ $K^{-1}$ and an effective thickness of 0.03 m, one finds heat flow balance for an ambient temperature of -15.3 °C. Should the temperature be below that, our Neanderthal person would keep his day-clothes under the hide cover. With 0.01 m of clothes (made from some animal skin) having the same 0.035 W $m^{-1}$ $K^{-1}$ heat conductivity (also valid for wool), one gets the acceptable cave temperature down to -29.3 °C. The skin of the animals employed by the Neanderthals has varying length of hair, and departures from the assumed average length of 0.03 m will affect insulating properties.

A number of things could affect these estimates. If the cave is sufficiently deep, the temperatures far inside would be roughly equal to the annual average temperatures for the region (i.e considerably colder than outside temperatures during summer and correspondingly warmer during winter). In northern Europe, such cave temperatures would typically be 8–9 °C, and this would be the temperature to consider indoor survival conditions for. If, say, 10 adult Neanderthals and 15 children are sleeping in the cave, the metabolic heat production would be around 1100 W. If, additionally, there would be embers from 5 fireplaces having been used to prepare dinner at 5 × 1000 W, the glowing embers may still produce some 2000 W for a couple of hours but then diminish towards zero as the night proceeds. If a quarter of the heat production is heating the air in the cave (the rest lost through the cave entrance and heating the cave walls), and the cave volume is 1000 $m^3$, then it would take 4 hours to raise the temperature by some 10 °C. After that, heat production would have fallen to the metabolic 1100 W and the further temperature rise would be about 1 °C in 4 hours, considering the same losses.

Alternatively, consider the family of 2 adults and 3 children living in a small hut of floor area 7 $m^2$ and producing 220 W while sleeping. The materials of the hut being assumed (as above) to be plant materials, wooden poles and possibly hides to cover against precipitation, the main effect would be to reduce the heat loss to the surroundings caused by wind. Assuming 12 $m^3$ of air in the hut and the embers of a single fireplace, the hut would offer practically no heat insulation against losses through its 120 $m^2$ air-exposed surface, and its inside temperature would be raised by a mere 1–3 °C, depending on how much of the night the embers will remain effective. The use of such huts requires a suitable solution to the problem of leading the smoke from the fire away. A chimney opening should be conceived in such a way that a minimum of cold air can get in, and such that rain and snow is kept from entering too easily.

Should the Neanderthal male try to spend the night sleeping outside on the ground (e.g. during a multi-day hunting trip; see below), the heat loss in an average wind of 1.5 m $s^{-1}$ would make the minimum endurable temperature equal to 31.8 °C (naked), 17.8 °C (one layer of clothes), or -10.3 °C (blanket made from 0.03 m of animal fur). Carrying such heavy gear as whole animal hides on hunting trips seems more than unlikely, but the estimate can be used to evaluate the need for a shelter at the base camp. At 5 m $s^{-1}$ wind speed, the hide blanket case would correspond to a minimum acceptable temperature of –8.6 °C. In situations with precipitation, the thermal resistance offered by clothes and fur blankets would be substantially reduced, and shelter thus required in most seasons of the year. While 5 m $s^{-1}$ is roughly the annual average wind at a height of 10 m given by the circulation models, and not unreasonable as the wind encountered in an open landscape on a typical day, the appropriate wind speed to use for a Neanderthal sleeping in a forest or heavy growth area may well be considerable lower (e.g. 1–2 m $s^{-1}$, as used in the estimate above), particularly if the male person has been able to find a sheltered slope in the terrain, away from the main wind direction (west).

The numbers discussed above are summarized in Table 1.1, including the corresponding estimates for female Neanderthals. The two lower rows in Table 1.1 give average temperatures for Bispingen (Fig. 1.12) and variations derived from modern statistics 1874–1971 (Danmarks Statistik, 1976). The lowest and highest values inside the parentheses are the extremes measured somewhere in the region during this 100-year period, and the middle ones are 30-year low and high values for the month, averaged over the region (Denmark). The amplitudes of variation are considered to be applicable also to the Eem period. Comparing the endurable temperature estimates with the actual temperatures that would be encountered during the Eem period in northern Germany or Denmark, it is clear that, in average winter cold-spells (-9

°C), the Neanderthal male could survive sleeping outside (away from the wind), provided he had a cover made from animal skin. The female, however, would need additional insulation in order to survive outside. Inside a structure or a cave, both sexes could survive with a 0.03 m animal blanket (taken as mammoth in the table), but not with just a single layer of clothing.

*Table 1.1. Estimates of ambient minimum endurable temperatures for Eem Neanderthals during sleep, and actual temperatures (°C).*

| *Degree C if no other unit mentioned* | male | female |
|---|---|---|
| Body mass (kg) | 80 | 75 |
| Body height (m) | 1.65 | 1,55 |
| Body surface area ($m^2$) | 1.87 | 1.74 |
| Metabolic rate of energy conversion at sleep (W) | 92 | 77 |
| Naked on dry ground (at wind speed $v$ = 1.5 m/s) | 31.8 | 32.3 |
| On dry ground with one layer of clothes ($v$ = 1.5 m/s) | 17.8 | 19.6 |
| On dry ground with mammoth skin cover ($v$ = 1.5 m/s) | -10.3 | -5.7 |
| On dry ground with mammoth skin cover ($v$ = 5 m/s) | -8.6 | -4.1 |
| Temperature rise from dying wood fire ($v$ = 0) | 1–2° for a few h | |
| Naked in cave or small hut | 26.9 | 27.8 |
| In cave or small hut with one layer of clothes | 12.9 | 15.2 |
| In cave or small hut with mammoth skin cover | -15.3 | -10.1 |
| In cave or small hut with mammoth skin cover and one layer of clothes | -29.3 | -22.7 |
| Temperature rise from body heat and embers from dying wood fire in small hut | 1-3° for some 4 h | |
| Temperature rise from body heat and embers from dying wood fires (5) in cave | up to 10° for 4 h | |
| Eem (Bispingen, 125 ky BP) July temperatures (average and ranges; see text) | 17.4 (0, 8, 27, 36) | |
| Eem (Bispingen, 125 ky BP) January temperatures (average and ranges; see text) | 1.0 (-30, -9, 7, 12) | |

In extreme weather situations (-30 °C according to Table 1.1), one more layer of clothes makes the male cope, but additional heat (such as from embers of fires) is required for the female (the sex difference being primarily due to differences in basic metabolic rate). In conclusion, clothes and bed covers are essential for winter survival, as are shelter in the form of caves or built structures, including wood-burning facilities.

In examining daytime energy use, one notes that the Neanderthals were as efficient hunters of their selected prey as modern hunting people (Sorensen and Leonard, 2001), and it has been suggested that Neanderthal women perhaps took part in the hunt (Kuhn and Stiner, 2006). The total time spent on hunting might have been similar to that of modern hunting tribes, averaging 5–7 hours per day. However, because of the size of the Neanderthal prey, dominated by large animals such as horse, deer and, in open landscapes, rhinoceros and mammoth, it is likely that hunting expeditions were conducted only at intervals, and that each hunting trip was a multi-day endeavour. Time would have been spent in other periods on replenishing weapon arsenals and tending to camp maintenance (with much less energy expenditure, except for tree felling and cutting up firewood). The comparison with modern hunters is not totally fair, since modern hunters–gatherers mostly live in Equatorial parts of the world, where collection of edible plant roots, berries or fruit is possible most of the year, and also because the hunting range is lower than the 50–100 km estimated for Neanderthals.

Let me attempt to describe a hunting expedition in the Eem period. If Hassan's estimate (1981) of each Neanderthal group comprising 25 individuals is correct, a hunting party could not have comprised more than 5–6 adult people: some adults would have to stay at the dwelling site to attend the 15 children, including breast-feeding of babies. I use mammoth as a prey example because of its size and mass (some 8 metric tons for a fully grown individual) and it therefore presents a maximal challenge for the hunters. However, mammoth was rare in northwest Europe during the Eem period, but common both before and after.

Spotting a mammoth herd, the Neanderthal hunting party would then track the herd and single out individual animals for attack (weak or not fully grown), and possibly throw a spear into such an animal in order to have it lag behind the group. Relatively little time would be used to actually attack and kill the prey. The biggest challenge appears after having killed the prey. I take again the most demanding example, that of a fully grown woolly mammoth, 3–4 m high and with a mass of 6–8 tons. Carrying just the best meat parts back to home (perhaps 50 km or more away) is a formidable challenge. Each adult Neanderthal might carry a maximum of 25 kg, and probably less for so long a distance. Just 2 tons of meat would then take 100 adult Neanderthals each carrying 20 kg. The hunting party of 6 would need to leave two people at the slaughter site to guard the remains against other predators, leaving 4 people 25 round trips to carry the meat home. What did they actually do? Did they take only a smaller fraction of the meat provided? Did they neglect pieces of the prey animal that could have been used for other purposes than food? The hunters probably feasted immediately on the bone marrow, which could be eaten without cooking, and then started to make plans. With a heavy weight on the shoulders, the 50 km trip home may have taken 10 hours, plus rest periods (i.e. a whole day). If they really did 25 round trips, it would have taken 50 days. On the other hand, the reward in terms of food would have been large: two tons of mammoth meat has a food energy value of some 16 GJ (judged from similar meat products in use today). The 25-member Neanderthal base group would have required an energy input from food of some 300 MJ per day (see below), implying that the single mammoth kill could have fed the small community for 53 days, just breaking even. But add to this the problem of food preservation: during the coldest period, 2 months of meat storage might have been feasible, but certainly not during the warmer periods of the year (and spending 50 days to get the meat back into camp would also not work in summer without access to preservation methods). So something has to be changed in the assumptions made above. Could it be drying the meat at the slaughter site, with use of solar energy in summer and wind energy with low air moisture in winter? This would reduce the meat mass to be carried home to about a third (FAO, 1990) and require only simple techniques of placing meat slices hanging on a set of horizontal wooden sticks for a couple of days. Similar techniques have been used by recent American Indians and by African hunter–gatherers. When the time for consuming comes, the dried meat is just cooked with water and retains nearly full nutritional value. Storage times could be prolonged by smoking and salting the meat. Smoking seems fully accessible to Neanderthal societies. We do not know if they used salt.

I assume that the meat was dried at the slaughter site, with a couple of Neanderthals defending the drying scaffold against scavenging animals, and that bringing the food back to home base now took some 7 round trips. In this way, two weeks of work by 6 people would supply more than 7 weeks of enough food for the entire 25-person group. So, once more, considering the energy expenditure of various behaviours has taught us something that the artefacts found have not given any clues to: the immediate drying of meat as a condition for making big-game hunting feasible. For other prey types, such as deer varieties, the techniques used could be similar, although for the smallest kinds of prey, it may have been feasible to carry them back to camp before processing.

As concerns energy use, tracking down and following a group of large animals would mostly involve walking, running and standing, with rates of enhanced metabolism in the range of 1.5 to 3 times the basic one (Leonard and Robertson, 1997), and only the actual kill activity would have used energy at levels as high as 3–4 times the basic metabolism. The muscle work performed while walking is dispersed as friction against the ground or contributes to changing potential energy, and would have been adding 0.3 to 1.0 times the basic metabolic rate, depending on the weight of meat carried. Doubling the basic rate of energy conversion constitutes the maximum work a human can deliver for an extended period. The difference between total metabolic conversion and work performed is heat, used to maintain body temperature in cold weather or otherwise transmitted to the environment by evaporation or conduction (Sørensen, 2004). The maximum work a human being can deliver for a short time (e.g. 1 minute) is as high as 20 times the basic rate (e.g. tree-felling using axes, of which several have been found at Neanderthal sites).

Considering that members of the hunting team would have to sleep overnight at the site of meat drying and other processing, this would cause problems during cold periods, assuming that no insulating cover was carried along on the hunting expedition. The skin from the mammoth killed could have been separated as a high priority, but it could hardly have been scraped and readied in time for immediate use. Alternative solutions would be to build a small igloo, in cases of snow/ice cover on the ground, or a temporary hut of twigs with leaves, if available. It is debatable whether the Neanderthals would have had tools for igniting fires with them on hunting trips, but considering the light weight of a small polished wooden pole and some woolly hair, it would appear quite probable.

Daily average energy expenditure (determining food energy requirements) for Neanderthals has been estimated as in the range of 163–217 W (Steegmann *et al.*, 2002); however, this estimate assumes that Neanderthals had a built-in thermoregulation in their body, based on conversion in cold weather of fat from a particular type of deposit in the body (in brown adipose tissues). This makes the heated body environment come ultimately from food energy rather than from insulating clothes (Sørensen, 2009). The even higher estimates (266 W) by Sorensen and Leonard (2001) appear unrealistic, because roughly a third of the hours in the day would be spent sleeping at basic metabolic rate, another third sitting, tending fires etc. (110–200 W), and the final third would therefore have to be beyond sustainable levels to give an average above 200 W. Estimates for modern human hunters and gatherers are around 125 W (Sørensen, 2004).

Some data are available on the energy use when carrying weights (such as bringing killed prey back to camp, infant carrying; Watson *et al.*,

2008) and, as assumed above, this could bring the "walking" estimate up to the maximum of three times basic in terms of food energy intake. Note that there is a further argument for use of heavy footwear: helping to avoid being wounded by sharp stones or plant thorns during periods of carrying heavy prey animals (presumable cut into pieces of convenient size, possibly transported in sacks), where it is also more difficult to look out for such perils. Good footwear is indispensable when having to make several trips of possibly over 50 km to and from the site where the prey animal was killed. So for the hunters, energy use when carrying a load of killed prey would be estimated as around 285 W, some 160 W when following the prey, 140 W during cutting and arranging the temporary site, and 92 W during sleep. The average is 152 W.

At home, the hardest work for the men would probably be wood cutting, averaging 5 workdays a month at 350 W, with the rest of the day at 140–150 W. The average would then be around 140 W, and for the females about 120 W, giving an energy conversion (and hence food requirement) of 1300–1400 W for the ten adult group members. The children would have added some 400 W. Vertebral bones found with *osteoarthritis* damage may be taken as support for the assumption that Neanderthal hunters had to carry large burdens of meat long distances back to their settlements (Dawson and Trinkaus, 1997).

The shorter lower legs of the Neanderthals may have reduced energy spending while walking at a pace adapted to the limb proportions, but a slight reduction in convenient speed may have been accompanied by a reduction in maximum walking and running speeds, and may thus be an explanation for Neanderthals not hunting swift animals. Energetic expenses for childbirth and lactation have been investigated and found to be small (Aiello and Key, 2002). Kaplan *et al.* (2000) have noted a significant variation in energy use with age in modern hunter–gatherer societies. This would probably also apply to Palaeolithic communities.

Table 1.2 summarizes the remarks made above and completes the picture by estimating the energy from fires, giving the amount of meat that would have to be consumed at a 70% efficiency to support the daily energy use. The table assumes that one hunting trip is carried out every month. If the return from a mammoth kill is two tons, as assumed in the example, there would be meat for some 8 weeks, so the assumption is that not all hunting trips are fully successful (maybe bringing back a deer rather than a large animal), and that meat preservation by drying and possibly smoking may be less than 100% successful, leading to occasional losses of stored meat.

*Table 1.2. Neanderthal energy use.*

| *Example of time and corresponding rate of energy use (W/cap)* | *4 males* | *1 male* | *1 female* | *4 females* |
|---|---|---|---|---|
| Hunt: tracking down prey (8 h, 1 day in month) | 2.04 | 2.04 | 1.71 | |
| Hunt: prey killing (1 h, 1 day in month) | 0.45 | 0.45 | 0.37 | |
| Hunt: parting mammoth, drying (3 h, 1 day in month) | 0.96 | 0.96 | 0.80 | |
| Hunt: eat, sleep, rest at hunt site (12 h, 1 day in month) | 1.69 | 1.69 | 1.41 | |
| Hunt: eat, watch, cut, scrape, sleep (24 h, 12 days in month) | | 44.16 | 36.96 | |
| Hunt: carrying meat back (10 h, 7 or 1 days in month) | 26.83 | 3.83 | 3.21 | |
| Hunt: sleep, rest at home (14 h, 7 or 1 days in month) | 13.77 | 1.97 | 1.65 | |
| Hunt: returning to hunt site from home (8 h, 6 days in month) | 12.27 | | | |
| Hunt: eat, sleep, rest at hunt site (16 h, 6 days in month) | 14.72 | | | |
| Home: wood cutting (8 h, 5 days in month) | 15.33 | | | |
| Home: stone flaking, tools construction, clothes making (8 h, 11, 16 or 30 days in month) | 22.49 | 32.71 | 27.38 | 61.33 |
| Home: fire attention, child rearing, food prep., leisure, eat (8 h, 16 or 30 days in month) | 22.90 | 22.90 | 19.16 | 42.93 |
| Home: sleep (8 h, 16 or 30 days in month) | 16.36 | 16.36 | 13.69 | 25.67 |
| **Monthly average energy expenditure, adult humans (W/cap)** | **149.81** | **127.06** | **106.35** | **129.93** |

| *Summary:* | *male* | *female* |
|---|---|---|
| Average adult minimum energy requirement (W/cap) | 145.26 | 125.22 |
| Total average adult energy requirement, W for whole group of 10 | 1352.37 | |
| Children's average energy requirement (W for whole group of 15) | 400.00 | |
| Equivalent meat intake (loss 30%), W and GJ per month for whole group of 25 | 2503 W or 6.5 GJ/month | |
| Equivalent meat intake (loss 30%), in kg per month for whole group of 25 | 813 kg/month | |
| Fires: 5 cooking fires 8 h, 30 d (346 kg dry wood), average over month | 1667 W | |
| Fires: large outdoor fire 8 h, 30 d (622 kg dry wood) | 3000 W | |
| Fires: Possible fire at hunt site 12 h, 14 days in month (168 kg) | 810 W | |

The hours of maintaining fires indicated in Table 1.2, as well as their equivalent energy conversion, should be taken as indicative only, and realistic time sequences would show highly varying energy conversion, from flame-based burning to just glows from embers. The Neanderthal metabolic energy expenditures come at 145 W for males and 125 W for female adults, slightly lower than some of the other estimates quoted above but consistent with later (and slightly smaller) *Homo sapiens* hunters' rates of energy conversion. The table also shows that the use of wood energy is about twice as big as the 2500 W of food energy consumed on average by the 25-person group. The estimate is based on availability of hardwood for felling, which is the most convenient source. It would be available in mixed tundra and scattered tree biomes, as well as in the boreal forest areas, but in some areas only twigs from minor trees and other plant material would be present. In this case, the Neanderthals would have to spent more time on collecting firewood and would possibly lower the power of the fires or the length of time keeping them going.

An important factor in determining the way in which a Neanderthal society worked is that of mobility. Did Neanderthals spread quickly over Europe, did individual communities trade with each other and were there sexual relations outside the (on average) 25-person groups mentioned? Neanderthals did use local materials when available, but at least towards 40 ky BP there is evidence for stone material having been collected more than 100 km from a given settlement and ornamental materials such as seashells collected more than 500 km from the base (Weaver and Steudel-Numbers, 2005). Furthermore, to avoid inbreeding problems and fluctuations in female populations, several 25-person groups would have had to exchange sexual favours at intervals (not implying that Neanderthals were aware of the former connection). That this was nevertheless successfully achieved follows from the genetic studies recently made (and with many more on their way; Hawks *et al.*, 2007; Weaver and Roseman, 2008). Further, there is the formation of new settlements, often involving mobility over distances exceeding 1000 km for each event of resettlement (cf. the case of the Finnish Wolf Cave). Changing climatic conditions have forced entire groups to move, and disease may have exterminated groups and left their hunting territory open to newcomers. A hunting territory for a group is estimated (using as above either data from modern hunters or ecological models of prey–herd size stability) to have comprised at least 10 000 $km^2$. Many modern hunters–gatherers have less area at disposal per capita, but they derive additional food from plant material (Hassan, 1981).

How big was the entire European population during the Eem period?

At 25 people per $10^{10}$ $m^2$ (or 0.025 cap/$km^2$), the populated part (excluding the far north and far east) could have sustained some 12 000 big-game hunters. This is consistent with the upper limit population suggested by genetic considerations (Harpending *et al.*, 1998). According to Hassan (1979; 1981), a mating network needed to comprise nearly 500 people (or 20 groups). This would have allowed some 24 (possibly overlapping) mating networks in populated parts of Europe, organized in smaller groups willing to walk up to 130 km to search for a spouse. Palaeolithic population estimates have had a tendency to get smaller with time: out of a world population of 1 million around 100 ky BP (Hassan, 1981), less than 20 000 persons may have lived in Europe. The 10 000-person estimate above for Europe is repeated by Jorde *et al.* (2001), using genetic data. These are supposed to count only the part of population in the fertile age, which according to the age distribution proposed by Hassan (1981) implies a total population of 20 000. Other estimates are based on actual archaeological sites uncovered (Bocquet-Appel *et al.*, 2005). They arrive at about 5000 individuals at 40 ky BP (but with an uncertainty interval of 1700 to 37 700). This is a lower estimate, with a centre value that would agree with estimates of an order-of-magnitude size of 10 000, if half the European sites have been destroyed by subsequent events or simply not found (some are presently underwater). The problem is that going further back than 40 ky BP, only some 10 European sites for each 10 ky time span have been found to contain Neanderthal remains (skulls, bones or artefacts found), and the lower-limit European population estimate from excavations is therefore only about 50 persons, too few for sustainability.

Because the Neanderthals survived from before the Eem to after 40 ky BP, the population must have exceeded the minimum sustainable level, which would be at least one mating network of about 500 persons, and by some margin to take care of unexpected setbacks. There is more on this in Chapter 2. The number of sites lost or not found also increases with time passed, so I shall assume that the Eem Neanderthal population was rather at or above 1000 people. The wide geographical dispersal of sites found suggests that the number of mating networks was at least 3 and more likely 5, corresponding to a European Neanderthal population of 1500–2500. Given the area of 472 000 $km^2$ occupied by the European Neanderthals (omitting a few outside excursions, e.g. the mentioned ones to the Nordic countries), it is clear that at such a population level, the Neanderthals never experienced systematic food shortage and never overexploited their environment or put any of their prey species at risk of extinction. But why did the population not grow during the long presence in Europe? To understand this, one has to look at the mortality of the Neanderthal people.

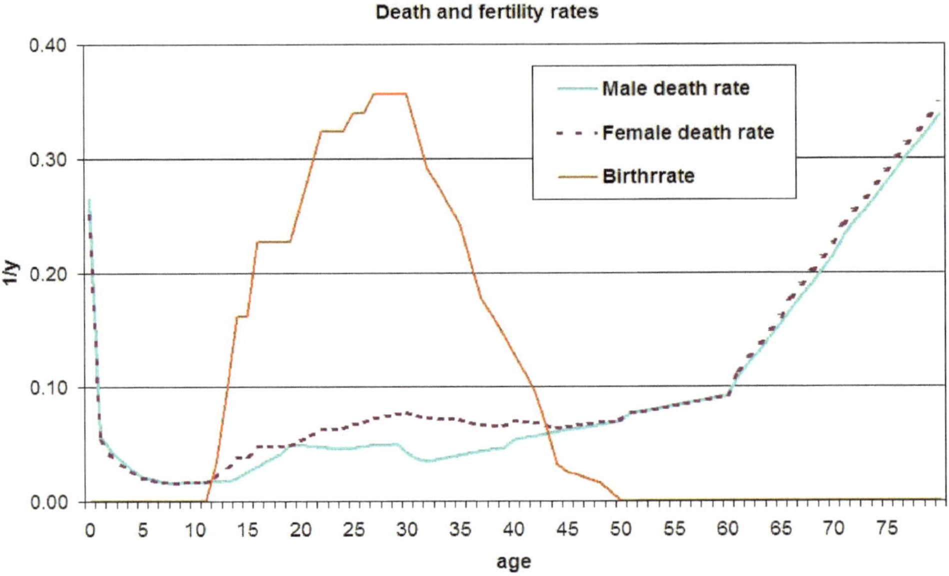


*Figure 1.18. Estimated death and birth rates of Eem Neanderthals. There are few data above age 55–60. However, this is of little importance due to small numbers surviving to such ages (cf. Fig. 1.20).*

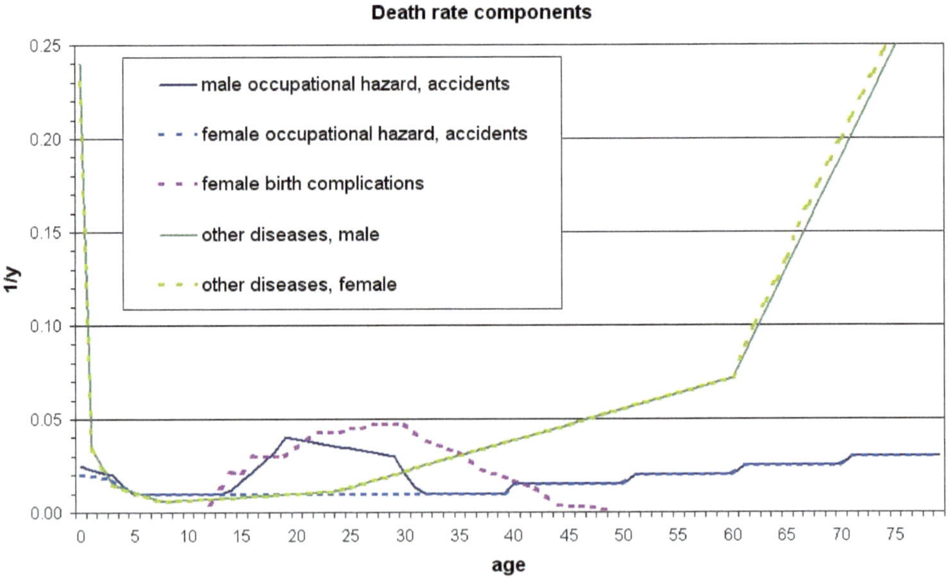


*Figure 1.19. Estimated components of Neanderthal death rates.*

Based on individual age determinations for finds of Neanderthal skeleton parts with age determination, Hassan (1981) has constructed a survival curve from which can be extracted the death rate (probability that an individual at a certain age will die within one year), as shown in Fig. 1.18. The death rates have been refined, with differences in male and female death rates having been added, because of their obvious importance. This is not based on finds, because gender could not be determined with certainty for all bones around 1980, when Hassan constructed his death rates. Fig. 1.19 shows a split of the death rates between the two categories. Most deaths at medium ages are attributed to two specific reasons: women dying during childbirth (in 13% of birth cases), and young men dying in accidents while participating in hunting and other activities with danger. The total agrees with Hassan's estimates, as well as with more recent ones (Trinkaus, 1995; Sørensen, 2011). The lump number of deaths in Hassan's 0–5 year age group has been distributed in such a way that the infant deaths (age 0–1) agree with numbers for modern hunter groups without access to medical facilities. The Hassan reconstruction has one peculiar anomaly: a very high death rate for the age group 40–60 years. An explanation could be that many remains of dead Neanderthals in this age group are found because they would have been more likely to die within the base settlement than younger people, who might die during hunting trips and be buried or left where they died, in places unlikely to have been identified for archaeological investigation. For this reason a more smoothly rising death rate has been used for the age group 40–60, but in any case it does not make much difference, due to the short overall lifespan of most Neanderthals.

The birth rate curve (number of live births by a women of a given age), also shown in Fig. 1.18, is derived from the estimate by Hassan (1981) and is consistent with generic curves for comparable groups in recent times. The overall rate is adjusted to yield a total population that is growing modestly (roughly doubling in 500 years). This involves increasing the total fertility from Hassan's 4.5 to 7.75 children during a woman's total reproductive period. The corresponding number of females produced by each female is 3.78, considering a sex ratio at birth of 0.488/0.512. The relatively high number of offspring is required for avoiding extinction, given the death rates that are high for infants and elevated for females giving birth, and for everyone during daily chores.

Using the data from Fig. 1.18, the implied population distribution can be calculated. It is shown in Fig. 1.20, including a 1000-year integration, which partly has the effect of correcting the normalized input distributions in case of any inconsistency, and partly indicates the population increase over 1000 years with unchanged birth and death rates. The increase is about a factor of three. This number is very sensitive to the assumed total fertility (overall birth rate scaling) and the growth is probably the lowest that could have allowed long-term survival of the Neanderthals, given the periods with rougher climate than that of the Eem. The effect of changing environment will be considered in the discussion of the Weichselian ice age in Chapter 2, but also outbreaks of new diseases could have led to extinction, if there were no surplus to take from. The death rates shown in Fig. 1.18 are consistent with at least some of the similar numbers for modern hunter–gatherer societies (Hill *et al.*, 2007), but the latter show large variations, associated with size of territory, diet composition and belligerent behaviour, for example.

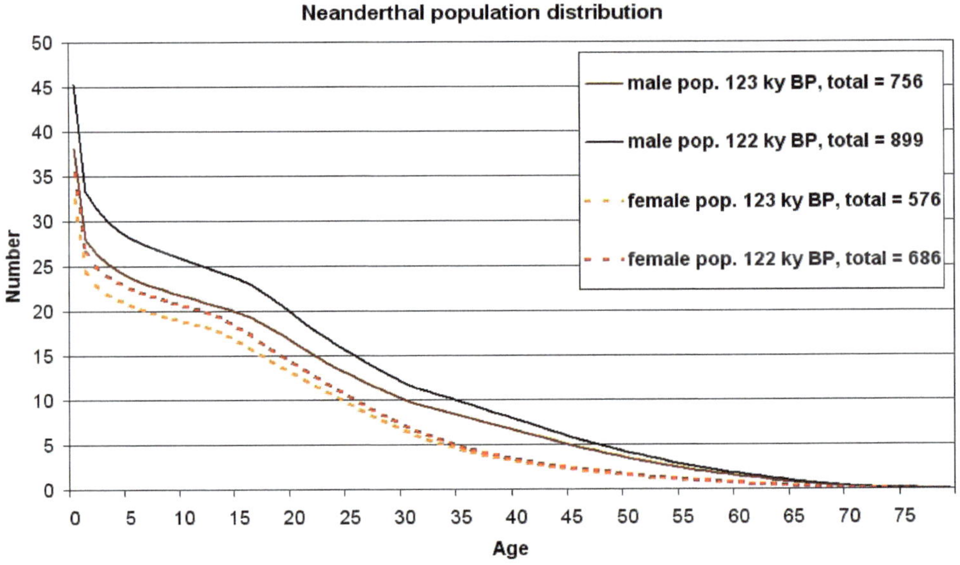


*Figure 1.20. Population age distribution and growth in the European Eem Neanderthals, based on the rates given in Fig. 1.18.*

It is sometimes claimed that the low overall population density of the Neanderthals would make epidemic diseases unimportant (Armelagos *et al.*, 2005). However, as noted, the Neanderthal survival depends on fairly frequent visits to other groups in the mating network, so the total mortality rate may after all reflect transmission of severe diseases, as well as the obvious risk of fatalities during hunt or attack by wild animals (plus childbirth risks for the female population). It is also evident that the high mortality is a key factor in preventing the Neanderthals from reaching high population numbers. The total fertility could in principle be as high as 12, a number seen in Europe in the 19th century. The increase would have to be at ages away from the peak in Fig. 1.18.

However, this requires shorter intervals between births and, in consequence, better conditions for the women, which was not possible for the Neanderthals, because of additional miscarriages and stillbirths (neither of which are included in the death rates of Fig. 1.19), as well as the need to participate in other activities, such as foraging or shelter maintenance. This discussion will be essential in dealing with survival during the following, colder conditions of the Weichselian period.

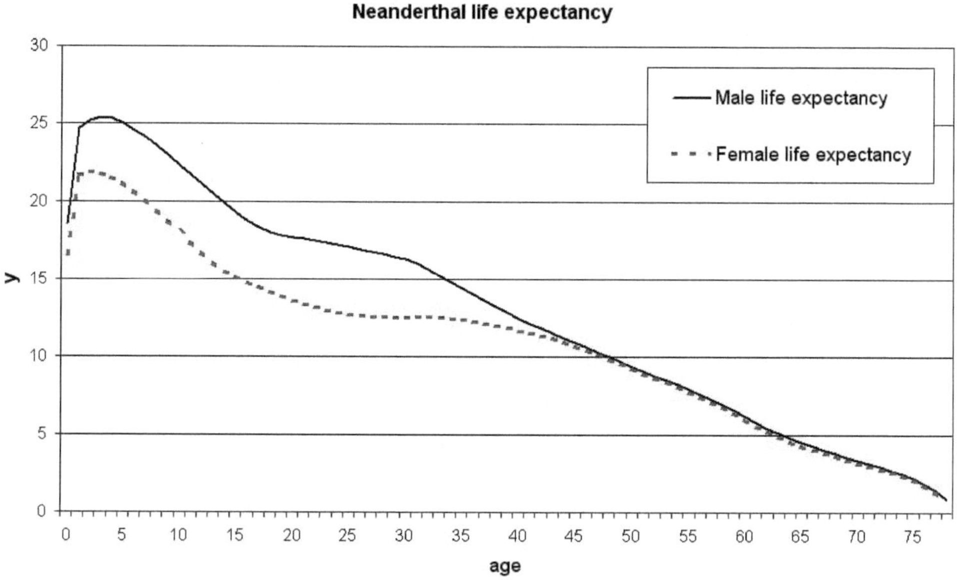


*Figure 1.21. Neanderthal life expectancy (unchanged over the 1000-year period of population numbers shown in Fig. 1.20, because the rates determining life expectancy are assumed to be time-independent).*

The Neanderthal life expectancy based on the calculated population curves is shown in Fig. 1.21. At birth, the male life expectancy is about 19 years and the female life expectancy 17 years. For the mid-life period, the female life expectancy is 4 years lower than the male one, a feature known from many more recent societies with a high mortality beeing associated with childbirth. If the woman survives to age 40, her life expectancy is roughly the same as for men. The life expectancy at age 18 is about 18 years for men, but only 14 years for women.

One may wonder why the Neanderthals never developed spear-throwing, given that they totally depended on spears for their big-game hunt. Presumably the manufacture of an efficient spear was considered a major effort, explaining why the Neanderthal hunter did not want to lose the spear, and why he did not carry an arsenal of more than the one used to repeatedly stab the mammoth. The Neanderthal's reluctance to throw the spear was most likely reflecting his fear of rendering himself defenceless in front of a big and ferocious animal. To kill a large animal like the mammoth, the small group of Neanderthal hunters would have to stab the animal many times, and each hunter thus had to retain his spear for the next attack. The design was therefore one that allowed easy retraction. Given this circumstance of their very specialized hunting strategy, it becomes easier to understand that they did not start to hunt smaller (and quicker) animals (by throwing spears or inventing bow and arrows) during the Eem period, where forests invaded many of their conventional hunting ranges. The reason may also simply be that the remaining big game was sufficient for the small population of a few thousands. An alternative explanation could be that the Neanderthals were afraid of entering the large forests. Maybe one of them had experienced of losing his or her way in the woods. A widely held view is that they systematically avoided the forests.

In the future, additional Neanderthal sites may be discovered. Perhaps a new approach should be used in the excavation of such sites. First, any handling of artefacts by excavating personnel without protective gloves and clothing should be avoided, in order to be able to make more use of DNA analysis without the risk of contamination by modern humans. In particular, this would enable weak surface DNA traces on tools to become analysed. Second, new non-destructive excavation methods or supplements should be employed once a layer expected to contain artefacts from a relevant period has been identified, and covering layers studied and removed. Systematic 3D X-ray photography using a range of frequencies and hardness of radiation could enable the identification of soft remains (such as rush mats and other building materials, wooden tools having already been reduced to flimsy fibre arrays) that are overlooked with traditional excavation techniques. First steps in this direction have been taken by the Las Caldas Cave laser project (González-Aguilera *et al.*, 2009).

***References:***

Aiello, L., Key, C., 2002. Energetic consequences of being a *homoerectus* female. *American J. Human Biology* **14**, 551-565.

Aiello, L., Wheeler, P., 2004. Neanderthal Thermoregulation and the Glacial Climate. *Neanderthals and Modern Humans in the European Landscape of the Last Glaciation–Archaeological Results of the Stage 3 Project* (van Andel, T., Davies, W., eds), Ch. 9, pp. 147–166. The McDonald Institute for Archaeological Research, Cambridge. Reprinted with corrections. The article is based on two assumptions: 1) Neanderthals always performed maximum sustainable physical work, even at sleep; 2) a subjective wind chill index is inserted into heat loss expressions instead of the real temperature and no explicit modelling is made of convection by wind.

Armelagos, G., Brown, P., Turner, B., 2005. Evolutionary, historical and political economic perspectives on health and disease. *Social Science and Medicine* **61**, 755–765.

Bard, E., Hamelin, B., Fairbanks, R., 1990. U-Th ages obtained by mass spectrometry in corals from Barbados: Sea level during the past 130 000 years. *Nature* **346**, 456–458.

Berger, A., Loutre, M., Kaspar, F., Lorenz, S., 2007. Insolation during interglacial. *The climate of past Interglacials* (Sirocko, F., Claussen, M., Goñi, M., Litt, T., eds), *Developments in Quaternary Sciences* **7, 3**–12 (Elsevier).

Bintanja, R., Wal, R. van de, Oerlemans, J., 2002. Global ice volume variations through the last glacial cycle simulated by a 3-D ice-dynamical model. *Quaternary Int.* **95–96**, 11–23.

Bocquet-Appel, J-P., Demars, P-Y., Noiret, L., Dobrowsky, D., 2005. Estimates of upper Paleolithic meta-population size in Europe from archaeological data. *J. Archaeological Sci.* **32**, 1656–1668.

Calov, R., Ganopolski, A., Claussen, M., Petroukhov, V., Greve, R., 2005a. Transient simulation of the last glacial inception. Part I: glacial inception as a bifurcation in the climate system. *Climate Dynamics* **24**, 545–561.

Calov, R., Ganopolski, A., Petroukhov, V., Claussen, M., Brovkin, V., Kubatzki, C., 2005b. Transient simulation of the last glacial inception. Part II: sensitivity and feedback analysis. *Climate Dynamics* **24**, 561–576.

Danmarks Statistik, 1976. *Statistisk Årbog 1976*. Danmarks Statistik, Copenhagen.

Dawson, J., Trinkaus, E., 1997. Vertebral Osteoarthritis of La Chapelle-aux-Saints 1 Neanderthal. *J. Archaeological Sci.* **24**, 1015–1021.

Defleur, A., White, T., Valensi, P., Slimak, L., Crégut-Bonnoure, É., 1999. Neanderthal cannibalism at Moula-Guercy, Ardèche, France. *Science* **286**, 128–131.

d'Errico, F., 2003. The invisible frontier. A multiple species model for the origin of behavioural modernity. *Evolutionary Anthropology* **12**, 188–202.

Donat, M., Kaspar, F., 2006. Simulation of the last interglacial and the subsequent glacial inception with the Planet Simulator. *Climate of the Past Discussions* **2**, 1347–1369.

FAO, 1990. Manual on simple methods of meat preservation. Animal Production and Health paper No. 79, FAO, Rome, http://www.fao.org/docrep/003/x6932e/X6932E00.htm

Frisch, R., 2004. *Female fertility and the body fat connection*. University of Chicago

Press, Chicago.

Funder, S., Demidov, I., Yelovicheva, Y., 2002. Hydrography and mollusc faunas of the Baltic and the White Sea seaway in the Eemian. *Palaeogeography, Palaeoclimatology, Palaeoecology* **184**, 275–304.

González-Aguilera, D., Muñoz-Nieto, A., Gómez-Lahoz, J., Herrero-Pascual, J., Gutierrez-Alonso, G., 2009. 3D Digital Surveying and Modelling of Cave Geometry: Application to Paleolithic Rock Art. *Sensors* **9**, 1108–1127.

Harpending, H., *et al.*, 1998. Genetic traces of ancient demography. *Proc. Nat. Acad. Sci. USA* **95**, 1961–1967.

Hassan, F., 1979. Demography and archaeology. *Ann. Rev. Antropology* **8**, 137–160.

Hassan, F., 1981. *Demographic Archaeology.* Academic Press, New York.

Hawks, J., Cochran, G., Harpending, H., Lahn, B., 2007. A genetic legacy from archaic Homo. *Trends in Genetics* **24**, 19–23.

Hill, K., Hurtado, A., Walker, R., 2007. High adult mortality among Hiwi hunter-gatherers: Implications for human evolution. *J. Human Evolution* **52**, 443–454.

Huber, C., Leuenberger, M., Spahni, R., Flückiger, J., Schwander, J., Stocker, T., Johnsen, S., Landais, A., Jouzel, J., 2006. Isotope calibrated Greenland temperature record over Marine Isotope Stage 3 and its relation to $CH_4$. *Earth and Planetary Science Letters* **243**, 504–519. Data available from World Data Center for Paleoclimatology at http://www.ncdc.noaa.gov/paleo

Jansen, E., *et al.*, 2007. Palaeoclimate. Ch. 6 in *Climate Change 2007: The Physical Science Basis.* Contribution of Working Group I to the Fourth Assessment Report of the Intergovernmental Panel on Climate Change. (Solomon, S., *et al.*, eds). Cambridge University Press, Cambridge. http://www.ipcc.ch/

Johnsen, S., *et al.*, 1997. The d18O record along the Greenland Ice Core Project deep ice core and the problem of possible Eemian climatic instability. *J. Geophys. Res.* **102**, 26397–26410.

Jorde, L., Watkins, W., Bamshad, M., 2001. Population genomics: a bridge from evolutionary history to genetic medicine. *Human Molecular Genetics* **10**, 2199–2207.

Kaplan, H., Hill, K., Lancaster, J., Hurtado, M., 2000. A theory of human life history evolution: Diet, intelligence, and longevity. *Evolutionary Anthropology* **9**, 156–185.

Kaspar, F., Cubasch, U., 2007. Simulation of the Eemian interglacial and the subsequent glacial inception with a coupled ocean-atmosphere general circulation model. *The climate of past Interglacials* (Sirocko, F., Claussen, M., Goñi, M., Litt, T., eds), *Developments in Quaternary Sciences* **7**, 499-515

(Elsevier).

Kinder, H., Hilgemann, V., 1964. *Atlas zur Weltgeschichte, Band I.* Deutscher Taschenbuchverlag, Stuttgart.

Kühl, N., Litt, T., 2007. Quantitative time-series reconstructions of Holsteinian and Eemian temperatures using botanical data. *The climate of past Interglacials* (Sirocko, F., Claussen, M., Goñi, M., Litt, T., eds.), *Developments in Quaternary Sciences* **7**, 239-251 **(Elsevier).**

Kühl, N., Litt, T., Schölzel, C., Hense, A., 2007. Eemian and Early Weichselian temperature and precipitation variability in northern Germany. *Quaternary Sci. Rev.* **26**, 3311–3317.

Kuhn, S., Stiner, M. (2006). What's a mother to do? *Current Anthropology* **47**, 953–966.

Lahr, M., Foley, R., 1998. Towards a theory of modern human origins: Geography, demography, and diversity in recent human evolution. *Yearbook of Physical Anthropology* **41**, 137–176.

Leonard, W., Robertson, M., 1997. Comparative primate energetics and hominid evolution. *American J. Phys. Anthropology* **102**, 265–281.

Lisiecki, L., Raymo, M., 2005. A Pliocene-Pleistocene stack of 57 globally distributed benthic $\delta^{18}O$ records. *Paleoceanography* **20**, PA1003, doi:10.1029/2004PA001071. Data available from World Data Center for Paleoclimatology at http://www.ncdc.noaa.gov/paleo

Lohmann, G., Lorenz, S., 2007. Orbital forcing on atmospheric dynamics during the last interglacial and glacial inception. *The climate of past Interglacials* (Sirocko, F., Claussen, M., Goñi, M., Litt, T., eds), *Developments in Quaternary Sciences,* **7**, 527-545 (Elsevier).

Manning, P., 2006. *Homo sapiens* populates the Earth : A provisional synthesis, privileging linguistic evidence. *J. World History* **17**, 115–158.

Møhl-Hansen, U., 1958. Første sikre spor af mennesker fra interglacialtid i Danmark. *Årbøger for Nordisk Oldkyndighed*, 101–126.

Mounier, A., Marchal, F., Condemi, S., 2009. Is *Homo heidelbergensis* a distinct species ? New insights on the Mauer mandible. *J. Human Evol.* doi:10.1016/j.jhevol.2008. 12.006.

Ng, J. Ying Chi, 2006. Automatic measurements of human subcutaneous fat with ultrasound. University of British Columbia, Canada (Master of Applied Science thesis).

Noe-Nygaard, N., 1989. Man-made trace fossils on bones. *Human Evolution* **4**, 461–491.

Patou-Mathis, M., 2000. Neanderthal subsistence behaviours in Europe. *Int. J. Osteoarchaeology* **10**, 379–395.

Peltier, W., 1994. Ice age paleotopography. *Science* **265**, 195–201.

Petit J., *et al.*, 1999. Climate and Atmospheric History of the Past 420,000 years from the Vostok Ice Core, Antarctica. *Nature* **399**, 429–436. Data available from World Data Center for Paleoclimatology at http://www.ncdc.noaa.gov/paleo

Rhodes, J., Churchill, S., 2009. Throwing in the middle and upper paleolothic: inferences from an analysis of humeral retroversion. *J. Human Evol.* **56**, 1–10.

Richards, M., *et al.*, 2008. Isotopic dietary analysis of a Neanderthal and associated fauna from the site of Jonzac (Charente-Maritime), France. *J. Human Evol.* **55**, 179–185.

Rousseau, D-D., Hatté, C., Duzer, D., Schevin, P., Kukla, G., Guiot, J., 2007. Estimates of temperature and precipitation variations during the Eemian interglacial : New data from the Grande Pile record (GP XXI). *The climate of past Interglacials* (Sirocko, F., Claussen, M., Goñi, M., Litt, T., eds), *Developments in Quaternary Sciences*, **7**, 231-238 (Elsevier).

Schulz, H-P., 2001. The lithic industry from layers IV-V, Sisiluola cave, Western Finland, dated to the Eemian interglacial. *Préhistoire Européenne 2000–2001*, **16–17**, 43–56.

Schulz, H-P., 2006. Artefact-geofact analysis of the lithic material from the Sisiluola cave. *Fennoscandia Archaeologica 2005* **XXIV**, 64–75, references at back of volume.

Shackleton, N., 1987. Oxygen isotopes, ice volume and sea level. *Quaternary Science Reviews* **6**, 183–190.

Siddall, M., Chappell, J., Potter, E-K., 2007. Eustatic sea level during past interglacials. *The climate of past Interglacials* (Sirocko, F., Claussen, M., Goñi, M., Litt, T., eds), *Developments in Quaternary Sciences*, **7**, 75-92 (Elsevier).

Skinner, L., Shackleton, N., 2006. Deconstructing terminations I and II: revisiting the glacioeustatic paradigm based on deep-water temperature estimates. *Quaternary Science Reviews* **25**, 3312–3321.

Sørensen, B., 2004. *Renewable Energy* 3rd edition. Elsevier/Academic Press, Burlington, San Diego and London.

Sørensen, B., 2009. Energy use by Eem Neanderthals. *J. Archaeological Sci.* **36**, 2201–2205.

Sørensen, B., 2011. Demography and the extinction of European Neanderthals *J. Anthropological Archaeology* **30**, 17–29.

Sorensen, M., Leonard, W., 2001. Neanderthal energetics and foraging efficiency. *J. Human Evolution* **40**, 483–495.

Speth, J., 2006. Housekeeping, Neanderthal-style. *Transitions before the transition* (E. Hovers and S. Kuhn, eds), 171-188. Springer Science & Business Media, New York.

Spitzer, H., 1954. *Hütte, Maschinenbau A*, 556-566. W. Ernst und Sohn, Berlin.

Steegmann, A., Cerny, F., Holloday, T., 2002. Neanderthal cold adaptation: Physiological and energetic factors. *American J. Human Biology* **14**, 566–583.

Stiner, M., 2006. Middle paleolithic subsistence ecology in the Mediterranean region. *Transitions before the transition* (E. Hovers and S. Kuhn, eds), 213-231. Springer Science & Business Media, New York.

Svendsen, J. *et al.*, 2004. Late quaternary ice sheet history of northern Europe. *Quaternary Science Reviews* **23**, 1229–1271.

Tarasov, L., Peltier, R., 1997. Terminating the 100kyr ice age cycle. *J. Geophys. Res.* **102**, 21 665–21 693.

Trinkaus, E., 1995. Neanderthal mortality patterns. *J. Archaeological Sci.* **22**, 121–142.

Wadley, L., 2006. The use of space in the late middle stone age of Rose Cottage Cave, SA. *Transitions before the transition* (E. Hovers and S. Kuhn, eds), 279-294. Springer Science & Business Media, New York.

Waelbroeck, C., Frank, N., Jouzel, J., Parrenin, F., Masson-Delmotte, V., Genty, D., 2008. Transferring radiometric dating of the last interglacial sea level high stand to marine ice core records. *Earth and Planetary Science Lett.* **265**, 183–194.

Watson, J., Payne, R., Chanberlaim, A., Jones, R., Sellers, W., 2008. The energetic costs of load-carrying and the evolution of bipedalism. *J. Human Evolution* **43**, 675–683.

Weaver, T., Roseman, C., 2008. New developments in the genetic evidence for Modern Human origins. *Evolutionary Anthropology* **17**, 69–80.

Weaver, T., Steudel-Numbers, K., 2005. Does climate or mobility explain the difference in body proportions between Neanderthals and their Upper Paleolithic successors? *Evolutionary Anthropoplogy* **14**, 218–223.

Wenzel, S., 2007. Neanderthal presence and behaviour in Central and North-western Europe during MIS 5e. *The climate of past Interglacials* (Sirocko, F., Claussen, M., Goñi, M., Litt, T., eds), *Developments in Quaternary Sciences*, **7**, 173-193 (Elsevier).

# 2

# Weichselian stone age: Coping with the cold north

## 2.1 Last ice age: A cold but variable climate

Our knowledge of the climate during the Weichselian ice age is fairly modest except for the final deglaciation period (from 21 ky BP). The reason is that one cannot (as in the Eem interglacial) just use current climate models, modifying only solar radiation inputs. The shorelines have changed due to melting of ice, land depression and later rebound; and when shorelines and underwater topography change, so does oceanic circulation; and when oceanic circulation changes, so does atmospheric circulation. A few attempts have been made to model the mechanisms responsible for general sea level changes, including not only the volume of water from melting ice, but also the land depression and rebound, by considering (in addition to the ingredients of conventional climate modelling) the compressibility of the continental plates in response to ice loads. Still, the most reliable shoreline information comes primarily from sediment analysis at specific locations.

The Greenland temperature record shown in Fig. 1.1 is, as mentioned, representing temperatures somewhere in Eurasia or North America, but not necessarily the same locations at different times. The temperatures indicate a 15 °C cooling from 120 to 115 ky BC, then highly varying temperatures ending in a further 15 °C cooling during the period 75–60 ky BP, followed by some 7 °C warming, again with huge variations, and once more falling back to temperatures some 30 °C below the Eem temperatures, around 30 ky BP. Finally, the period is terminated by a warming to present temperature levels, reached around 10 ky BP.

The world total ice volume in different periods is inversely proportional to the general sea levels shown in Fig. 1.5. The sea levels (and hence the ice build-up) change gradually during the entire period from 120 to 18 ky BP, with minor relative sea level minima at 110 ky BP, possibly at 90 ky BP and at 65 ky BP. This behaviour is generally in agreement with the simplified calculations of Tarasov and Peltier (1997) and of Bintanja *et al.* (2002), but not in detail, and in particular the latter calculation underestimates the early ice build-up events. There is general agreement that neither the 110 ky BP nor the 90 ky BP relative ice maximum caused glaciation in Denmark. The 90 ky BP event may have caused some ice cover in northern Scandinavia, as estimated by Svendsen *et al.* (2004) and illustrated in Fig. 2.1. The theoretical estimates mentioned have ice only at quite high latitudes of Norway and Sweden.

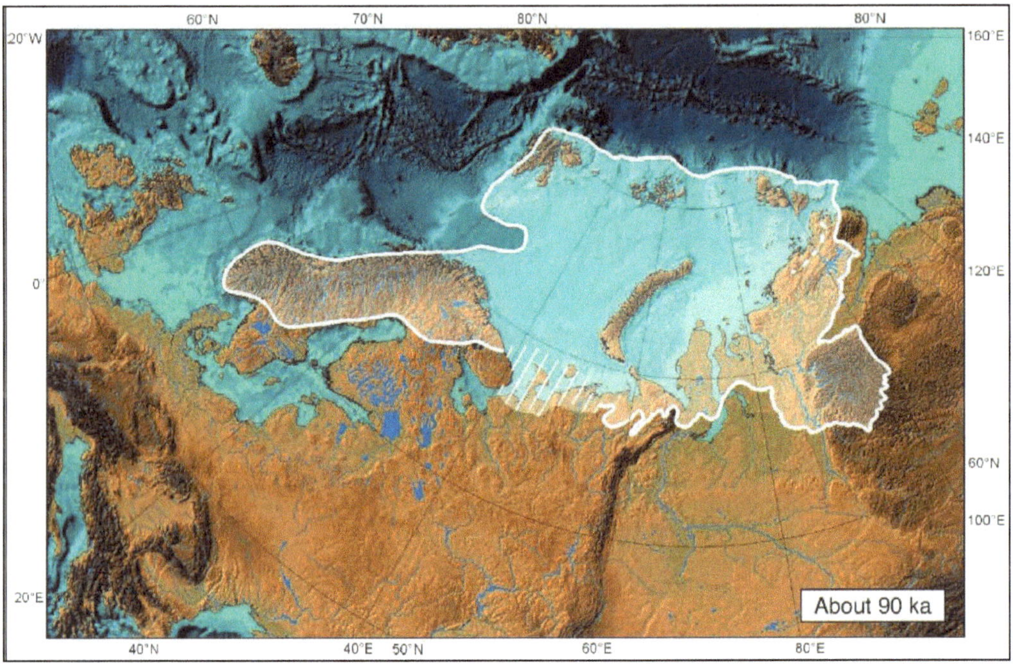


*Figure 2.1. Estimated extent of Eurasian ice sheet at the relative maximum around 90 ky BP (Svendsen et al., 2004).*

Vegetation shifted between warm periods (with forests and tree clusters) and colder periods (with more open vegetation and tundra). During the middle Eem, trees in northern Germany included elm, ash and hazel, but during the cooling from 120–110 ky BP, first spruce and then pine and birch became the dominant trees (Seelos and Sirocko, 2007). Findings of fauna depending on open plains (such as mammoth) from these periods indicate that the forest cover was never complete, but left enough open range areas to allow the favoured Neanderthal prey animals to survive, while at the same time offering plenty of wood for construction or firewood uses. Similarly, the cold periods (called "interstadials"), such as the one from 96 to 85 ky BP, were never devoid of trees, although their numbers, as estimated from pollen analysis, were reduced to about half (Kühl *et al.*, 2007). The split between tree and non-tree pollen is shown in Fig. 2.2 for the period 127 to 81 ky BP and for the site of Gröbern in northern Germany, along with variations in precipitation, summer and winter temperatures, and solar radiation. It is seen that winter temperatures dropped more than summer temperatures during the cold periods, and that Neanderthal settlers would probably have had to put on one additional layer of clothes when hunting during the coldest winter periods.

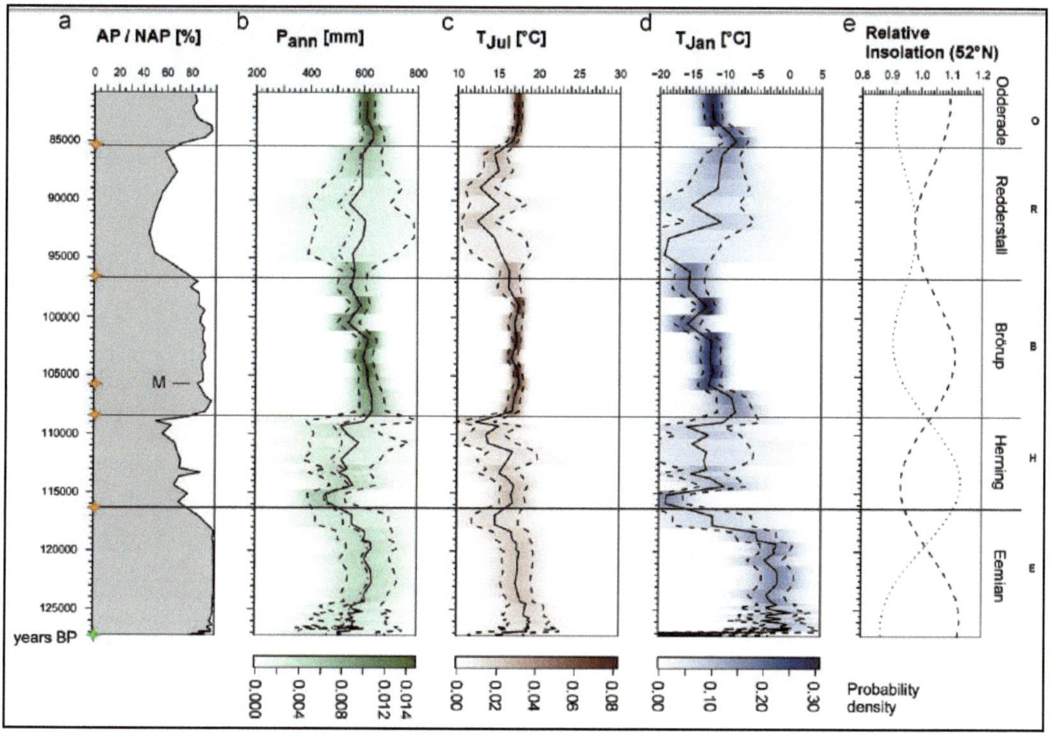


*Figure 2.2. Tree pollen, precipitation, temperatures and radiation at Gröbern, northern Germany, 127–81 ky BP (Kühl et al., 2007).*

At the 60 ky BP lowering of sea levels, most estimates have a Fenno-scandian ice sheet extending into part of Denmark from the East. Again, the Svendsen *et al.* (2004) estimate shown in Fig. 2.3 is slightly larger than other estimates, where the ice only reaches the Eastern parts of Sealand. The uncertainty is due to the often dramatic influence on sediment cores exerted by ice scraping against the underlying sediments and by sediments moved about by the ice-melting water.

During the entire period between the Eem and the start of the final Weichselian deglaciation, there are no signs of human presence in

Denmark, but – particularly from around 40 ky BP – abundant evidence for presence of prey animals such as mammoth and various deers (Aaris-Sørensen *et al.*, 1990). While it cannot be excluded that hunting parties may have visited Denmark, by analogy with the Eem Hollerup indication, the likely conclusion is that Denmark had no permanent human inhabitants during this period. Still, known settlements exist not far away in northern Germany throughout the Weichselian period.

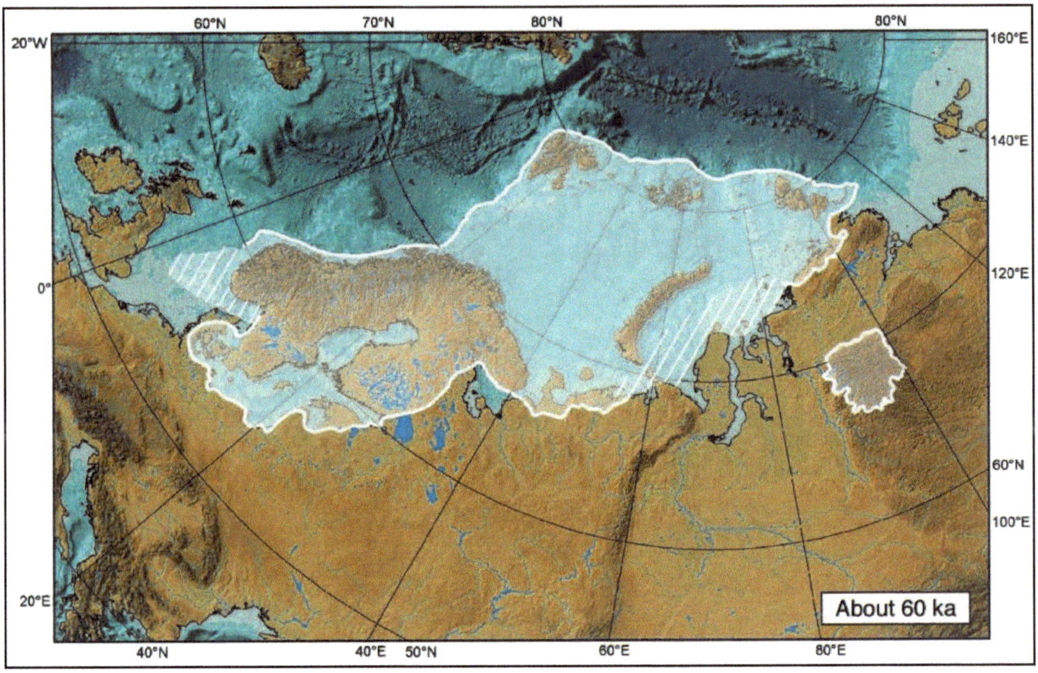


*Figure 2.3. Estimated extent of Eurasian ice sheet at the relative maximum around 60 ky BP (Svendsen et al., 2004).*

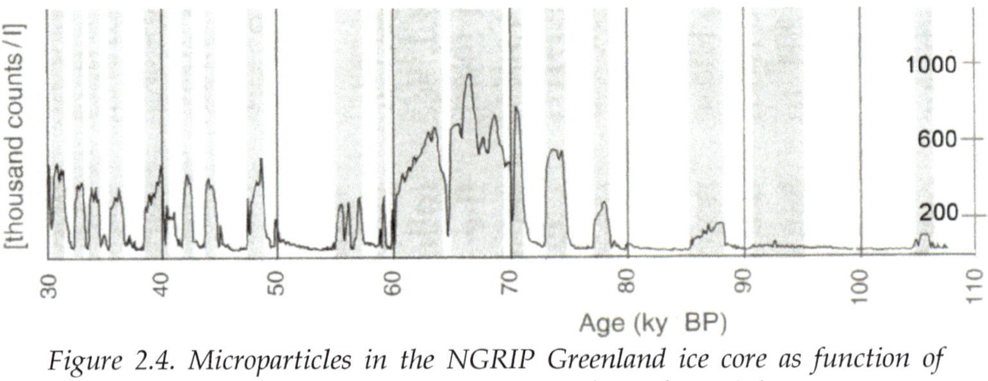


*Figure 2.4. Microparticles in the NGRIP Greenland ice core as function of reconstructed time (Ruth et al., 2003; Seelos and Sirocko, 2007).*

The causes of the frequently dramatic variations in temperatures during the mid-Weichselian period are not completely understood. In part there are the variations in solar radiation (Fig. 1.2 and right-hand panel of Fig. 2.2), but changes in oceanic circulation could also play a role, as could other events such as volcanic eruptions. The latter is a very probable cause, at least for the cooling period at 70–60 ky BP, where counts of micro-particles in the atmosphere over Eurasia show order-of-magnitude increases persisting for 10 000 years, according to the NGRIP ice core data from Greenland (Ruth *et al.*, 2003), shown in Fig. 2.4. The culprit seems to have been an extremely large volcanic eruption on Toba, Sumatra, 75–70 ky BP (Ambrose, 1998).

The conditions of Neanderthals south of Denmark between 120 and 60 ky BP are difficult to assess, due to the small number of sites found. Population sizes were probably very small, but they seem to have regained their position from 60 ky BP, with an increasing number of identified settlements (Andel *et al.*, 2003). The fraction of older settlements found is presumably lower than for more recent settlements. One should also bear in mind that costal settlement sites from the glaciation periods are mostly below water today.

In order to gain more insight into the development of Neanderthal population size, the simple population model used above for the Eem interglacial period will now be extended to run for the entire period of possible Neanderthal presence (i.e. from 260 ky BP to the recent extinction). The exponential behaviour of population models makes the choice of parameters particularly important for such long periods of calculation. Requiring that Neanderthal population never got above 100 000, but on the other hand did not dwindle to zero during the cold period near the end of the previous (Saalian) period of glaciation, will seriously narrow the possible parameter choices of the model. The basic form of the model based only on the fertility and death rates of Figs. 1.18 and 1.19 would imply a long-term growth to billions of people, because of the lack of two important features: prey ecology restraints and the progressive cooling of climatic conditions.

Climate considerations can be added to the model by making the death rate depend on temperature. The age distribution of health impacts from extremely hot or cold weather is known from current data. It involves increased sensitivity to temperature for the young and the old, and leads to significant increase in deaths only at the extremes of the temperature range (Sørensen, 2011a). The population model constructed here uses the age and temperature distributions shown in Figs. 2.5 and 2.6, with a simplified temperature dependence on only the annual average temperature. Prey ecology enters when the preferred food animals of the Neanderthal population become scarce from over-hunting, without immediate substitution options. The hunting area available to Neanderthals may (cf. Chapter 1) be large enough to support a population of about 12 000 people feeding exclusively on mammoth. Although this was one of the preferred food animals, the Neanderthals also hunted various types of deer and several other animals (including fish and shellfish at coastal sites), and presumably would have increased the share of these if a decline in mammoth availability occurred. This may have happened in southern Europe, but certainly not in northern Europe, where the mammoth lived healthily until after the most recent glacial maximum. Neanderthals preferred open landscapes and rarely moved into the forests to find prey, even during periods (e.g. Eem) where the forest had advanced at the expense of open ranges (Bocherens and Drucker, 2003).

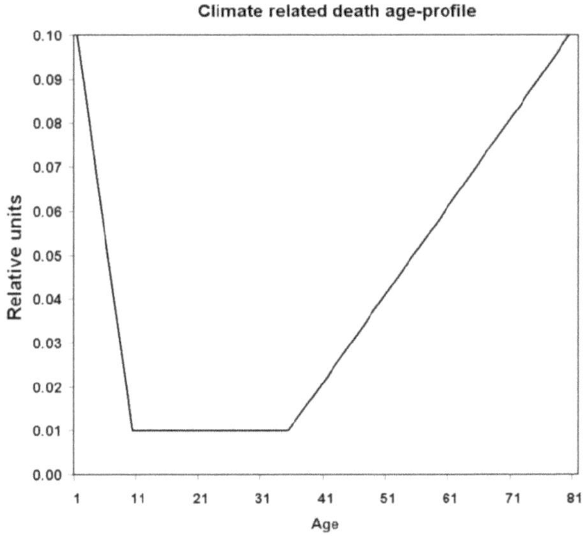


*Figure 2.5. Age profile of temperature-dependent death rate (Sørensen, 2011b).*

Only if the Neanderthal population in Europe had been substantially larger than suggested by finds (e.g. near 100 000 individuals) might it not have been possible to find enough substitution prey animals, because the Neanderthals seemed unwilling to hunt the smaller and more abundant swift animals that were later utilized by modern humans (see the discussion in Chapter 1 on the absence in Neanderthal remains of shoulder bone asymmetries that would have signalled activities of spear–throwing). Again, this could have relevance for Neanderthals in the south of Europe, where other conditions for rapid growth were better, and where the mammoth stock would have been declining because the mammoths migrated to the north. It is not known if this was because of over-hunting or because of more suitable open landscapes in the north. The prey ecosystem dependence is added to the population model in the form of a population-dependent addition to the death rate, which sets in above a population level of 12 000 (for the entire European area modelled) and increases as shown in Fig. 2.7.

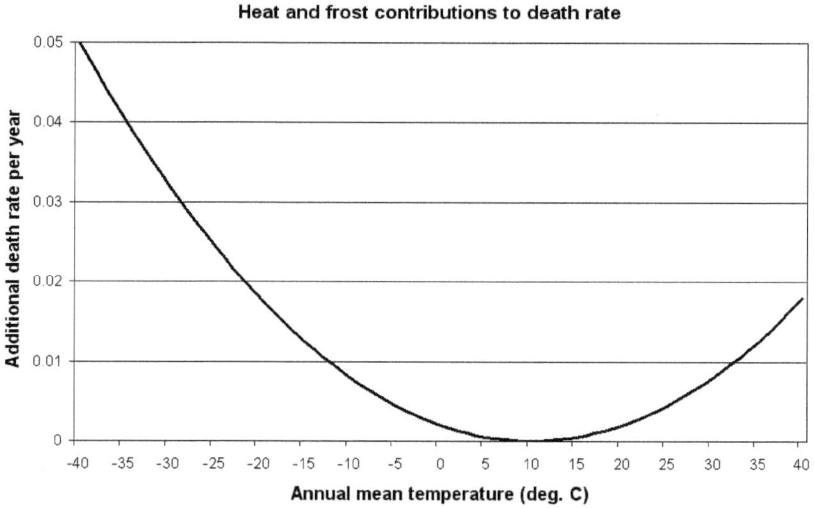


*Figure 2.6. Temperature dependence of climate-related death rate (Sørensen, 2011b).*

The population model thus has three main parameters to vary:
$\beta$: amplitude of birth rate for fixed death rate profile.
$\gamma$: amplitude of additional temperature-dependent death rate.
$\delta$: amplitude of additional death rate related to prey overkill.

In fact, these parameters are largely determined by the requirements that the Neanderthal population never exceeded 50 000 people (as judged from settlements and remains found, even assuming that only part of the actual sites have been unearthed) and that Neanderthals survived the worst cooling at the end of the previous glacial period (around 160 ky ago). This is illustrated in Figs. 2.8–2.10. The climate dependence ($\gamma$) selected just about saves the population from going extinct in northern Europe at the end of the Saalian glaciation, despite a drop of three orders of magnitude (Fig. 2.10), while in southern Europe population drops only by a factor of 20 (Fig. 2.8, curve *a*). A doubling of the climate-related death rate also makes the southern European population drop to extinction (Fig. 2.8, curve *c*, leaving no survivors beyond

160 ky BP), and a halving makes the population frequently hit the ceiling set by food availability. Fig. 2.9 compares the selected reference case *a* with calculations assuming different impacts ($\delta$) of prey availability for southern Europe: weakening the influence of prey overkill causes the population to become too large during the early Weichselian glaciation (around 100 ky BP), while increasing it makes little difference, as the population in the reference case *a* is already below the ecologically sustainable upper limit (Sørensen, 2011b).

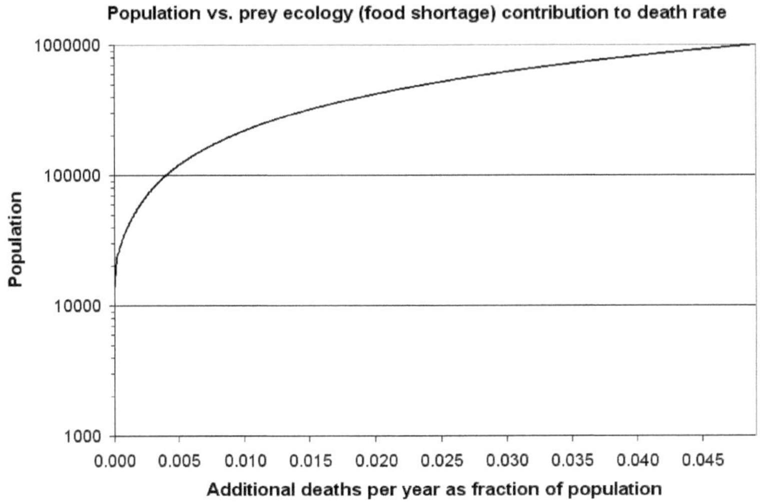


*Figure 2.7. Effect of food shortage on death rate, for large populations limited by prey availability with assumed Neanderthal hunting methods (Sørensen, 2011b).*

The effect of minute changes in birth rate ($\beta$) is illustrated in Fig. 2.10 for northern Europe. Again, the selected reference case is denoted by *a*. A one per cent decline in the overall fertility makes the population die out at the end of the Saalian period (*b*), and a similar increase makes the 80 ky BP population reach an unrealistic half million (*c*), in the absence of famine-related restraints. With such restraints, as in Fig. 2.9, the upper curve changes to a maximum population of 100 000 (*d*), while cases *a* and *b* are unaffected. Clearly, the initial population of two times 300 at the beginning of the calculation (260 ky BP) is arbitrary, and the curves can collectively be moved up and down to accommodate other guesses for an initial population. However, due to the constraints posed by consistency with our general knowledge of particularly the period from the late Saalian to the glacial maximum of the Weichselian period, this would not influence the course of the reconstruction, but only cause small changes in the amplitudes. The choice of allowing the late Saalian population in northern Europe to drop to a handful, while that in southern Europe stays viable, can be changed slightly to make the Neanderthals die out in the north, to become repopulated from the south during the following Eemian period. The subsequent population behaviour during the Weichselian period will be only marginally affected.

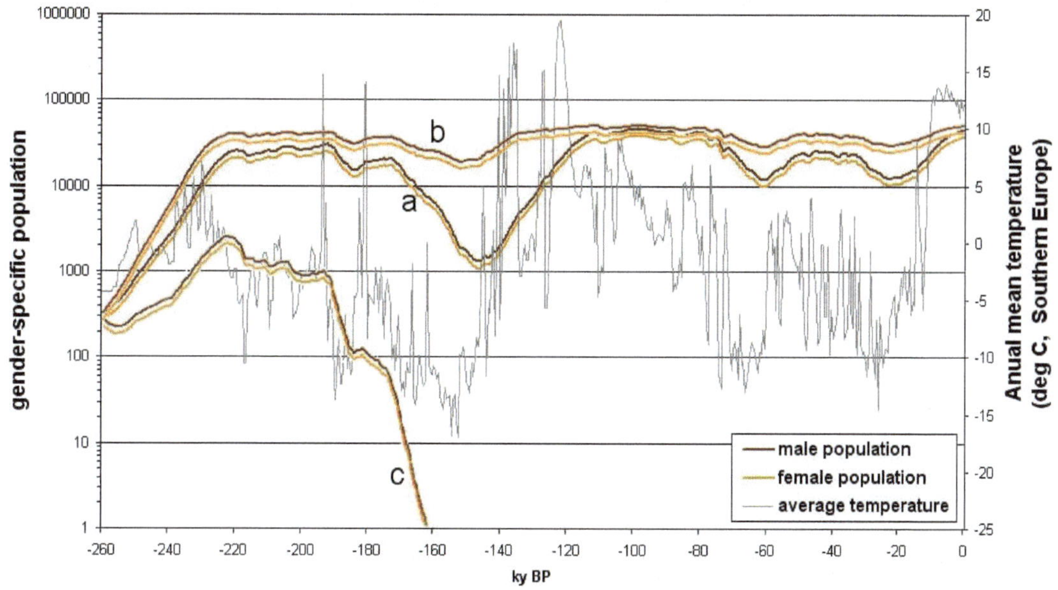


*Figure 2.8. Population model reconstruction of Neanderthal population using climatic impact data for southern Europe (some 4 °C higher temperatures than for northern Europe; the full time series used was reconstructed from the GRIP and NGRIP data of Figs. 1.1 and 1.8). The amplitude of the climate-related death rate ($\gamma$) is in the reference case (a) 0.00002, in (b) 0.00001 and in (c) 0.00004. The amplitude is multiplying the age and temperature dependence given in Figs. 2.5 and 2.6. The food availability restraint is in effect (corresponding to the reference case of Fig. 2.9) and confines the population to below 100 000 individuals at all times (Sørensen, 2011b).*

A feature of the population model results is a considerable lag between population lows in cold periods and the following growth during warmer periods. After the lowest northern European population (around 140 ky BP, some 15 ky after the Saalian low in temperature), it takes until 90 ky BP to reach the next population maximum, made possible by the Eem temperature peaks (ending at 120 ky BP). The explanation is of course that the near–extinction at 140 ky BP requires a thorough rebuild of population one generation at a time. The similar build-up in southern Europe is considerably faster. The temperature drop after 120 ky BP coexists with population rise until 90 ky BP.

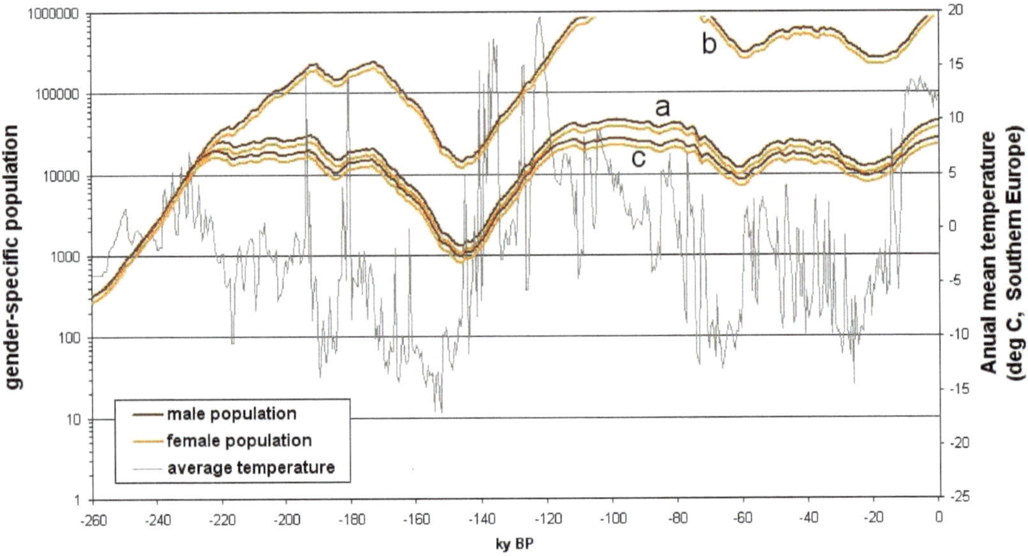


*Figure 2.9. Population model reconstruction of Neanderthal population using climatic impact data for southern Europe (cf. Fig. 2.8). The amplitude of the food-shortage death rate ($\delta$) is in the reference case (a) 0.00000005, in (b) 0.00000001 and in (c) 0.0000001.*

Following this line of features in the Neanderthal population development, both for northern and southern Europe, it is clear that the model calculations that allow Neanderthals to survive the glacial maximum of the late Saalian period do not bring them even close to extinction when approaching the subsequent glacial maximum at the end of the Weichselian period, nor at the temperature depression from 75 to 60 ky BP.

A similar conclusion has been reached by a different approach by Banks *et al.* (2008). They use machine learning software to describe ecological niche populations. Feeding known Neanderthal-occupied sites at a given moment in time into the software, along with primarily climate data, the model learns which environments the Neanderthals can survive in, somewhat like a neural network program. The problem is that the ecological niches found are underdetermined, so that a stochastic approach involving many models must be used if results are to emerge. At each point in time, the model in this way indicates likely sites other than those used for input that may be suitable for Neanderthal occupancy. However, there is no point in comparing niche models for different times, as the sites with climatic conditions identified as suitable at one time may not be so with identical conditions at a different time, because the software would then have its "suitable conditions" determined by different input data. What Banks *et al.* do is to use the models derived from earlier data to predict Neanderthal presence at around 37 ky BP. They find a strong increase in "potential sites" allowing presence, despite the fact that the model derived by using data specifically for 37 ky BP just confirms the near–absence of Neanderthal population at that time. They conclude that it cannot be the climate around 37 ky BP that made the Neanderthals disappear and, furthermore, based upon similar model calculations for modern humans, that the conditions for modern humans were neither better nor worse than those of the Neanderthals (i.e. they occupy precisely the same ecological niches).

The simple demographic model is now expanded to include two important aspects (Sørensen, 2011b). One is the migration from warmer to colder climates to improve hunting (taking advantage of less populated areas and better conditions for the mammoth), or migration in the opposite direction, fleeing south if the temperatures become too cold and permafrost spreads to the occupied areas. The other addition is prey competition with modern humans. The latter have a slightly lower mortality, due to hunters' spear-throwing being less dangerous than close-range spear-piercing of the mammoth, and because a wider range of possible food prey implies a higher threshold for food shortage (assumed onset at a population of 20 000 rather than 12 000).

Figure 2.11 shows the results of the expanded model, now dividing Europe into three climate zones (with average temperatures 0, 2 and 5 °C below the northern European ones, and fixing the model parameters (fertility, climate sensitivity and initial population) so that Neanderthals survive the Saalian maximum glaciation and have the same observed population of about 250 in each of the three regions (northern, middle and southern Europe) during the Eem interglacial period. Food shortage never occurs with these parameters, and the Neanderthal population during the period 40–20 ky BP remains between 4000 and 3000. A recent study of the variation in mitochondrial DNA from five Neanderthals estimates a similar population (Briggs *et al.*, 2009). Again the conclusion is that, as Neanderthals survived the Saalian glaciation, they should also survive the Weichselian one, unless something lying outside the model happened (Sørensen, 2011b).

It is seen from Fig. 2.11 that in certain colder periods, neither middle or northern Europe is populated at all. This holds for Neanderthal as well as modern populations and is not contradicted by the distribution of excavated sites over the regions.

Having argued that the Neanderthal disappearance from Europe at 38–

35 ky BP (except for a small enclave near Gibraltar, possibly surviving to 28 ky BP) is not likely to have been due to climatic conditions, nor to shortage of food, then it must have been due to some other factor that could affect mortality. There is no reason that the total fertility or the probability of death from accidents or from the diseases included in the long-term death rate levels of Fig. 1.19 should have changed abruptly at the time of the sudden Neanderthal extinction. In fact, any emergence of insights on medical treatment would have worked the opposite way. So what other possibilities exist?

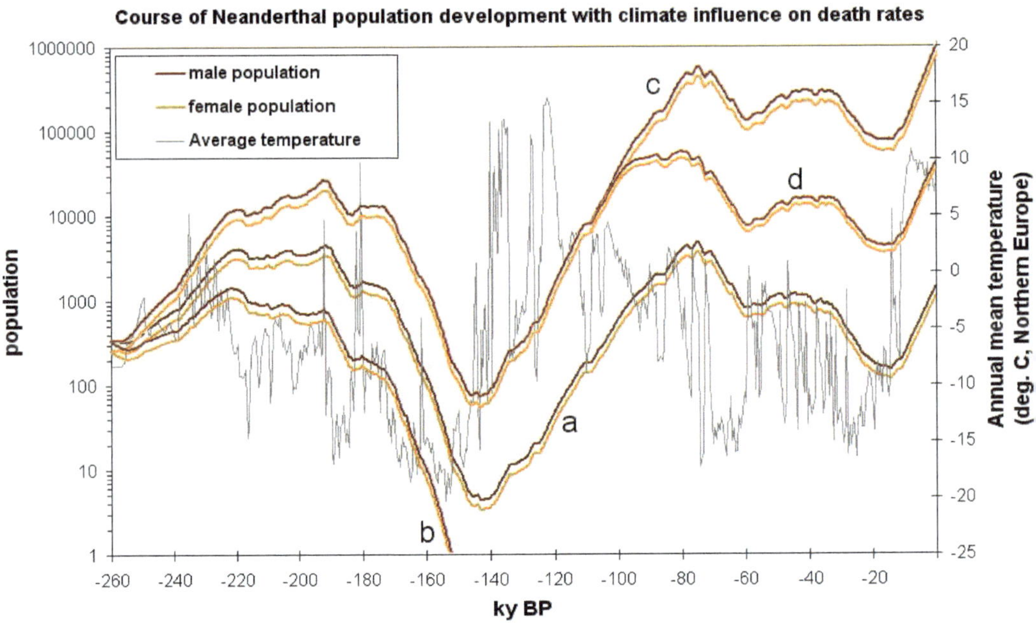


*Figure 2.10. Population model reconstruction of Neanderthal population using climatic impact data for northern Europe (with temperature reconstruction from the GRIP and NGRIP data of Figs. 1.1 and 1.8). The amplitude of the total fertility β (for fixed death rate profile) is in the reference case (a) 1, in (b) 0.9994 and in (c) 1.0006. The amplitude β is multiplying the fertility age-distribution curve given in Fig. 1.18. The climate-dependent death rate is as in the reference case (a) of Fig. 2.8. The food availability restraint is added in (d), corresponding to the reference case of Fig. 2.9, again leading to confining the population to below 100 000 individuals.*

I can think of two such possibilities, both of which are linked to the invasion of modern humans from the east or southeast, reaching southern Germany and France after 38 ky BP: violence or new diseases brought in by the intruders. I shall return to the second possibility.

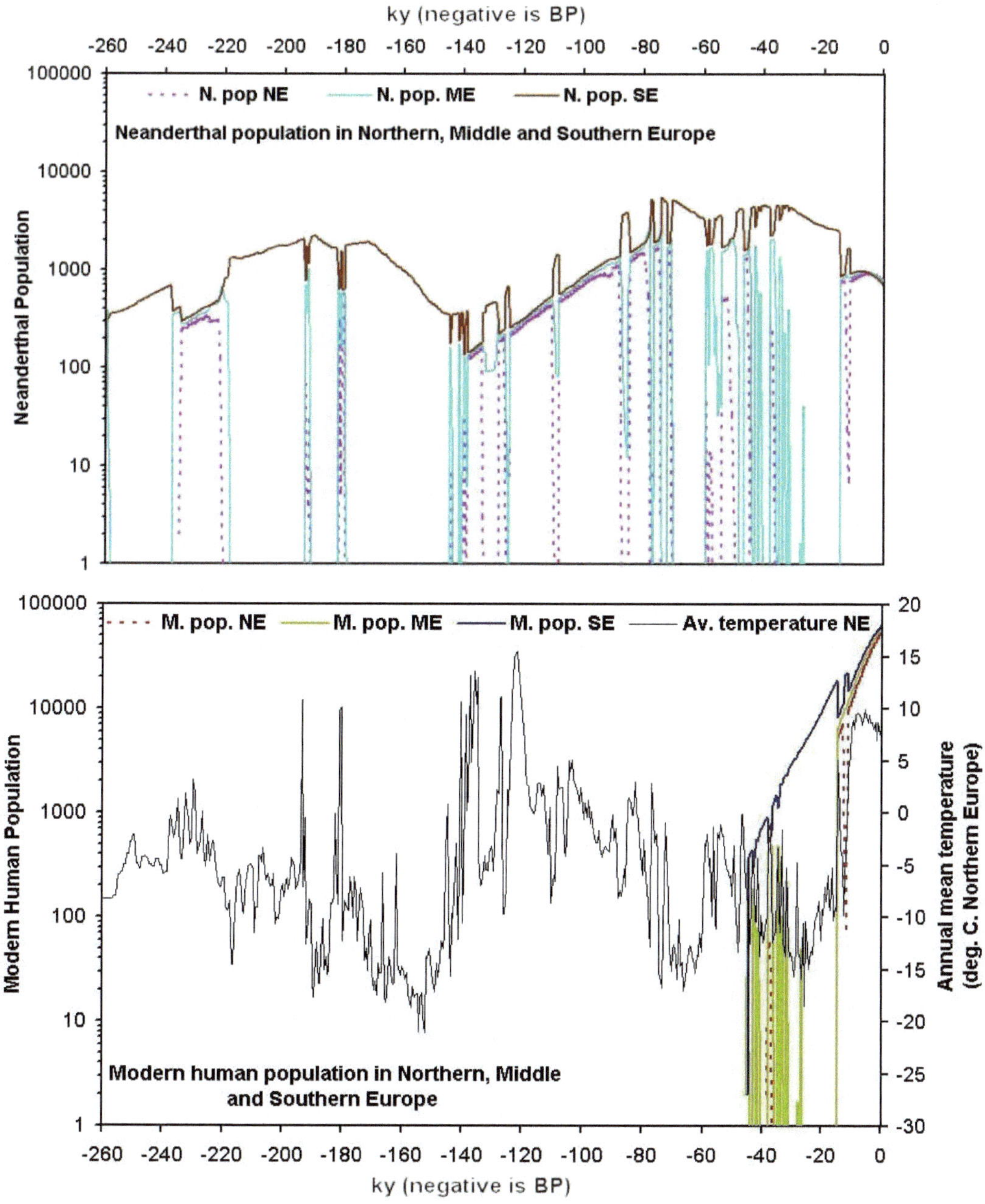


*Figure 2.11. Neanderthal (N) and modern human (M) population development model with possibility of migration between regions (SE is southern Europe, ME is middle Europe and NE is northern Europe; Sørensen, 2011b).*

Violence and aggressive behaviour are common in human societies. The history of *Homo sapiens sapiens* is full of accounts of violent fights and wars, all the way up to the most recent century, with two devastating world wars and hundreds of local conflicts. Few Neanderthal or early modern human bones show marks of violence (disregarding the signs of wounds inflicted during hunting or as part of burial rituals). Evidence for belligerent behaviour increases after the appearance of modern humans on the European scene (Walker, 2001), but, as the following chapters will detail, it only becomes severe after the societies have become much larger (as a result of the population growth allowed by agriculture) and have created layers of social structure. So, did the intruders simply kill the European Neanderthals? There is no evidence that it happened. A surge in dead Neanderthal remains with signs of violence is not found at the time in question, and one is led to suggest that the hospitable attitude of friendly relationships that is also a feature common among modern humans must have prevailed in this case.

The more than 10 ky overlap between Neanderthals and modern humans in Europe would have given ample opportunity for coexistence and exchange of ideas and technology. Neanderthal technology shows remarkable little development during the period from their first known European presence in the early Saalian period to about 45 ky BP. This could be due to the absence of challenges related to food availability, but also quite possibly due to the low population density, which despite the mobility related to mating networks, left relatively little opportunity for discussing new ideas. Trips between different groups would have allowed visitors to take home new technologies seen in the foreign group, had there been any, but the conversations over nightly campfires that could have generated new technology or art do not seem to have taken place. It has therefore been suggested that the interaction with modern humans, bringing with them at least slightly different technologies and some symbolic behaviour, made the Neanderthals change their habits. On the other hand, the new technologies and the artwork appearing in Neanderthal settlements during the last 10 ky of survival are sufficiently distinct from those of contemporary modern humans to be considered a parallel development and not merely a copying activity.

Most palaeolithic sites, in the open, in rock shelters or most notably in caves, have had human occupancy spanning a very long time, without necessarily being continuous. A good example is Grotte du Renne at Arcy-sur-Cure in mid-eastern France, where personal ornaments have been found in the form of finger rings and pierced animal teeth for a necklace or some similar kind of jewellery. The stratigraphy has top layers with Aurignacian (modern human) finds, several intermediary layers that are Châtelperronian (late Neanderthal), and bottom layers with Mousterian (earlier Neanderthal) artefacts. As in most cases, there has been debate over possible shuffling of layers by later flooding, etc. (White, 2001), but in the Grotte du Renne case, artefacts from the middle layers have been unambiguously identified as of Neanderthal origin through analysis of the ear labyrinth of a young person's skeleton found there: the posterior canal is lowered in a way unique to Neanderthals and never found in modern humans (Hublin *et al.*, 1996). This does not, however, exclude that other objects could have been moved between layers, but certainly reduces the likelihood that this would be true for any large fraction.

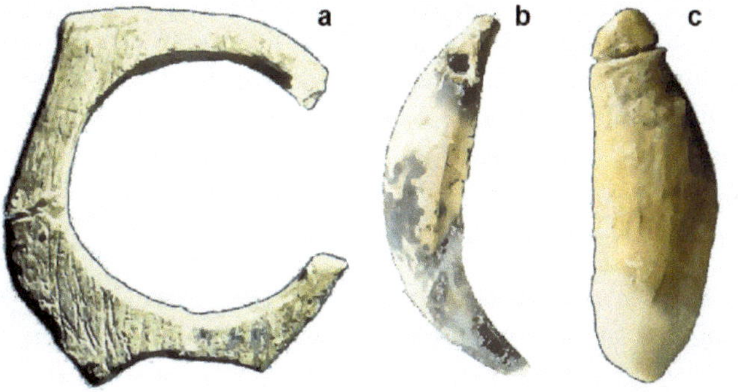


*Figure 2.12. Late Neanderthal ornaments: finger ring (a) and perforated canine tooth (b) for a necklace, both from the lowest Châtelperronian layer X at Grotte du Renne. To the right, another tooth, but here for a necklace with string fastening (c), found in the top Châtelperronian layer VIII (White, 2001; Conard and Bolus, 2003). For a dissenting view on the dating of these artefacts, see Balter (2010).*

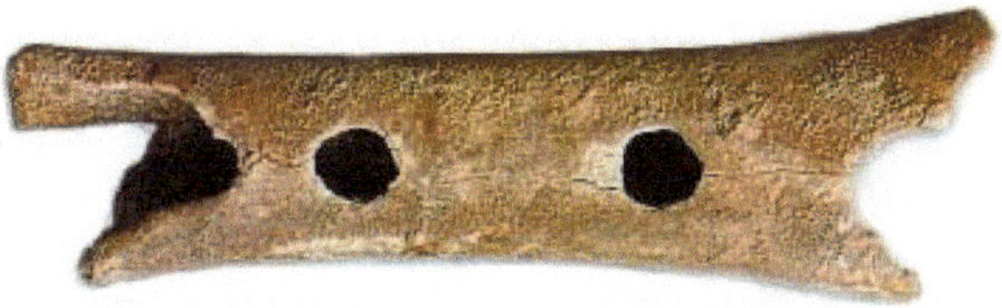

*Figure 2.13. Flute-like perforated bone artefact found in a Neanderthal-occupied cave layer at Divje Babe I in Slovenia, and dated to earlier than 40 ky BP (Turk, 1997; Turk et al., 2006). The possibility of such early cultural activity by Neanderthals has made many sceptical about the flute interpretation, and suggestions have been made that cave bears gnawed the holes (d'Errico et al., 1998; 2003). The probability of animals producing three holes on one side of the bone and aligned with the axis of the bone is roughly one in 10 000, so perhaps a reopening of the case is due.*

Figure 2.12 shows examples of the personal ornaments found in the intermediary Neanderthal layers at Grotte du Renne, with two different necklace techniques used for the tooth in the lower and in the upper Châtelperronian layer, both belonging to the late Neanderthal period dated to around or just before 35 ky BP (reconsideration of $^{14}$C-dating calibration has pushed many previously published dates from this period backwards; Bard, 1998). In Fig. 2.13, a disputed Neanderthal flute is depicted. Recent studies of the Neanderthal vocal tract and the discovery that modern humans and Neanderthals share the special FOXP2 gene responsible for language makes it clear that Neanderthals had the ability to communicate by spoken language (Boë *et al.*, 2007; Krause *et al.*, 2007). The Neanderthal artwork and ceremonial burials strongly suggest that they used this capability. d'Errico *et al.* (2003) quote over 40 Neanderthal sites where black pigment has been found, and in some cases facets created by rubbing against soft material or tips intended for line drawing. We do not know if the drawing was on the human skin or on other canvas, as such painting has not survived. Maybe it resembled the cross-patched patterns carved in stone that have been found in the early human settlement at Blombos, South Africa.

The age distribution of modern human sites across Europe indicates that *Homo sapiens sapiens* probably arrived through eastern Europe and within some 10 ky spread throughout middle and southern Europe, finally reaching the Iberian peninsula (Bocquet-Appel and Demars, 2000). Another possible route would be from the eastern Mediterranean shores through the Balkans. Evidence that seemed to support this idea has lost strength (Conard *et al.*, 2004a). The advancing ice sheet over northern Europe prevented new human appearance in this region until the warming periods associated with deglaciation. The population growth would have been fast and the total population soon exceeded the maximum Neanderthal population.

Having made it plausible that the coexistence between the last Neanderthals and the modern humans was probably peaceful and that there was a possible exchange of ideas on technology, jewellery and other beauty parlour stuff, it seems unlikely that there should not be some measure of sexual activity between the two populations (Zilhão, 2006). Sequencing of DNA, first from Neanderthal mitochondria and later genomic DNA from one particular individual living around 38 ky BP in Vindija Cave (Croatia), suggests separate genetic development of Neanderthals and modern humans for 370 ky or more (Krings *et al.* 1997; comments in Gutiérrez *et al.*, 2002; Green *et al.*, 2006; Noonan *et al.*, 2006; the first full draft sequencing finally made by Green *et al.*, 2010). Evidence for mixing of Neanderthals and modern humans is suggested but also is very difficult to establish from a genomic comparison: The relatively brief period open for admixture would leave a very infrequent trace in the genome of modern humans, and although a large sample of modern humans can be screened in the comparison, the number of different Neanderthals from which useful DNA can be extracted is likely to remain minute. The genes of humans and other living things consist of active and inactive sequences, the inactive being thought to carry information on previous development all the way back to the first unicellular creatures. As a result, differences between different species show only in a small portion of the DNA, and quantitative differences between, for example, humans and apes, are extremely small, although detectable due to the considerable databases derived from (mostly) modern DNA sequencing of various species.

To this comes the fact that we are currently unable to read the language of the genetic code. We have noted that it consists of sequences of certain molecule pairs (base pairs) spanned between two backbone spirals. We can move chunks of code from one animal to another and see what happens, but nobody can write the sequence of code to insert, that will produce a creative scientist, painter or composer. Our knowledge is limited to identifying general chunks of code as probably related to given diseases or to specific organ functions in the body. Using genetics to discuss whether we inherited, say, musical ability from the Neanderthals is beyond our present capability. Finally, it should be said that the primitive archaeological excavation methods used in the past (and unfortunately often also at present) implies that all artefacts have been heavily handled by current modern humans, leaving their DNA all over them. The sequencing teams claim that they have indicative evidence that the contamination of the DNA used for sequencing Neanderthals is below 6% (Pennisi, 2006). It would clearly be important to conduct more DNA-analyses on Neanderthal remains where possible.

While genetic approaches remain cautious in providing answers to questions related to mixing between Neanderthals and modern humans, a more direct study of individuals showing traits of both (sub-) species may be more amenable. The first clear finding was that of a four-year–old boy from Portugal, with several Neanderthal traits, but living some 20 ky ago in a modern human family and being buried with all the ornaments typical of modern human burial style at the epoch (Duarte *et al.*, 1999). Since then, several other examples of hybrid humans have been found, including a recent one in Romania (Soficaru *et al.*, 2007). These morphological findings, dating to some 10 ky after the Neanderthals is presently believed to have been extinct, constitute a very strong evidence for interbreeding.

Returning now to the last remaining explanation for the Neanderthal extinction, this is an area where DNA analysis may be of real assistance. If the modern humans brought with them a disease for which the Neanderthals did not have any immediate defence, this could have increased the Neanderthal mortality and have caused the demise of an already fragile population of low population density. The ability of the modern humans to hunt smaller and faster animals (e.g. by spear–throwing) that were inaccessible to the Neanderthals might have been an additional factor, but unlikely to be decisive, because there should have been enough food for the remaining Neanderthals, as long as the modern human population had not grown to over 100 000 (cf. the population model used above for Neanderthals). The modern humans probably caused similar problems much later (e.g. during colonization of South America in the 16th century), because they brought diseases along that the indigenous populations initially had no defence against (although one of them, tuberculosis, has been proven to have existed before the invasion; Salo *et al.*, 1994). However, the Amazonian Indians did not completely disappear, possible due to presence of remote communities outside the reach of the Europeans. The Neanderthals may not have used this avenue, but may actively have sought peaceful interaction with the newcomers. The disease transmission hypothesis is becoming testable, in view of the recent identification of concrete diseases by DNA traces in several populations, including the classical Egyptian one around 5–6 ky BP (Crubézy *et al.*, 1998; Gụhl *et al.*, 1999; Zink *et al.*, 2001; Fournier *et al.*, 2007). The culprit in the case of Neanderthals should be a transmittable disease (not necessarily epidemic) already in reasonable control by the early modern humans, but unknown to the Neanderthal population (e.g. a parasitic one; Sørensen, 2011b).

A volcanic eruption, large in comparison with eruptions of the recent millennium but small compared to the Toba eruption around 75 ky BP, took place in southern Italy around 40 ky BP (Fedele *et al.*, 2008). According to Fig. 2.4, the Greenland ice core particle contents indicate a five times shorter duration of fallout and over 10 times lower accumulated particle fallout for the Italian *Campanian Ignimbrite* eruption, compared with the Toba one. The main plume travelled northeast (as indicated by the ash layers found in sediments), but with subsequent changing wind left traces also in other regions, including Greenland. The demographic calculation presented above takes into account the increased mortality due to ash cooling, but other effects (e.g. on vegetation) could cause additional stresses on the human populations. However, hunters are not very directly affected and, anyway, the main impact is probably limited to the south Italy, Balkan and Ukrainian regions where the ash fallout is largest. In any case, the geography of the climatic changes around 40 ky BP does not suggest that they caused the extinction of the Neanderthal population across Europe. The explanation of a contagious disease thus remains more probable.

Despite inhospitable climatic conditions, the modern humans already reached a larger population size than the maximum Neanderthal one during the first 10 ky of their European presence, as judged from the number of settlement sites (Jöris and Street, 2008; Banks *et al.*, 2008). Part of the explanation could be the ability of the modern humans to hunt a larger variety of prey than the Neanderthals, making use of precision throwing of spears to hit hares and foxes, perhaps. The population model investigations will be continued in the following chapters for the modern humans subsequently populating Europe alone.

*Figure 2.14. Aurignacian art. Above: flute, Geissen-klösterle, made of swan wing bone, 13 cm, 35 ky BP. Below: mammoth, Vogelherd, made of ivory, 7 cm, 31 ky BP. Right: lionman, Hohlenstein-Stadel, made of ivory, 28 cm, 33 ky BP (Conard and Bolus, 2003, Conard et al., 2004b; photos courtesy of N. Conard and H. Floss, University of Tübingen; K. Wehrberger, Stadt Ulm Museum).*

Several of the Aurignacian sites (from about 35 ky BP presumed populated by modern humans, with slight doubts expressed in Conard et al., 2004a) contain new expressions of art and culture, as well as refinements in hunting tools and fishing equipment. Among items found are sophisticated musical instruments (notably flutes), figurines and cave wall paintings, many of which show an artistic excellence that has only been matched in a few periods of more recent history. Figure 2.14 shows three-dimensional artwork from mid-German sites.

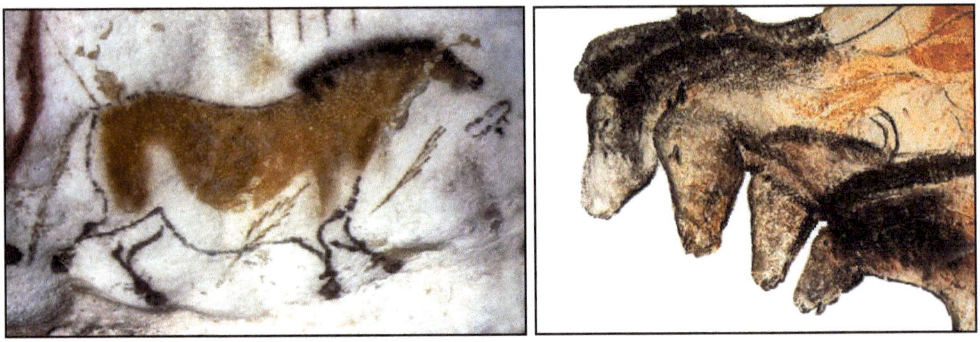

*Figure 2.15. Cave paintings of horses. Left: Lascaux Cave, 17 ky BP (Delluc and Delluc, 2006). Right: Chauvet Cave, occupied 32 to 20 ky BP (Feruglio, 2006). Public domain pictures from Wikipedia Commons.*

In France and Spain, the most impressive form of art from this period and more recently was painting, examples of which are shown in Fig. 2.15. The style is naturalistic but with increasing idealization and artistic simplification. The younger paintings found in the Lascaux Cave use abstraction in the contour lines to convey an impression of animals in motion, and colours that do not fill the space expected from realism. Such techniques are not seen in more recent art until the impressionism 17 ky later. But even the earliest cave paintings, such as those from Chauvet or Altamira caves, are not fully realistic: they use the style based on black contour painting subsequently filled with colours. Real animals do not have black contours, and thus already the first paintings had a level of abstraction. It is interesting that the black contour outlining is a style invention that has played an important role in painting and drawing ever since.

It is also remarkable that art at a very high level appeared in several places, from Germany through France to the Iberian peninsula. Allen (2006) is inclined to think that this artistic renewal appeared instantly, rather than as the result of a gradual development. While progress usually involves leaps, I think that there is also a development aspect with interaction and mutual learning among these artists. After all, the figurines and cave paintings found are dated to various points in time spanning a period of more than 10 ky, which may be short in relation to human development, but is much longer than the time span within which we have in the most recent 2 ky experienced classic Egyptian and Greek sculptures, renaissance paintings and impressionism. We would not lump these style changes together as instant and simultaneous manifestations, and nor should we in describing the appearance of new art forms during the late period of glaciation and ice withdrawal.

While the range of materials required for paint brushes and other tools related to the novel activities of late Neanderthals and early European modern humans increased substantially from 40 to 30 ky BP, few activities that required additional energy can be discerned. Hunting, maintaining fires and other facilities at the dwellings were still the dominant energy uses, directly or through food intake. The slowly emerging trend seen at the end of the period is increased use of particular rock and other materials not just found "lying around", such as the mineral compounds used for paint. Searching for such more exotic materials was the first step towards a raw materials industry, with mining activities later developed to the level required for the Bronze and Iron ages. Travelling far away to obtain raw materials created a novel need for transportation energy. Grinding mineral pigments such as haematite (iron-based, red), pyrolusite (manganese-based, black) and goethite (iron-based, yellow) with traces of magnesium and titanium, for use in paints, would have been done in mortars (Molerus, 1996). Black pigments were not heated over fires, as this would have left detectable changes in the manganese compounds (Chalmin *et al.*, 2006), but ochre must be heated to go from yellow to red.

The energy use per capita largely remains at the values presented in Table 1.2, but with modifications caused by climate variations and by increase in industrial use of fire towards the end of the period. This is illustrated in Fig. 2.16, based on Tables 2.1, 2.2 and 3.1, giving the energy "delivered" to the final user, a term indicating that losses in the end-use phase are not subtracted (e.g. human muscle power is typically under one-third of the metabolic heat associated with food intake, and medium- to high-temperature heat from open fires delivers only a modest fraction as comfort heat to people trying to sleep near the fire). The word "delivered" is used because of its importance in later periods, where preceding treatment and conversion of energy (say in oil refineries or fossil power plants) makes it necessary to distinguish between "primary" and "delivered" energy. By dividing energy use over the many tasks considered in describing the life conditions of Stone Age people, a more reliable energy usage estimate is obtained than would be possible with aggregated categories (summing up individual contributions, as in Fig. 2.16). This is called a "bottom-up" approach, as opposed to the "top-down" approach. The bottom-up approach is a better tool, as long as the risk of "forgetting" some energy-use activities is low, which is believed to be the case for the Stone Age society.

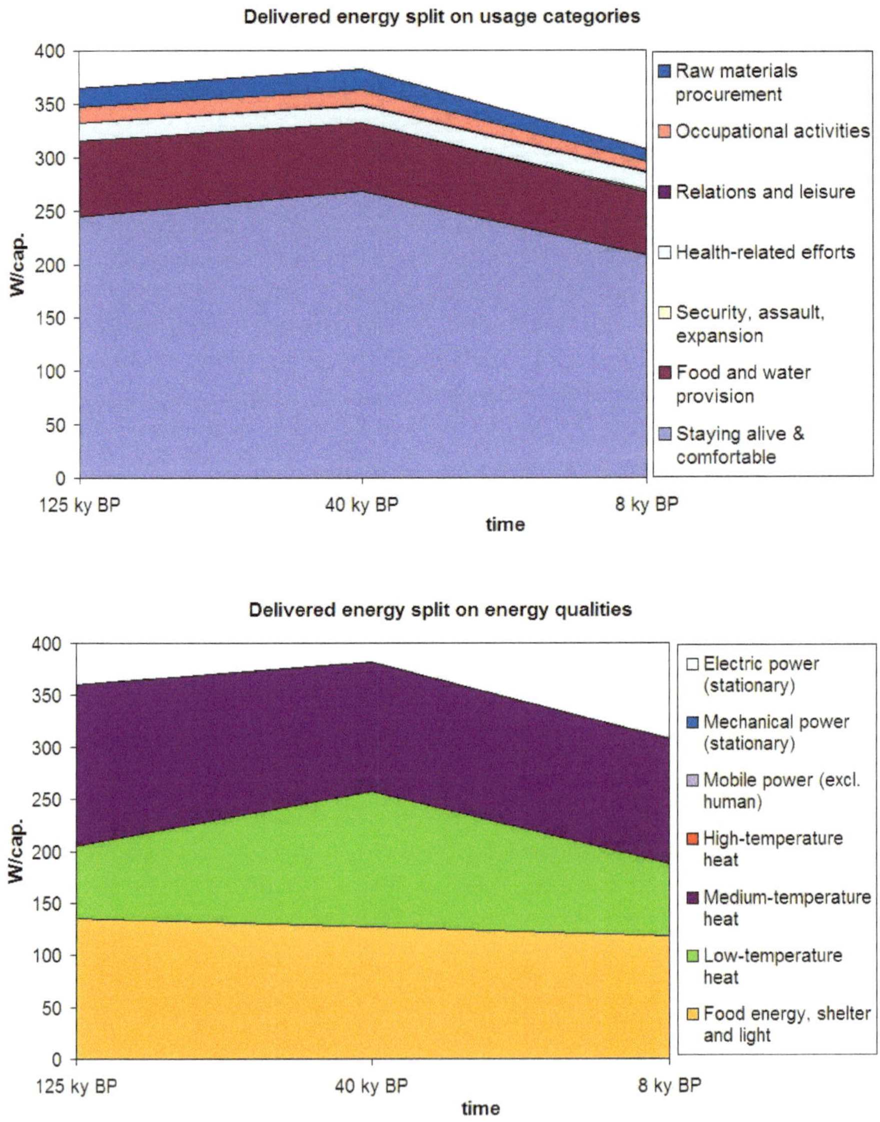


*Figure 2.16. Estimated energy use (delivered) 125 to 8 ky BP, distributed on categories of usage (a) and energy type (b; see text).*

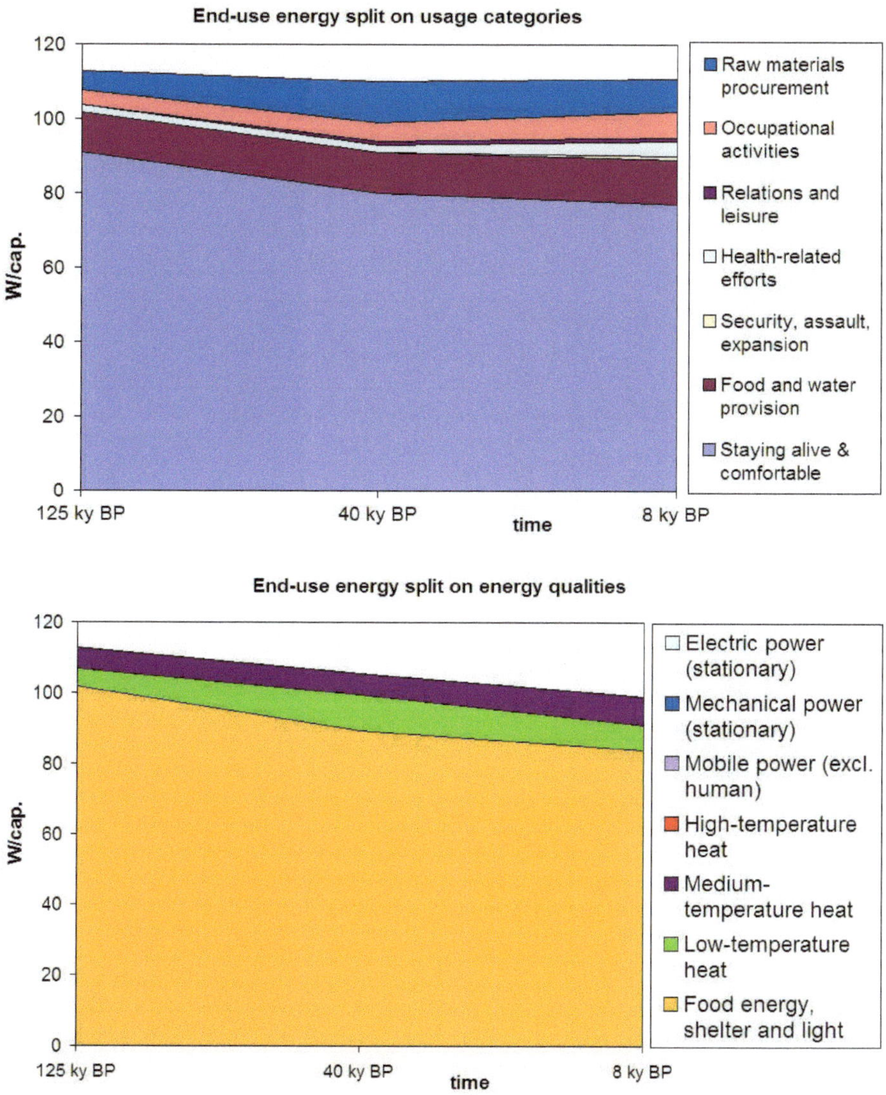


*Figure 2.17. Estimated end-use energy 125 to 8 ky BP, distributed on categories of usage (a) and energy type (b; see text).*

Figure 2.17, based on Tables 2.3, 2.4 and 3.2, estimates the ***end-use*** energy efficiency of the tasks listed in the Tables 2.1, 2.2 and 3.1, and on that basis gives the final end-use energy, after including all conversions and omitting all losses up to the final service or concrete item delivered by the act of energy conversion. The efficiency of energy conversion chains will be an important consideration in the following chapters. However, this "end-use energy" concept should be used with care in order not to hide that some goods and services can be delivered in many different ways, using use of different technologies, of which there could even be some that do not involve energy use at all.

*Table 2.1. Estimated energy use (delivered) to adults, 125 ky BP.*

| *Northern Europe 125 ky BP.* | *Human energy* | *Energy from fuels and flows* | | |
|---|---|---|---|---|
| *Rate of energy use (average W for an adult person)* | Food energy | Low-temperature heat | Medium/high-temperature heat | TOTAL |
| **Liveable surroundings** | | | | **244** |
| Basic metabolism | 84 | | | **84** |
| Heating | | 60 | | **60** |
| Light | | | 100 | **100** |
| **Food and water provision** | | | | **72** |
| Hunting | 15 | | | **15** |
| Gathering | 2 | | | **2** |
| Preparation, cooking | 4 | | 50 | **54** |
| Water provision | 1 | | | **1** |
| **Health-related efforts** | | | | **16** |
| Food preservation | 1 | 10 | 5 | **16** |
| **Occupational activities** | | | | **12** |
| Construction | 5 | | | **5** |
| Manufacture | 4 | | | **4** |
| Transportation of goods | 3 | | | **3** |
| **Raw materials procurement** | | | | **16** |
| Materials for construction and industry | 10 | | | **10** |
| Energy provision | 6 | | | **6** |
| TOTAL | **135** | **70** | **155** | **360** |

The changes in energy expenditure between 125 ky BP and 8 ky BP have a large component from the basic metabolic rate, which on average is 84 W/cap. for an adult Neanderthal person but only 68 W/cap. for an average adult modern human. The number used at 40 ky BP is for a modern human (Table 2.2) and the additional metabolic energy used for further activities thus increases from 58 to 75 W/cap. between 125 and 40 ky BP. The basic metabolic rate is considered as also being the end-use energy, while useful muscle energy is taken to be about a third of the increased metabolic rate (Tables 2.3 and 2.4), making the corresponding muscle activities increase more modestly from 20 to 26 W/cap. This is less than the decline in basic metabolic rate, and the 40 ky BP peak in delivered energy thus becomes a decline in end-use energy from 125 to 40 ky BP (Fig. 2.17b).

Between 40 and 8 ky BP, the amount of muscle energy made useful again increases slightly (Fig. 2.17b), but not the delivered energy (Fig. 2.16b). This is seen from Tables 2.4 and 3.2 to be caused by an increase in activities such as fishing and tool manufacturing, combined with a slight decline in efforts needed to obtain the raw materials (wood and flint), due to increased forest cover. The beginning of a shortage of flint stone will be felt in the subsequent period.

*Table 2.2. Estimated energy use (delivered) to adults, 40 ky BP.*

| *Northern Europe 40 ky BP.* | *Human energy* | *Energy from fuels and flows* | | |
|---|---|---|---|---|
| *Rate of energy use (average W for an adult person, Modern Human)* | Food energy | Low-temperature heat | Medium/high-temperature heat | TOTAL |
| **Liveable surroundings** | | | | **268** |
| Basic metabolism | 68 | | | **68** |
| Heating | | 120 | | **120** |
| Light | | | 80 | **80** |
| **Food and water provision** | | | | **64** |
| Hunting | 14 | | | **14** |
| Gathering | 4 | | | **4** |
| Preparation, cooking | 5 | | 40 | **45** |
| Water provision | 1 | | | **1** |
| **Health-related efforts** | | | | **16** |
| Food preservation | 1 | 10 | 5 | **16** |
| **Relations and leisure** | | | | **1** |
| Toys, artwork | 1 | | | **1** |
| **Occupational activities** | | | | **14** |
| Construction | 6 | | | **6** |
| Manufacture | 5 | | | **5** |
| Transportation of goods | 3 | | | **3** |
| **Raw materials procurement** | | | | **19** |
| Materials for construction and industry | 12 | | | **12** |
| Energy provision | 7 | | | **7** |
| TOTAL | **127** | **130** | **125** | **382** |

The efficiency of open fires (for providing light or heat) is generally low, probably under 5% for heat. Indoor fires for cooking would gradually become more efficient, reaching around 10% at the end of the period, with three-stone stove types using heat-accumulating stones. For space heating, glowing embers provide higher efficiency than bonfire flames. The division of energy from fires between low and medium temperature entries reflects the uses made of the fire, not the flame or ember temperature. The categories are defined by the boundary between low and medium temperature being set at about 100 °C, and that between medium and high temperature around 500 °C. High temperature energy is not required during the early and middle Stone Age, but will be as when use of metals becomes common (actually slightly before the Bronze Age and Iron Age).

*Table 2.3. Estimated energy use (end-use) to adults, 125 ky BP.*

| *Northern Europe 125 ky BP.* | *Human energy* | *Energy from fuels and flows* | | |
|---|---|---|---|---|
| *Rate of energy end-use (average W for an adult person)* | **Food energy** | **Low-temperature heat** | **Medium/high-temperature heat** | **TOTAL** |
| **Liveable surroundings** | | | | **91** |
| Basic metabolism | 84 | | | **84** |
| Heating | | 4 | | **4** |
| Light | | | 3 | **3** |
| **Food and water provision** | | | | **10.7** |
| Hunting | 5 | | | **5** |
| Gathering | 0.7 | | | **0.7** |
| Preparation, cooking | 1 | | 3 | **4** |
| Water provision | 1 | | | **1** |
| **Health-related efforts** | | | | **2** |
| Food preservation | 1 | 1 | 0 | **2** |
| **Occupational activities** | | | | **4** |
| Construction | 1.7 | | | **1.7** |
| Manufacture | 1.3 | | | **1.3** |
| Transportation of goods | 1 | | | **1** |
| **Raw materials procurement** | | | | **5** |
| Materials for construction and industry | 3 | | | **3** |
| Energy provision | 2 | | | **2** |
| **TOTAL** | **101.7** | **5** | **6** | **112.7** |

The energy use of children below 12 years has not been included, due to its strong age dependence and lack of archaeological indications. As noted in Table 1.2, it is likely to be small.

Figures 2.16a and 2.17a divide the energy use on demand categories. "Liveable surroundings" include basic metabolic conversion, as well as space heating and light. The energy used by the building (hut, tent) itself is not estimated. It would comprise heat gains (negative energy use) from absorption of solar radiation and from reducing heat loss by wind. An idea of the size of such contributions can be inferred from the changes in tolerable temperatures on the human skin (multiplied by a heat conductivity), as discussed in Table 1.1. The imprecise term "energy use" denotes energy carried away or converted per person and per unit of time (hence the unit W/cap; cf. Sørensen, 2004).

"Food and water provision" covers energy spent during hunting, fishing, gathering and later farming. For water, this is mainly muscle energy for walking to a source of water and carrying it back. During most of the Weichselian period, northern Europe would have been quite wet, and small lakes and streams would usually have been available nearby, even on inland plains or tundra. In winter, melted snow could have been used. From the Maglemose period (Chapter 3), human settlements were preferably placed at freshwater reservoirs. The water from such sources could occasionally be infected and contribute to disease. There is no indication of cleaner water from rainwater collection being used, but it cannot be excluded. Concentrating funnels have not been found, but modest collection in pots could have taken place.

*Table 2.4. Estimated energy use (end-use) to adults, 40 ky BP. Corresponding tables for 8 ky BP are in Chapter 3.*

| *Northern Europe 40 ky BP.* | *Human energy* | *Energy from fuels and flows* | | |
|---|---|---|---|---|
| *Rate of energy end-use (average W for an adult person, Modern Human)* | Food energy | Low-temperature heat | Medium/high-temperature heat | TOTAL |
| **Liveable surroundings** | | | | **80** |
| Basic metabolism | 68 | | | **68** |
| Heating | | 9 | | **9** |
| Light | | | 3 | **3** |
| **Food and water provision** | | | | **11** |
| Hunting | 4 | | | **4** |
| Gathering | 1 | | | **1** |
| Preparation, cooking | 2 | | 3 | **5** |
| Water provision | 1 | | | **1** |
| **Health-related efforts** | | | | **2** |
| Food preservation | 1 | 1 | 0 | **2** |
| **Relations and leisure** | | | | **1** |
| Toys, artwork | 1 | | | **1** |
| **Occupational activities** | | | | **4.7** |
| Construction | 2 | | | **2** |
| Manufacture | 1.7 | | | **1.7** |
| Transportation of goods | 1 | | | **1** |
| **Raw materials procurement** | | | | **6.7** |
| Materials for construction and industry | 4 | | | **4** |
| Energy provision | 2.7 | | | **2.7** |
| TOTAL | 89.4 | 10 | 6 | 105.4 |

Tools specifically aimed for use as arms are found only near the end of the period, around 6 ky BP. Food preservation was important, and drying of meat has already been mentioned. This would take solar energy or a fire. There is no indication of using salt until later, when it could be imported from European regions further south.

Activities such as manufacture of toys and creation of art begin around the last glacial maximum. In Denmark, two types of arts and crafts can be distinguished: one decorates tools with abstract patterns, the other is figurative and depicts prey animals and hunters, as well as sexual attributes of women (perhaps suggesting that the artists were male?). Energy-related techniques are the production of colours, involving heating of metal-containing materials. The motives of art suggest that religious rituals were not conducted during the period. Of course, human beings would have speculated over the workings of the sun, thunder and lightning, rain and so on, and perhaps about the meaning of life, but there was no institutionalized religion, as this would have influenced motives for the art produced. Such motives are a very strong indicator of the things that occupied people's minds. Religious art and other manifestations appear later, in conjunction with the mediators of institutionalized religion, often called *shamans*. Their origin may be in the functions of "wise men" or "wise women", proposing to relieve pain and cure diseases, sometimes with justification in the form of functioning herbal medicine. The wise men became priests or doctors, and the wise women became midwives or witchdoctors, suggesting that such developments did not occur before the societies had become considerably more stratified, with division of labour and with class and gender structures supported by invention of norms that could keep the less distinguished in place. This inter-relationship between institutionalised religion, division of power and enforcement of laws and codes of conducts (originating in the need to encourage just or moral behaviour, but also to become used to subdue parts of the population) did not become generally apparent until after 6 ky BP (in the north), a time characterized by a fast increase of population, by the introduction of agriculture and by the emergence of substantial amounts of social layering and the associated violence. Institutionalized religion seems to serve as a tool for maintaining the social disparity and permitting exploitation of those not holding the reins of power, and therefore it did not appear before the "need" arose.

"Occupational activities" comprise construction of huts and fireplaces, cleaning of hides, skins and bones, making clothes, and production of axes (for felling trees), hunting equipment (such as spears, and later bows and arrows), and tools (from truffle-searching poles to toothpicks). Behind these activities is the "procurement of raw materials", primarily wood and suitable stone material such as flint. During the colder periods, there would have been few trees and sometimes long walks involved in securing wood. Permafrost would have made the search for flint more difficult, and the inhabitants of cold regions would presumably have needed to put raw materials in stock during the warmer months. Later, with the deglaciation and increasing temperatures, the forests expanded and made wood resources easy to access. Flint resources were fairly evenly distributed, although in Denmark the largest deposits were in particular places (Fig. 3.29). For energy provision, not only hardwood but also twigs and other small-sized wood resources would have been exploited.

### *References:*

Aaris-Sørensen, K., Petersen, K., Tauber, H., 1990. Danish finds of mammoth (*Mammuthus primigenius* (Blumenbach)). Stratigraphical position, dating and evidence of late Pleiostene environment. Geological Survey of Denmark, DGU Series B, No. 14. Copenhagen.

Allen, B., 2006. Prometheus and the muses. *Common Knowledge,* **12**, 354-378.

Ambrose, S., 1998. Late Pleistocene human population bottlenecks, volcanic winter, and differentiation of modern humans. *J. Human Evolution* **34**, 623–651.

Andel, T. v., Davies, W., Wenninger, B., 2003. The human presence in Europe during the last glacial period I: Human migration and the changing climate. *Neanderthals and Modern Humans in the European Landscape of the Last Glaciation. Archaeological Results of the Stage 3 Project* (van Andel, T., Davies, W., eds), pp. 31–56. The McDonald Institute for Archaeological Research, Cambridge. Reprinted 2004 with corrections.

Balter, M., 2010. Reanalysis of French cave could deal setback to Neandertal smarts. *Science* **330**, 439.

Banks, W., d'Errico, F., Peterson, A., Kageyama, M., Sima, A., Sánchez-Goñi, M-F., 2008. Neanderthal extinction by competitive exclusion. *PloS ONE* **3**, e3972, 1–8.

Bard, E., 1998. Geochemical and geophysical implications of the radiocarbon calibration. *Geochimica et Cosmochimica Acta* **62**, 2025–2038.

Bintanja, R., Wal, R. van de, Oerlemans, J., 2002. Global ice volume variations through the last glacial cycle simulated by a 3-D ice-dynamical model. *Quaternary Int.* **95–96**, 11–23.

Bocherens, H., Drucker, D., 2003. Reconstructing Neanderthal diet from 120,000 to 30,000 BP using carbon and nitrogen isotopic abundances. *Le role de l'environnement dans les comportements des chasseurs-cueilleurs préhistoriques* (M. Patou-Mathis, H. Bocherens, eds), 1–7. *British Archaeological Reports.* Int. Series 1105. Archaeopress, Oxford.

Bocquet-Appel, J-P., Demars, P., 2000. Neanderthal contraction and modern human colonization of Europe. *Antiquity* **74**, 544–552.

Boë, L-J., Heim, J-L., Honda, K., Maeda, S., Badin, P., Abry, C., 2007. The vocal tract of newborn humans and Neanderthals: Acoustic capabilities and consequences for the debate on the origin of language. A reply to Lieberman (2007a). *J. Phonetics* **35**, 564–584.

Briggs, A., *et al.*, 2009. Targeted retrieval and analysis of five Neanderthal mtDNA genomes. *Science* **325**, 318–321.

Chalmin, E., Menu, M., Pomiès, M-P., Vignaud, C., Aujoulat, N., Geneste, J-M., 2006. Les blasons de Lascaux. *L'anthropologie* **108**, 571–592.

Conard, N., Bolus, M., 2003. Radiocarbon dating the appearance of modern humans and timing of cultural innovation in Europe: new results and new challenges. *J. Human Evolution* **44**, 331–371.

Conard, N., Grootes, P., Smith, F., 2004a. Unexpectedly recent dates for human remains from Vogelherd. *Nature* **430**, 198–201.

Conard, N., Malina, M., Münzel, S., Seeberger, F., 2004b, Eine Mammutelfenbeinflöte aus dem Aurignacien des Geissenklösterle. *Archäologisches Korrespondenzblatt* **34**, 447–462

Crubézy, E., Ludes, B., Poveda, J-D., Clayton, J., Crouau-Roy, B., Montagnon, D., 1998. Identification of *Mycobacterium* DNA in an Egyptian Pott's disease of 5400 years old. *Comptes Rendus Acad. Sci. Paris III. Life Sciences* **321**, 941–951.

Delluc, B., Delluc, G., 2006. Art paléolithique, saisons et climats. *Comptes Rendus Palevol* **5**, 203–211.

d'Errico, F., Villa, P., Llona, A., Idarraga, R., 1998. A middle paleolithic origin of music? Using cave-bear accumulations to assess the Divje Babe I bone "flute". *Antiquity* **72,** 65–79.

d'Errico, F., *et al.*, 2003. Archaeological evidence for the emergence of language, symbolism, and music – an alternative multidisciplinary perspective. *J. World Prehistory* **17**, 1–70.

Duarte, C., Maurício, J., Petitt, P., Souto, P., Trinkaus, E., Plicht, H., Zilhão, J., 1999. The early Upper Paleolithic human skeleton from the Abrigo do Lagar Velho (Portugal) and modern human emergence in Iberia. *Proc. Nat. Acad. Sci. USA* **96**, 7604–7609.

Fedele, F., Giaccio, B., Hajdas, I., 2008. Timescales and cultural process at 40,000 BP in the light of the Campanian Ignimbrite eruption, Western Eurasia. *J. Human Evolution* **55**, 834–857.

Feruglio, V., 2006. De la faune au bestiaire – la grotte Chauvet-Pont-d'Arc, aux origines de l'art pariétal paléolithique. *Comptes Rendus Palevol* **5**, 213–222.

Fournier, P., Drancourt, M., Raoult, D., 2007. Bacterial genome sequencing and its use in infectious diseases. *Lancet Infectuous Diseases* **7**, 711–723.

Green, R., *et al.*., 2006. Analysis of one million base pairs of Neanderthal DNA. *Nature* **444**, 331–336.

Green, R., *et al.*, 2010. A draft sequence of the Neandertal genome. *Science* **328**, 710–722.

Guhl, E., Jaramillo, C., Vallejo, C., Yockteng, R., Cárdenas-Arroyo, F., Fornaciari, C., Arriaza, B., Aufderheide, C., 1999. Isolation of *Trypanosoma cruzi* DNA in 4000-year-old mummified human tissue from northern Chile. *American J. Phys. Anthropology* **108**, 401–407.

Gutiérrez, G., Sánchez, D., Marín, A., 2002. A reanalysis of the ancient mitochondrial DNA sequences recovered from Neanderthal bones. *Molec. Biol. Evolution* **19**, 1359–1366.

Hublin, J-J., Spoor, F., Braun, M., Zonneveld, F., Condemi, S., 1996. A late Neanderthal associated with Upper Paleolithic artefacts. *Nature* **381**, 224–226.

Jöris, O., Street, M., 2008. At the end of the $^{14}C$ time scale – the middle to upper paleolithic record of western Eurasia. *J. Human Evolution* **55**, 782–802.

Krause, J., *et al.*., 2007. The derived *FOXP2* variant of modern humans was shared with Neanderthals. *Current Biology* **17**, 1908–1912.

Krings, M., Stone, A., Schmitz, R., Krainitzki, H., Stoneking, M., Pääbo, S., 1997. Neanderthal DNA sequences and the origin of modern humans. *Cell* **90**, 19–30.

Kühl, N., Litt, T., Schölzel, C., Hense, A., 2007. Eemian and Early Weichselian temperature and precipitation variability in northern Germany. *Quaternary Sci. Rev.* **26**, 3311–3317.

Molerus, O., 1996. History of civilisation in the western hemisphere from the point of view of particulate technology. *Advanced Powder Technology* **7**, 71–78.

Noonan, J., Coop, G., Kudaravalli, S., Smith, D., Krause, J., Alessi, J., Chen, F., Platt, D., Pääbo, S., Pritchard, J., Rubin, E., 2006. Sequencing and analysis of Neanderthal genomic DNA. *Science* **314**, 1113–1118.

Pennisi, E., 2006. The dawn of stone age genomics. *Science* **314**, 1068–1071.

Ruth, U., Wagenbach, D., Steffensen, J., Bigler, M., 2003. Continuous record of microparticle concentration and size distribution in the central Greenland NGRIP ice core during the last glacial period. *J. Geographical Res.* **108** (D3) ACL1-1

Salo, W., Aufderheide, A., Buikstra, J., Holcomb, T., 1994. Identification of *Mycobacterium tuberculosis* DNA in a pre-Columbian Peruvian mummy. *Proc. Nat. Acad. Sci. USA* **91**, 2091–2094.

Seelos, K., Sirocko, F., 2007. Abrupt cooling events at the very end of the last interglacial. *The climate of past Interglacials* (Sirocko, F., Claussen, M., Goñi, M., Litt, T., eds), pp. 207-229. *Developments in Quaternary Sciences* **7** (Elsevier).

Soficaru, A., Petrea, C., Dobos, A., Trinkaus, E., 2007. The human cranium from the Pestera Cioclovina Uscata. *Current Anthropology* **48**, 611–619.

Sørensen, B., 2004. *Renewable Energy*, 3rd edition. Elsevier Academic Press, Burlington (4th edition 2010).

Sørensen, B., 2011a. *Life-Cycle Analysis of Energy Systems. From Methodology to Applications.* RCS (Royal Society of Chemistry) Publishing, Cambridge.

Sørensen, B., 2011b. Demography and the extinction of European Neanderthals. *J. Anthropological Archaeology* **30**, 17–29.

Svendsen, J. et al., 2004. Late quaternary ice sheet history of northern Europe. *Quaternary Science Reviews* **23**, 1229–1271.

Tarasov, L., Peltier, R., 1997. Terminating the 100kyr ice age cycle. *J. Geophys. Res.* **102**, 21665-21693.

Turk, I., 1997. Description and origin of the suspected bone flute. *Mousterian bone flute & other finds from Divje Babe I cave site in Slovenia* (Turk, I. ed.), pp. 157–175. Znanstvenoraziskovalni Center Sazu, Ljubljana.

Turk, I., Blackwell, B., Turk, J., Pflaum, M., 2006. Résultats de l'analyse tomographique informatisée de la plus ancienne flûte découverte à Divje Babe I (Slovenie) et sa position chronologique dans le contexte des changements paléoclimatique et paléoenvironmentaux au cours du dernier glaciaire. *L'anthropologie* **110**, 293–317.

Walker, P., 2001. A bioarchaeological perspective on the history of violence. *Ann. Rev. Anthropology* **30**, 573–596.

White, R., 2001. Personal ornaments from the Grotte du Renne at Arcy-sur-Cure. *Athena Rev.* **2**, 41–46.

Zilhão, J., 2006. Neanderthals and moderns mixed, and it matters. *Evolutionary Anthropology* **15**, 183–105.

Zink, A., Haas, C., Reichl, U., Szeimies, U., Nerlich, A., 2001. Molecular analysis of skeletal tuberculosis in an ancient Egyptian population. *J. Med. Microbiol.* **50**, 355–366.

# 3

# Late Stone Age: More settlements and transitions

## 3.1 Bølling and Allerød periods, and the end of glaciation

The modern humans had come from Africa to Europe by way of the Danube corridor, or perhaps more probably via a route East of the Black Sea. They found plenty of interesting prey to satisfy their food requirements, and the colder climate did not seem to bother them too much. In summer it was more pleasant than further south, and the challenge of the cold winters was eventually overcome. That the modern humans did not see low temperatures as an insurmountable barrier is evident from the fact that they ventured north as soon as the ice sheets withdrew. Before the glacial melt became visible (after 20 ky BP), they stayed south of Hamburg, but as soon as the first ice-free temperature excursion towards warmer weather occurred, which was during the Bølling period (14.6–14.2 ky BP), they came to Denmark.

Figure 3.1 shows the temperatures in Denmark from 15 ky BP to the current time. The figure is simply an expansion of the relevant part of Fig. 1.1. For a few hundred years, starting at 14.5 ky BP, the average temperature was only 4 °C below the present one. However, the shorelines in the Nordic region were substantially different from the simple eustatic levels (i.e. those derived from the volume of ice melted; Fig. 1.5). Calculations of topography, including the effect of ice depressing the underlying sediments and subsequent raising of the land after the ice disappeared, are presented below in Fig. 3.2.

Figures 3.2a and 3.2b show the first signs of melting ice, at the Shetland Islands and at the Barents Sea. The height of the latter at 20 ky BP is some 50 m above the mean sea level, including the ice sheet, whereas at 19 ky BP it is below zero, also including the height of the ice, indicating that a substantial depression is manifesting itself both before and during the melting of the ice (heights shown include ice but not inland waters).

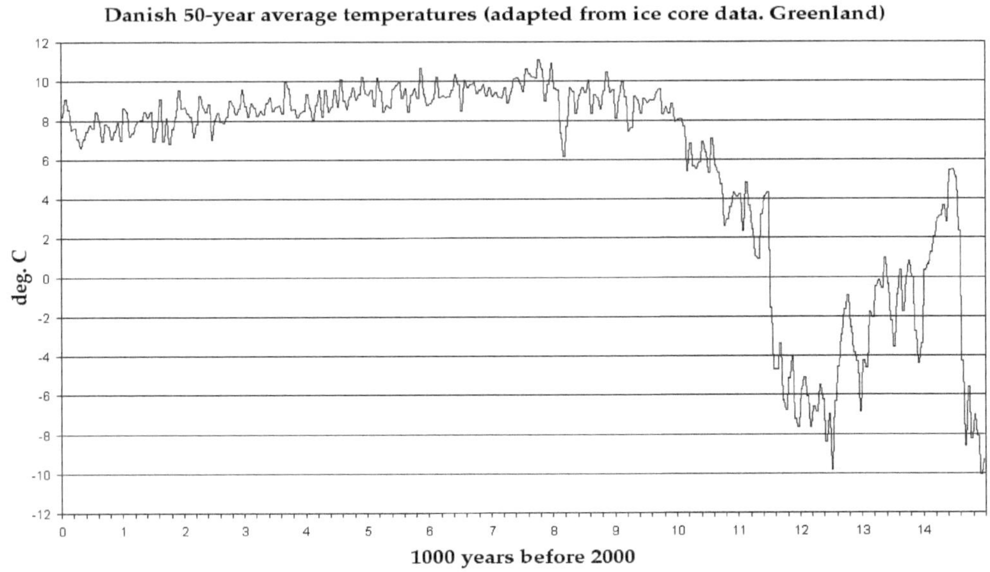


*Figure 3.1. 15 ky of temperatures adjusted from NGRIP ice-core data to fit current Danish average temperature (based on Fig. 1.1).*

At 18 ky BP (Fig. 3.2c), there is further thaw in the Barents Sea, and at 17 ky BP (Fig. 3.2d) it is ice-free. The same is true for nearly all of (present) Denmark, but the temperatures (Fig. 1.1) suggest presence of extensive permafrost.

---

*Figure 3.2. (Panels appearing on the following pages are labelled by letter from a to n, and by time before present). Height map (m above global mean sea level) for the Nordic region calculated with consideration of global sea level change due to ice melting, as well as local phenomena due to depression and upheaval (based on the model of Peltier, 1994). The ice coverage is calculated with a resolution of 1°, as indicated by the white grid-lined ice areas, and so is the topography both in ice-covered areas and outside these. Grids indicate ice somewhere inside, not necessarily extending to the edges. However, height variations inside these grid cells are taken from modern data with a higher resolution of 0.08333° (National Oceanographic and Atmospheric Administration, 2004). The spacing in time between each panel of the figure and the following one is 1 ky, except for the last two panels. The sequence of times is 20, 19, …, 10, 9 ky BP and then 6 ky BP, 3 ky BP. A more recent model (Peltier, 2004) gives somewhat different shorelines, but no clear improvement relative to available sediment data.*

a: 20 ky BP

b: 19 ky BP

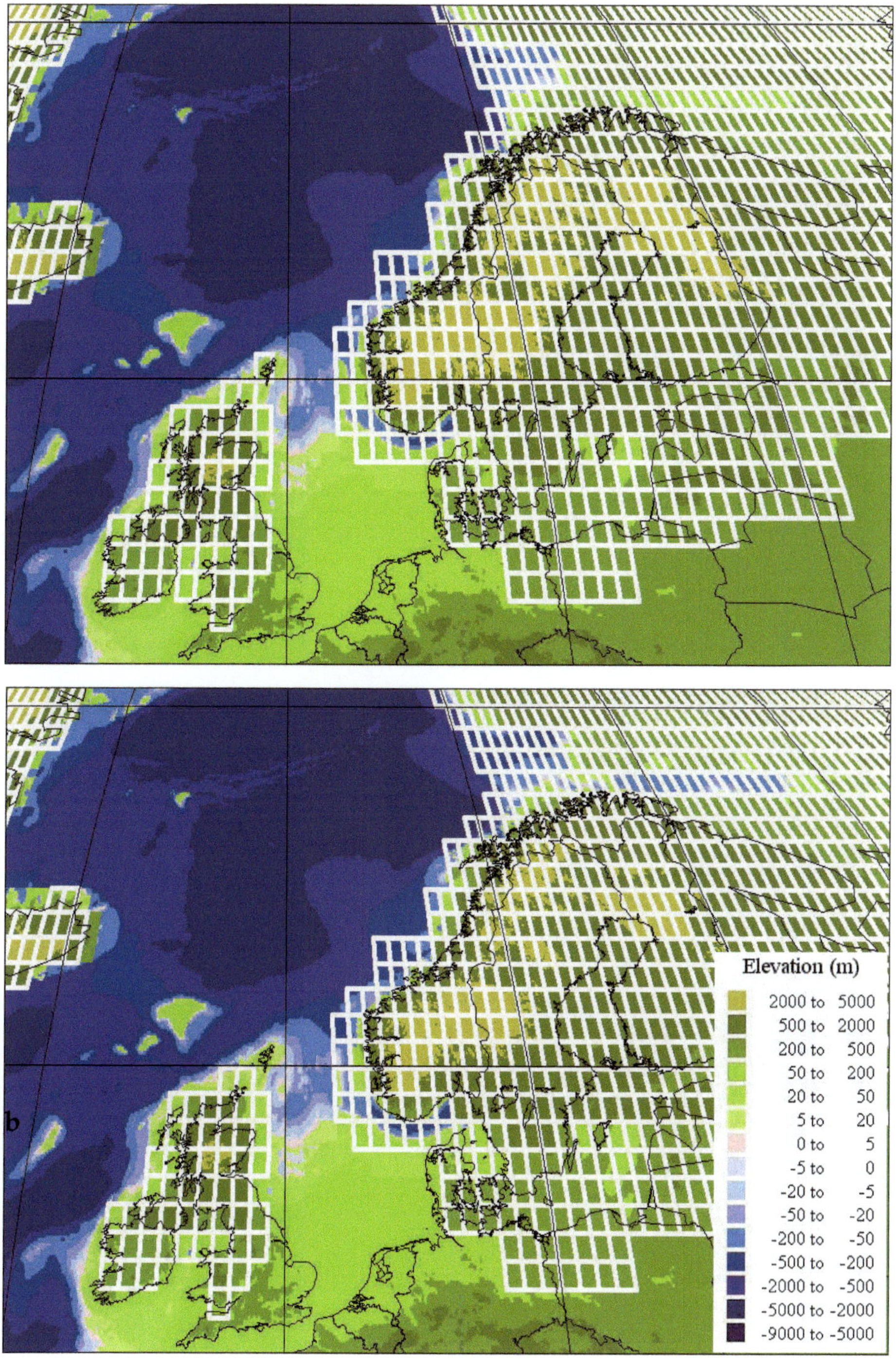

c: 18 ky BP

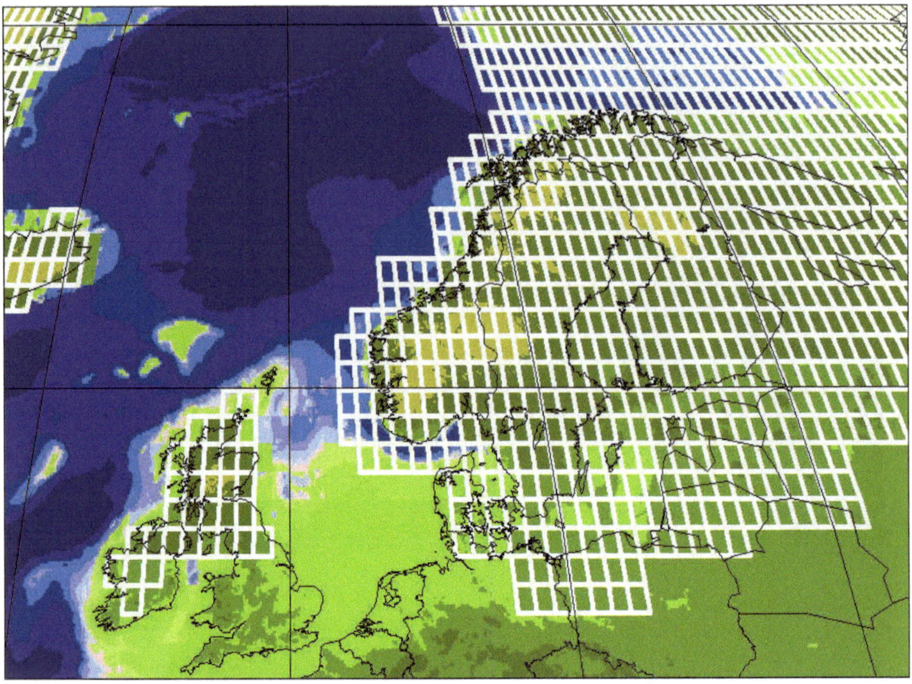

d: 17 ky BP

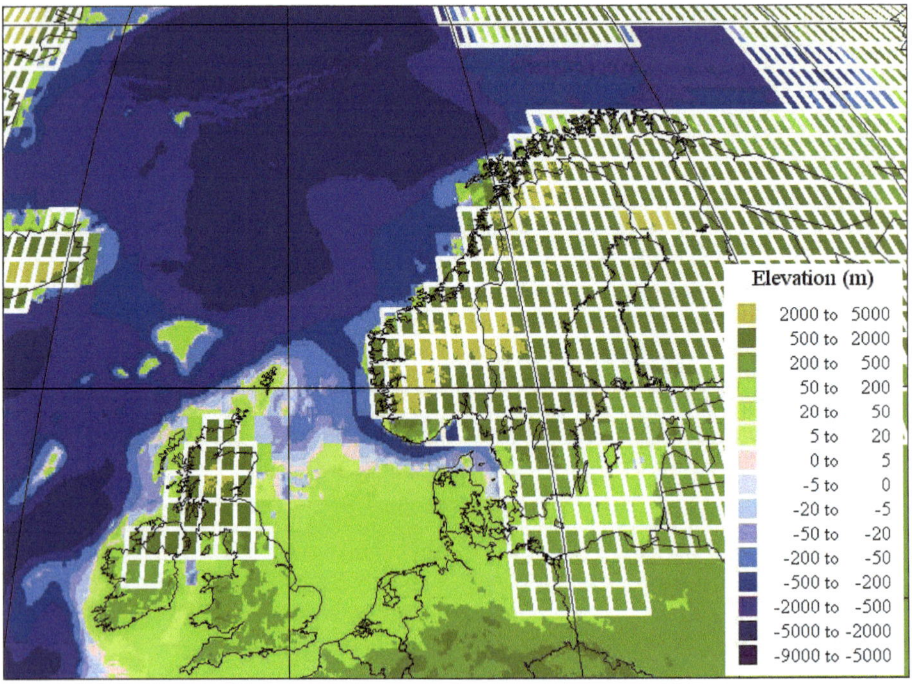

e:
16
ky
BP

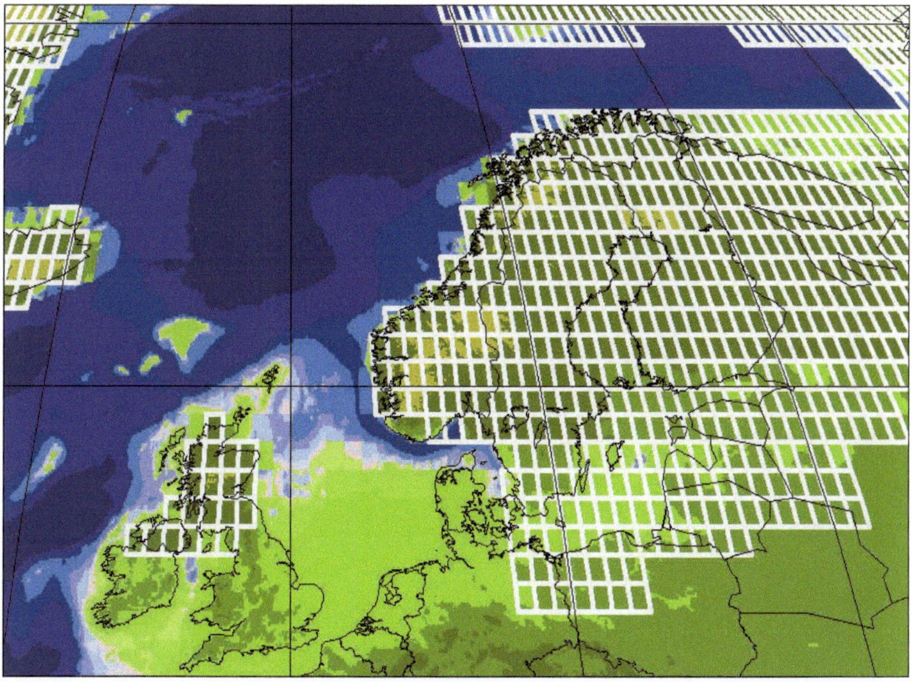

f:
15
ky
BP

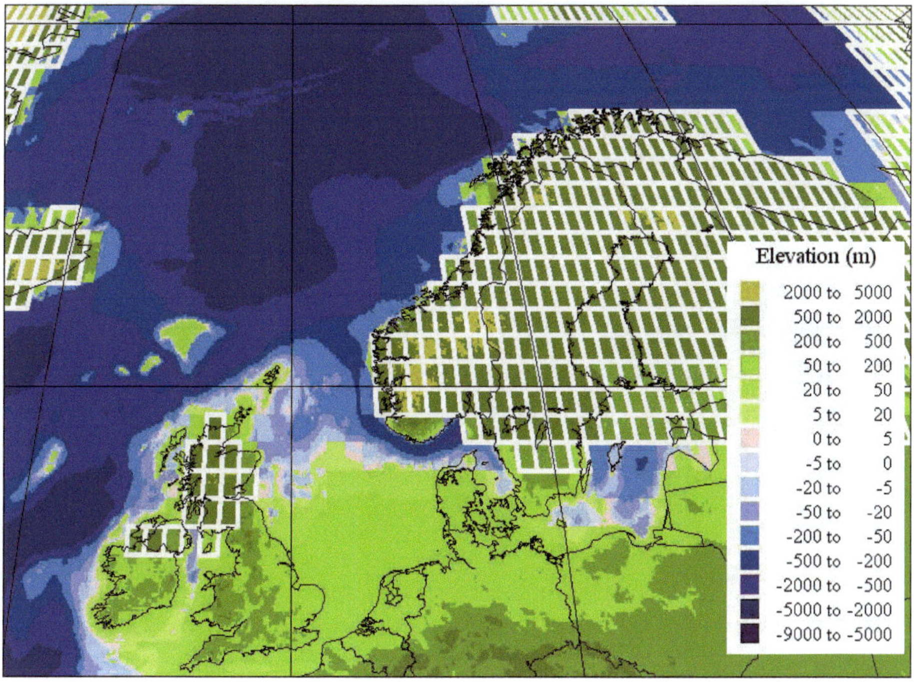

g:
14
ky
BP

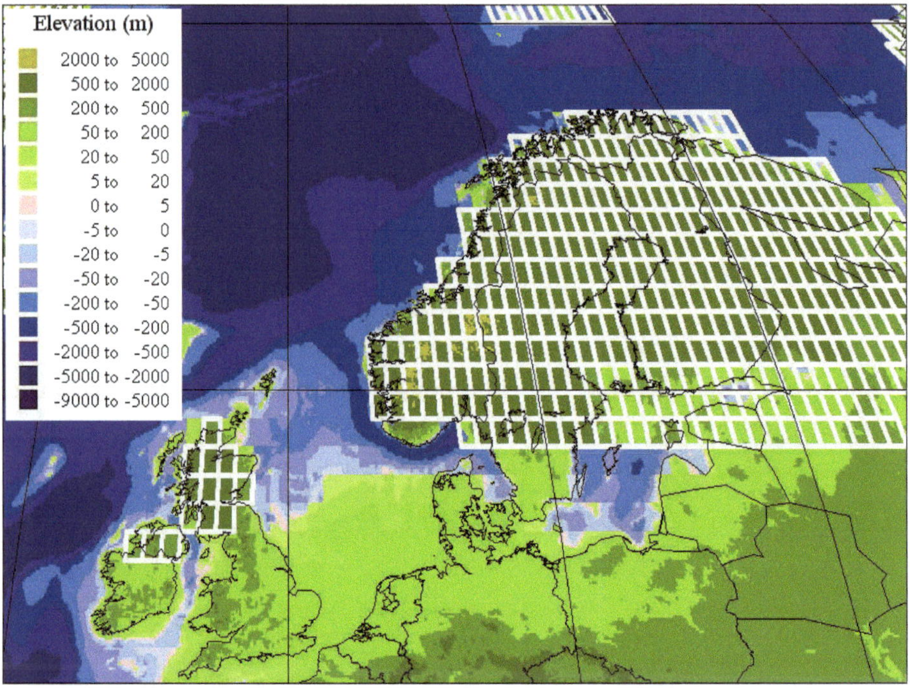


h:
13
ky
BP

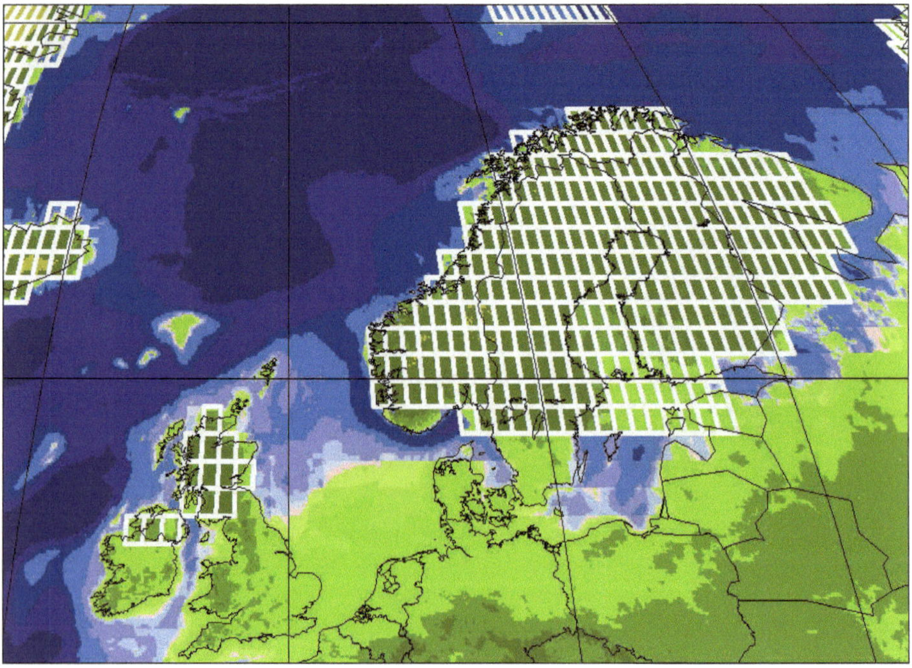

i: 12 ky BP

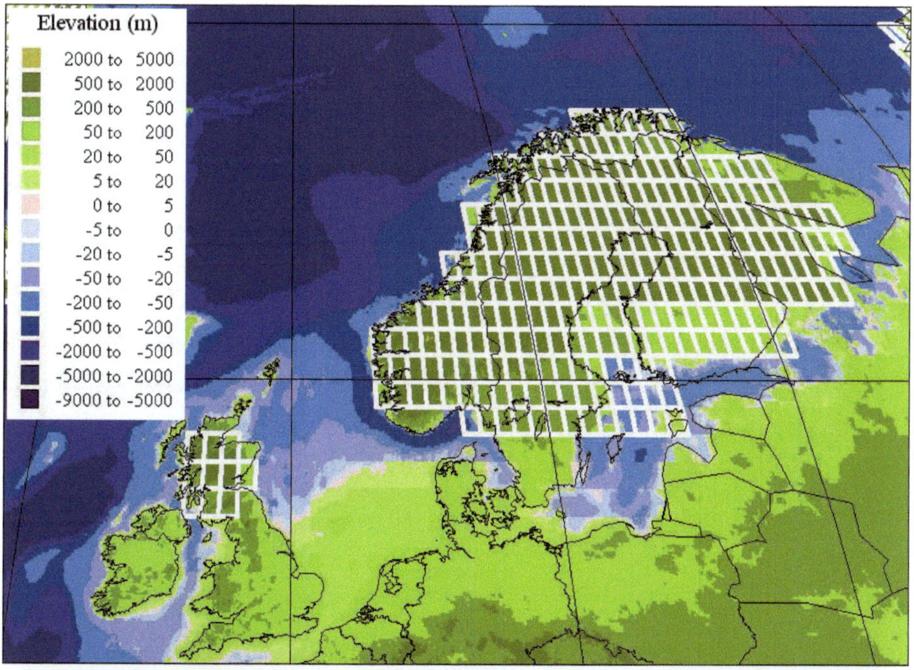


j: 11 ky BP

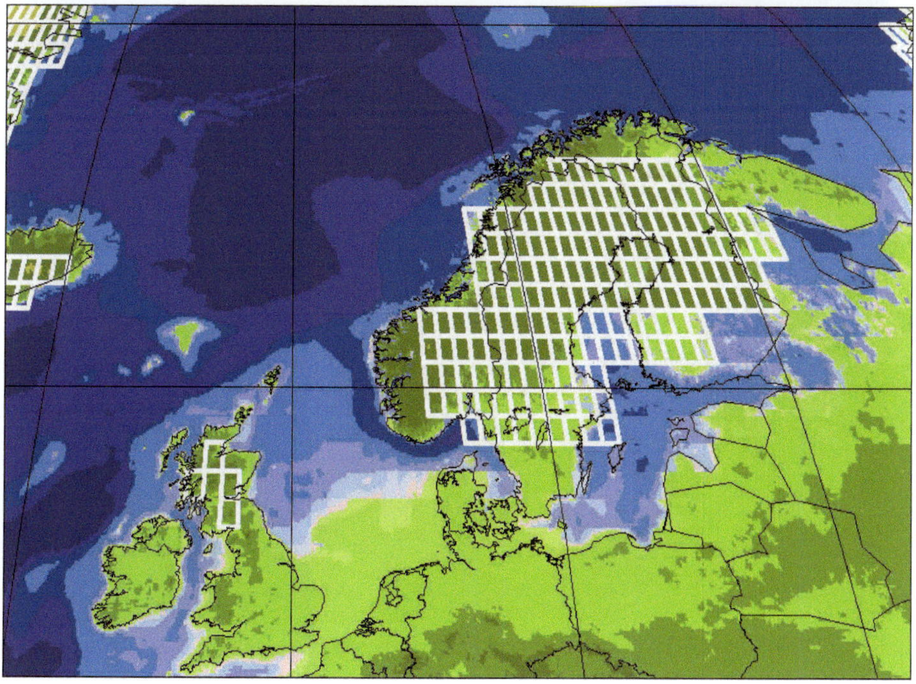

k:
10
ky
BP

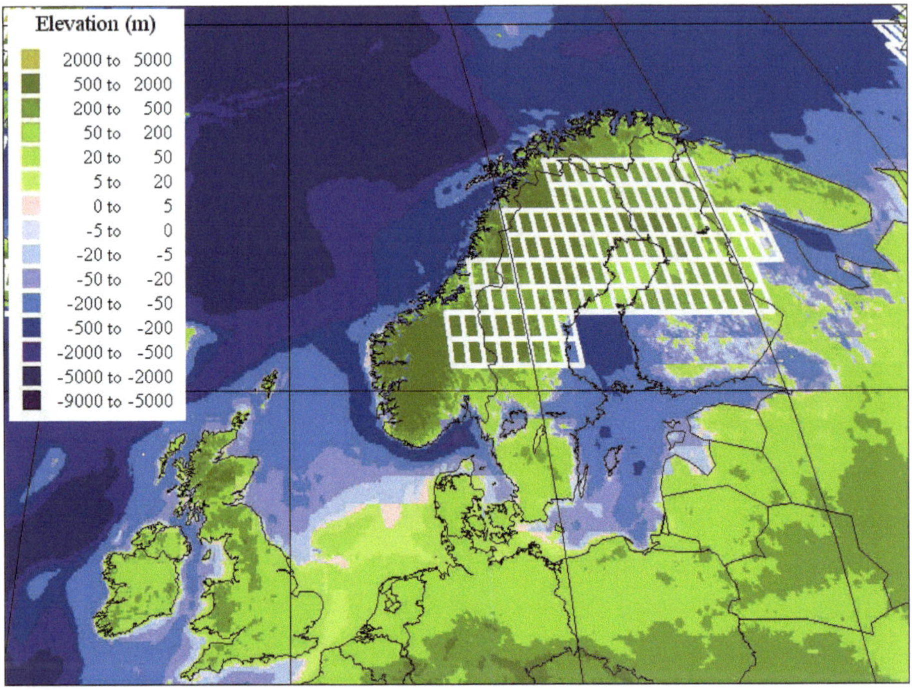


l:
9
ky
BP

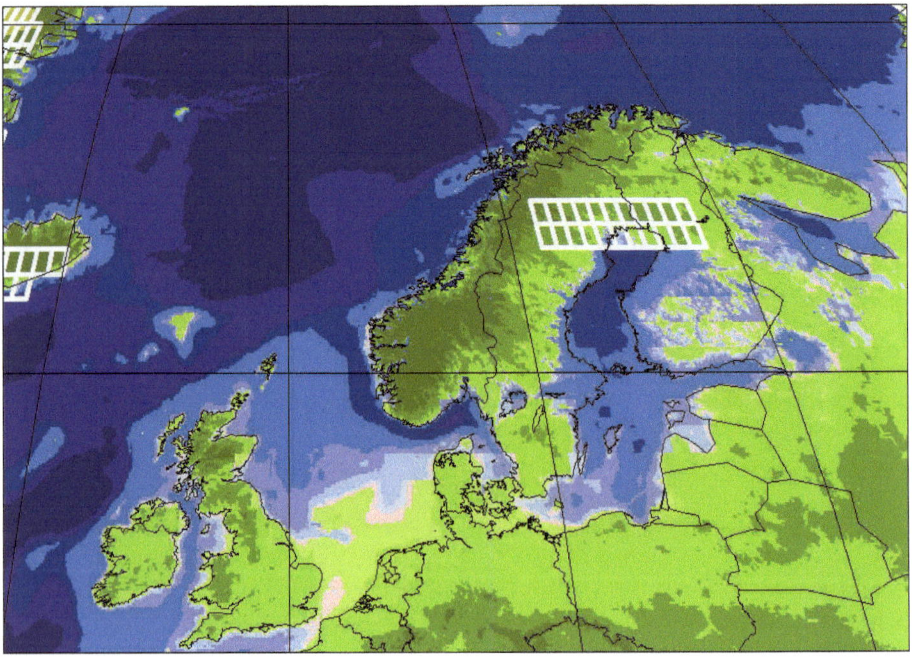

m: 6 ky BP

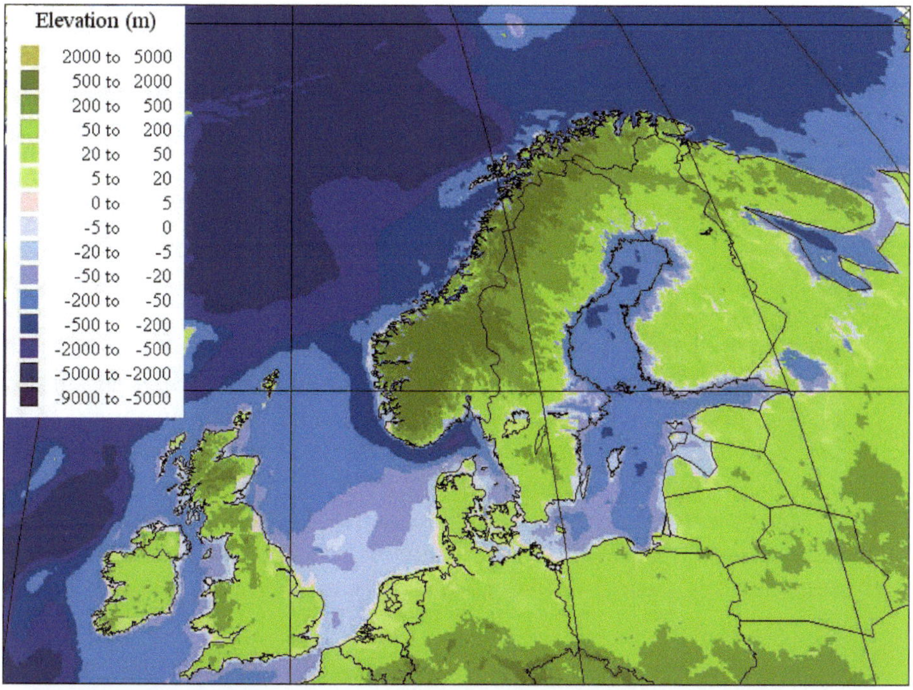


n: 3 ky BP

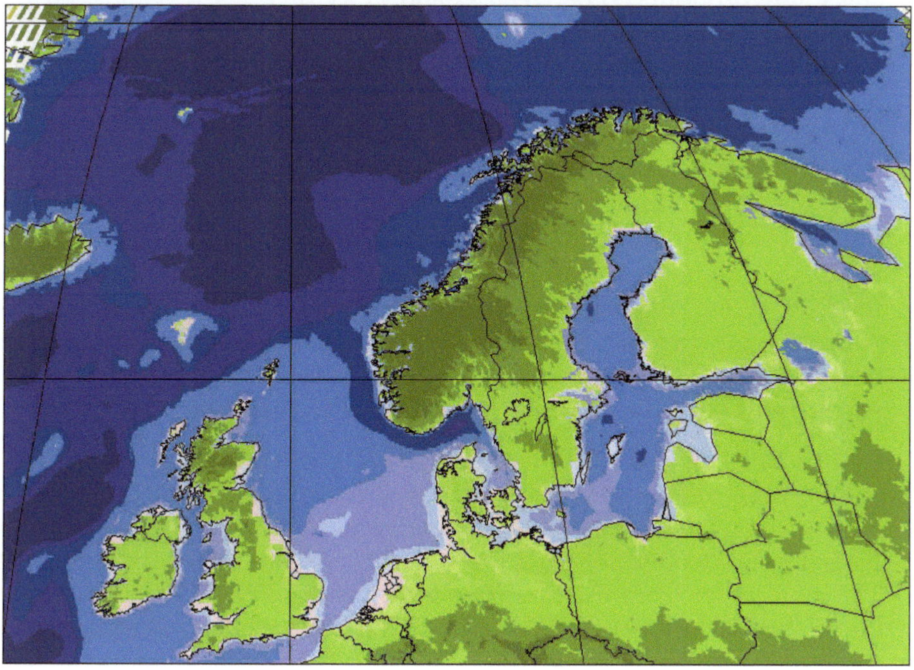

Between 16 and 15 ky BP (Figs. 3.2e and 3.2f), the last ice disappears from eastern Denmark, but the average temperature is still around –10 °C, leaving the ground frozen during most or all of the year. The Arctic Ocean becomes increasingly open and ice-free, allowing acceleration of the warming through circulation of water from the Equatorial Atlantic Ocean into the far north.

By 14 and 13 ky BP (Figs. 3.2g and 3.2h), the average temperature in Denmark is first increasing to 5 °C but then dropping to near zero. The warm Bølling period around 14.4 ky BP is when people from the Hamburg region visit Denmark and perhaps settle down. The map also shows that one can walk to England, which people eventually did. The Baltic south of the ice rim started to fill with water at 15 ky BP, and by 13 ky BP a string of water bodies extended nearly continuously to the White Sea in present Russia (as in the early Eem period). The reason here is also the depression, caused by the ice, by more than 200 m, which takes several additional millennia to disappear by uplifting. The fact that Denmark and the plain stretching to England are above water throughout this period indicates that the ice thickness in Denmark was never very large; Fig. 3.2a suggests it was always below 200 m.

The waterway from the southern Baltic to the White Sea becomes increasingly pronounced between 12 and 11 ky BP (Figs. 3.2i and 3.2j), but there remains a land barrier at the SE corner of Finland (possibly intercepted by river streams). The ice sheets are now covering only Norway, Sweden and Finland (minus the Kola peninsula) plus Scotland, leaving plenty of options for oceanic circulation. Throughout this time you can still walk from Denmark to England without getting wet shoes, but Ireland has been an island since 14 ky BP. After having experienced a cooling down to an average temperature near –10 °C, from 11.5 ky BP the Danish temperature again rises abruptly to about 4 °C, signalling the warm Allerød period, again with indications of human settlements. Although this presentation focuses on northern Europe, I might mention that a similar development is found in North America, where the warm Twin Peaks period coincides with the European Allerød period.

By 10 ky BP (Fig. 3.2k), the ice-sheet has left Scotland, western Norway, southern Sweden and southern Finland. However, the released area of Finland, as well as the Stockholm area of Sweden, is totally underwater, having been deeply depressed by an ice sheet of some 2 km thickness. The melt-water from the ice forms a freshwater Baltic Sea considerably larger than today's brackish sea (with a complex two-way exchange of fresh and salty water through the Øresund Belt), still expanding at 9 ky BP (Fig. 3.2l), where only a small ice sheet is remaining in the north. A possible outlet from the Baltic across Sweden at the level of Stockholm had already reached the North Sea before 10 ky BP (Jensen, 2001). It is still possible to walk from Denmark to England, and uplift makes southeast Finland start to come above water, while the Stockholm area is still deeply under water.

The footpath between Denmark and England disappeared into the sea sometime between 8 and 7 ky BP, and by 6 ky BP (Fig. 3.2m) Scandinavia begins to look almost as today, except for the Stockholm area and some Finnish Baltic costs still being below water. The water corridor from Estonia to the White Sea is now closed in more than one place. By 3 ky BP (Fig. 3.2n), the North Sea had deepened to nearly its present depth, and the Baltic shores had taken approximately the form they now have.

From 11 ky BP, Danish temperatures kept increasing, and by 9 ky BP they had reached the levels still prevailing today, with excursions confined to ± 1–2 °C (Fig. 3.1).

From the warm Bølling Period roughly 14.6 to 14.2 ky BP (see Fig. 3.1), four or five human settlements have been found in Denmark, with flint tools and bones indicating a diet of deer (reindeer—the large, now extinct *Dama dama*, and elk). Tηε tool techniques resemble those of settlements near Hamburg, and occupation of the sites in what is currently Denmark seems to have excluded at least the coldest winter months (Aaris-Sørensen *et al.*, 2007). Seasonal occupation may also apply to some 30 sites dated to the following 2000 years through the Older Dryas cold spell to the Allerød period of 13.8–12.6 ky BP (Mortensen *et al.*, 2008). In particular, a dead-ice hole at Slotseng in the south of Jutland has provided both pollen for assessment of plant life and bones with clear marks from human tools (Fig. 3.3).

*Figure 3.3. Reindeer breastbone, penetrated by arrowhead, found at Slotseng and dated to 14.1 ky BP (photo Jørgen Holm, Danish National Museum).*

The number of sites indicates a rapidly increasing population of modern humans (which is the case all over Europe), in contrast to the 200 ky of nearly stable Neanderthal population shown in Fig. 2.11. This also explains the need for expanding the food supply by introducing new components into the diet. The subsequent Bromme (Allerød period) and Ahrenburg cultures (to 11 ky BP) increasingly used fish resources and sophisticated harpoon points were developed (Fig. 3.4). The cooling period at 12.5 to 11.4 ky BP has been suggested to originate from cold water released to the Atlantic Ocean by a huge North American sea of melted dead-ice breaking its way through to the ocean. A hypothesis for the substantial migration from northern Germany into Denmark around 13 ky BP is that people were fleeing from the fallout from the Laacher See (near Bonn) volcanic eruption. With an eruption column reaching an estimated height of 40 km, the airborne fragments (tephra) dispersed over a sizeable area of northern Europe (Fig. 3.5), which could have caused people in the affected areas to migrate either north or south (Riede and Bazely, 2009). The 13 ky BP lowering of temperature (Fig. 3.1) could also be connected to this event.

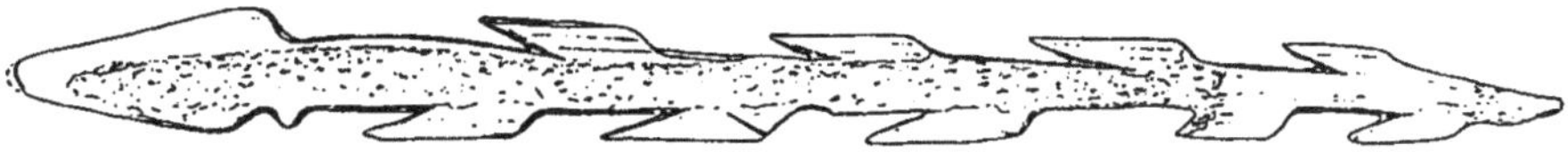

*Figure 3.4. Harpoon made from reindeer point, found at Skaftelev, Sjælland and dated to about 11 ky BP (Danish National Museum; Jensen, 2001).*

Material for flint tools is abundantly available in Denmark (Fig. 3.29), and so is wood for construction. Pollen analysis finds evidence during the period from 14.5 to 11.5 ky BP for grass–covered ranges and trees such as willow and dwarf birch, followed by regular birch trees in the later part of the period (Mortensen *et al.*, 2008).

Among the sites used by hunters of the late Bromme culture (12.6 ky BP, the Bromme site itself being on Sealand) are base sites for longer occupancy and various specialized sites (e.g. for parting prey and for flint industries). Near one of these (in Vendsyssel, north Jutland) a hut of approximately 9 $m^2$ has been found, with fireplace and signs of occupancy by both children and adult persons. Thus, although the use of most sites may have been temporary, the hunters did not leave their families somewhere else during months-long hunting expeditions in Denmark, but seem to have brought them along.

The colder period from 12.5 to 11.4 ky BP may have made the majority of the hunters move south, but the following warming brought them back and gave rise to more permanent settlements in greater and greater numbers. This required new ideas and skills. Some of the key prey animals, the mammoth and the giant deer (*Dama dama*), disappeared after 12 ky BP. Some authors think that overkill by humans was the reason for the extinction of these species, after several hundred thousand years of presence (Bulte *et al.*, 2006). The declining human occupation of the north in the colder period speaks against this, unless the overkill took place a little earlier while the north was still habitable and the population of, for example, woolly mammoth may have been too small to survive both the cold period and human hunting.

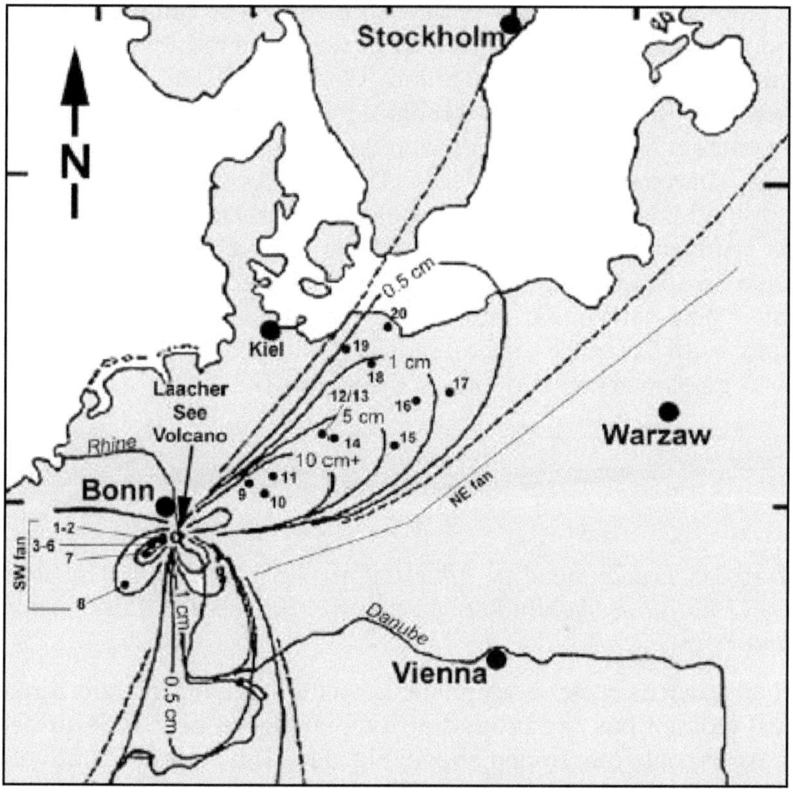


*Figure 3.5. Estimated fallout from Laacher See volcanic eruption at 13 ky BP. Numbers indicate sample sites and heights of tephra depositions in cm (Riede and Bazely, 2009).*

The human population returning to settle in Denmark around 11 ky BP is called the Maglemose culture. Because permafrost had a strong grip on the land during the period of final ice age low temperatures, subsequent warming caused the environment around 11 ky BP to become extremely wet. Settlement in the moors and bogs of the wetland was difficult, and a number of houses built on floating rafts have been found (such as those used today in some parts of Korea). Here, humans and mosquitoes (identified in large numbers in sediments) shared life.

Shellfish and freshwater fish (such as perch or carp), small animal prey, roots, berries and nuts were the main diet for the human inhabitants. Quite possibly, the early generations of settlers would have moved away from the floating platforms in winter, but no alternative winter residences have been found in Denmark (Jensen, 2001).

The swamp-based society faced conditions quite different from those of the Bølling or Allerød periods. The reduced megafauna presence forced development of a number of new sources and techniques for procuring food. In a settlement at Alkestrup in west Sealand, two huts with bark-insulated floors have been found. One had a size of 33 m$^2$, sleeping places with fern bedding, a fireplace built up by clay and sand (to separate embers from the wood-covered parts of the floor), and lots of nutshells, charcoal and flint pieces. The door opening faced a rush- and reed-lined lake (part of the Åmose) and a 120-cm-long paddle for a boat was found in the hut (Fig. 3.6). At the rear of the hut, an alder forest with a swamp-like floor prevented access, making the use of boats essential. Several similar huts from the Maglemose period have been found, with evidence for both adults and children living there. Fig. 3.7 shows a reconstruction of one of these huts.

*Figure 3.6. Paddle found in Alkestrup hut from the Maglemose cultural period, 11 ky BP or slightly later (drawing by J. Jensen, Danish National Museum).*

On an aurochs bone, a simplistic drawing of two men and three (pregnant) women has been found at Ryemarksgård near Jyderup (south of the Åmose site mentioned above; Fig. 3.8). The bone is tentatively dated to the early Maglemose period. The style of feather-like (for males) and cross-hatched (for females) engraving has some similarity to that of much older human drawings found in South Africa, and the drawing is quite different from the sophisticated drawings and paintings found in French and Spanish caves dating to 35–15 ky BP (cf. Fig. 2.15). Does this mean that the modern human groups in various parts of Europe had no contact and no collective recollection of their common past? Migration moves in response to climate variations suggest differently, but perhaps interest in art was not considered important in most social encounters between groups, which focussed on trade or spouse seeking.

The most likely explanation for the apparent lack of art in northern Europe until after 11 ky BP (and lack of music even later) is that the creative talents witnessed at Grotte Chauvet or Geissenklösterle represent exceptional skills available to only a small fraction of the population (probably far below 1%). The same is largely true today, but the fraction of current populations capable of enjoying the art and music of others is much larger. If that was so in the Stone Age is not known. Another argument could be that the northern population was small and conditions of life more harsh, leaving less time for creative artwork. This argument also appears not very convincing, considering that some of the French cave art was made at times when the climate in southern Europe was no better than in Denmark during the younger Dryas (the cold period between the Allerød and Maglemose cultural periods).

*Figure 3.7. Reconstruction of early Maglemose-style hut (at Skånes Djurpark; photo: Danish National Museum).*

The Maglemose cultural period continued to around 8.4 ky BP (covering geological periods denoted Boreal and Atlantic). Towards the end of the period, the bow and arrow became a favoured hunting tool aimed at swift animals (including birds) and animals difficult to get close to. The bow and arrow technique expanded the range of hunter–prey distances relative to that of spear–throwing, and had a positive effect on food procurement.

The arrow points of the late Maglemose period are quite interesting. Some of them seem to have had microlites (small points) inserted near the arrowhead and pointing backwards to act as a barb (hurting the animal if it tried to get the main arrow point out), while others are blunt like a club, presumably in order not to destroy the skin of an animal where use of the skin was intended. Figure 3.9 shows a 154cm-long bow made of elm and Fig. 3.10 shows some examples of arrowheads from this period. Typical arrow lengths were 60–90 cm (Becker, 1945).

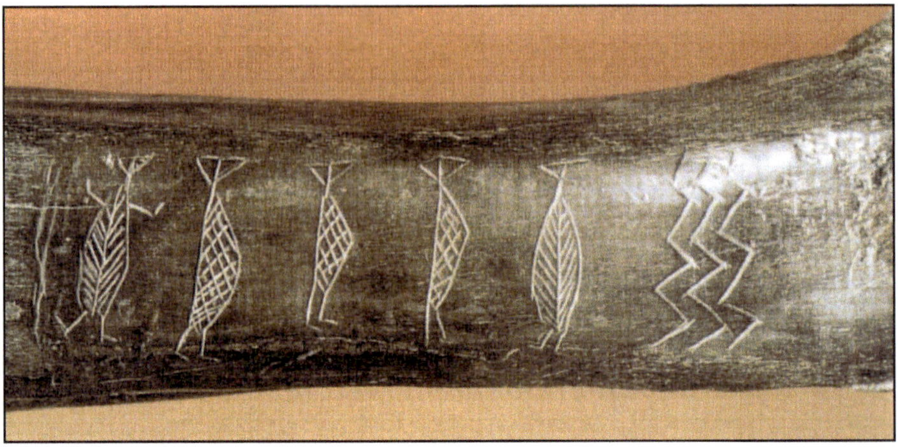

*Figure 3.8. Engraved human figures on aurochs bone from the Maglemose site, Ryemarksgård (Brøndsted, 1940; Kjærum and Olsen, 1990). Probable date is around 11–10 ky BP (photo by Danish National Museum).*

*Figure 3.9. Bow found at Holmegaard site IV (southern Sealand), late Maglemose period (Becker, 1945; photo by Danish National Museum).*

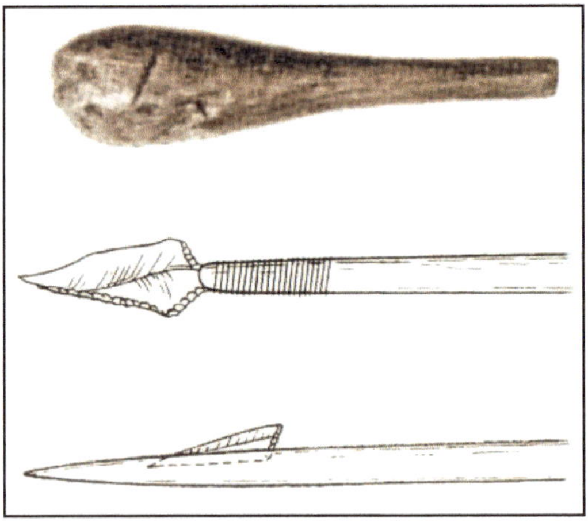

*Figure 3.10. Late Maglemose arrowheads: wooden with microlite barb inserted (bottom), with flint point (middle) and designed as a blunt club (top) (Becker, 1945; photo and drawings by Danish National Museum).*

Fastening the microlite in the groove made in the arrow was done using a resin extracted from the bark of birch. This resin ("beg" in Danish) was widely used in the Stone Age for fastening points to spears or heads for arrows, for fixing the lines (made of bast) used to wrap around or tie things together, and for mending holes in wooden equipment supposed to be water-tight (bowls, boats, etc.). The birch bark is turned over a slow fire, thereby making the resin bubble up at the surface or drip from it, enabling collection. Once back to ambient temperatures, the dark-coloured resin gets hard, but the Maglemose people learned that by chewing it for a while, it would soften and stay pliable long enough to be applied to its gluing task. A piece of birch bark resin has been found at Barmose, with tooth-marks from both a child and an adult ("I will show you how"; Fig. 3.11).

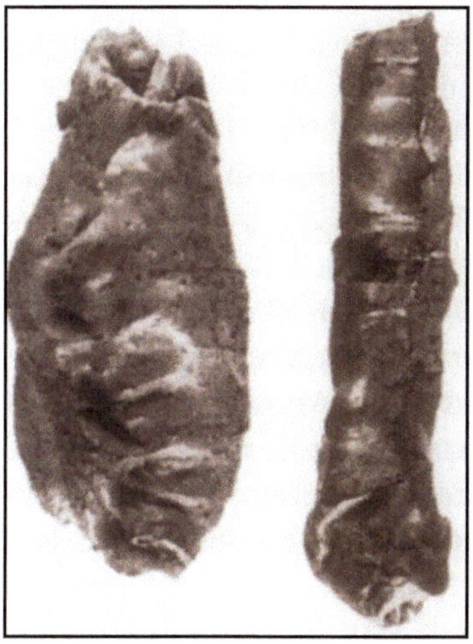

*Figure 3.11. Birch bark resin chewed by a child and an adult, found at Barmose (south Sealand) and dated to about 10.5 ky BP (Andersen, 2001; photo by Sydsjællands Museum).*

The substantial increase in variety of the activities did not cause much change in dedicated energy use. Muscle power and fire were still the main sources of energy. Of course, solar energy played a role in drying food and washed clothes, helped by ventilation caused by wind. It is conventional to disregard environmental energy flows in energy analysis, but this can lead to strange results, if passive energy use is replaced by active energy input, producing an apparent jump in energy usage. Furthermore, the large energy input associated with the greenhouse effect, making the environment at the surface of the Earth much more suitable for human habitation than it would be without this assistance from constituents such as water and carbon dioxide in the atmosphere, is excluded from all conventional analysis. The magnitude of such environmental heat would be more than two orders of magnitude larger than even the human energy use today (Sørensen, 2004). Because of the uncertainty of such estimates, the convention applied here is to include active energy use from food, fuels and environmental flows, to the extent they can be estimated, but not the basic global greenhouse warming of the lower atmosphere.

The paddle shown in Fig. 3.6 indicates that boats were used during the Maglemose period. By the following Kongemose (8.4 to 7.4 ky BP) and Ertebølle (7.4 to 5.9 ky BP) periods, increasingly more sophisticated fishing methods came into use. Boats with one or two fireplaces have been found, presumably used to attract fish that could be caught with a spear, a harpoon or a fork. Fishing rods would also have been in widespread use, as witnessed by the large number of fish hooks found from late Maglemose and into the following periods (Fig. 3.12). A boat found at Tybrind Vig is made from lime wood and dug out with the use of axes (Jensen, 2001). Wooden and later ceramic basins with traces of being used to burn train oil or fat have been found (Fig. 3.13). Because these fuels were more rare than the fuelwood extensively used for outdoor or indoor fires, these basins called "lamps" were probably not used for lighting but rather for flare fishing, presumably of eels (Hulthén, 1980). Specially shaped eel forks (called "leisters" or "eel spears") have been found. They could have been used with the lamps to catch eels (Fig. 3.14). Other fishing tools were traps ("ruses") made of bast (plant fibre threads) weaved into double tubes, allowing the fish easy access but difficulty in getting out (Fig. 3.15). Another technique is to construct an underwater fence, for example at a river or inlet, where the water is moving in or out so that the fence will catch fish (Fig. 3.16). These fences were used throughout the Ertebølle period as well as later (in fact until the 20th century), both in fresh- and saltwater environments. A common species to fence in for fish farming was the eel, and the areas fenced in are called eel weirs ("ålekister"). Fish were the dominant diet for Danes in the Ertebølle period.

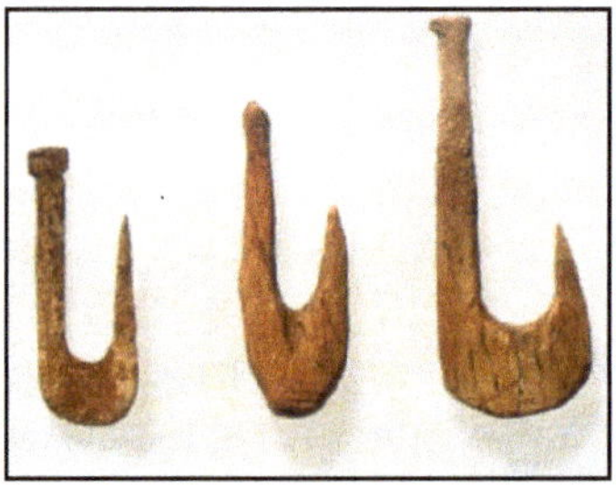

*Figure 3.12 (left). Ertebølle period fish hooks found at Tybrind Vig (Andersen, 2001; photo by P. Dehnholm, Danish National Museum). Similar hooks exist from the Maglemose period.*

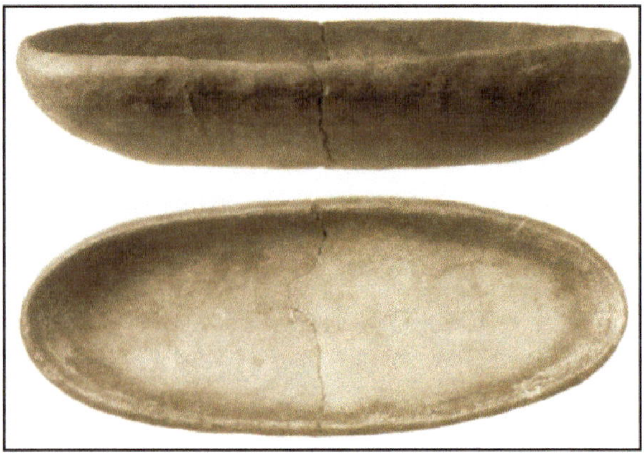

*Figure 3.13 (right). Lamp basin made of wood, used in early Ertebølle period and found at Gudsø Vig (Jutland) (Andersen, 1997; photo by L. Larsen, Danish National Museum).*

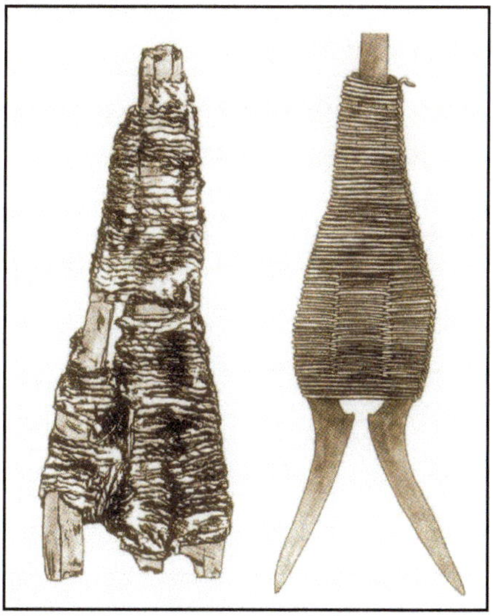

*Figure 3.14 (left). Ertebølle eel spear found at Næbbet (Ærø; left side) and a reproduction (right side; Skaarup, 1993; drawing by J. Kortermann Jauch, Danish National Museum).*

*Table 3.1 (below). Estimated energy use (delivered) in Denmark around 8 ky BP.*

| *Denmark, 8 ky BP.* | *Human energy* | *Energy from fuels and flows* | | |
|---|---|---|---|---|
| *Rate of energy use (average W for an adult person)* | **Food energy** | **Low-temperature heat** | **Medium/high-temperature heat** | TOTAL |
| **Liveable surroundings** | | | | **208** |
| Basic metabolism | 68 | | | **68** |
| Heating | | 60 | | **60** |
| Light | | | 80 | **80** |
| **Food and water provision** | | | | **59** |
| Hunting | 8 | | | **8** |
| Gathering | 5 | | | **5** |
| Fishing, aquaculture | 5 | | 5 | **10** |
| Preparation, cooking | 5 | | 30 | **35** |
| Water provision | 1 | | | **1** |
| **Security, assault, expansion** | | | | **2** |
| Weapons, armies | 2 | | | **2** |
| **Health-related efforts** | | | | **17** |
| Food preservation | 1 | 10 | 5 | **16** |
| Doctors, lay or professional | 1 | | | **1** |
| **Relations and leisure** | | | | **1** |
| Toys, artwork | 1 | | | **1** |
| **Occupational activities** | | | | **9** |
| Construction | 4 | | | **4** |
| Manufacture | 3 | | | **3** |
| Transportation of goods | 2 | | | **2** |
| **Raw materials procurement** | | | | **12** |
| Materials for construction and industry | 6 | | | **6** |
| Energy provision | 6 | | | **6** |
| TOTAL | **118** | **70** | **120** | **308** |

*Table 3.2. Estimated end-use energy in Denmark around 8 ky BP.*

| ***Denmark, 8 ky BP.*** | ***Human energy*** | ***Energy from fuels and flows*** | | |
|---|---|---|---|---|
| *Rate of energy end-use (average W for an adult person)* | **Food energy** | **Low-temperature heat** | **Medium/high-temperature heat** | TOTAL |
| **Liveable surroundings** | | | | **77** |
| Basic metabolism | 68 | | | **68** |
| Heating | | 6 | | **6** |
| Light | | | 3 | **3** |
| **Food and water provision** | | | | **11.9** |
| Hunting | 2.5 | | | **2.5** |
| Gathering | 1.7 | | | **1.7** |
| Fishing, aquaculture | 1.7 | | 1 | **2.7** |
| Preparation, cooking | 1.7 | | 3 | **4.7** |
| Water provision | 0.3 | | | **0.3** |
| **Security, assault, expansion** | | | | **0.7** |
| Weapons, armies | 0.7 | | | **0.7** |
| **Health-related efforts** | | | | **2.6** |
| Food preservation | 0.3 | 1 | 1 | **2.3** |
| Doctors, lay or professional | 0.3 | | | **0.3** |
| **Relations and leisure** | | | | **0.3** |
| Toys, artwork | 0.3 | | | **0.3** |
| **Occupational activities** | | | | **3** |
| Construction | 1.3 | | | **1.3** |
| Manufacture | 1 | | | **1** |
| Transportation of goods | 0.7 | | | **0.7** |
| **Raw materials procurement** | | | | **4** |
| Materials for construction and industry | 2 | | | **2** |
| Energy provision | 2 | | | **2** |
| TOTAL | **84.5** | **7** | **8** | **99.5** |

Considered in the energy use estimate of Table 3.1 (cf. Tables 2.1 and 2.3) is use of indoor and outdoor fires to create space heating, light and process heat (e.g. for cooking or melting whale fat or birch resin). The table also reflects muscle power used for walking and running, for hunting, fishing and gathering, for felling trees and collecting branches, twigs and stones, for wood cutting and manufacture of hand tools from flint or wood, and for constructing boats, landing places and huts with floors. The end-use figures corresponding to Tables 2.2 and 2.4 are shown in Table 3.2, and a summary of delivered and end-use energies are included in the time-development displays Figs. 2.16 and 2.17. The development during the presence of modern humans is one of continuous improvement in energy efficiency by use of better techniques. Primary wood usage did not increase all the time, but depended on availability. Forest cover of Denmark diminished before the Allerød period but grew later. This is seen from the development from 40 to 8 ky BP, where the delivered energy drops substantially (Fig. 2.16a) while the end-use energy (the benefits from energy use) increases (Fig. 2.17a).

*Figure 3.15. Fish trap found at Lille Knapstrup (Sealand), presumed to be from the late Maglemose period (Becker, 1941; Danish National Museum).*

*Figure 3.16. Fish fence found at Oleslyst (Great Belt), dated to about 5 ky BP. Fish traps may have been placed at the ends of the fence. Another possibility is the creation of a delimited fish enclosure for fish farming (e.g. of eels). The oldest fish fence remains found in the Great Belt area (at Halsskov) are from 7.4 ky BP (Pedersen, 1997; Danish National Museum).*

Society continued to develop in many respects. The monotonous diet of the big-game hunting society was, starting in the early bog settlements, being replaced by small game (brought down by bow and arrow) and considerable amounts of fish. Dwellings became more comfortable and dogs were domesticated (first Danish evidence about 10 ky BP; Jensen, 2001). Dogs could have helped by retrieving small game. Increasingly sophisticated ways of fishing came into use (with fishing rods or extensive fixed structures) and so did specialized tools, including axes (e.g. for wood cutting), and fireplaces with stones placed to ensure more constant heating of food, by their energy-storing capacity. Capacity to drill holes in wood, teeth and amber was developed (Fig. 3.17). Similar drills could be used to create new fires, as well as for other drilling purposes (Figs. 3.18–3.20).

*Figure 3.17. Drill consisting of a mouth-held stick where a flint tip is attached at the end (with resin), rotated by a string operated by a bow. The photo is from a recent Eskimo society. Drills have been found from Maglemose (the one shown here at Mullerup) and Ertebølle periods, and may have been used for several purposes: holes for connecting parts in hand tools, jewellery such as animal tooth necklaces, and for starting fires (replacing the flint tip by a wooden pole lighting wood chips by friction heat; Andersen, 2001; photo by L. Larsen, Danish National Museum).*

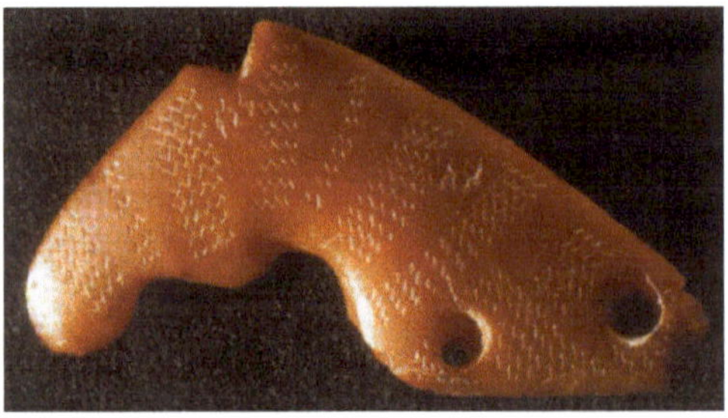

*Figure 3.18. Amber elk from Maglemose period, found at Egemarke (north Sealand), with drilled holes (Jensen, 2001; photo by L. Larsen, Danish National Museum).*

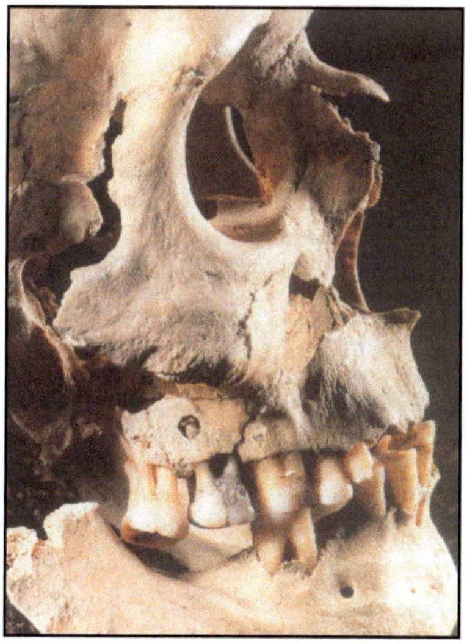

*Figure 3.19. Man with indication of caries treatment by drilling in the upper part of a molar (no. 7 in upper jaw). Found at Hulbjerg (Langeland) and dated to peasant period (5.9 to 3.7 ky BP). Familiarity with drilling in animal teeth (for jewellery) would have made it natural to think of this way of relieving toothache (Bennike, 1984; photo from Bennike et al., 1990).*

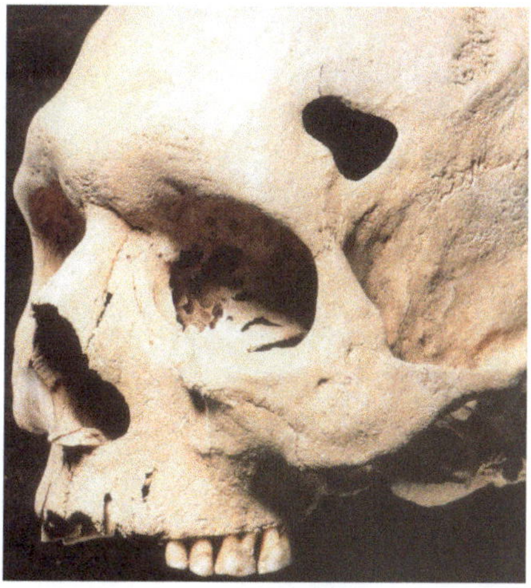

*Figure 3.20. Trepanning hole in skull of man from Ertebølle period (dated to around 6 ky BP). Trepanning was often used on skulls with previous damage from weapons and is supposed to have diminished pain by "diminishing the internal pressure" (this explanation obviously more recent; photo from Bennike et al., 1990).*

Personal ornaments have been found in considerable numbers, along with art pieces and objects with unknown use (suggested by some as ceremonial or religious). Women's jewellery moved through several fashions, from the simple teeth with holes drilled for a string, to amber and fanciful combinations of snail shells and teeth (Figs. 2.12b, 3.18, 3.21 and 3.24).

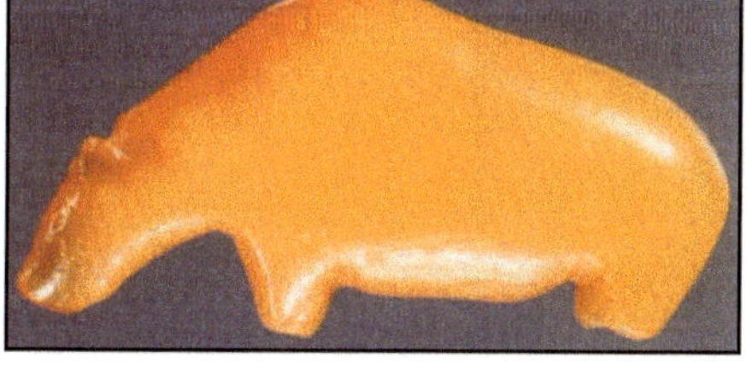

*Figure 3.21. Amber bear found at Fanø Vesterstrand, from Ertebølle period (Petersen, 1998; photo by K. Weiss, Danish National Museum).*

*Figure 3.22. Picture of man and deer carved into a deer antler, found at Åmose (western Sealand) and dated to the Maglemose period (Petersen, 1990; Moesgaard Museum).*

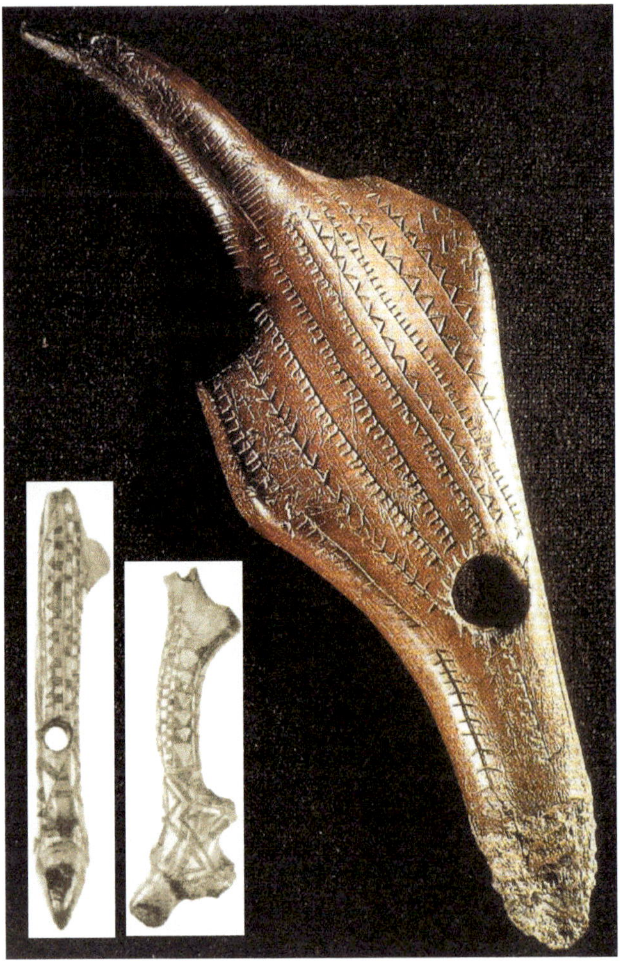

*Figure 3.23. Engraved piece made from the antler of an elk, found at Ugerløse at Åmose (Sealand) and ascribed to the late Maglemose period (Andersen, 2001). Inset: Double-edged axe from the Kongemose period found at Værebro, Sealand (top and side view; Petersen, 1971). For both artefacts, the hole is for a shaft, and it is suggested that the item in the main picture is some kind of pickaxe or similar weapon (all photos from Danish National Museum).*

The Kongemose period from 8.4 to 7.4 ky BP (i.e. between the Maglemose and Ertebølle periods) represents a continuous transformation, first of all characterized by the same settlements being in permanent use through the entire period.

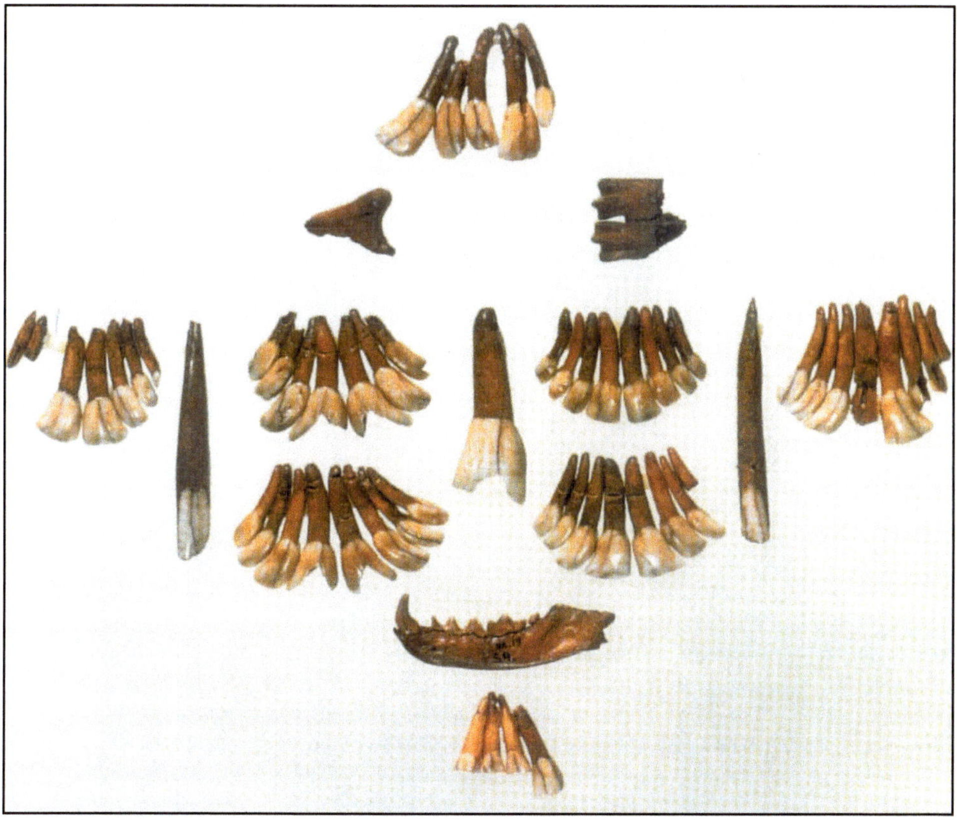

*Figure 3.24. Ertebølle jewelery found at Bøgebakken (Vedbæk), made from prey teeth and snail shells (*Neritina fluvatilis balthica*; Andersen, 2001; photo by L. Larsen, Danish National Museum).*

The Ertebølle period (7.4 to 6.0 ky BP) sees larger specialization, evidenced by the variety of artisan work that cannot have been based on skills available to everyone. During the Ertebølle period, improved harpoons were developed and large sea animals such as seal and porpoise (of the whale family) were hunted. This again required large boats carrying several hunters and options for getting the prey back to the shore. The topography of the Maglemose period, where one could walk from England through current Denmark to Skåne (cf. Fig. 3.2l), changed so that during the Ertebølle period, the North Sea became similar to today's and the sound (Øresund) between current Denmark and Sweden was beginning to appear (cf. Fig. 3.2m). While the early period had many settlements now under water, the late period topography looked much like today, and settlements such as Vedbæk excavated at current shores were also shore localities when in use.

Whether the specialization also implied an increasing social layering is not known, but is quite possible and supported by two apparent traits: greatly increased population size and beginning signs of violence playing a more significant role in human interaction.

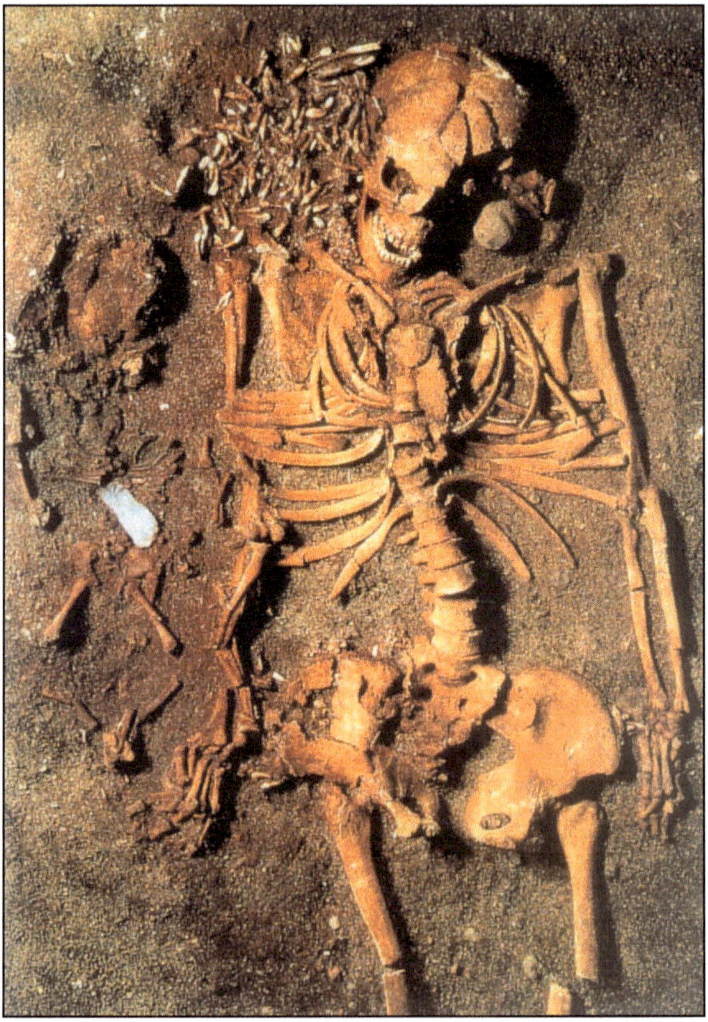

*Figure 3.25. Woman (aged 18 years) buried with her 8–9-month-old foetus at Bøgebakken/Vedbæk (Sealand). Ertebølle period (Jensen, 2001; photo by Danish National Museum).*

Figure 3.26 gives an idea of the population development in this period, based on the model used earlier (e.g. in Fig. 2.11). It has been assumed that mortality remains at the level used for modern humans since their arrival to Europe. Figure 3.27 shows the results assuming 1% lower disease mortality for Danes after the diet shifted from meat to predominantly fish at about 11 ky BP. Mortality from hunting accidents would also have gone down, but in compensation, drowning deaths during bad weather at sea (or pulling porpoises) would have increased. The model parameters for food shortage as a limiting factor have also been adapted to Danish conditions, starting the reduction given by Fig. 2.7 at a population of 1000 for the big-game hunters, rising to 2000 when fish takes over as the main diet by 11 ky BP, and then to 20 000 when agriculture is introduced, and further to 50 000 when livestock appears, both around 6.0 to 5.5 ky BP. The last two additions are successively illustrated in Figs. 4.2 and 4.3.

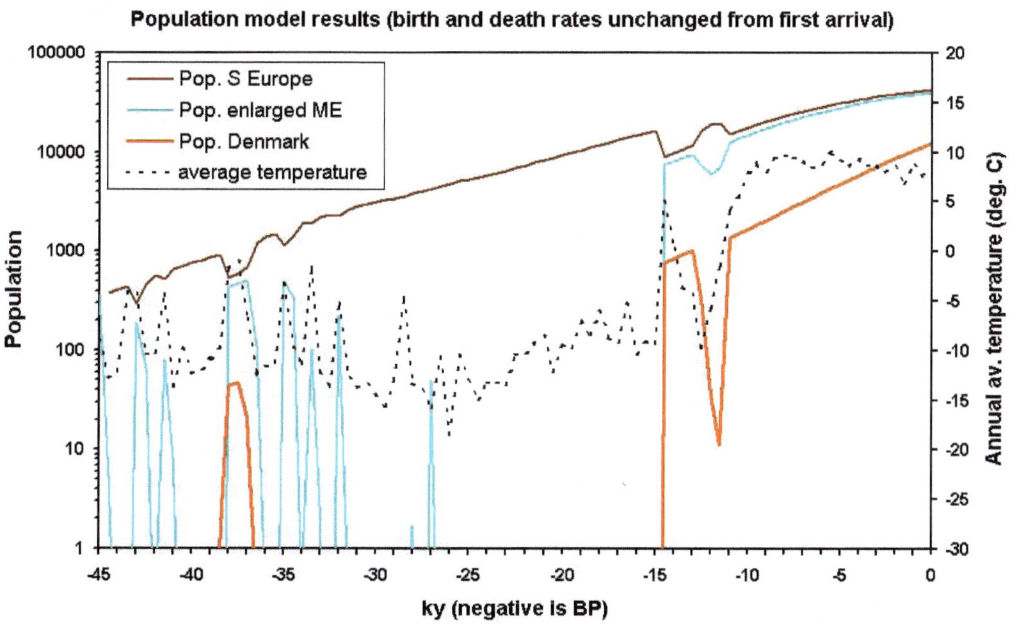


*Figure 3.26. Detail of the development in modelled modern human population in Europe from 45 ky BP. It differs from Fig. 2.11 by singling out the Danish population, through letting northward migration be limited to situations where the population in central Europe (denoted ME) stays more than ten times higher than in Denmark, and as before where the Danish average temperature is above 0 °C (and lumping the rest of northern Europe together with central Europe). Fertility and death rates are so far unchanged. Data are plotted only at 0.5 ky intervals. Note the theoretical possibility of a brief human presence in Denmark around 38 ky BP, not confirmed by physical finds.*

In all the model calculations, population development is extrapolated to present days, showing what the result would have been if the Danes had stayed hunters, fishing buffs or pioneer agriculturists, with the associated implications for the components of male and female death rates. The greater variety of tasks in society, and a possible extension of the breast-feeding period (or at least larger amounts of time devoted to small children) could change fertility; however, this possibility is deferred to the following chapters.

The mortality of women during childbirth remained high, as witnessed by finds such as the one shown in Fig. 3.25, from a cemetery at Vedbæk (Sealand), being part of a large conglomeration of settlements (Albrethsen and Petersen, 1976).

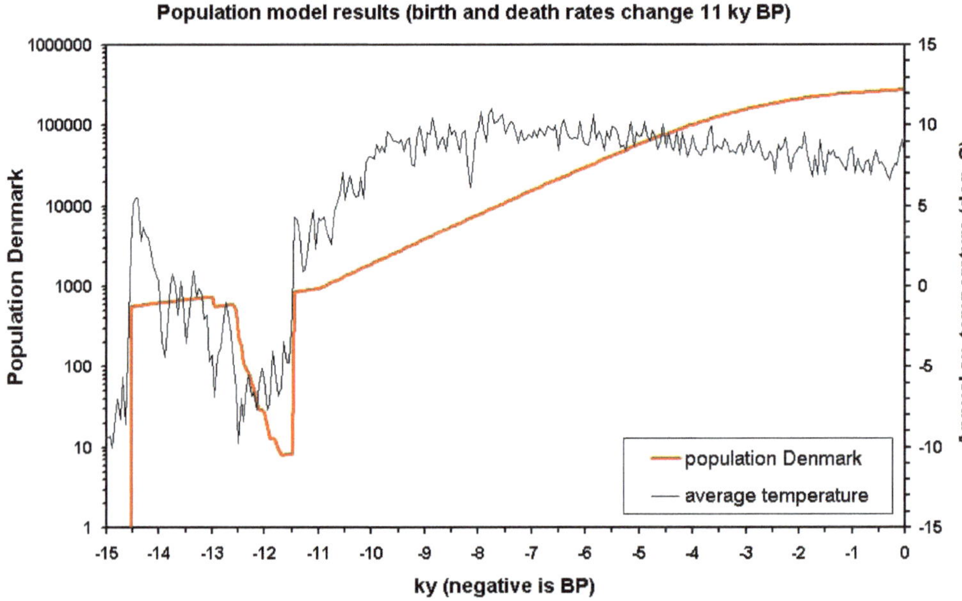


*Figure 3.27. Danish population development from Bølling period and onwards. A 1% reduction in death rate is assumed from 11 ky BP (Maglemose period) to reflect the positive health impact of increased amounts of fish in the diet (also supported by the caries-free teeth found). Assumptions regarding prey availability are discussed in the text. The effect of later changes in birth and death rates, when agriculture was introduced, are added in Figs. 4.2 and 4.3.*

The average height of men (1.66 m) and of women (1.54 m) remained unchanged to mid-Neolithicum (Bennike, 1993). As in the case of 20th century Japanese, the fish kept people in good health, but it was not until livestock with milk appeared on the menu that heights started to increase (cf. Fig. 4.4).

The common way of preparing meat and fish was to boil it. Meat was rarely fried, as this is a more wasteful method. The favourite way of preparing fish was in a clay pot with water and possibly some vegetables, placed on three stones, the heat capacity of which provided smooth cooking. Such cooking stones have been found and dated to a thousand years before the first pottery use (7.6 ky BP; Andersen, 2001) in Denmark. Because the stones get hot and stay warm for a while (e.g. placed in a cavity in the soil), one could have cooked meals in this way by use of wooden containers or even leather bags. Several clay pots found bear evidence of burnt fish stuck to sides and bottom (Andersen, 2001). This does not mean that the gastronomic value was low (cf. French onion soup).

Water availability and quality are of fundamental importance to any society. Little is known about the provision of drinking water in the Danish hunting and fishing societies. The same is true for the European Neanderthal and early modern human societies, many of which settled in places for which the availability of water is unknown. Presumably the number of freshwater lakes, streams and water holes was large throughout the Weichselian period and final withdrawal of the ice, and presumably the water used was nearly all surface water. In a very few places there would have been springs providing ground water. Some such springs in Denmark are known to have been in use and revered for a long time. The use of wells seems to have emerged much later, and equipment for rainwater collection has not been found, although such techniques may have been in operation. The location of settlements in the period 11 to 6 ky BP is predominantly at lake or river shores (Jensen, 2001), and access to streams and water holes in other locations would have been easy, due to the sheer number of such water resources. In this respect, Denmark differs greatly from the settlements in the Mediterranean or Near East areas, where access to water resources often required ingenious engineering solutions. The use of surface water would have caused occasional infections and thereby contributed to the disease component of the death rate.

In Chapter 2 it was pointed out that surviving in Europe required clothes designed to keep the wind out, and particularly tight (and preferably water-resistant) footwear, both during the warmer interstitial periods and during extended glaciation. Neanderthals solved the problem with clothes made of skin from prey animals. Such clothes would have been heavy, and as the end of the hunting society was drawing near during the Ertebølle period, a number of lighter materials for clothes-making were developed. Hides were chewed (as evidenced by wear on teeth, particularly of women) to make them softer, and weaved textiles emerged. Fig. 3.28 shows small pieces of textile found at Tybrind Vig (Funen). They are made from willow bast or lime fibres. Suggestions of underwear made with swan down have also been made (6.2 ky BP; Jensen, 2001; Jørgensen, 1990).

*Figure 3.28. Textile pieces found at Tybrind Vig. The weaving pattern is needle binding (Andersen, 2001; photo by P. Dehlholm, Danish National Museum).*

The use of fish soup with vegetables indicates an increasing role played by food of plant origin. The gathering of wild vegetables was increasingly systematized, employing in particular women and children. They would use a wooden stick (specimens have been found) to unravel edible roots, and probably considered laying out vegetable gardens near the dwellings, constituting a sign of a gradual transition to an agricultural society. There could hardly have been an understanding of the ability of different diet compositions to provide necessary proteins and vitamins, so each period, with different diets, would have had its imprint of diseases caused by cases of insufficient supply of needed food components. In the meat-dominated era, fruit-related vitamins might have been insufficient, while in the agricultural period, certain proteins may have been insufficiently supplied.

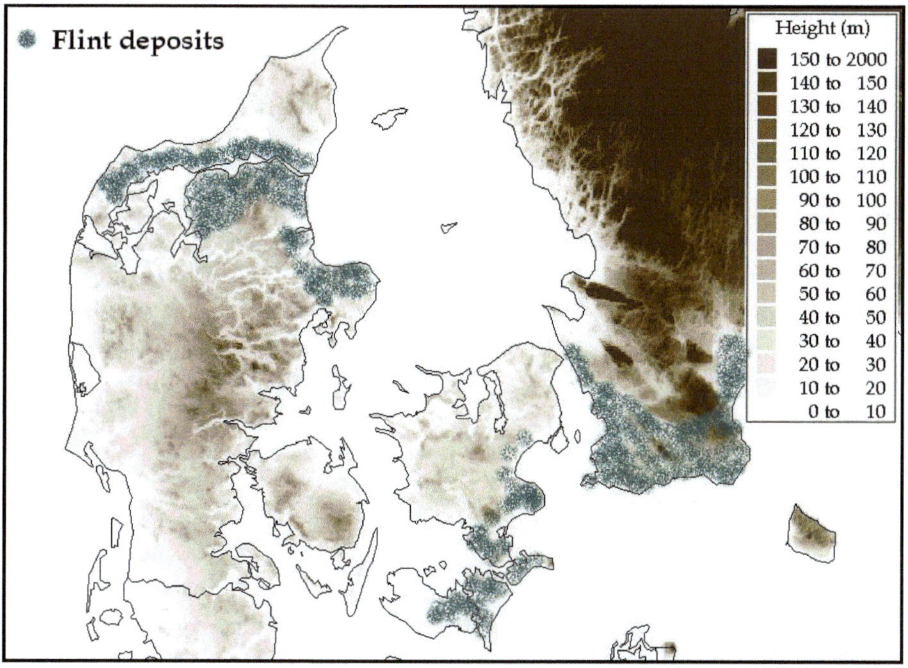


*Figure 3.29 (above). Locations of large Danish flint deposits accessible from the surface (with use of information from Jensen, 2001; base map from Sørensen et al., 2001).*

*Figure 3.30 (right). Flint mine at Skovbakken, North Jutland (Becker, 1951).*

Trade must have acquired an increasing importance during the Ertebølle period. As population increased, local flint resources would have been insufficient for the growing number of tools demanded, and the large flint deposits south of Limfjorden and along the southeastern shores of Sealand (Fig. 3.29) would have been collected by travelling members of human groups, and used elsewhere. A rudimentary mining effort was started, probably around 6 ky BP, aimed at getting at the flint resources below the land surface, by digging holes or shafts and using ladders and rush baskets to bring the stone material hacked out to the surface (the mine shown in Fig. 3.30 is more recent, from the late Neolithic period). Hacks that have been found at Hov (north Jutland) were made from dear antlers (Nielsen, 2001); not a good choice for efficiency or durability.

Flint of probable Danish origin has been found in Norway and Germany. Other signs of trade are the ornamented artefacts found in modest numbers (around 50) at Danish sites, which use rock material specific to other regions. One item is the "shoe last axe" (Fig. 3.31), the rock material of which is known to have its origin in the Balkans or in the European Alps (Schwarz-Mackensen *et al*, 1983). It is debated whether trade was in a long chain of exchanges between neighbouring communities, or if traders travelled long distances through Europe to acquire exotic tools or jewellery. The Danes could pay for their purchases (for instance with amber, which was highly valued in southern Europe).

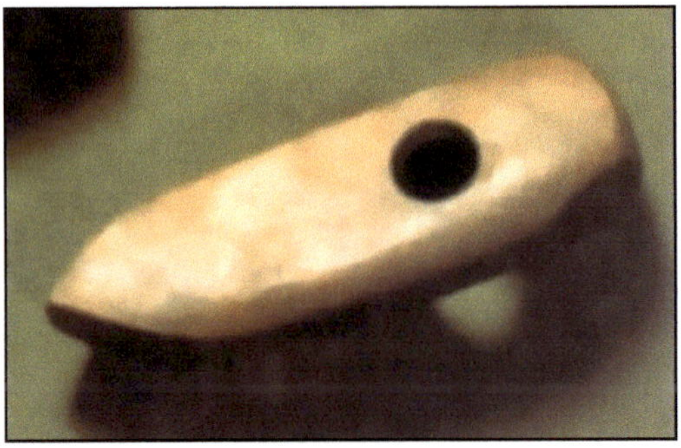

*Figure 3.31. Imported ornamental axe (late Ertebølle period; Jensen, 2001; Danish National Museum).*

While the Neanderthals were nomadic, moving around Europe according to shifts in climate and prey availability, and the early modern humans likewise, with time people became more stationary. Obvious causes for this emerged when agriculture and husbandry of livestock became part of the social structure, but also earlier, when fairly open steppe and tundra were converted into dense forests, travel on land became increasingly difficult. During the following millennia, travel at sea became more and more the preferred mode, at least until roads and wagons had been developed.

Graves found from all the periods demonstrate a high mortality of females at childbirth. Often, the foetus or child was buried with the mother and both were sometimes accompanied by grave gifts. Young women were buried in their best clothes (of which often no more remains than an impression of leather or a spot of ochre). Better conserved were ornaments such as the jewellery shown in Fig. 3.24. Men are found buried with their knives and axes, hunting trophies such as bear teeth, and the head could be laid to rest on a set of red deer antlers. Suggested uses of ochre are for curing animal skins used for clothing, for dying clothes, for painting the human skin, or in graves for ritual decoration. Graves were often marked with large stones.

Signs of violent death are rare in the early and middle Stone Age skeletons, as compared with the situation in the subsequent farming communities. A few of the 64 graves found at Skateholm in Skåne and dated to about 7.2 ky BP contain skeletons with arrow-shot wounds, indicating that attacks on one settlement by another did happen (Larson, 1986). Similarly, a male skeleton with an arrow shot through the neck was found at Bøgebakken, Vedbæk (Petersen, 2008). A mass grave with 8 skeletons at Strøby Egede (Sealand) could hold the victims of drinking infected water. The number of graves found is greater than the number of people estimated to have lived in the dwellings found, but numbers are not so large as to suggest that graves were not reused after a time.

### *References:*

Aaris-Sørensen, K., Mühldorff, R., Petersen, E., 2007. The Scandinavian reindeer (*Rangifer tarandus* L.) after the last glacial maximum: time, seasonality and human exploitation. *J. Archaeological Sci.* **34**, 914–923.

Albrethsen, S., Petersen, E., 1976. Excavation of a Mesolithic cemetery at Vedbæk, Denmark. *Acta Archaeologica* **47**, 1–28.

Andersen, S., 1997. Tybrind Vig: a submerged site and aspects of the late Mesolithic coastal exploitation. *The built environment of coast areas during the stone age* (Król, ed.), Gdansk.

Andersen, S., 2001. *Oldtiden I Danmark: Jægerstenalderen*. Sesam, Copenhagen.

Becker, C., 1941. Fund af ruser fra Danmarks stenalder. *Årbøger for Nordisk Oldkyndighed og Historie,* 131–149.

Becker, C., 1945. En 8000-aarig stenalderboplads I Holmegaards mose. *Fra Nationalmuseets Arbejdsmark,* 61–72.

Becker, C., 1951. Late-Neolithic flint mines at Aalborg. *Acta Archaeologica* **22,** 135–152.

Bennike, P., 1984. Tandlægebehandling i oldtiden. *Dansk Medicinhistorisk Årbog,* 45-55.

Bennike, P., 1993. Menneskene. *Da klinger i muld* (S. Hvass, B. Storgaard, eds), pp. 34–39. Kgl. Nordisk Oldtidsselskab, Jysk Arkæologisk Selskab, sold through Aarhus Universitetsforlag.

Bennike, P., Nielsen, P., Skaarup, J., 1990. Figures 9 and 10 in *Faces of the Past* (P. Kjærum, R. Olsen, eds). Kgl. Nordiske Oldskriftselskab and Jysk Arkæologisk Selskab.

Brøndsted, J., 1940. Human figures on a Danish Mesolithic Urus bone. *Acta Archaeologica* **11,** 207–212.

Bulte, E., Horan, R., Shogren, J., 2006. Megafauna extinction: A paleoeconomic theory of human overkill in the pleiostone. *J. Economic Behavior & Organisation* **59**, 297–323.

Hulthén, B., 1980. Ertebøllekulturens lampor. *Ale,* 1–5 (Lund).

Jensen, J., 2001. *Danmarks oldtid: stenalder*. Gyldendal, Copenhagen.

Jørgensen, L., 1990, Stone age textiles in North-Europe, *Textiles in northern archaeology* (Walton, J. and Wild, J., eds). Proc. North European Symposium held in Edinburgh. Archetype Publications Ltd, London.

Kjærum, P., Olsen, R. (eds), 1990. Figure in *Oldtidens ansigt*. Kgl. Nordiske Oldskriftselskab and Jysk Arkæologisk Selskab.

Larson, L., 1986. Skateholm. *Skalk* (4), 21–31.

Mortensen, M., Holm, J., Christensen, C., Aaris-Sørensen, K., 2008. De første mennesker i Danmark. *Nationalmuseets Arbejdsmark,* 69–82.

National Oceanographic and Atmospheric Administration (US), 2004. NOAA NGDC ETOPO5 elev. Dataset available from http://iridl.ldeo.columbia.edu

Nielsen, P., 2001. *Oldtiden i Danmark. Bondestenalderen.* Sesam, Copenhagen.

Pedersen, L., 1997. Bebyggelse og erhverv i sen jægerstenalder og tidlig bondestenalder. *Storebælt i 10000 år* (L. Pedersen, A. Fischer, B. Aaby, eds), pp. 109–115; 124–143. AS Storebæltsforbindelsen, Copenhagen.

Peltier, W., 1994. Ice age paleotopography. *Science* **265**, 195–201.

Peltier, W., 2004. Global glacial isostacy and the surface of the ice-age Earth: The ICE-5G (VM2) model and GRACE. *Annual Rev. Earth Planetary Sci.* **32**, 111–149.

Petersen, E., 1971. En ornamenteret hjortetaksøkse. *Nationalmuseets Arbejdsmark,* 188–189.

Petersen, E., 1990. From *Acta Archaeologica* **23**, 167ff, reproduced as Fig. 3 in *Faces of the Past* (P. Kjærum, R. Olsen, eds). Kgl. Nordiske Oldskriftselskab and Jysk Arkæologisk Selskab.

Petersen, E., 2008. Et 7000 år gammelt mordvåben. Danish National Museum, webpage www.natmus.dk (June 2008).

Petersen, P., 1998. Rav, hjortetak og mesolitisk magi: danefæ fra Jægerstenalderen. *Nationalmuseets Arbejdsmark,* 87–100.

Riede, F., Bazely, O., 2009. Testing the `Laacher See hypothesis': a health hazard perspective. *J. Archaeological Sci.* **36**, 675–683.

Schwarz-Mackensen, G., Schneider, W., 1983. Wo liegen die Hauptliefergebiete für das Rohmaterial Donauländischer Steinbeile und Äxte in Mitteleuropa? *Archäeologisches Korrespondenzblatt* **13**, 305–314.

Skaarup, J., 1993. Submarine bopladser. *Da klinger i muld* (Hvass, S., Storgaard, B., eds) pp. 70–75. Det Kgl. Oldskriftselskab and Jysk Arkæologisk Selskab, Aarhus.

Sørensen, B., 2004. *Renewable Energy*. 3rd edition. Elsevier Academic Press, Burlington (4th edition 2010).

Sørensen, B., *et al.*, 2001. Hydrogen as an energy carrier. *IMFUFA Text* 390, Roskilde University. Available at http://rudar.ruc.dk/handle/1800/3500 as file IMFUFA_390.pdf.

# 4

# Agriculture, trade and metals

Bt 6 ky BP, the successful Ertebølle culture had reached a total population in Denmark of nearly 30 000 (Fig. 3.26). At this level, provision of enough food was not easy. The conventional hunter's prey plus roots and berries that could be gathered were at their limits, and so was shore-based fishing, and perhaps fresh water access. The limits of sustainability had been reached, and the additional possibilities of ocean fishing had its limitation with the tools available. That there was a food pressure is evident from the effort to catch whales ("marsvin") in open sea, an enterprise that must have been very dangerous, even if the boats available were among the world's best at the time.

Trade with Mediterranean countries (of goods such as the ornamental axe shown in Fig. 3.31) would have brought knowledge of the Middle East's successes in agriculture to the Danes, directly or through middlemen if the trade was made through chains of exchanges between closer neighbours. Cereal plantation started in Mesopotamia before or around 10 ky BP (Zohary and Hopf, 2000). Plant domestication contains two components. One is the transfer of wild plant species to plantations (and later yield optimization by handmade genetic selection–i.e. keeping only the plants with highest yield, not yet employing interbreeding processes). The other is developing the technologies for transforming cereal grains into edible food, such as grinding and milling. Interestingly enough, evidence for grinding cereal has been found in the form of a grinding stone (in Israel; Piperno *et al.*, 2004) as early as around 23 ky BP, suggesting that the two components are independent inventions: first learning how to transform gathered wheat and barley into porridge and bread, and much later domesticating wild types of cereals.

In the Middle East, evidence for agriculture and animal husbandry appear at approximately the same time. The cereal types were typically wheat and barley, and the domesticated animals goat and sheep. None of these were native to Europe.

*Figure 4.1. Spread of agriculture in a direction mainly northwest across Europe. Indicated are 940 archaeological sites with agriculture. A gradient of shading in 250-year intervals from 10 ky BP (darkest) to 5.5 ky BP (lightest) illustrates the time of introduction. Areas unsuited for agriculture are white with a black outline (Bocquet-Appel et al., 2009).*

The introduction of farming in Europe, on the basis of knowledge available in the Near East, has been subjected to modelling by diffusion models, based on the above-mentioned assumption that transfer was mainly between adjacent regions, rather than having Danes travel to Mesopotamia to learn about the newest advances in farming, or Middle Eastern people travel to Denmark to sell wheat and goats (Hazelwood and Steele, 2004; Davison *et al.*, 2006; Bocquet-Appel *et al.*, 2009). Fig. 4.1 shows the results of such a diffusion model, adjusted to roughly agree with the known dates of agriculture appearing in different parts of Europe. Agriculture is assumed to enter Europe from the Levant (where it was established around 10 ky BP), through Turkey and the Danube corridor, reaching Denmark, Sweden and Ireland at 6.0 to 5.5 ky BP.

Since wheat and barley did not grow indigenously in Europe, they would initially have to be transported from the Middle East. After harvesting, some grain would have to be saved as seeds, just as it is done today. This allowed the cultivated area to grow gradually. However, before sowing, the land had to be prepared, particularly in countries such as Denmark, where forest cover at the time was substantial. In other regions, such as the mid-German one, there were areas of loess having prevented the entrance of trees, but still being most suitable for agriculture (Nehlich *et al.*, 2009). In forests and other areas with bushes or trees, one first had to create open ranges for planting. This was typically done by setting fire to the existing vegetation (Iversen, 1941). Burning a forest leaves fertile soil only for a limited time, and after a few generations of crops, the cereal yields will go down, and more quickly with no ploughing. The first ploughs were simple irons (or sharp stones on a shaft), not turning the earth but just carving a linear furrow. Such furrows have been found in Denmark and dated to about 5.4 ky BP; the ploughs themselves (named "ard") have only been found from later periods (Nielsen, 2001). In Denmark, agriculture seemed stable over long periods, probably due to use of animal manure to provide the necessary replenishment of nutrients, along with a practice of rotation (cf. Williams, 2008). This is another argument in favour of livestock introduction roughly at the same time as the cultivation of plants.

In any case, introduction of agriculture in Denmark was, as mentioned, made necessary by increasing population. The opposite inference–that agriculture allowed population to increase–although possible, does not seem to be correct, as the arguments put forward in Chapter 3 suggest that the food supply situation was already stressed 1–2 ky before the appearance of cereal cultivation in Denmark. The effect of agriculture on population can, in the demographic model already employed for the preceding periods, be incorporated through a shift in the food-shortage threshold from 2000 to 20 000 people (taking place not instantly but over a period long enough to raise the productivity of agriculture to the levels required at each level of population), and a modification of mortality (1% increase) as described below.

The results are illustrated in Fig. 4.2. Now the population at 4 ky BP would be more than twice that predicted by the model without agriculture, despite the increase in mortality.

Until a few decades ago, the conventional interpretation was that new immigrants from the Middle East brought the novel cereals and domesticated animals to Europe in the gradual process depicted in Fig. 4.1, and hence that the introduction of agriculture in Denmark was also associated with an influx of foreign people. The obvious alternative is for the residents to have learned of the new possibilities, maybe by inviting a few "instructors", or by themselves travelling to regions where the secrets of agriculture were already mastered. An influx of foreign people into a society already facing shortage in food provision is unlikely to have been peaceful, so buying cereal and sheep in the way in which trade had already been established is more likely. The suggestion that the changes could not have been produced within the society, but needed invasion of people from the outside, is weakly founded, because it just shifts the introduction of innovation from Europe to the Near East, where it could not have been learned from invaders. It is more likely that the option of agriculture and animal husbandry was already fully appreciated in Europe by 10 ky BP, but that the climatic differences in growing seasons, solar radiation and soil quality required additional motivation and adaptation before the agricultural technology would function in Europe. The suggestion that it only happened when the food supply situation in relation to a rapidly growing population made it necessary therefore stands out as the most plausible one.

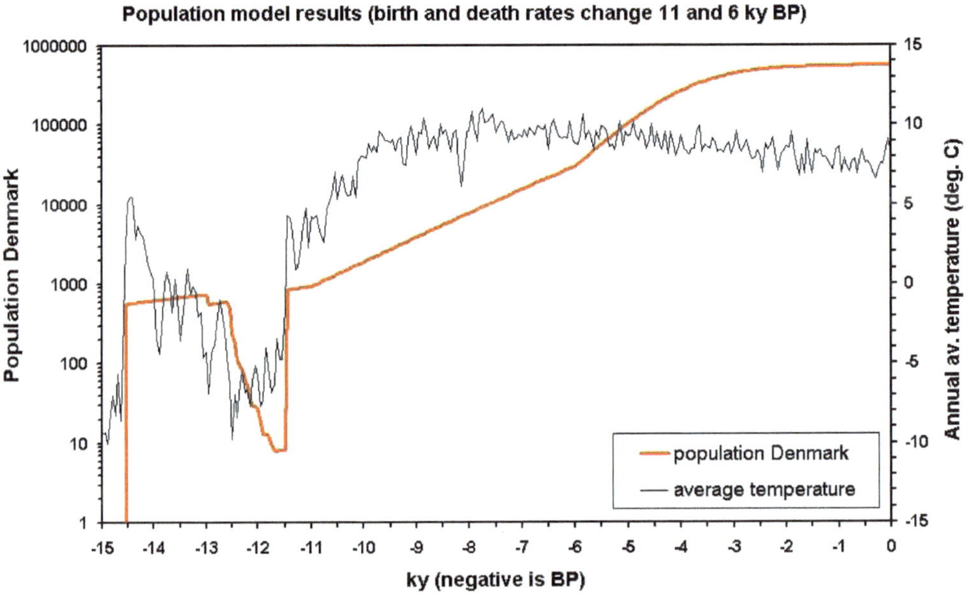


*Figure 4.2. Modelled Danish population development. It differs after 6 ky BP from Fig. 3.26 by a tenfold relaxation in the mortality threshold associated with food shortage and an 8% decrease in the rate of female deaths at childbirth, but a 1% increase in disease mortality.*

While the first signs of agriculture and domesticated livestock in Denmark appears around 6 ky BP, more massive use of these technologies is only seen from around 5.5 ky BP (Nielsen, 2001). Hunting and gathering continued to play a role, except that the fish component in the diet declined, in Denmark as well as in other European countries (Fischer *et al.*, 2007).

Raising livestock could have proceeded through domestication of prey animals, or by importing animals from farmers in regions where they had already been domesticated. Danish cattle seem to have been imported rather than derived from the existing aurochs (Noe-Nygaard and Price, 2005). DNA analysis suggests that the same is true in other European countries (Edwards *et al.*, 2004; Zeder *et al.*, 2006), although some contrary examples do exist (Beja-Pereira *et al.*, 2006). The food-shortage threshold may now be increased further, to set in at a population of 50 000, but with mortality increased by another 1 or 2% (see below). This is illustrated in Fig. 4.3, where the animal husbandry is added from 5.5 ky BP. The result is that population growth is nearly stopped. The continuation of the population curve to present time again only shows what would have been happened if there were no further changes in fertility or mortality.

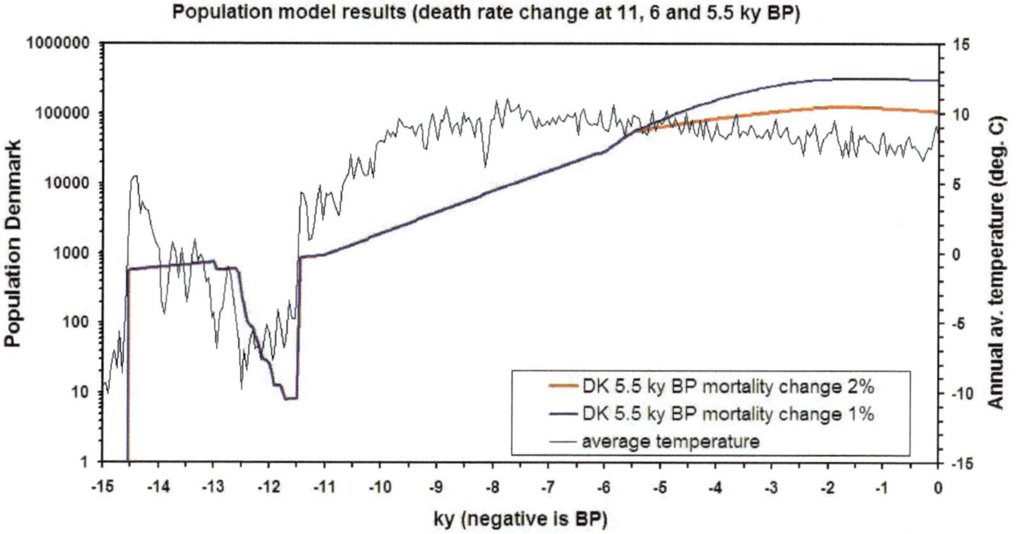


*Figure 4.3. Modelled Danish population development. It differs after 5.5 ky BP from Fig. 4.2 by an additional 2.5-fold relaxation in the mortality threshold associated with food shortage, and a tentative 1 or 2% increase in mortality from diseases. This will be further discussed in connection with Fig. 4.20.*

Although the general effect of hunters–gatherers transforming into farmers with agriculture, horticulture and animal husbandry was to allow greatly increased populations to survive on a fixed territory, there were also drawbacks, although numerically smaller than the advantages. As a first step, the transition from hunters and possibly gatherers to pure plant-growing agriculturists (no livestock) is considered below, although this is not what actually happened in Europe. After that, domesticated animals are added.

Meat provides practically all the proteins and vitamins required by the human body, in addition to its energy content. A vegetarian diet does not, as there are proteins and vitamins missing in nearly all plant food, and particularly in cereals. Modern vegetarians know that the missing ingredients can be obtained from special fermented plant food and a few exotic plants imported from equatorial regions. The early agriculturists in Europe or elsewhere cannot be expected to have known this. As argued, the early introduction of domesticated plants in the European diet may in places have been combined with a continued reliance on meat and fish, in proportions that were nutritionally sound. The relative vegetable/meat proportions in the diet can be inferred from carbon and nitrogen isotope ratios in human bone collagen, but care should be exercised when fertilizer based on animal manure is utilized (as it presumably would have been in Denmark), because the derived meat fraction in the diet will become overestimated (Bogaard *et al.*, 2007).

It was not just the nutritional deficiencies of plant food that could cause problems. The fact that mainly cereal monoculture-based agriculture led to a less diversified diet would also have increased the incidence of deficiency symptoms. This is demonstrated by skeletal evidence revealing a number of problems, relative to people living on the former meat diet: anaemia signs in female remains, bone loss, caries in teeth, loss of teeth, as well as impaired bone growth in infants and young children (Barrett *et al.*, 1998). In addition to the valuable teeth studies, which reveal both dietary composition and changes in wear patterns, genetic studies such as DNA analysis of both humans and livestock animals are playing an increasing role in revealing the relation between diet and health in ancient populations (Zeder *et al.*, 2006).

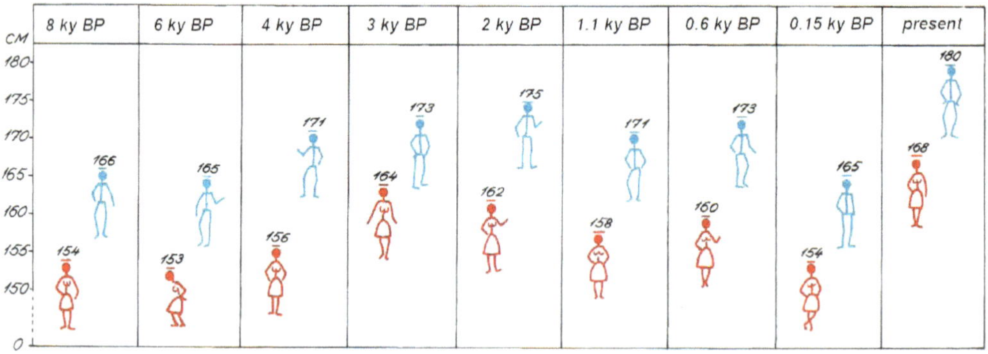


*Figure 4.4. Typical body height of Danish males and females, from middle Stone Age to present (Bennike, 1993). The effect of milk in the diet shows up very markedly from 6 to 3 ky BP.*

Domestication of goats and cows allowed milk to be added to the human diet, which, apart from an increased frequency of allergies, would cause a number of positive changes, including an increasing body height (Fig. 4.4). The same effect appeared in post–World War II Japanese youth, where the addition of cows milk to a primarily fish and rice-based diet caused height increases of around 25%.

DNA analysis of a different kind, involving only samples from current populations, has also contributed to the discussion of whether farmers immigrated into Europe from Mesopotamia and other Middle Eastern locations ("demic diffusion", as proposed by Ammerman and Cavalli-Sforza, 1984), or if the agricultural knowledge was learned and put into practice by the people already living in Europe ("cultural diffusion"). The latter seems to be the more likely answer, regarding the small admixture into the DNA of modern Europeans (Richards *et al.*, 2000; Haak *et al.*, 2005). However, migration has certainly taken place extensively over the past 10 ky, just as it did earlier when the modern humans arrived in Europe. Examples are the many sweeps of warriors raiding central Europe from the regions now called Asia, Russia or Ukraine (e.g. invasions by Kurgans in the early period, Mongols and Huns more recently; Rosser *et al.*, 2000; Renfrew, 2000). A series of visits from the Middle East between 10 and 6 ky BP is a possibility, but more likely by traders and not invaders. Isotopic studies suggest such local admixture (Bentley *et al.*, 2003).

Another line of argument has suggested invasion as the reason for the introduction of what we now call Indo-European languages into most regions of Europe. The Indo-European languages can, of course, be traced back only some 5 ky from writing, and it seems that formation of the underlying common origin happened after the introduction of agriculture (because words like "plough" and "cow" are in the common vocabulary) and in an area far from the sea (because no words related to oceans or ships are present in the proto-language; Wolfe and Stanley, 1980). An alternative explanation is that the mobility and transfers that left traces in the current genetic material happened much earlier, during the Weichselian deglaciation (some 15 ky BP or earlier). One line of thinking claims that the European population at some point withdrew to a refuge in the Pyrenees and that all later Europeans derive from this enclave (Torroni *et al.*, 1998; Achilli *et al.*, 2004). Archaeological evidence does not support the refuge idea, as healthy societies elsewhere were, before and around 15 ky BP, painting in French caves or making excursions to Denmark from their homes in Germany.

Generally speaking, the study of current mixing of genetic traits (grouped in entities called haplogroups) does not seem to be a viable method for studying migration routes in the Stone Age, because current distributions are extremely mixed and reflect numerous movements at different times of people between regions, leaving only the possibility of carrying out certain statistical analyses, but not to pinpoint the timing of individual movements (cf. Barbujani *et al.*, 1998; Richards, 2003). A high concentration of a given genetic group (e.g. in the Pyrenees) could indicate the origin of a spreading population, or it could be just the results of a single family withdrawing to this isolated place and, over the following millennia, creating a group of individuals with little genetic variation, due to lack of outside contacts.

Infectious diseases increased with the introduction of plant farming, because of higher population densities, but also more specifically due to bites during field work by insects operating as vectors for micro-organisms carrying viral or other infections, as well as direct contraction by touch or inhalation, and by contact with parasites (also during work in fields, especially during harvest). Further diseases would be transmitted by use of faeces as fertilizer, and from stored plant food that might harbour contamination related to a number of gastro-intestinal diseases (Barrett *et al.*, 1998).

It has been suggested that the significant decrease in life expectancy at birth, observed after the introduction of agriculture, could be the effect of rising fertility and the ensuing preponderance of children, and therefore also preponderance of children in mortality statistics revealed through skeleton remains (Larsen, 1995). Although women may well have responded to increased child mortality by having children more frequently, the death statistics would after a little while be free of the transient behaviour and reveal the true age distribution of mortality, which therefore was indeed higher in young persons after the introduction of agriculture.

The demographic model used in constructing Fig. 4.2 includes the negative impact on life expectancy after 6 ky BP by a modest 1% increase in mortality at all ages, not enough to upset the positive effect of being able to feed more people. In this parameter choice there is consideration of the actual number of sites found as a function of time, as well as the fact that meat and fish did not entirely disappear from the Danish diet. That changes in fertility and mortality were at the 1% level or lower is clear from the sensitivity of demographic models, and has also been noted by Armelagos and Harper (2005).

Turning now to the parallel domestication of animals, the close contact with animals in addition to the dogs (domesticated long before agriculture and livestock raising) would induce a number of new pathways for disease transmission. There would be diseases such as tuberculosis or anthrax, transmitted through milk, skin contact or ambient dust from animal hair and stools (Barrett *et al.*, 1998). It is established that measles and tuberculosis are transmitted from cattle and influenza from pigs and ducks or chickens. None of these diseases are believed to have bothered humans before domestication of such animals (Diamond, 2002). Arrangements where livestock lived close to or in the same building as humans were common in many farming societies (Fig. 4.11), and the model calculations in Fig. 4.3 use a mortality increased by 1 or 2% at all ages, relative to that for pure agriculture, despite the amelioration of some of the nutrition-related health problems associated with agriculture, when a share of meat is present in the diet. Now the population curve stabilizes, in agreement with indications from data, and despite the new increase in the ceiling for starvation.

On the positive side, the introduction of animal husbandry increased the diversity of food, and ensured the proteins and other nutrients that were hard to derive from vegetarian food. Milk was a beneficial factor (Fig. 4.4), despite causing some allergies. Berries and roots retained their role as dietary supplement. Livestock further serves as a store of energy, providing a measure of insurance against the risk of failed harvests, at least on a timescale of a single year.

The livestock introduction did not appear in an identical way in different parts of Europe. In some places, it is believed that, for a long initial period, plant crops were grown only to feed domesticated animals. Use as food for humans required milling and baking or cooking. Securing food for domesticated animals during winter, where grazing options were reduced, particularly in northern regions, may have been avoided, at least for a time, by selecting animals such as goats, sheep and pigs. Pigs could often be left to find their own food, if there was a forested area nearby, while sheep and goats could graze even marginal open land. In Ireland, for example, sheep and goats are not believed to have been fed harvested produce during winter until around 800 (1200 y BP; Kerr *et al.*, 2009).

A disputed question is the distribution and nature of forests across Europe from 13 to 5 ky BP. Pollen analysis shows that there were forests, but also other plants (Huntley, 1990). In pre-agricultural Denmark, lime dominated in dry areas, elm and alder in wet areas, with some additional oak and hazel (Rasmussen, 2005). There are indications that forests at times must have been so dense that the mobility of humans in the area was reduced (making waterways the preferred way of travelling in Denmark during the Maglemose period), and big game would not have been able to penetrate into the forest areas. On the other hand, actual occurrence of some animals requiring open or semi-open landscapes has been established in most regions (Moore, 2005). After 6 ky BP, Danish elm declined rapidly (probably due to disease) and lime more slowly (but reappearing after 4.4 ky BP). Pollen counts for oak, hazel, black alder and birch increase. In reality, Danish forests may have become more open, because the change is from trees with low pollen production to trees with high production (inviting false deductions). After 5.5 ky BP, pollen counts from grasses and cereal increase, and in many areas dominate (Aaby, 1993). The change shows substantial effects of human interference by slash-and-burn cultivation in previous forest areas, causing formation of moors by soil nutrient depletion, unless animal manure was used to refertilize the soil.

An important implication of farming was the need to attend to animals year round, and fields at least during the sowing and harvesting seasons. This meant that the highly mobile hunters, maintaining contact both with mating groups several hundred kilometres away and with other populations or hunting territories more than 1000 km away were changed into sedentary settlers, many of whom spent their entire lives within an area of some 20 km × 20 km. This had a cultural effect, as well as a genetic one: negative effects of inbreeding must have multiplied and caused the population-wide genetic health to decay. It is sometimes claimed, that population increase and farming are connected through an increase in women's fertility, arguing that, in a hunting society with "nomadic" migrational moves, a woman had to restrict herself to one infant requiring carrying from settlement to settlement, whereas in sedentary agricultural settlements, she could carry two at a time!

I have argued that fertility was already quite high (7–8 children) in the Stone Age hunting societies, because it was required for survival due to the high mortality during the first year of life. Furthermore, in most of Europe, and certainly in Denmark, the hunting societies were already pretty sedentary from the late Maglemose period onwards, and my analysis above suggests that the population increase (doubling every 1000 years from 11 to 5 ky BP) started well before agriculture and was probably the reason that (the already long-known) agriculture was finally introduced. In fact, the model results suggesting a stagnation of population growth after the introduction of farming (in Fig. 4.3 the 2% mortality growth curve has 31 000 people by 6 ky BP, 67 000 by 5 ky BP, and only 88 700 by 4 ky BP and 111 000 by 3 ky BP) are broadly supported by archaeological data: Nielsen (2001) uses the Danish database of Stone Age settlements (Nationalmuseet, 2001) to compare the number of settlements in the Ertebølle period with that after introduction of agriculture. Apart from a significant move of settlements from rivers, lakes and seashores to inland locations, the overall number of settlements in northwest Sealand drops from 378 to 162. Compensating for an increase in the number of people in each settlement (to provide labour for the work in the fields and tending livestock), this would be in general agreement with the population increase mentioned above, dropping to only 32% from 5 to 4 ky BP, as shown in Fig. 4.3 (the 2% mortality increase case). The early hunter–settlers would have had a group size of around 25 (like the Neanderthals discussed in Chapters 1 and 2) or perhaps even lower during the lake and river fishing period, increasing when the team-work of ocean fishing made it necessary, and in agricultural settlements around 5.5 ky BP, villages with 100 or more inhabitants would presumably have been common.

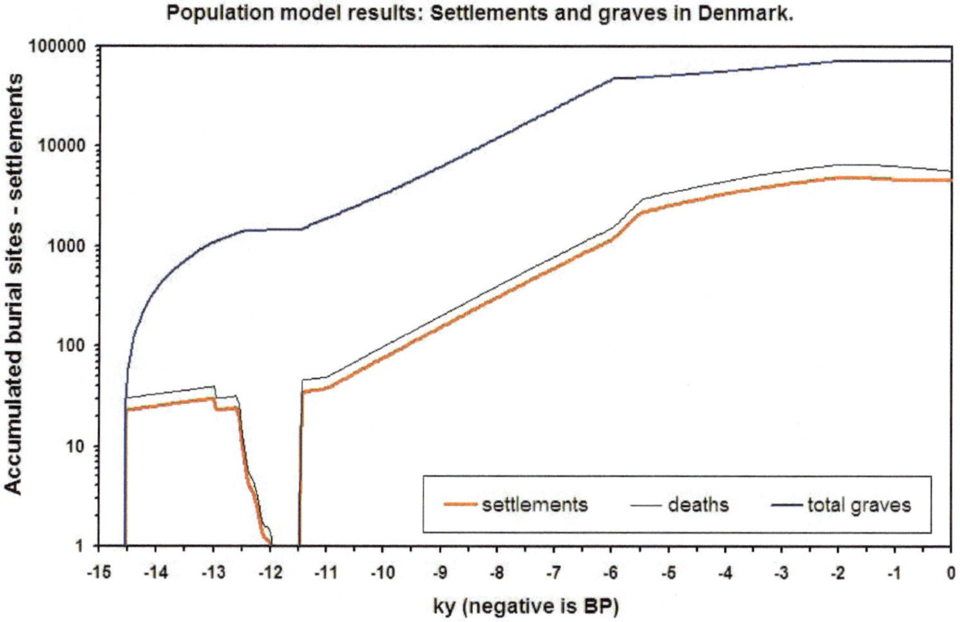


*Figure 4.5. Based on the population numbers of Fig. 4.3 (high mortality case), settlement numbers are calculated by dividing by the number of inhabitants in each settlement at a given time. This is assumed to be 25 until 6 ky BP, then increasing to 150 by 1 ky BP. Graves contain one body until 6 ky BP, and 15 after that. The number of deaths indicated are per year.*

The number of burial places (first single graves or dolmens, later passage graves) found is 5–10 times the number of settlements, in fair agreement with the results derived from the population model (Fig. 4.5). Stone graves were more durable than the wood-pole and (presumably) clay-wall houses used for living, and the number of graves suggests that they were starting to get reused only towards the end of the Ertebølle period. The graves of the farmer society had an increasing number of people buried in each grave, and an acceleration of reuse (Jensen, 2001a). This is built into the model underlying Fig. 4.5 as an assumption. Fig. 4.6 to 4.8 show the succession of grave styles, and a housing reconstruction from the early farming period around 6 to 5 ky BP. Cultural subdivisions of the periods are made according to the style of pottery and its decoration, an example of which is shown in Fig. 4.10.

*Figure 4.6. Dolmen with chimney-type grave from the Funnelbeaker cultural period, about 5.4 ky BP (Dæmpegård, Sealand). The super-ellipse-shaped ring of stones encircles two graves (photos by author).*

Skeletons found in the passage graves are rarely complete and they show evidence of considerable amounts of treatment, in contrast to the early single burial of mostly complete skeletons. The explanation proposed is that dead bodies were kept in a temporary place while being readied for final burial (Jensen, 2001a). So-called cult houses and special gathering places identified as neither dwelling nor burial place could have played a role in the associated rituals. Cleaning crania and omitting certain organs might aim to avoid attracting unwanted animals (whether rodents, maggots or other decomposers) and minimize any foul smell in a grave intended for reuse and re-entering. Grave gifts continue to appear (in the Fig. 4.7 passage grave, a 4.2 ky BP one-piece stone dagger was found), perhaps indicating a conception of a possible afterlife.

*Figure 4.7. Passage grave from the Danish agricultural period (5.2 ky BP), at Ølshøj (Sealand). There are two passages leading to two rather spacious grave chambers that were in use for more than 1000 years. An external site lining with two elliptical rows of stones has later been lost through use as raw material for newer houses (photos by author).*

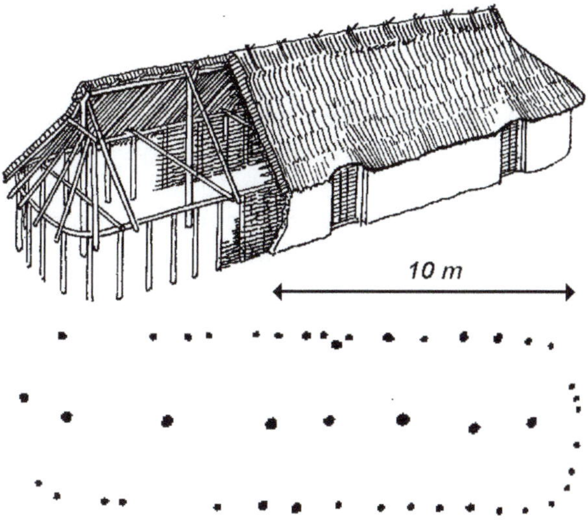


*Figure 4.8. Reconstruction of dwelling from early agricultural period, 5.5 ky BP (Limensgård, Bornholm; Nielsen, 1999; drawing by F. Bau, Danish National Museum).*

*Figure 4.9. Sickle found at Christiansminde Mark, Sealand (Nielsen, 2001; photo by Danish Nation al Museum.*

*Figure 4.10. Bowl from 5.3–5.2 ky BP, in the "Troldebjerg" style, found at Lille Knapstrup, Sealand (Jensen, 2001a; photo by J. Weng, Danish National Museum).*

The Danish farmhouses developed from the type shown in Fig. 4.8 to the larger one in Fig. 4.11. The site of the latter contains layers of successive buildings on top of previous collapsed ones, and towards the end of the agricultural period (4.3–3.7 ky BP, called the "dagger period" after elaborate daggers found; see Fig. 4.12) one finds 16 large houses of the type illustrated in Fig. 4.11. The direction of most of the houses is from west-northwest to east-southeast, ensuring maximum solar passive heating and wind shielding (Nielsen, 2001). In each house, there would be grain storage (barley and wheat, or the early forms emmer and einkorn, with 4, 2 and 1 rows of grains), grinding stones, fireplaces, presumably sheds (for cattle and pigs, and perhaps goats, sheep and ducks, in periods with temperatures too low to leave them outside) and living quarters (in one or both ends). This arrangement could be compatible with large family holdings, but also with beginning delegation of tasks outside the family, to domestics or labourers.

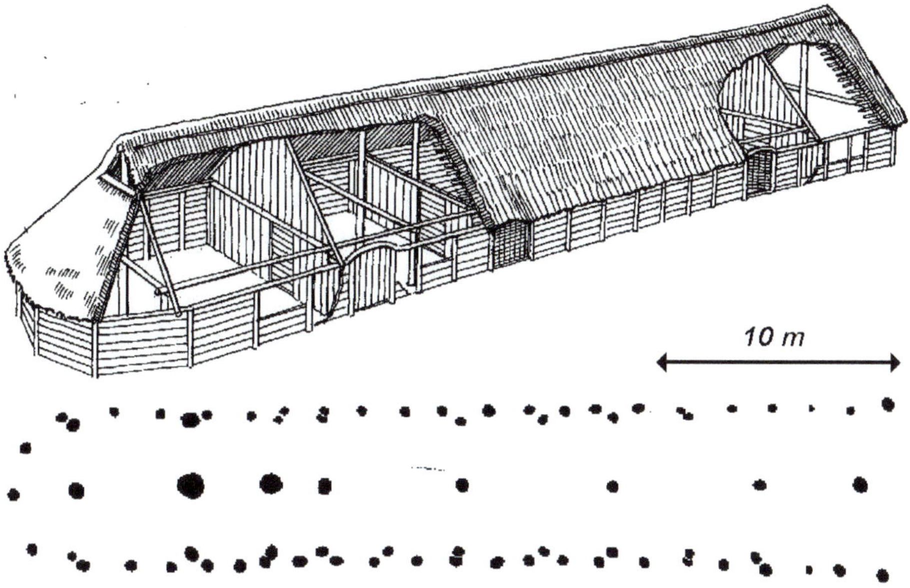


*Figure 4.11. Reconstruction of dwelling from the late agricultural period, around 4 ky BP (Limensgård, Bornholm; Nielsen, 2001; drawing by F. Bau, Danish National Museum).*

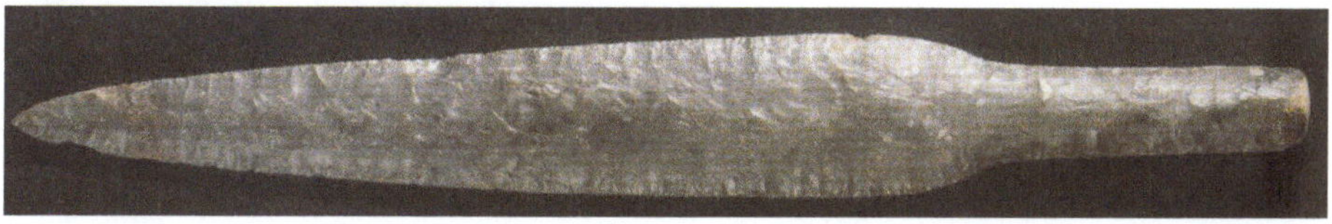

*Figure 4.12. Neolithic refinement of stone tool manufacture (cf. the early technology stage in Fig. 1.16). Daggers like the one shown were produced from around 5 to after 4 ky BP (into the Bronze Age; Nielsen, 2001; photo by L. Larsen, Danish National Museum).*

The increasing land areas cultivated, and occasionally abandoned due to diminishing returns, led to an increase in heathland starting at 5.5 ky BP. A few moors were present earlier (Aaby, 1993), but the substantial new bog areas must be seen as ecological damage caused by humans. Only in much later times (after 1400) did it become common to try to reclaim heathland for agriculture (Odgaard and Nielsen, 2009).

Much thinking has been devoted to explaining why the early egalitarian societies were rather suddenly transformed into stratified social structures, with increasing levels of disparities and power struggles, from the beginning of the agricultural period (O'Sullivan, 2008). Was it due to population size passing some threshold, or was it the tasks of organizing farming that entailed a need for a new social order? Malinowsky (1922) suggested that the main reason was that, for the first time, more material wealth became available than the minimum needed for survival. On the other hand, differences in grave goods suggest that earlier Stone Age societies in northern Europe were not totally egalitarian. Still, there may have been a transition from group or village "leaders" *administering* any required redistribution of goods or services to such leaders *controlling* the redistribution by coercive authority. In any case, this trend continued with larger and larger units controlled by chiefs, and then by feudal or royal monarchs. Authority may have acquired its basis by the invention of land ownership and inheritance, enabling owners to claim rent or other compensations from farmers and to build up intergenerational accumulation of wealth, thereby forming of a stratified social structure with classes of people (Earle, 2000). A natural consequence was the formulation of rules or laws, and the establishment of juridical and military units to enforce the rules (and as far is military forces are concerned, also to enlarge territories by conquest of neighbouring societies by assault or warfare).

*Figure 4.13. More weapons: battle-axe (left) found at the cult site Sarup II (Funen), dated to 5280 y BP. Similar axes from the early Neolithic have been found as grave gifts elsewhere. The three-sided arrowheads shown at right use flint technology like that of the dagger in Fig. 4.12. Flint cannot be polished as the rock used for the axe (Andersen, 1999; Jensen, 2001a; left photo by P. Dehlholm, right photo by L. Larsen, both at Danish National Museum).*

Religion was probably created at the same time as social stratification, in an effort to reinforce the authority of the rulers. As mentioned at the end of Chapter 2, there is no concrete Mesolithic evidence for religion, except if the grave goods late in the period are interpreted as a belief in afterlife, rather than just an exhibition of reverence for the deceased. I define religion as different from personal beliefs, as an institutional package of convictions that are guarded over by some custodian authority. The laws imposed on the stratified societies were believed to carry more weight if they were claimed to have been conceived by a god rather than by a squire or shaman. The fruit of religious institutions is intolerance, and ultimately war, often interwoven with political quest for more power, although both religious and territorial wars, as well as wars aimed to increase political power, have been fought. Such events dominate most conventional history tales from the last 5 to 6 millennia, but are still far from the whole story of the development of human societies.

Evidence for violence on different levels is plentiful in the period just before (as well as after) the introduction of agriculture in Denmark some 6 ky ago. At first, the violence seems connected with local struggles for control, and only in the Bronze Age (from 4 ky BP in Denmark) can one speak more broadly of warfare (Beyneix, 2007). Most Stone Age weapons are similar to hunting or domestic tools (spears, arrows, axes, clubs, knives or daggers), but it is often possible to distinguish the two applications by the design. For example, the battle-axe shown in Fig. 4.13 has a design different from the straight axes used to fell trees in the forest for use as timber to build houses and boats, and to sustain fires in connection with cooking or space heating.

Until a few decades ago, archaeologists and anthropologists painted a picture of the Danish Stone Age as non-violent, amicable and slowly evolving, despite the large numbers of weapons found (Vandkilde, 2006). The weapons were thought to be just status symbols, never actually used. This view is difficult to maintain after Bennike (1985) presented studies of a substantial number of skeletons with lesions that could only have been induced as a result of fights or assaults. Fig. 4.14 shows how many skulls have been found in Denmark for each time period, and how many of these show signs of violence, such as trepanation (as a remedy for mace damage), fracture or deformation of the skull, and arrow wounds in skeletons (complete with crania, but in some cases also wounds in other places). The data are separated for male and female crania. First one notices that the number of crania available is not proportional to the time-span of each period. This is partly due to changing burial practice (with cremation increasingly common in the Bronze to Viking periods), and partly because of differences in skeleton survival with location. Scarcity of finds makes conclusions less reliable for the Bronze Age and the early agricultural period.

On the other hand, particularly the early Neolithic period has such a high percentage of crania with signs of violence that it does not appear coincidental, despite the basis of only 3 male and 4 female skulls.

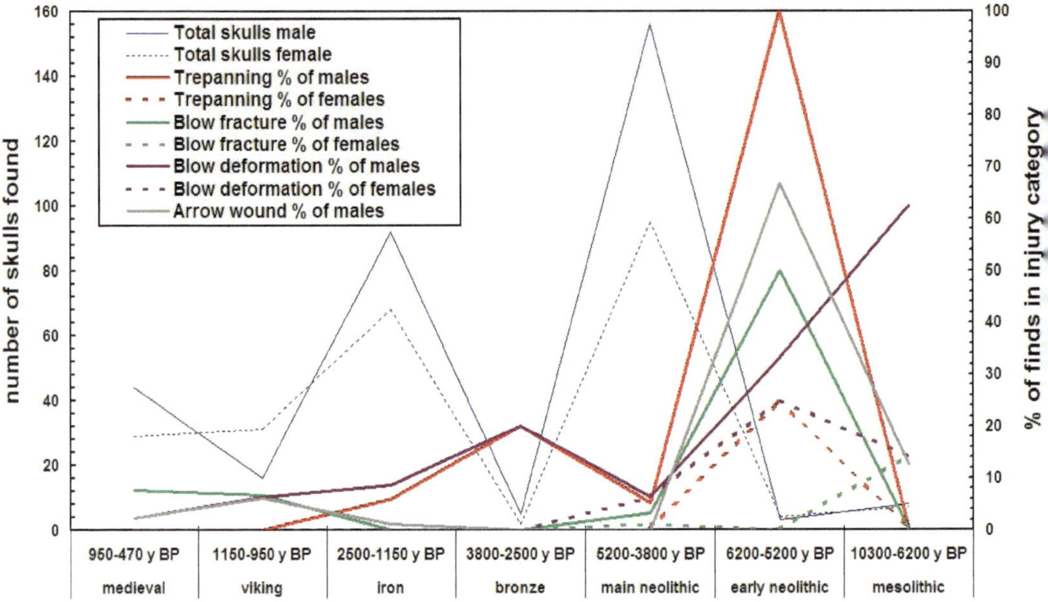


*Figure 4.14. Time variations in signs of violence in Danish crania, from trepanation, fracture, deformation and from arrows (including some found remaining, for example in ribs). The scale on the right gives percentages of different categories of cranial injury with respect to the total number of crania found in Denmark (the latter indicated in blue using the left scale). The data are divided on sexes, omitting a few where sex determination was not possible (based on Bennike, 1985).*

Use of trepannation is seen to start in the agricultural period and largely disappear as one approaches the most recent two millennia. The method is applied to reduce the effects on the brain of deformation wounds, and particularly those induced by maces (a sword in one case; cf. Fig. 3.20). A surprisingly high survival rate of 79% is deduced from the signs of bone healing around the bore-holes. Fig. 4.15 shows a case of an arrow stuck in the nasal region.

The alleged association of increase in violence with introduction of agriculture could be due to an influx of foreign farmers to the area, or (perhaps more likely) due to stratification within the society, but it is in any case hinging on rather weak evidence. Singling out the early Neolithic, as in the data behind Fig. 4.14, rests on dividing data in a way that emphasizes violence in the crucial time period, which only has data of dubious statistical significance, and the figure also shows a high incidence of violence during the previous (Mesolithic) period, where the statistics are slightly better. All are mace-induced lesions without signs of trepannation, and, strangely, there are no examples of skull fracture. Perhaps the clubs/maces were less effective during that period?

*Figure 4.15. Man killed by arrow by somebody shooting from an elevated position, found at Porsmose (Sjælland) and dated to 5.4 ky BP (Becker, 1952; Bennike, 1985; Madsen, 1990; photo by Danish National Museum).*

The scant data for the early Neolithic period indicate a sex difference in incidences of violence, supporting the notion that the violent fight with use of weapons is mainly an activity of men (all three men in the sample have wounds, compared with only one of the four women). One may ask if the incidences result from individual attacks or war-like exchanges. The fact that the combatants in the Neolithic were taken in for subsequent treatment such as trepannation, and that many of the wounds, such as skull deformations, were not fatal, may suggest that at least some of the violence was conducted in fairly controlled form, such as by settling disagreements in staged duels. During the main part of the Neolithic period, where statistics are reliable, 16% of all skulls found show signs of violence, and of these about 30% are women (conventionally not supposed to have practiced duels). The data could also be interpreted in terms of occasional raids by one village on another.

In the Bronze Age, statistics are too poor to allow firm conclusions, but in the Iron Age, violence is at a level similar to that of the Neolithic, except that now the victims are nearly 100% male. The same appears to be true in subsequent Viking and medieval periods, but here the nature of violence has changed and much of the action has been displaced to areas outside Denmark, implying that the skull data cannot be directly compared with those of the earlier periods. It is interesting that Tacitus (98) notes with surprise that German women took part in fighting the Romans, or at least performed a Red Cross-style nursing function. Could Danish women also have joined the Neolithic fights, but not the later ones due to a change in social order?

From the main Neolithic period, mass graves have been found that could be interpreted as a sign of more systematic warfare. However, we are still considering encounters between fairly small groups of people, rather than the territorial power struggles appearing after a further stratification of the society into entities ruled by ambitious sovereigns with standing armies (and associated requirements, such as tax collection and monetary lending facilities). At which level we define the conflict as war rather than assault is largely a matter of convention (Keeley, 2004).

The changing social order and change in the nature of violence in society needs to be incorporated into the demographic models used here to trace population developments, which again form the basis for evaluations of energy use in the variable societies. The basic age distribution of death rates has so far been kept fixed, and only overall changes in mortality or fertility have been introduced as a function of time since the Neanderthal societies. To describe the Danish society emerging after the agricultural development, such as at the beginning of the Bronze Age, the death and fertility rates of Figs. 1.18 and 1.19 are changed into those of Figs. 4.16 and 4.17. The resulting age distribution and life expectancies then take the form given in Figs. 4.18 and 4.19, instead of those shown in Figs. 1.20 and 1.21. The change in population development predicted as a consequence of adopting the new birth and death rates at the beginning of the Bronze Age (4 ky BP) is shown in Fig. 4.20.

The main differences in the mortality profiles come from changes in female death during childbirth (assumed to have decreased through the Ertebølle and Agricultural periods by a total of 8%) and the change in male occupational death. The argument for the decline in death during childbirth is indirect–otherwise, the increased male death risk would have led to a more significant drop in total population than the one observed. While the male risk associated with hunting large game or fish far offshore would have largely disappeared, the rising risk of violent death was extending to men of higher age, as the hostilities assumed a more war-like character (Fig. 4.16). In Fig. 4.17, total female fertility drops by 1% after 4 ky BP, partly as a reaction to reduced child death and partly due to the more complex tasks carried out by both men and women in the farming and metal ages. The effect of this on population development is explored by the two alternatives in Fig. 4.20. In Chapter 5, this analysis will be continued.

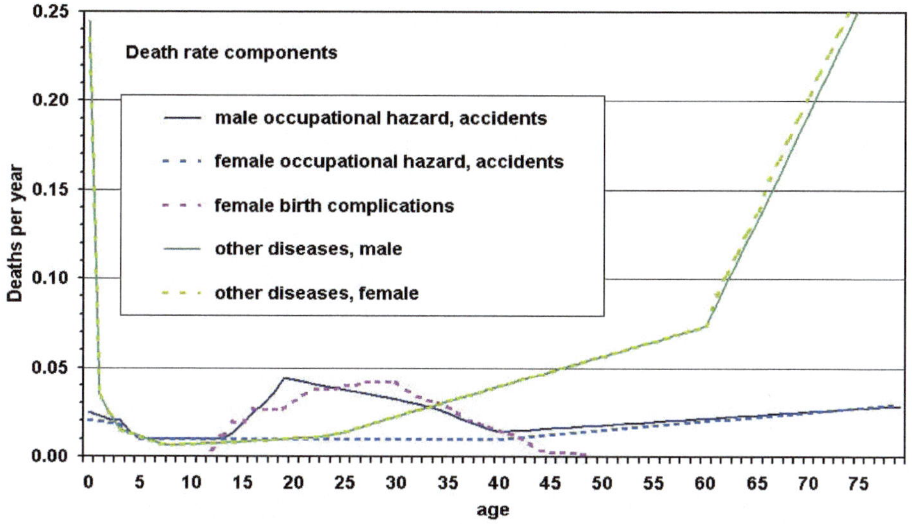


*Figure 4.16. Gender-specific mortality components for Danes used in the demographic model from the Bronze Age (4 ky BP).*

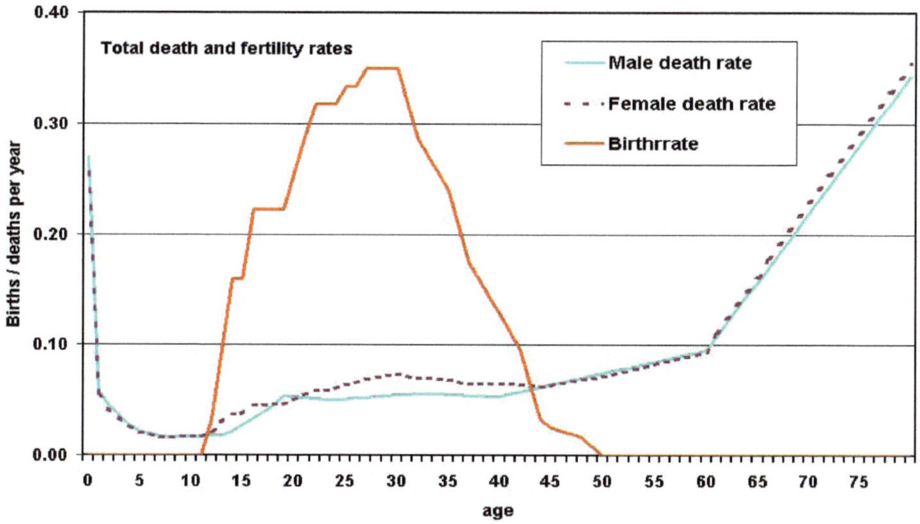


*Fig. 4.17. Overall fertility and mortality rates for Danes used in the demographic model from the Bronze Age (4 ky BP).*

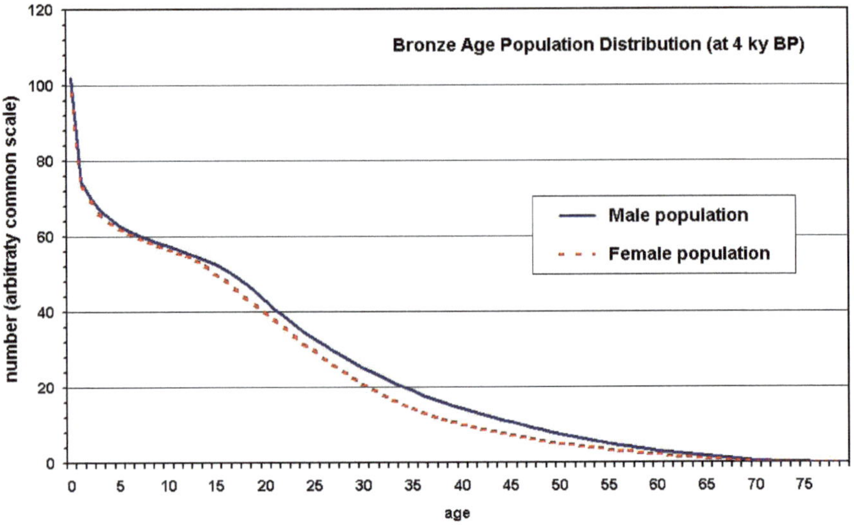


*Figure 4.18. Age distribution of Danish population at the beginning of the Bronze Age (4 ky BP), derived from birth and death rates in Fig. 4.17.*

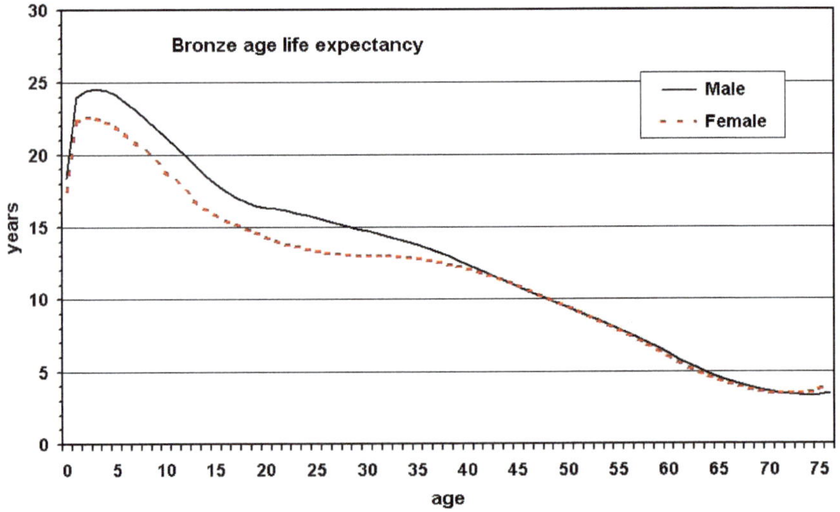


*Figure 4.19. Life expectancy of Danish population at the beginning of the Bronze Age (4 ky BP), derived from birth and death rates in Fig. 4.17.*

Compared with the life expectancies in Fig. 1.21, the new Bronze Age data used in Fig. 4.19 improve the situation for women, although the maximum expectancy of 22–25 years is not changed. The age distributions for men and women move a little closer together (Fig. 4.18).

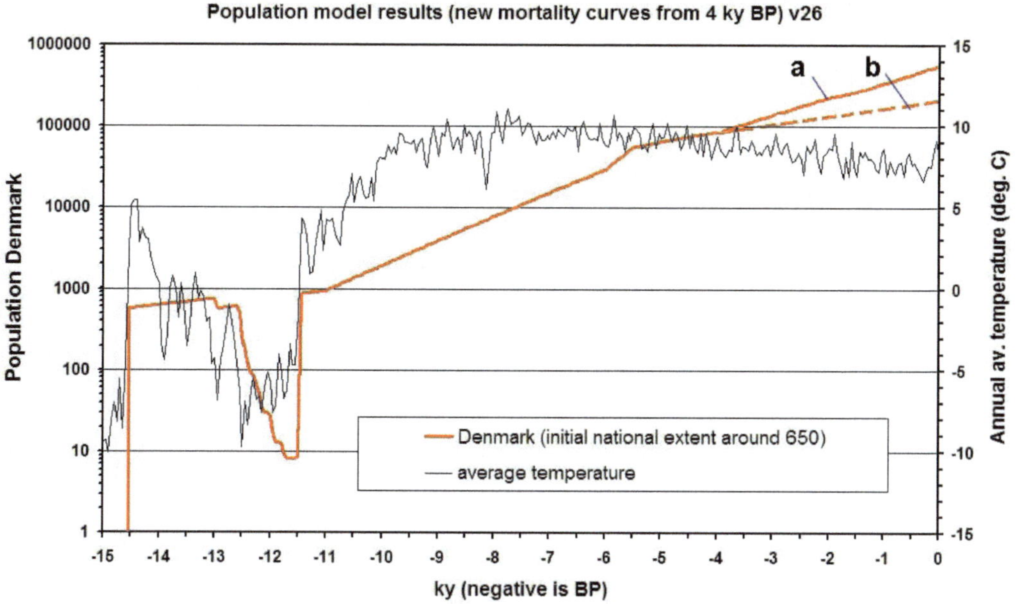


*Figure 4.20. Modelled Danish population development. The model is identical to the one used in Figs. 4.2 and 4.3 until 4 ky BP, and then switches to the one based on the fertility and mortality rates of Figs. 4.16 and 4.17 (case* b*). Case* a *omits the 1% reduction in fertility introduced by Fig. 4.17 (cf. text and further discussion in Chapter 5). No parameter change is made after 4 ky BP.*

The overall population development given in Fig. 4.20 shows that the new data introduced into the demographic model make it latch smoothly onto the ones used before 4 ky BP. The subsequent population development is similar to that of the old model (Fig. 4.3) in case *b*, despite the extension of the male surplus mortality to higher ages, due to violence. The case denoted *a*, where female fertility is not reduced by 1% (to 7.46), finds a somewhat higher population growth from 4 ky BP. The growth is in both cases exponential (linear on the logarithmic scale). Estimates for other parts of Europe confirm the general behaviour of the model used here, but in some regions have a decline in population at the beginning of the Bronze Age, again for reasons of increased violence, plus a less healthy diet previously introduced by agriculture, but with fewer mitigating circumstances than in Denmark (Biraben, 2003, considering the global population). The reason for the lesser population reduction in Denmark relative to the global situation at the start of the Bronze Age is the improvement of the Danish diet by the addition of fish, a commodity missing in most parts of the world during the time period considered. There is a modest reduction in long-term average temperature after 8 ky BP, continuing to quite recent times (some 100 y ago), also having a slight negative influence on population size. "Long-term average" here means averaging over a 300-year period of variability, in contrast to the 30-year averaging procedure meteorologists have traditionally used to define "climate".

Support for the occurrence of an increased male mortality during the metal ages comes from excavations of a cemetery at Vikletice (Vencl, 1999), finding a 15% increase in male mortality in the 15–30 year age group, presumably relative to smooth interpolation between adjacent age groups. The suggestion that average life expectancies remain fairly constant all the way from the Neanderthal period to medieval times was made already by Hassan (1981), who finds variations below 15% with lowest values at the end of the Neolithic period and again later at the time of the bubonic plague pandemic in 14th century Europe (based, however, on very limited data). Regarding the impact of dietary changes, there are hypotheses beyond the disease-related ones mentioned here, such as noting the differences in proportions of fat and carbohydrates and invoking them in explaining a higher mortality during the Neolithic, as well as in other (recent) periods of high-sugar diets. Lutz (1995) suggests that altered fat/sugar ratios at least require several centuries for adaptation.

*Figure 4.21. Imported gold necklace found at Grevinge Mark (Sealand). The half-moon* (lunula) *form was appreciated all over western Europe (Nielsen, 2001; photo by L. Larsen, Danish National Museum).*

The period of Danish history from about 4 to 2.5 ky BP is usually referred to as the Bronze Age, although flint tools continued to play an important role. Some archaeologists take the beginning of the Bronze Age as 3.7 ky BP, because the casting and processing in Denmark had been well established by that time. But some imported bronze goods came to Denmark earlier, and metals other than bronze were imported, notably gold (Fig. 4.21). The interesting thing is, of course, that none of these metals can be mined in Denmark, and yet importing the metal and manufacturing the tools and artefacts in Denmark reached a quite sizeable magnitude during the bronze period.

Raw materials such as wood and stone had been in use for as long as the presence of humans is known. Also, clay had been in use well before the introduction of metals. Clay may first have been employed as building material for huts and houses, using wooden rafts and poles, and plastering rush mats with clay for tightening. However, people would soon learn that clay became permanently hard after being heated to about 500 °C, a temperature level available at the fireplaces of settlements. Simple clay bricks could thus be produced and used as building material. Clays are formed by erosion of rock material, and would typically be $Al_2O_3$–$SiO_2$ compounds with crystal water and various impurities, forming a sheet-like structure (Hummel, 2004). The next step in development would be pottery based on earthenware techniques of heating to 1000 ± 300 °C, at which temperature clay fuses into a coarse-surfaced red, grey or black ceramic structure later called terracotta. The oldest example found is a female figurine (fertility charm; Fig. 4.22) from Dolní Věstonice in the present Czech Republic and dated to 21–25 ky BP. This and several figurine fragments found at the site have been produced from loess-based clay and constitute earthenware fired at the lower part of the temperature range given above. Also found are hearths or kilns suggesting that firing would possibly have employed charcoal (Vandiver *et al.*, 1989). Re-examination of the figurine has revealed what has been interpreted as fingerprints placed by a child on the artefact before firing (Kralik *et al.*, 2002). The more than 10 000 fragments (supposedly mainly from rejects) found at the site cluster indicate a sizeable industry, and the fingerprints may indicate that children helped in the production. The artist contribution would have been to form the artefacts (most of which depict animals) from a mixture of loess and water (similar to raw materials used in present kindergartens), before firing.

*Figure 4.22. Věstonice "Venus", dated to 25 ky BP (photo by Petr Novák, Wikipedia Creative Commons).*

Reaching earthenware firing temperatures usually requires the use of charcoal in closed or semi-closed kilns with draft control, rather than simple fuelwood hearths, and the extensive production of pottery evidenced by Chinese finds starting at around 18 ky BP (Boaretto *et al.*, 2009) are associated with charcoal remains found close to the artefacts. As mentioned in Chapter 3, Danish fired pottery appears around 8 ky BP. The porous surface of earthenware causes water in the container to seep through and evaporate, thereby cooling the content, a property presumably appreciated then as well as now (wine coolers).

The next advance in clay pottery technology is glazing. It consists of a second firing (at a temperature lower than the first) of the earthenware after covering the areas to be glazed by a ground sand–salt mixture suspended in water. Glazed clay-ware is found in Egypt some 5 ky ago (Hummel, 2004). An alternative technique using tin was used in Mesopotamia some 3–4 ky ago, and much more recently on Majorca Island and in Faenza (Italy) in the Mediterranean region. Other advances include the production of ultra-hard stoneware (China, 3.4 ky BP, requiring firing at temperatures of 1200–1300 °C) and later porcelain (also China, early attempts before 900, brought to Europe by Marco Polo around 1295). Each clay-ware type requires a different composition of the raw clay material. Another silicate-based material is glass, an amorphous solid with short-range ordering of molecules that over time may increase its crystalline character (fast for metal-based glasses, very slowly for silicon-based glasses). Pottery glazing and glazed beads may be considered the earliest examples of man-made glass, as opposed to the glass-type called *obsidian* that can be formed naturally in volcanoes. Vessels of glass are found from around 3.5 ky BP, or at least described in tablets, in Mesopotamia and Egypt (Shortland, 2007). The earliest Danish glassware found at Himlingøje (from 1800 y BP; Jensen, 2003) is clearly imported, and probably manufactured by the large glass-making industry of the Roman state.

Copper was first used as hammered sheets made directly from quality ore. A pendent found in Shanidar, northern Iraq from about 11.4 ky BP is probably the earliest find (Smith, 1994). Native copper ores are not very abundant, and larger-scale use of copper requires mining of suitable minerals such as malachite, as well as complex processing involving ore melting above some 1100 °C in a carbon monoxide atmosphere without molecular oxygen, and ideally with an iron catalyst (which could be present in the ore used). The kilns would have been built on the experience from pottery firing, and placed, for example, on the top of hills, to allow wind to carry $CO_2$ away. Later, bellows were incorporated into the design to do this. The technology may have emerged at

10–8 ky BP, but only from 8–7 ky BP is there concrete evidence (weapons and tools, Egypt; Hummel, 2004). The first evidence of mining operations are from 6 ky BP in the Balkans. Adding some 5% tin would make the much more useful alloy of bronze; it is harder, requiring lower firing temperature (1000 °C), and without the brittleness of cast and hammered copper.

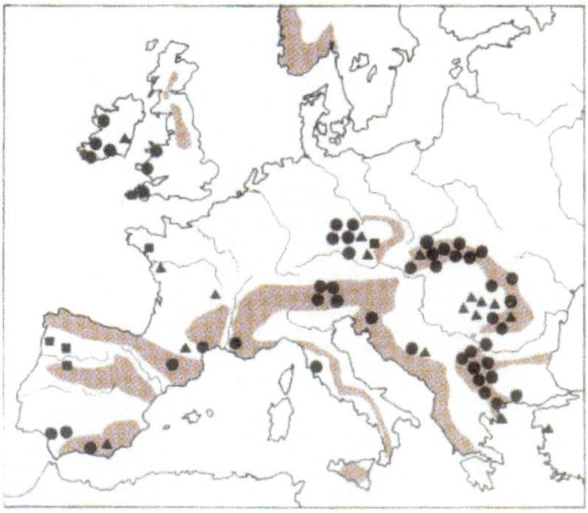

*Figure 4.23. European geological deposits of copper (shaded areas) and of tin (dots). Triangles indicate early locations of manufacture (Jensen, 2001b; Danish National Museum).*

Figure 4.23 shows the locations of deposits of copper and tin in Europe. While the main centres of ore extraction and melting were initially in the Balkan area, the technology spread to the western Mediterranean and to the north. Suppliers to Denmark are believed initially to have been from England (due to its large tin deposits, thought to be a key reason for the later Roman invasion) and then from the mid-German–Czech cultural area referred to as Únetiče. A large mining, manufacturing and export industry developed there, taking advantage of easy transportation to salt and bronze customers by way of large rivers. In Denmark, this shifted the manufacturing centres from northern Jutland to Sealand, which for a while became the main seat for an indigenous Danish metal industry.

Iron was at first (some 6 ky BP) derived from meteorites. Melting iron directly (requiring 1538 °C) was not possible, but at 1000 °C iron ore can be reduced to *sponge iron*, which can be hammered into *wrought iron* tools. As mentioned, small amounts may also have been used in bronze production, as indicated from sponge iron residues found. By 3.4 ky BP, the Hittites (in present Turkey) learned to apply several cycles of heating at 1200 °C, incorporating carbon from the charcoal CO emissions. This made the iron harder. Iron with 2% carbon is called *steel*. Iron is not better than bronze, but it is far more abundant and often can be retrieved from surface minerals, without mining.

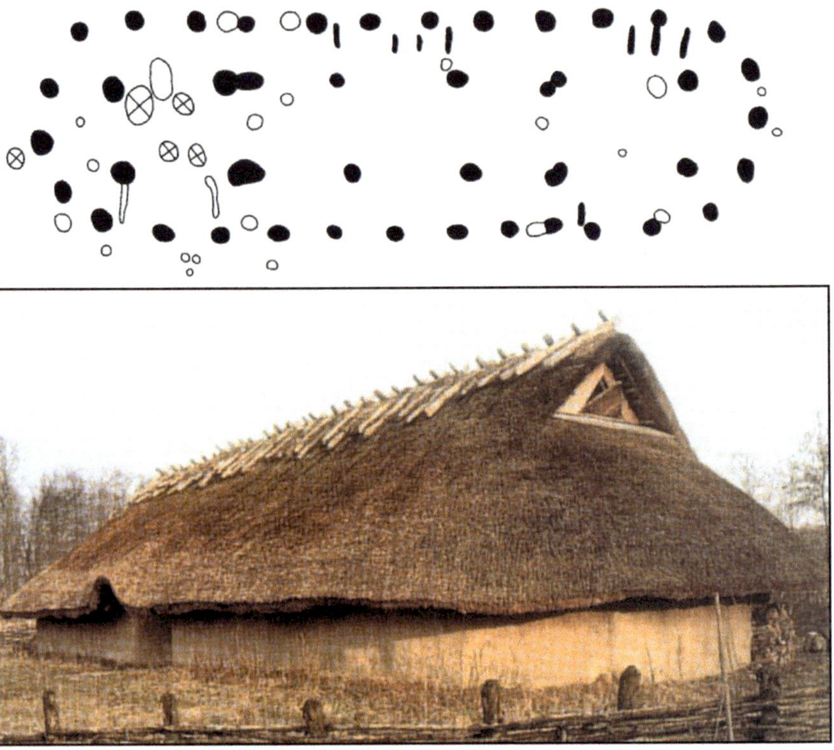

*Figure 4.24. Plan and reconstruction of three-aisled Bronze Age house (Rasmussen and Adamsen, 1993; Jensen, 2001b; Danish National Museum).*

Dwellings changed at the entrance to the Bronze Age. The two-aisled long-houses (Fig. 4.11) were replaced by three-aisled houses as the one shown in the upper panel of Fig. 4.24. It gives the floor plan, interpreted as giving better room for stall partitions for cattle or pigs (Rasmussen and Adamsen, 1993; the plan depicts a house found at Spjald, Jutland). Because only the positions of the vertical poles have survived, there is a considerable amount of guesswork involved in the reconstructions, and neither in the agricultural-period houses of Figs. 4.8 and 4.11, nor in the Bronze Age one in Fig. 4.24, can one be quite certain that the domesticated animals actually lived in the same houses as people. Below the plan is a museum reconstruction of a similar house, at Hollufgård (Funen).

As mentioned, simple ploughing has presumably taken place throughout the agricultural period, but the first ploughing tool found, a "hook ard", dates from 3.5 ky BP (Fig. 4.25a). A vertical steering stick was attached (to the right-hand side in the figure), and a team of two oxen would have pulled the ard, as shown in the petroglyph (rock carving) shown in Fig. 4.25b. Note that the farmer carries a bag, presumably with seed grain. An improved version is the "bow ard" shown in Fig. 4.26a (2.7 ky BP) and on Fig. 4.26b from the Iron Age (2.3 ky BP).

*Figure 4.25 (a, top) Plough of the hook ard type, found at Hvorslev, Jutland (Jensen, 2002), and (b, below) the same rendered on a petroglyph, showing its use in connection with sowing (found at Listleby, western Sweden; Glob, 1951). Both photos by Danish National Museum.*

*Figure 4.26 (a, above) Plough of the bow ard type, found at Døstrup, Jutland (Glob, 1951; Danish National Museum), and (b, right) rendering on petroglyph, showing its use for sowing, found at Aspeberget, west Sweden (Sharpe, 1992; photo by B. Almgren, Svenskt Hällristnings Forsknings Arkiv, id=222).*

The use of the ox for ploughing made it natural for the Neolithic or Bronze Age farmer to think of using the ox for carrying the harvest home, or for other tasks such as removing stones from newly cleared land. The notion of carts (and later carriages) would have been transmitted from the Mediterranean or Near East region, so one just had to build two wheels and connect them to a V-branched pole and a wicker basket. Probable wheel tracks have been found dating to the beginning of the agricultural period, but the earliest actual wheels found in Denmark (at Kilderis) are the pair of which one is shown in Fig. 4.27a (Jensen, 2001a; Rostholm, 1977). It is solid, and therefore breaks easily on rough roads, as the specimen witnesses. Varying numbers of wheels have been found at several locations, so it is not clear if the first carts had 2 or 4 wheels. The reconstruction shown in Fig. 4.27b has four solid wheels, and so have many petroglyph renderings of carts found in western Sweden (Bradley, 2006; Coles, 2003).

*Figure 4.27. (a, left) Solid wheel for ox-driven cart, found near Viborg (Jutland), dated to around 5.9 ky BP (Jensen, 2001a), and (b, right) reconstruction of 4-wheel cart, of which parts were found at Jørlunde (Sealand) and dated to about 2 ky BP. However, the design may well have been used in rural areas much earlier (Schovsbo, 1987). Photos by Danish National Museum.*

A better wheel, with spokes that enable a springy damping of road roughness, was introduced during the early Bronze Age, but probably first used only by the more wealthy parts of the now quite stratified population. A number of Iron Age chariots with Celtic (base: Munich region) ornamentation from the period around 2 ky BP have been found, for example at Dejbjerg Bog (Jutland), one of which is shown in

Fig. 4.28. It is contemporary with the less pretentious one in Fig. 4.27b.

*Figure 4.28. Chariot from Dejbjerg (Jutland), dated to early Iron Age (about 2 ky BP). Inset: decoration on the sitting box, presumed to have been made by a Celtic artist (Hvass, 2001; Jensen, 2002; photos by L. Larsen, Danish National Museum).*

*Figure 4.29. Excerpt of Kivik grave picture (Skåne) from early Bronze age (Strömberg, 1990; Coles, 2003). The drawing appears on one of a set of stone tablets. Photo by C. Bertilsson, Svenskt Hällristnings Forsknings Arkiv, id=201.*

That wagons or chariots with spoked wheels were produced in Denmark during the Bronze Age is evident from the petroglyph shown in Fig. 4.29. Another representation of a spoked-wheel vehicle is found in the solar carriage shown in Fig. 4.30, dating from 3.4 ky BP. It indicates not only the existence of a wealthy segment of society, but also the practising of a cultic worship of the mystery associated with the solar orbiting during the day (probably associated with further philosophizing on the course of seasons as well).

Wagons need some kind of road. The Neolithic farmers cleared vegetation to create local dirt paths where they needed to go, and roads between villages or between regions were probably no better. The solid wheels of the carts carved deep grooves in the soil. Such pairs of Neolithic or Bronze Age wheel grooves have been found in several places (Jensen, 2001b). As a matter of fact, they are still fairly common today on private (e.g. farm) land, where it has not been considered economical to get rid of the annoying middle ridge and upgrade to a dirt road with gravel. In wet areas, the heavy carts were not easy to move forward, and the passage was often ensured by creating a wooden floor from trunks of young trees, as illustrated in Fig. 4.31. A more primitive solution was to spread twigs and small branches on the trail, which could help passage at least for a while. Roads built with stones or cobbles appeared from the late Bronze Age. Denmark is claimed to be the first place to use such stone roads (Jørgensen, 1993).

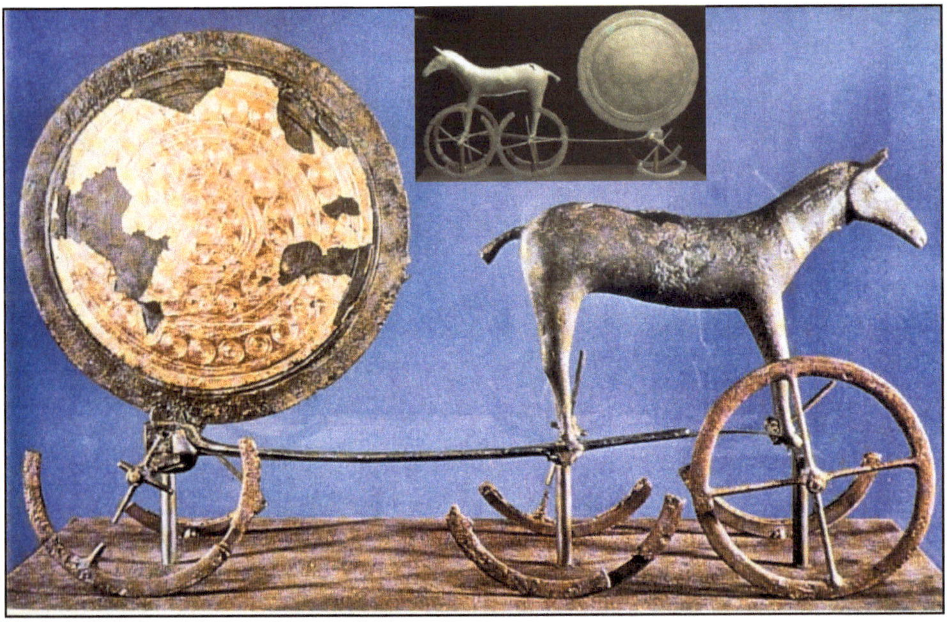

*Figure 4.30. Cult object depicting a horse rolling the sun across the sky, from Trundholm Bog (Sealand), dated to 3.4 ky BP (Jensen, 2002). The main photo by Sandro Prato-Previde is from the 1950ies, when damage from Copenhagen automobile pollution was still modest. The insert shows a recent backside photo taken after a restoration by the Danish National Museum. The sun-hauling story is retold in later accounts of Nordic mythology (Sturleson, ca. 1200).*

Boats must have played an important role for transporting people and goods, as witnessed by the many representations in petroglyphs such as the one depicted in Fig. 4.32. Artefacts manufactured in one place but found in another, often far away, and particularly the raw materials imported into Denmark starting in the Bronze Age, strongly suggest that the number of boats or ships must have been substantial. Yet relatively few actual boat remains have been found from either stone or metal periods. One fairly late example is shown in Fig. 4.38.

*Figure 4.31. Part of a 3 × 300 m² wooden road from 3450 y BP, found at Speghøje Bog (Jutland; Jørgensen, 1982, 1993). The cut planks lie on two oak battens and are adapted to lie nearly horizontally, with stop poles at intervals marking the road (photo by M. Jørgensen, Danish National Museum).*

Considering the varied tasks creating demand for sailing the waters of Danish streams and fiords, the oceans around Denmark and the Oder and Neisse rivers leading to the bronze centres near Prague, it is surprising not only that no Bronze Age boat remains have been found, but also that all the boats depicted in the Scandinavian petroglyphs, on rock where available (but in Denmark also on loose stones), are fairly similar. Perhaps they were equally suited for river and ocean use. The boats in the pictures all have a forcefully drawn keel, both ends of which rise above the water. All drawings are side views, and most of them have a second horizontal line that could represent the railings. This line rises at both ends, the front one on early petroglyphs (about 3.6 ky BP) bending first back and then forward with what looks like an animal head. Variations on this theme are found in the later drawings. The sides of the ship (between the keel and railing lines) are sometimes solidly carved out (Fig. 4.32), but in many of the petroglyphs detailed with drawing the individual planks bend from the keel up along the sides of the ship (but the bending is of course not seen in the side view; see Fig. 4.39). Finally, several of the pictures have a vertical line-stylized crew of 5 to 30 men sitting in the boat, indicating that boats of different sizes were essentially built according to the same master plan. More than 1000 petroglyphs have been found in Scandinavia, and the boat is by far the most common subject (other subjects are wheel crosses, men with erect penises, weapons, various animals, wheeled carts and sun-like disks).

*Figure 4.32. Bronze Age petroglyph from Tanum Heath (west Sweden). Note the dotted path (photo from Svenskt Hällristnings Forsknings Arkiv, id=62; cf. Swedish Rock Art Research Archives, 2010).*

The Danes sailed to centres such as those of the Únetiče culture to buy bronze, not the other way round, as the central Europeans did not seem to have had as keen a relation to sailing and shipbuilding as the Scandinavians. The purchases were paid primarily with amber, a commodity highly valued in central and southern Europe. More than 80% of graves with grave-goods in the Únetiče region contain amber jewellery (Gimbutas, 1963). Trace-impurity analysis shows that the amber came from the Baltic region, from Denmark to Lithuania. Not only was amber regarded as a nice jewellery stone, it was also considered to have magic power, because of the static electricity formed when rubbing it with a soft skin (the classical Greek name for amber is actually *ηλεκτρον*). Trade was eased by the introduction of a common Scandinavian weight system (Malmer, 1992).

The petroglyph in Fig. 4.32 (as well as the one in Fig. 4.35) may show a convoy of ships sailing down the river Oder to buy bronze. The dots regularly signposting the route of the ships in Fig. 4.32 may represent the river. Although such dots (semi-spherical indents) are common on petroglyphs, it is rare to find them ordered along a string.

The societies from which Scandinavia bought bronze are believed to have been considerably more socially stratified than those in the north, with strong rulers controlling the trade. However, mining had a limited time horizon before at least the easily extractable minerals had been depleted, and after 3.1 ky BP much of the mining activity moved to the Alpine region. Danish amber is then found in considerable quantities in Italy (e.g. in Verucchio near present Rimini), indicating strong trade ties (Jensen, 2002). The imported bronze was melted and cast into merchandise such as weapons in facilities back in Denmark (Fig. 4.33).

*Figure 4.33. Left: Bronze Age furnace, probably for cooking, found at Skamlebæk (Sealand; Jensen, 2002; photo by E. Lomborg). Larger ones (fire-pits) were used for pottery manufacture and probably similar ones for bronze melting. Right: Mould for bronze casting, found at Bregnemose, with axe-head found nearby (Funen; Jensen, 2002; photo by L. Larsen). Danish National Museum.*

The mythical thinking of the Bronze Age people, as first deduced from the singular find of Fig. 4.30, receives support from related topics depicted on a petroglyph found in Sweden (Fig. 4.34). Again the picture can be interpreted as a horse pulling a solar disc, here with some people surrounding it. A person in Denmark facing south would see the sun as rising to the left and moving over the sky along an arc towards the right. The person must have wondered how the sun got back east to be ready to rise next day. A solution would be to have another horse (the first one being now tired) move the sun back, perhaps along the ground or switching the light off, so that people could not see it return during the night. The back of the object in Fig. 4.30 is without shining bronze (see inset), so from the side where the horse would move left, the sun would not be glimmering. Such a cosmological hypothesis appears in slightly altered form in the more than 2 ky later Icelandic Eddas (Sturleson, ca. 1200), where it is proposed that two horse-drawn carriages driven by the son Day (Dagr) and the daughter Night (Nott) of a giant called Njorvi, and pulled by two horses named Skinfaxi and Hrymfaxi, take care of the transportation of the sun "around the world".

*Figure 4.34. Petroglyph from Askberget (in current western Sweden, previously Norway). A solar disc hauling in progress? (Photo by C. Bertilsson. Svenskt Hällristnings Forsknings Arkiv, id=206).*

The Eddas may be seen as a last–minute attempt to provide written evidence of an ancient Norse mythology, written at a time when Christianity was being introduced and its priests did what they could to suppress the old way of thinking. Of course, writing was also by 1200 fairly new in Iceland, where the Edda manuscripts were found (but in Scandinavia known from late Iron Age times, around 400; cf. Vebæk, 1980). It is remarkable (but not unlikely) that the philosophical and mythological thinking of the Nordic people might have stayed more or less the same for nearly 3 ky, at least in the fairly isolated Iceland.

*Figure 4.35. Petroglyph from Ekenberg (mid-Sweden). The spiral discs placed above or below the ships may be interpreted as suns, and the groove across the picture, dividing the view of the ships turned by 180°, could represent a river or a horizon (rubbing by Coles, 2003; he interprets the human figures along the groove as a cultic procession).*

*Figure 4.36. Bronze Age razor found at Neder Hvolris (Jutland), exhibiting a ship entering from the left, an S-shaped horse pulling the sun, and what looks like a bird (Bro-Jørgensen, 1996; photo by Danish National Museum).*

While the theme of Fig. 4.34 is rare in Scandinavian petroglyphs, there are frequent circular objects, some of them with a cross of spokes, others without, or filled in. It has been speculated that they could represent the sun in an alternative way of returning it to the eastern position during the night, making use of ships rather than horses (Bradley, 2006). The mechanisms are unclear, although on some pictures, the placement of discs relative to the ships suggests interpretations related to whether the discs are below or above the ship (Fig. 4.35). The most convincing argument in favour of a connection between ships and sun-pulling comes from pictures carved into men's razors (Kaul, 1998). Ship designs are common on razors, and 11 have been found with what appears to be a ship and a horse involved in the dawn appearance of the sun (Fig. 4.36). The interpretation is that the ship has transported the sun during the night and is now delivering it to the horse for the sun's spectacular journey over the daytime sky (Jensen, 2002).

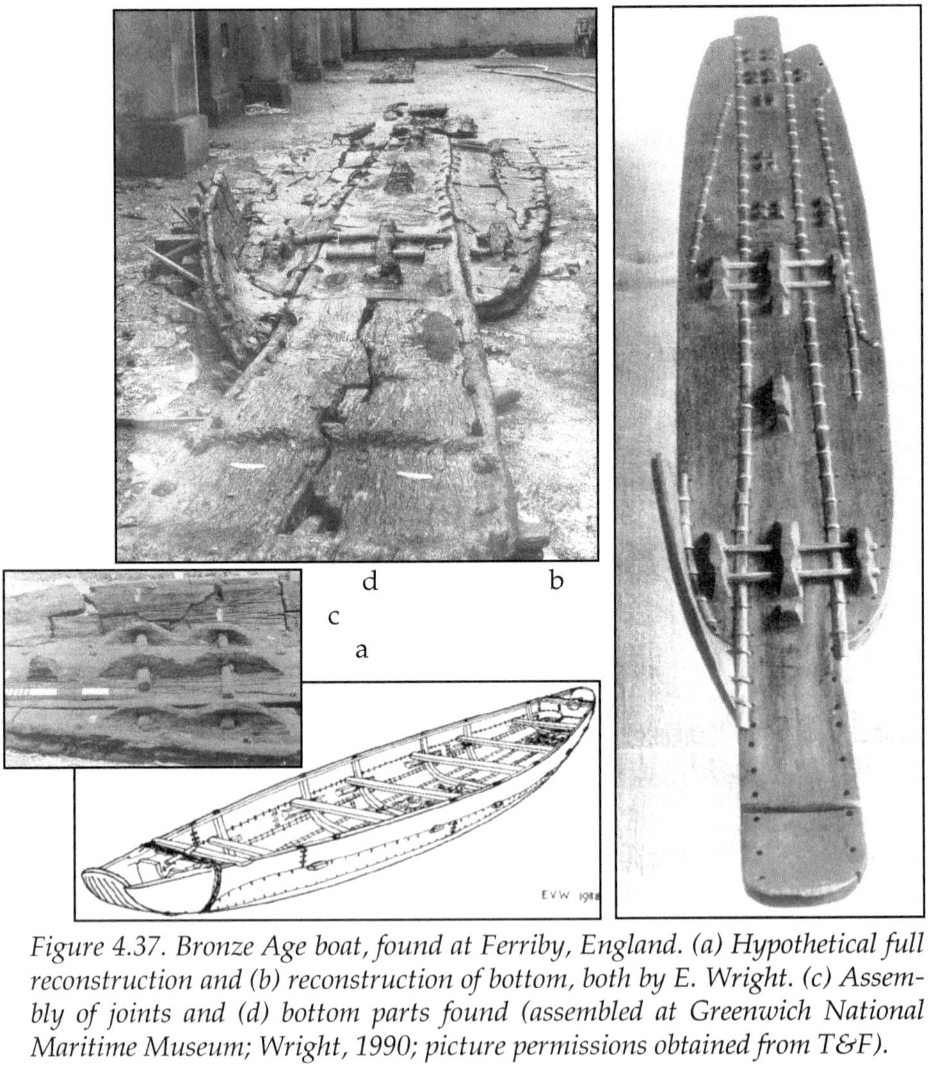


*Figure 4.37. Bronze Age boat, found at Ferriby, England. (a) Hypothetical full reconstruction and (b) reconstruction of bottom, both by E. Wright. (c) Assembly of joints and (d) bottom parts found (assembled at Greenwich National Maritime Museum; Wright, 1990; picture permissions obtained from T&F).*

While no Danish boat remains from the Bronze Age are known, there are several in England. Fig. 4.37 shows one find, dated to 3780 ± 100 y BP, of three boats at Ferriby (northern England). It has a number of advanced features, including a broad bottom (of a size similar to the diameter of a large tree) from which the side structure extrudes. Details of elegant connections and plank overlaps to improve watertightness can be seen from the remains. The upper structure is lost. The bottom plank has ends bent upwards, suggesting that the makers had mastered the technique of span bending. This requires fixing the wet plank with clamps, drying, and applying, for example, the wax extracted from birch bark. This glue-like resin could also have been used to tighten the joints of the ship's sides. These techniques are quite different from those used earlier in the Mediterranean region (Digby, 1954; Sørensen, 1995).

Scandinavian Bronze Age boats seem to have looked rather different from the English ones, judging from the vessels depicted in many of the petroglyphs found (e.g. Figs. 4.32, 4.35 and 4.39), but actual boats of this type exist only from the early Iron Age (2350 y BP), notably the one found at found in Hjortespring Bog (southern Jutland; Fig. 4.38). In contrast to the more barge-like Ferriby boat, it has a swung keel and structural railings, between which the bent side-planks are fixed.

*Figure 4.38. Iron Age boat sunk in Hjortespring Bog after a battle taking place around 2.35 ky BP. Top right: assembly of parts found. Bottom: reconstruction (Jensen, 2003). Top left: layout of design (Nelson, 2009). Photos by F. Kaul and K. Petersen, Danish National Museum.*

Although the Hjortespring boat is similar to the Nordic ones depicted, it actually came from the Hamburg region (but use of labels such as Danish and German make little sense for this region at the time). It was part of a small fleet of 4–5 boats, carrying at least 80 warriors seeking to raid the island of Als in southern Jutland. A regular battle was fought, as Als spies had spotted the attackers early enough to allow for assembling a small army of maybe 100 men (out of a settlement of some 3000 people). It was in 2350 BP, and the people of Als won. Surviving attackers fled on some of the boats, but one was left behind. This is the one that was carried to a nearby bog and sunk, together with the abandoned weapons of the attacking tribe (with spears and daggers bent, just to make sure that they were now useless; Jensen, 2003). Iron Age Danes had become warriors, no longer just farmers and blacksmiths.

*Figure 4.39. Warriors in Bronze Age petroglyph from Tanum Heath (western Sweden; photo by Svenskt Hällristnings Forsknings Arkiv, id=90).*

The development of Danish society to include a set of warrior clans may also be deduced from the petroglyphs. Just as the majority of them contain pictures of ships, there is also a majority depicting men fighting or men with weapons (Fig. 4.39). Of course, this does not mean that every Dane took part in the hostilities. The estimate for the battle of Als was, as noted, that 6% of the male inhabitants took part in the battle against the aggressors. The warriors may have had much influence on society, but there were certainly other groups in society, such as the people working in the fields or tending to livestock. They may not have made much imprint on the relics passed on to the present, but ignoring them would distort the picture of this society, its living conditions for the average person, and its resource consumption.

*Figure 4.40. Decorated stone piece (jar cover) found at Maltegårdens Mark (Sealand; Kaul, 2004; photo by Danish National Museum).*

The very abundant rendering of men with large penises on the petroglyphs has been interpreted as related to a worship of fertility (Kaul, 2004). Figure 4.40 (from the cover of a jar) shows such a male together with a female, both stretching out to embrace each other. The suggestion by Kaul that the cover may belong to a grave jar-offering and have something to do with a fertility cult seems unlikely. The phallus motive is not otherwise seen on grave ornaments, and fertility would be of little importance for someone already dead. In petroglyphs such as the one shown in Fig. 4.39, even the fighting men are equipped with an erected penis*. This may suggest that there is no mythological significance intended and that the penis stroke is just a convenient way to

* One wonders if the use of paint (white, yellow or red) on petroglyphs reflects actual discovery of paint remains in cavities of the carved stones, before such traces were hidden or destroyed by contemporary repainting.

identify a stick figure in the picture as male. The stylized iconography of the petroglyphs may already suggest a first step towards a (hieroglyph-type) written language.

Current interpretation of artefacts from the Stone Age is, of course, influenced by the framework set by the structure and ideologies of present societies, despite scientific efforts to maintain objectivity. It is therefore hardly surprising that religious interpretations have been offered for many of the finds made. As argued earlier in this chapter, religion probably did not exist before the change in social stratification taking place sometime into the agricultural period. It is important in this discussion to distinguish between mythology and religion. I use the concepts as follows:

*Mythology* is speculation on the origin of the things we see around us, and on life and death. Myths are the results of human reflections over these issues, of having discussed them with friends over the settlement fire, and having passed the thoughts from generation to generation (e.g. in the form of legends). For this reason, myths may be transmitted to later generations by way of a substantial number of similar artefacts. Still, such fables and beliefs are basically creations and possessions of individual human beings.

*Religion* is a set of beliefs that may be of mythological origin, but which has been canonized into rules that can be imposed by an institutional authority. While mythological thoughts are open to change if further reflections suggest so, religious beliefs are set packages, which are not open to amendment or scientific inquiry: they must be accepted as they are, possibly with a threat of repercussions if questions are posed. Religion has a priesthood, an institution capable of collecting fees and of punishing people who do not comply, whether by decapitation (crusades, inquisition) or by psychological manipulation (creation of guilt complexes, social exclusion). The most important distinction between personal beliefs (the right to hold such beliefs is currently "guaranteed" by the United Nations Declaration of Human Rights; cf. Chapter 12) and religion is the presence in the latter case of a power-exercising institution, overseeing that rules and dogmas are not altered or neglected, except by decree from the institution.

Possible early signs of the introduction of religion in Denmark are the so-called cult houses. When they were first introduced in the agricultural period they may have served primarily practical purposes, made necessary by an increased population that no longer permitted the building of a complete dolmen for each dead person. The dead bodies were prepared for the stay in a collective passage grave or larger mass grave, by removal of body parts that would develop odour or attract decomposing animals, making reuse of the grave chamber unpleasant. This meant creating a profession similar to the Egyptian embalmers and mummification specialists. Such professions usually introduce rituals as a way to boost their fees, and ceremonies taking place in the cult houses where the undertakers performed their preparation of the dead bodies could have been the precursors of church services. Then as later, the cult structures were often stone constructions standing out from the wooden houses where people spent the active part of their lives. Creating solemn rituals and demanding high fees have always been cornerstones of the funeral profession, and employment of shamans or priests would fit nicely into the cult house activities. Clearly, this fable of a business-driven introduction of cult institutions is only one possibility, the obvious other one being religion created by a power elite needing to prop up its image in order to get the "common" people to accept slave work without compensation. In the Bronze Age, cult houses such as the one found at Sandagergård (Sealand; Kaul, 2004) became even more impressive. The line of rationalizing goes like this: recognizing that farming requires hard labour that is unlikely to be offered voluntarily, and wanting a layered society where a few have a rich life, there must either be others persuaded or forced to contribute without sharing the rich life, or the totality of wealth must grow in absolute terms and be shared. The first case would be served by introducing a religious institution promising rewards such as a second life, with better conditions after death, in compensation for toil in the first life, and adding some kind of punishment for those who do not buy the idea.

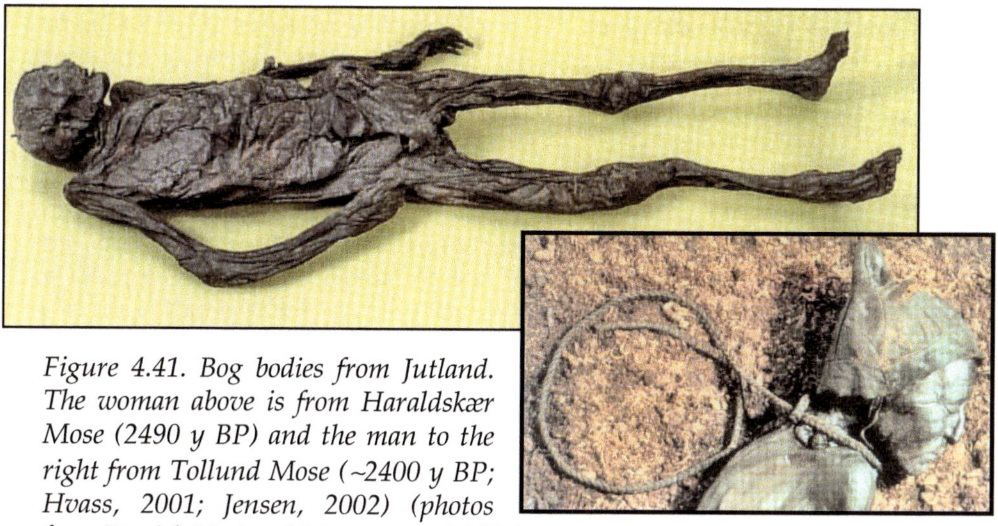

*Figure 4.41. Bog bodies from Jutland. The woman above is from Haraldskær Mose (2490 y BP) and the man to the right from Tollund Mose (~2400 y BP; Hvass, 2001; Jensen, 2002) (photos from Danish National Museum and Silkeborg Museum & Library).*

Evidence for introduction of punishment for those not conforming to the rules set by the rulers and their religious institutions are becoming abundant from the Bronze–Iron Age transition. Having noted the rise of a warrior class it is natural to think of the rulers being either the military leaders and people supported by them, or lay leaders who could afford to buy military support. The body of a woman (perhaps tortured to a drowning death by being fastened to the bottom of a bog by four clamps over arms and legs) is shown in Fig. 4.41, as well as a man hanged and disposed of in a bog. It seems that bogs were not used to bury law-abiding citizens. For the woman, an alternative explanation is that she was killed before being placed in the bog, and that the clamping was only to prevent the body from floating away (Hvass, 2001). In that case it is strange that no other bog corpses have been tied to the bottom. Tacitus (98) tells of the southern Germans that they (500 years later) had courts where traitors were condemned to hang and those "dishonouring their body" to drown. The possible torturing of the woman suggests a sexual motive (sleeping with the wrong man) or practice of witchcraft (such as herbal healing), rather than violation of a civil code, such as theft. The introduction of execution and other punishment signals a new type of society. However, the total fraction of people killed in this way can be assumed to be modest. That fairly many bog corpses are found may be because the bogs preserve bodies so well while other sites of burial do not, especially as cremation had become a common procedure in the Iron Age (2.5–1.2 ky BP).

In contrast to bronze, it was possible to mine iron within Denmark, using the bog iron ores abundant, as the name says, in bog areas. The extraction of iron involves a complex series of steps (Fig. 4.42), for which signs appear after 2.5 ky BP. As with bronze, import of iron tools started long before domestic production was mastered. Bog iron comes from low-quality deposits formed by biochemical processes on the basis of dissolved iron-containing substances carried by water streams.

In many ways, the Iron Age society became much more complex, exhibiting more division of labour and more insights into natural phenomena. The increased trade with the rest of Europe brought such ideas to Scandinavia, but often they were adapted to local conditions. The petroglyphs show some of these advances, but they have probably appeared in other, less durable forms. For instance, written number systems began to develop, as indicated in the calendar shown in Fig. 4.43. Groupings of simple counts point to the use of a lunar calendar with 4 × 7 days in a month. The choice of grouping counts in 7-day weeks is not based on any natural phenomenon or known Nordic mythology and so is most likely to have been imported from the Mediterranean calendars.

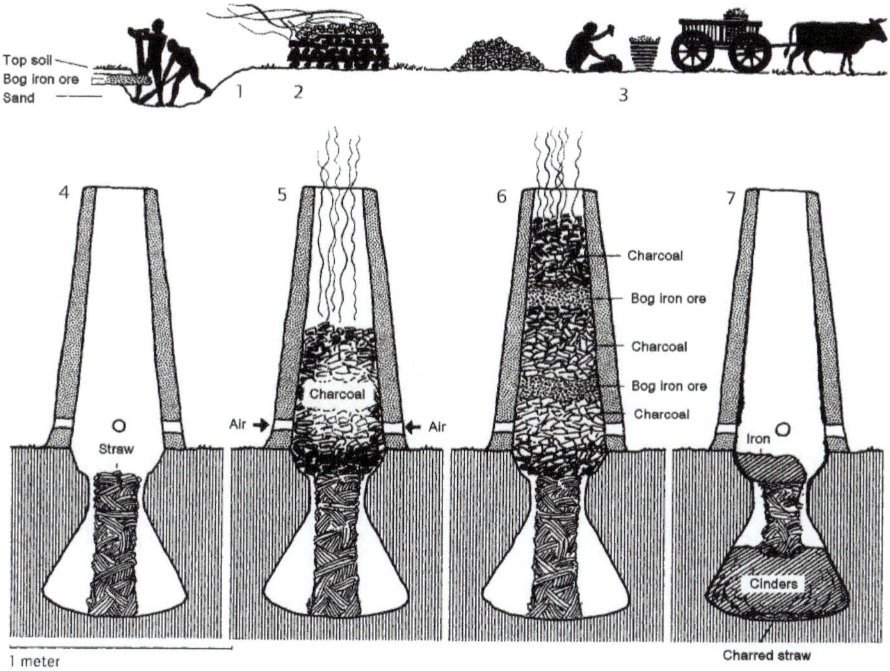


*Figure 4.42. Steps in transforming bog iron ore into useful iron. 1: mining; 2: ore roasting; 3: crushing; 4–7: oven operation. The top of the oven is then hacked away and the iron conglomerate removed. The remaining cinders in the bottom have been found in several places (Jensen, 2003; drawing by F. Bau, Danish National Museum).*

*Figure 4.43. Bronze Age petroglyph with calendar operated by a female (moon or menstruation cycle?), found at Aspeberget (western Sweden; Kaul, 2004, Danish National Museum).*

Most people were associated with agriculture and lived in villages, typically consisting of 10–40 rather similar houses, and from 2 ky BP most of them were inside a wall or fence. Some of the houses were occupied by craftsmen such as blacksmiths or potters, and the villages were often placed near deposits of raw materials such as sand, clay and bog iron (e.g. Hodde village in Jutland, occupied 2150–1600 y BP; Jensen, 2001b). A grave from an earlier settlement is preserved and respectfully integrated into the village layout by the Iron Age inhabitants.

*Figure 4.44. Clothes and hair fashions.* (a) *Bronze Age, Egtved grave (Munksgaard, 1974; Jensen, 2002; ~3.5 ky BP). The chord skirt is found in Bronze Age graves and on figurines. Open-air exhibit of the navel (during summer) seemed to have been as popular then as today.* (b) *Iron Age dress codes. Blue dress from grave at Lønne Hede, two grey dresses from Huldremose (woman killed and mutilated, 1945 y BP), man's dress from Søgård Bog, ca. 1650 y BP (reconstruction/photo by Silkeborg Museum & Library).* (c) *Decapitated woman's head from Stidsholt Bog. The hair is set by threads like in* (a) *(Jensen, 2003; photo by L. Larsen). Photos* (a) *and* (c) *from Danish National Museum.*

The Danish Iron Age was influenced by the presence of the Roman state (later empire), not very far from the southern border. A class of wealthy merchants, squires and warlords was established. They traded with like-minded people in many parts of Roman Europe, as evidenced by artefacts found (for example the Gundestrup kettle, made from 9 kg silver with relief plates of Thracian motives; Jensen, 2003). This new Danish upper class wore clothes more sophisticated than those of the farming people, and women seemed to have more elaborate hairstyles, although it is often difficult to judge to what extent the grave finds are representative (Fig. 4.44). Probably most Iron Age graves found represent upper class people. Whether people thrown into the bogs were lower class, or whether morality laws pertained equally or even more to upper class violators, are not known. Some of the bog bodies appear to come from upper classes (Fig. 4.44c). That remains uncertain, and the exact purpose of the many executions is not known. Grave gifts, where found, were equally splendid for male and female burials, possibly indicating an equal status of genders, again pertaining primarily to the upper class people. The sleeveless or short-sleeved women's two-part dresses found from both Bronze and Iron ages are not common further south in Europe. Since Danish temperatures were only about a degree higher than today (Fig. 4.2), one is tempted to conclude that Danish women were sturdy, but it is more likely that the dresses found are all summer or indoor wear, and that indoor temperatures were reasonably high as a result of large fireplaces and high activity levels. The dye used for the blue dress in Fig. 4.44b is extracted from the woad plant (*Isatis tinctoria;* Danish: vajd), at the time common in northern Europe.

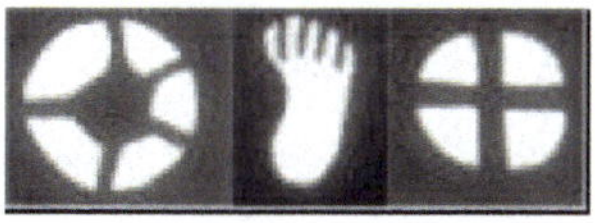

*Figure 4.45. Pottery marks (Jensen, 2003; Danish National Museum).*

Marks made on pottery (Fig. 4.45) and on rock carvings (e.g. Figs. 4.32 and 4.39) could, as suggested earlier, have been precursors for the construction of a picture-based written language. However, this did not happen, as writing based on the Roman alphabet became imported. The runes are basically Roman letters, adapted to carving into wood or stone (Fig. 4.46). Runic writing became the standard throughout the region from around the year 180 and was used mainly for short messages, such as noting the artist's name on a tool. A shortened version of the runic alphabet was subsequently in use, perhaps indicating that Danes were not equipped with any deep scholarly interest.

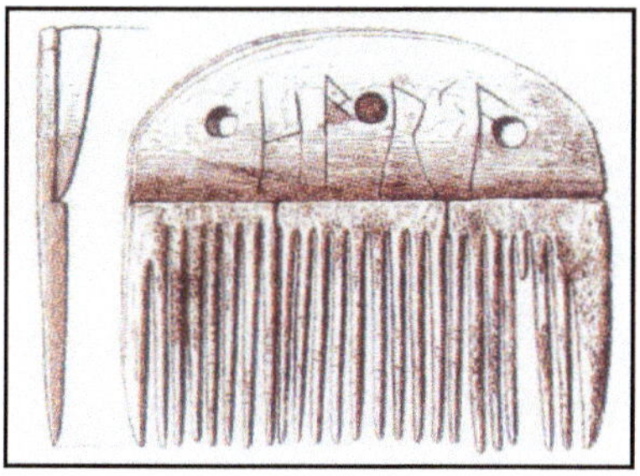

*Figure 4.46. The runic alphabet (above) and the oldest find with runic inscription ("Harja", Vimose, year 150, left). The inscription could be the name of the maker of the comb or its user (Jensen, 2003; drawings by L. Hammelev, Danish National Museum; J. Petersen, Moesgaard Museum).*

A more direct import from the Mediterranean region is the game set and dice shown in Fig. 4.47. Often found together with weapons, they were probably pastime tools for soldiers. Even today, many computer games have military connotations.

*Figure 4.47. Third century game-playing tools found at Vimose (Funen; Jensen, 2003; photo by L. Larsen, Danish National Museum).*

Additional findings consistent with the assumption of an increased militarizing of the Danish society are helms, horns, lures and horses with elaborate harnesses. The story of the Danish Iron Age will be continued in the following chapter, where the implications of the many new activities for population size and energy use will be analysed in a follow-up of the timeline perspective started in Chapters 2 and 3.

***References:***

Aaby, B., 1993. Menneske og natur. *Da klinger i muld* (S. Hvass, B. Storgaard, eds), pp. 16–19. Kgl. Nordisk Oldtidsselskab, Jysk Arkæologisk Selskab, sold through Aarhus Universitetsforlag.

Achilli, A., *et al.*, 2004. The molecular dissection of mtDNA haplogroup H confirms that the Franco-Cantabrian glacial refuge was a major source for the European gene pool. *American J. Human Genetics* **75**, 910–918.

Ammerman, A., Cavalli-Sforza, L., 1984. *The Neolithic transition and the genetics of populations in Europe*. Princeton University Press, Princeton, USA.

Andersen, N., 1999. Sarup bopladsen. *Jysk Arkæologisk Selskabs Skrifter* **33** (2), Aarhus.

Armelagos, G., Harper, K., 2005. Genomics at the origins of agriculture, part two. *Evolutionary Anthropology* **14**, 109–121.

Barbujani, G., Bertorelle, G., Chikhi, L., 1998. Evidence for Paleolithic and Neolithic gene flow in Europe. *American J. Human Genetics* **62**, 488–491.

Barrett, R., Kuzawa, C., McDade, T., Armelagos, G., 1998. Emerging and re-emerging infectious diseases: The third epidemiologic transition. *Annual Rev. Anthropology* **27**, 247–271.

Becker, C., 1952. Skeletfundet fra Porsmose ved Næstved. *Nationalmuseets Arbejdsmark.* Årsskrift, Copenhagen.

Beja-Pereira, A., *et al.*, 2006. The origin of European cattle: Evidence from modern and ancient DNA. *Proc. National Academy of Science USA* **103**, 8113–8118.

Bennike, P., 1985. *Palaeopathology of Danish skeletons. A comparative study of demography, disease and injury.* Akademisk Forlag, Copenhagen.

Bennike, P., 1993. Menneskene. *Da klinger i muld* (S. Hvass, B. Storgaard, eds), 34–39. Kgl. Nordisk Oldtidsselskab, Jysk Arkæologisk Selskab, sold through Aarhus Universitetsforlag.

Bentley, A., Chikhi, L., Price, T., 2003. The Neolithic transition in Europe: comparing broad scale genetic and local scale isotopic evidence. *Antiquity* **77**, 63–66.

Beyneix, A., 2007. Réflexions sur les débuts de la guerre au Néolithique en Europe occidentale. *L'Anthropologie* **111**, 79–95.

Biraben, J-N., 2003. The rising numbers of humankind. *Population & Societies* **394**, 1–4.

Boaretto, E., *et al.*, 2009. Radiocarbon dating of charcoal and bone collagen associated with early pottery at Yuchanyan Cave, Hunan Province, China. *Proc. Nat. Acad. Sci.* **106**, 9595-9600.

Bocquet-Appel, J-P., Naji, S., Linden, M., Kozlowski, J., 2009. Detection of diffusion and contact zones of early farming in Europe from the space-time distribution of 14C dates. *J. Archaeological Sci.* **36**, 807–820.

Bogaard, A., Heaton, T., Poulton, P., Merbach, I., 2007. The impact of manuring on nitrogen isotope ratios in cereals: archaeological implications for reconstruction of diet and crop management practices. *J. Archaeological Sci.* **34**, 335–343.

Bradley, R., 2006. Danish razors and Swedish rocks: Cosmology and the Bronze Age landscape. *Antiquity* **80**, 372-386.

Bro-Jørgensen, M., 1996. Rageknivskunst. *Skalk* (1), 30–31.

Coles, J., 2003. And on they went … processions in Scandinavian Bronze Age rock carvings. *Acta Archaeologica* **74**, 211–250.

Davison, K., Dolukhanov, P., Sarson, G., Shukurov, A., 2006. The role of waterways in the spread of the Neolithic. *J. Archaeological Sci.* **33**, 641–652.

Diamond, J., 2002. Evolution, consequences and future of plant and animal domestication. *Nature* **418**, 700–707.

Digby, S., 1954. Boats and ships. *A History of Technology* (C. Singer, E. Holmyard, A. Hall, eds), p. 735. Clarendon Press, Oxford.

Earle, T., 2000. Archaeology, property, and prehistory. *Annual Rev. Anthropology* **29**, 39–60.

Edwards, C., *et al.*, 2004. Ancient DNA analysis of 101 cattle remains: limits and prospects. *J. Archaeological Sci.* **31**, 695–710.

Fischer, A., Olsen, J., Richards, M., Heinemeier, J., Sveinbjörnsdóttir, Á., Bennike, P., 2007. Coast-inland mobility and diet in the Danish Mesolithic and Neolithic: evidence from stable isotope values of humans and dogs. *J. Archaeological Sci.*, **34**, 2125–2150.

Gimbutas, M., 1963. *The Balts.* Thames & Hudson, London. Available at http://www.vaidilute.com/books/gimbutas

Glob, P., 1951. Helleristninger I Danmark. *Jysk Arkæologisk Selskabs Skrifter* **7**, Århus.

Haak, W., *et al.*, 2005. Ancient DNA from the first European farmers in 7500-year-old Neolithic sites. *Science* **310**, 1016–1018.

Hassan, F., 1981. *Demographical Archaeology.* Academic Press, London.

Hazelwood, L., Steele, J., 2004. Spatial dynamics of human dispersal. Constraints on modelling and archaeological validation. *J. Archaeological Sci.* **31**, 669–679.

Hummel, R., 2004. *Understanding Materials Science.* Springer, Berlin.

Huntley, B., 1990. European post-glacial forests: compositional changes in response to climatic change. *J. Vegetation Sci.* **1**, 507–518.

Hvass, L., 2001. *Oldtiden I Danmark. Jernalderen.* Sesam, Copenhagen.

Iversen, J., 1941. Land occupation in Denmark's stone age. Dansk Geologisk Selskab, Serie II, No. 66. Copenhagen.

Jensen, J., 2001a. *Danmarks oldtid. Stenalder.* Gyldendal, Copenhagen.

Jensen, J., 2001b. *Oldtiden I Danmark. Bronzealderen.* Sesam, Copenhagen.

Jensen, J., 2002. *Danmarks oldtid. Bronzealder.* Gyldendal, Copenhagen.

Jensen, J., 2003. *Danmarks oldtid. Ældre Jernalder.* Gyldendal, Copenhagen.

Jørgensen, M., 1982. To jyske bronzealderveje – og en ny metode til arkæologisk opmåling. *Nationalmuseets Arbejdsmark* , 142–152.

Jørgensen, M., 1993. Bronzealderens veje. *Da klinger i muld* (S. Hvass, B. Storgaard, eds), p. 144. Kgl. Nordisk Oldtidsselskab, Jysk Arkæologisk Selskab, sold through Aarhus Universitetsforlag.

Kaul, F., 1998. *Ships on bronzes: a study in Bronze Age religion and iconography.* Studies in archaeology & history, vol. 3, from the National Museum of

Denmark, Copenhagen.

Kaul, F., 2004. *Bronzealderens religion.* Studier af den nordiske bronzealders ikonografi. Det Kongelige Nordiske Old-skriftselskab, Copenhagen.

Keeley, L., 2004. Archaeology of conflict and war. *Int. Encyclopedia of Social and Behavioral Sciences*, 2537–2542. Elsevier.

Kerr, T., Swindles, G., Plunkett, G., 2009. Making hay while the Sun shines? Socio-economic change, cereal production and climate deterioration in early medieval Ireland. Preprint.

Králík, M., Novotný, V., Oliva, M., 2002. Fingerprint on the Venus of Dolní Věstonice I, *Anthropologie* **40** (2), 107–113 (publisher: Moravské zemské muzeum, Brno, Czech Republic).

Larsen, C., 1995. Biological changes in human populations with agriculture. *Annual Rev. Anthropology* **24**, 185–213.

Lutz, W., 1995. The colonisation of Europe and our Western diseases. *Medical Hypotheses* **45**, 115–120.

Madsen, T., 1990. Manddrab. Picture in *Oldtidens ansigt* (Kjærum, P., Olsen, R., eds). Kgl. Nordiske Oldskriftselskab and Jysk Arkæologisk Selskab.

Malinowsky, B., 1922. *Argonauts of the Western Pacific. An account of native enterprize and adventure in the archipelagoes of Melanesian New Guinea.* Dutton & Co, New York.

Malmer, M., 1992. Weight systems in the Scandinavian Bronze Age. *Antiquity* **66**, 377.

Moore, P., 2005. Down to the woods yesterday. *Nature* **433**, 588–589.

Munksgaard, E., 1974. *Oldtidsdragter.* Danish National Museum, Copenhagen.

Nationalmuseet, 2001. Det Kulturhistoriske Centralregister. Transferred to Kulturarvsstyrelsen and available at http://www.dkconline.dk, although geared to serve construction contractors with information on specific locations, rather than general scientific searches and queries.

Nehlich, O., Montgomery, J., Evans, J., Schade-Lindig, S., Pichler, S., Richards, M., Alt, K., 2009. Mobility or migration: a case study from the Neolithic settlement of Nieder-Mörlen (Hessen, Germany). *J. Archaeological Sci.* **36**, 1791–1799.

Nelson, A., 2009. Fartyg före Vikingatiden. Webpage http:// axelnelson.com/skepp/prewiking.html

Nielsen, P., 1999. Limensgård and Grødbygård. Settlements with house remains from the early, middle and late Neolithic on Bornholm. Settlement and landscape. *Proc. Conf. of the Jutland Archaeological Society*, 149–165, Århus (reproduced in Jensen, 2001a).

Nielsen, P., 2001. *Oldtiden I Danmark. Bondestenalderen.* Sesam, Copenhagen.

Noe-Nygaard, N., Price, T., 2005. Diet of aurochs and early cattle in southern Scandinavia: evidence from $^{15}$N and $^{13}$C stable isotopes. *J. Archaeological Sci.* **32**, 855–871.

Odgaard, B., Nielsen, A., 2009. Udvikling i arealdækning i Danmark i perioden 0-1850. *Danske landbrugslandskaber gennem 2000 år* (Odgaard, B., Rydén Rømer, J. (eds), pp. 41–58. Aarhus Universitetsforlag, Aarhus.

O'Sullivan, P., 2008. The "collapse" of civilizations: what paleoenvironmental reconstruction cannot tell us, but anthropology can. *The Holocene* **18**, 45–55.

Piperno, D., Weiss, E., Holst, I., Nadel, D., 2004. Processing of wild cereal grains in the Upper Palaeolithic revealed by starch analysis, *Nature* **430**, 670–673.

Plutarch, M., ~100. *The Parallel Lives. Vol. 9: The Life of Marius.* English translation: the Loeb Classical Library edition, 1920. http://penelope.uchicago.edu

Rasmussen, F., 2005. Mid- to late-Holocene land-use change and lake development at Dallund Sø, Denmark: vegetation and land-use history inferred from pollen data. *The Holocene* **15**, 1116–1129.

Rasmussen, M., Adamsen, C., 1993. Bebyggelsen. *Da klinger i muld* (S. Hvass, B. Storgaard, eds), 136–141. Kgl. Nordisk Oldtidsselskab, Jysk Arkæologisk Selskab, sold through Aarhus Universitetsforlag.

Renfrew, C., 2000. At the edge of knowability: towards a prehistory of languages. *Cambridge Archaeological J.* **10**, 7-34.

Richards, M., 2003. The Neolithic invasion of Europe. *Annual Rev. Anthropology* **32**, 135–162.

Richards, M., *et al.*, 2000. Tracing European founder lineages in the Near Eastern mtDNA pool. *American J. Human Genetics* **67**, 1251–1276.

Rosser, Z., *et al.*, 2000. Y-chromosomal diversity in Europe is clinal and influenced primarily by geography, rather than by language. *American J. Human Genetics* **67**, 1526–1543.

Rostholm, H., 1977. Neolitiske skivehjul fra Kideris og Bjerregårde I Midtjylland. *Kuml*, 185–222.

Schovsbo, P., 1987. *Oldtidens vogne i Norden.* Bangsbomuseet, Denmark.

Sharpe, E., 1992. Scandinavia. *Mythology* (R. Cavendish, ed.), pp. 178–187. Barnes & Noble, New York.

Shortland, A., 2007. Who were the glassmakers? Status, theory and method in mid-second millennium glass production. *Oxford Journal Archaeology* **26**, 261–274.

Smith, F., 1994. Samples, species, and speculations in the study of modern human origins. ***Origins of Modern Humans*** (M. Nitecki and D. Nitecki, eds). Plenum, New York.

Sørensen, B., 1995. History of, and recent progress in, wind-energy utilization. *Annual Rev. Energy & Environment* **20**, 387–424.

Strömberg, M., 1990. Människobilder i Kivikgraven. Figure 17 in *Faces of the Past* (P. Kjærum, R. Olsen, eds). Kongelige Nordiske Oldskriftselskab & Jysk Archaeologisk Selskab.

Sturleson, S., ca. 1200. *The Younger Eddas,* 10–12. Translation from Old Norse: I. Blackwell, http://www.Gutenberg.org E-Book #14726.

Swedish Rock Art Research Archive, 2010. Web resource at http://www.shfa.se/bild

Tacitus, 98. *Germania.* Translation "Moeurs des ancien Germains" (l'Abbé Legendre), http://www.Gutenberg.org E-Book #19662.

Torroni, A., *et al.*, 1998. mtDNA analysis reveals a major late Paleolithic population expansion from Southwestern to Northeastern Europe. *Americam J. Human Genetics* **62**, 1137–1152.

Vandiver, P., Soffer, O., Klima, B., Svoboda, J., 1989. The Origins of Ceramic Technology at Dolni Věstonice, Czechoslovakia. *Science* **246**, 1002–1008.

Vandkilde, H., 2006. Archaeology and war: Presentations of warriors and peasants in archaeological interpretations. *Warfare and society* (T. Otto, H. Thrane, H. Vandkilde, eds), pp. 57–73. Aarhus University Press, Aarhus.

Vebæk, C., 1980. Trææske med runeindskrift. No. 21 in *Danefæ* (P. Glob, ed.). Det kongelige Nordiske Oldskriftselskab & Jysk Arkæologisk Selskab. Danish National Museum, Copenhagen.

Vencl, S., 1999. Stone age warfare. *Ancient warfare. Archaeological perspectives* (J. Carman, A. Harding, eds), pp. 57–72 (references at end of book). Sutton Publishing, Stroud.

Williams, M., 2008. A new look at global forest histories of land clearing. *Annual Rev. of Environment and Resources* **33**, 345–367.

Wolfe, S., Stanley, J., 1980. Linguistic problems with patriarchal reconstructions of Indo-European culture: A little more than kin, a little less than kind. *Women's Studies Int. Quarterly* **3**, 227–237.

Wright, E., 1990. *The Ferriby boats: Seacraft of the Bronze age.* Routledge, London.

Zeder, M., Emshwiller, E., Smith, B., Bradley, D., 2006. Documenting domestication: the intersection of genetics and archaeology. *Trends in Genetics* **22**, 139–155.

Zohary, D., Hopf, M., 2000. *Domestication of Plants in the Old World.* Oxford Science Publications, Oxford.

# 5

# Migration and exploration by land and sea

The Danish relation to the Roman empire formed during the final centuries before the currently used "year-zero" (2000 y BP) was not just one of trade. A major relocation of both warriors and their entire families took place in 2114 or 2113 y BP. The tribes participating were the Cimbrians (believed to come from Himmerland in mid-Jutland, because that name figures on a map of Jutland drawn by Ptolemaïos, 150), the Teutons, possibly coming from Thy in northwest Jutland or from the German Baltic coast; and the Ambrons (maybe from Schleswig south of Jutland; Fig. 5.1). Why they chose to move south is unknown. Natural catastrophes and famine have been suggested, but no concrete evidence is available. Temperature oscillations were fairly strong for a period of several hundred years preceding this time (cf. Fig. 4.2) and sea surface temperatures were 4 °C warmer than in adjacent times (Sarnthein *et al.*, 2003), with a corresponding peak in plankton production (Hebbeln *et al.*, 2006). Pollen in a Danish inland lake on the island of Funen does not show any anomaly (Rasmussen, 2005).

The migrating party itself said that it was fleeing from a tsunami, an explanation that the Turkish/Greek/Roman geographer Strabo dismissed by stating that "everybody knew" that flood waves were slow and predictable (Strabo, ~20, evidently only knowing of tides). Tsunamis would in fact not be uncommon along the shores of the North Sea, due to major Arctic ice blocks being occasionally dumped into the sea (rather than caused by underwater quakes). Impacts of tsunamis have been found in the Shetland Islands, at 5.5 and 1.5 ky BP (Bondevik *et al.*, 2005). The impacts of, say, Greenland ice shedding on other shores could have occurred at other times, such as that of the exceptional sea surface warming some 2.2 ky BP, mentioned above. Settlement evidence shows that Jutland was far from totally depopulated by the migrational exodus, but it appears reasonable to accept the tsunami explanation given by the people involved, suggesting that the tsunami had hit certain areas of Jutland and made life difficult.

The exodus from Jutland may have been joined by Germanic tribes on the way along the Elb, Oder and Danube rivers to Pannonia (northwest of Beograd). Here the party was about to found a major settlement, *Teutoburgium* (2113 y BP, at current Dalj), which, however, they left after repeated attacks and battles with local Scordisci tribes (Fig. 5.2).

They then moved east and met the Roman army at Noreia (in present Austria), still in 2113 y BP. The Nordic migrants sent an envoy (with gifts) to ask the Roman Senate for a piece of land where they could settle, but the head of the Roman army, Gnæus Papirius Carbo, double-crossed them and tried to lure them into a trap. Instead, he himself and his army were slain (Book 63 of Livius's *History of Rome*; Livius, -1). The Nordic alliance, said by the defeated Romans to count several hundred thousands, moved on to Gallia Transalpina (present east France*), saddened that the Romans were no longer their friends when they did not come to trade.

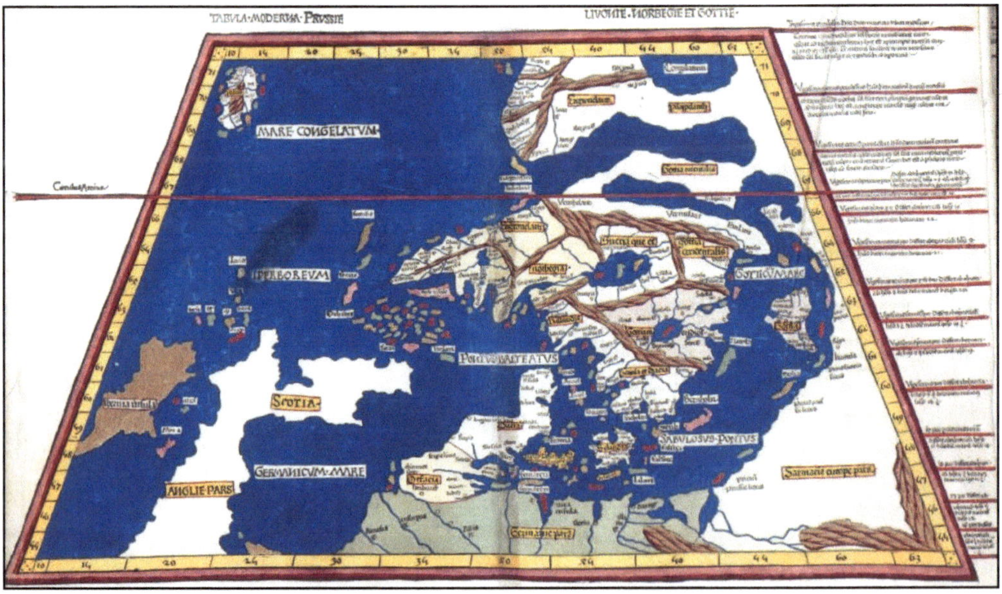


*Figure 5.1. A 1496 attempt at restoring Ptolemaïos's map (~150) of northern Europe (see Lang, 1955). The map shown appeared at Ulm in* Atlas Cosmografica *and was redrawn several times over the next century (this picture from a web exhibition of the Danish Royal Library in Copenhagen, 2010).*

In 2108 y BP, the Romans, led by consul Marcus Junius Silanus, were again defeated at a location west of present Lyon; another plea for permission to settle was sent to the Senate in Rome and again refused. The

* The Alpine part of present Italy was also called Gaul, but Gallia Cisalpina.

Senate was afraid to receive new settlers after a negative experience with earlier people settling in northern Italy. In fact they were so afraid that the Nordic tribes might march directly towards Rome that they now sent not just one but two of their armies to France to fight the migrants.

The battles took place in 2105 y BP at Arausio (present Orange), where Cimbrian leader *Boiorix* massacred both Roman armies, said to muster 80 000 soldiers and 40 000 servants, helped by a lack of co-operation between the two Roman leaders, Gnaeus Manlius and Quintus Servilius Caepio (Livius, -1, Book 67). The Roman armies were placed at two locations along one shore of the river Rhône, and when they were driven into the river it turned out that they could not swim.

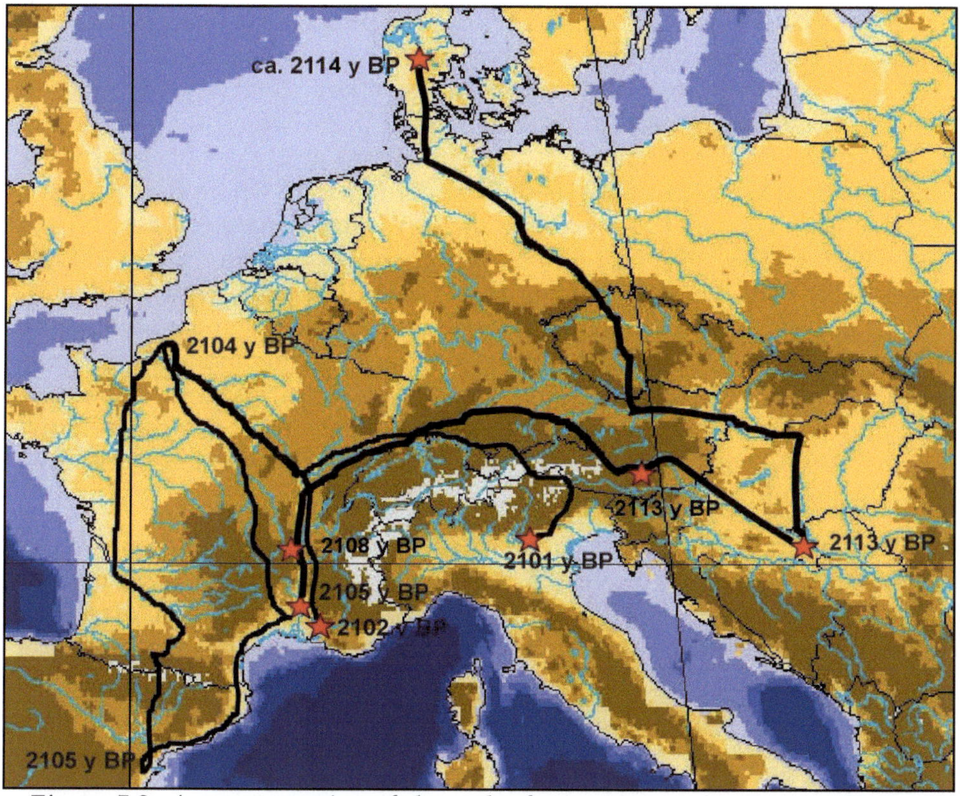


*Figure 5.2. A reconstruction of the path of migration undertaken by Cimbrians, Teutons and Ambrons from about 2114 to 2101 y BP, ending in defeat and survivors taken prisoners by the Romans.*

Consul Caius Marius (celebrated in Rome for having caught the North African commander Jugurtha) had the Romans spend the next years completely reorganizing their army structure, forming a new, professional army of mercenaries arranged in a hierarchical structure (legions, cohorts, centurions and maniples; Kinder and Hilgemann, 1964). This new army was trained during the following years, delighted that the Nordic warriors did not march straight against Rome, and probably being given swimming lessons. Meanwhile, the Nordic campaign had broken up to search for land accepting immigration. Evidently they did not want to settle in a place where they were not welcome. The Cimbrians moved to Spain (Barcelona area) and from there to northern France and Belgium, while the Teutons and Ambrons went directly to the latter areas (see Fig. 5.2). In 2102 y BP they all moved back to the Rhône area, maybe feeling that this was after all the most hospitable place they had been. It is highly likely that their numbers were substantially reduced after some 12 years of migrational travel, and strange that they again chose to split up: the Ambrons and Teutons marched to Massilia and were beaten by the new Roman army at Aquae Sextiae (present Aix). The Cimbrians went through Helvetiere (Switzerland/Austria) and crossed the Alps into present Veneto, from which they moved in the direction of Mediolanum (present Milan) and were beaten by the Romans at Vercellae (there is doubt about exactly where that was and whether it is related to a present village of similar name). According to Livius (-1, Book 68; repeated by Plutarch, ~100), there were 200 000 Ambrons and Teutons killed, and 160 000 Cimbrians, with another 90 000 and 60 000 taken prisoner. We do not know how much Livius may have exaggerated. Plutarch expands on the character and physiognomic description of the Nordic "barbarians": fierce, fearless, tall, blue-eyed, blond and behaving like children (e.g. sliding naked down the snow-clad Alpine hillsides on their shields). It may be folklore, but most Danes would readily accept this description.

The length of travel undertaken by the Nordic tribes is difficult to understand, unless, as suggested, they really did not want to be considered enemies. This fits very poorly with the views of Roman writers, who considered the Nordic campaign as a series of assaults on local people, stealing food and jewellery from them. Evidently far from everyone in the migrant group was up to the exhausting trip, and as mentioned the size of the group would have diminished along the way. The first stop at Teutoburgium would be consistent with them looking for a suitable place to settle, but the further long march appears increasingly aimless. In a longer historical perspective it has been suggested that they made several contacts across Europe that eased interaction through the following centuries (Jensen, 2003). Maybe north European bonds were strengthened, but there had been trade relations all over Europe long before, and people in the Alpine, Po and Rhône regions were probably at this time extremely tired of the many wars the Romans had fought on their territory (e.g. as part of the Punic wars). A direct effect of the large Nordic migrational expedition would naturally be lowering the population in the regions of origin, notably Jutland. On the other hand, the Cimbrians, Teutons and Ambrons captured by the Romans, although first used as slaves and prostitutes, would later have become integrated into the Italian population, which even today sports a remarkable number of blond and blue-eyed (a recessive trait) people. This perspective on the story is told in a novel by Jensen (1922).

Unfortunately, only one side of the tale is available, and the Roman authors seem more interested in battles and wars, so not much is learned about the motivations of the migrants from the north. It seems certain, however, that they came with the entire family as refugees forced from their homes in Jutland by events that are poorly known, and that their primary wish was to be given a new piece of land where they could settle. The Romans were not the least bit open to migrants from other parts of the world. Their own migrations were always with a sword in the hand (the same could of course be said of the Nordic tribes, but to be used only if diplomacy failed). Human history has many examples of massive relocations based on lack of satisfaction with the present lot, sometimes diffusely defined and sometimes based only on short-term setbacks, such as the tsunami proposal. Dreams of a new land with improved living conditions have often been a driving factor, although reality has not always lived up to the dreams. Migrants have rarely been welcome, but in some cases (such as Spanish and English settlements in South and North America, and later Australia), the indigenous populations were too weak to prevent their land from becoming colonized.

One may suggest an inherent desire for exploration and a longing for new horizons, probably associated with the nomadic life of early hunters. When agriculture became necessary, man must have felt restrained by the demands for staying around the farm and its livestock animals. As suggested in Chapter 4, agriculture also brought about social layering, the concept of "property" and military organization to protect one's property. The new waves of migration were in part assaults on land occupied by others, with battles as needed for incorporating such land into one's own "property" and raising status (plus laying the foundations for the next war). This was the reason Romans were constantly involved in several wars, and similar behaviour had characterized their predecessors (e.g. in Mesopotamia, Egypt, Greece and Carthage). However, another cause originated outside the traditional cultural centres, as a result of population pressure or environmental setbacks. Examples are the Cimbrian campaign and the later major migrational moves into central Europe by people from the east. From the collapse of the Roman Empire, one sweep of intruders after the other assaulted central parts of Europe, sometimes joined by earlier migrants, such as the Goths originally from Sweden (Kinder and Hilgeman, 1964), the Normans from Denmark and Norway, or the Germanic tribes coming from areas just south of Denmark.

Danish hunter–gatherers would have moved north through western Sweden into Norway as soon as the ice withdrawal allowed it, around 10 ky BP (see Fig. 3.2 j and k). The oldest Norwegian stone tool finds are from Fosna (near Trondheim), dated to 9.6 ky BP. They are similar to imported tools from Ahrensberg in northern Germany, appearing during the Allerød period (Bibby, 1980). A claim has been made that Norway was not populated from Denmark through western Sweden, but by Sami people from North Siberia arriving through the Murmansk region. As noted in Chapter 3, a major strait stretching from Stockholm to Kattegat may have made the passage from Denmark difficult. However, a similar major Barents Sea inlet towards the Baltic would have made the passage to the north of Norway as difficult as the one through Bohus Län in Sweden. Support for the pathway through southwest Sweden comes from evidence for the relevant area of Sweden being populated already by 11.5 ky BP (Schmitt *et al.*, 2009). Only when the climate had improved further did permanent settlement appear in Norway. During the subsequent development, even after establishment of separate kingdoms (which happened later in Norway than in Denmark, for obvious geographical reasons), the people of the two countries continued to feel closely related, often collaborating in conquests in England and France, for example, and being friendly as often as they were in mutual war over their own territories.

The share of warriors in late Iron Age and early medieval societies may have increased from the 6% found (for a particular case) in Chapter 4 to over 10% during the peak assault and colonization periods, with a ruling class of warlords and later kings, surrounded by an elite group of nobility, hardly larger than 1% of the total population. The remaining population would have included various skilled people serving the elite, but still some 90% were people working as slaves or citizens with limited rights, primarily in agriculture. How many people had employment as servants to the elite and as workers in raw materials or construction (in mines, smelters, tools and weapons manufacture, ships, cities, roads, fortifications) is difficult to estimate, but probably about 5% or less, except in special circumstances (similar to the Great Wall construction in China, to take a non-European example). The classical

Middle East and Mediterranean civilizations probably had a higher fraction of non-farmers, because of their large cities, which must have required a sizeable amount of manpower related to infrastructure (water supply, sewage, etc.), construction and maintenance work.

During the 4th century, the introduction of the Christian religion in southern Europe signalled the beginning of what has been called "the Dark Age" because of the high level of mind control exerted by the institutions of the Christian Church. The Roman Empire began to collapse and warrior clans from eastern Europe and beyond (notably the Huns) raided the rich Roman provinces, causing entire cities to become abandoned. Scandinavian elites realized, if they hadn't before, that armies and weapons were the basis of wealth, and that they could easily behave as ferociously as the eastern tribes. Why trade with foreign cities if you can just raid them and take what you need? Fortunately, humans are more complex than that, and even what began as a plundering expedition in many cases changed and developed into a fairly benevolent relationship, including settlement in the region concerned and integration with existing populations. Danish Vikings were not only attackers, but sometimes called in to help the local population in fighting a "worse" intruder. This happened, for example, in Ireland (Bugge, 1904), as well as in England (Jensen, 2003), in both cases to help existing populations fight Norwegian Vikings.

*Figure 5.3. Horse harness buckle in the refined ornamental style characterizing the period around 600 (found at Veggerslev, east Jutland; Jensen, 2004; photo by K. Klein, Danish National Museum).*

The late Danish Iron Age is usually taken to comprise the period from 2500 to 1230 y BP (770; some authors use 800), followed by the Viking age lasting to about 1050. The late Iron Age saw a number of developments. Farmers gathered in fenced villages with a number of common facilities. Veritable towns came into being; the first was Hedeby in south Jutland (present Schleswig), serving as an important hub for all trade along the Baltic shores, just as Dorestad (near present Utrecht in Holland) did for the trade along the European continental west coast. Ships became larger and increasingly sophisticated, but still based on rowing rather than using sails. The hardware for the warriors must have created a substantial blacksmith profession. Horse-riding improved the efficiency of personal transport over land (Fig. 5.3). The attractiveness of sailing rather than crossing land on foot or horseback remained significant: the road journey from Hedeby to the Alps has been estimated to take 9 weeks, even in the 12th century (Olrik, 1891).

In art, the Nordic people developed unique ornamental styles. From the possibly Greek–inspired spiral motive of the early Bronze Age (Fig. 5.4a; also on the rim of the sun-dial in Fig. 4.30 or the spiky belt pieces worn by women, Fig. 4.44a) developed intricately interwoven snake-like motifs (regaining popularity in the early 20th century Jugend style). Examples from the Iron Age are in Figs. 5.4b and c.

Figure 5.5 shows a buckle from the Viking age, used to fasten a woman's braced dress. The interesting development is that the face depicted here is clearly female, with arranged knots of hair. The background uses a layered material with holes in the upper layer to achieve a more stunning decorative style. The religious transition during the Viking age took Danes from the Greek–inspired, happy, and, in behaviour, very human, family of gods with Odin as the head (Wotan in Germany) to the stern Catholic version of Jesus. The Christian Church made the concession to Rome and its earlier religions not to choose monotheism (as did Egyptian Farao Akneton, the Jews and, later, Islam), but to add Jesus as a second god rather than a prophet, then adding a third elusive god, the holy spirit, and finally all the half-gods called saints. In this way, the Christian religion became familiar and acceptable to the non-monotheists, albeit in plain contradiction to the teaching of the historical Jesus (Thomas, ~75).

Figure 5.6 illustrates the transition from Nordic to Christian religion, by a small charm depicting Odin, and the runic stone by which King Harald III Blåtand (Bluetooth) announced that he converted the Danes (or at least some of them) to Christianity. The stone is nominally commemorating his parents Gorm and Tyre, but both the picture side with

Jesus Christ and the text side of the stone deals mostly with Harald's own accomplishments.

*Figure 5.4a-c. (a) above: Early Bronze Age motif (Vognserup Enge, Sealand; Poulsen, 1980). (b) upper right: Iron Age harness buckle from Vendel Church, Uppland (about 600; Jensen, 2004). (c) lower right: Iron Age brooch with inlaid garnet stones from Bornholm (ca. 650, Jørgensen, 1990). The face in the upper middle is a commonly found motif; Jørgensen turns the plate 180 ° and sees a man with a divided beard in the middle, surrounded by two animal-like creatures. Photos by J. Poulsen (a), L. Jørgensen (b), both Danish National Museum, and (b): © J. Ljungkvist, Swedish Historical Museum, Inventory No. 9785:XII, gallery No. 225873.*

*Figure 5.5. Woman's dress fastener from Viking age, found at Ågerup, Sealand (Fuglesang, 1990). Photo by Danish National Museum.*

*Figure 5.6. (a) left: lead necklace charm depicting Odin, made in a Ribe (Jutland) workshop, likely used to make a production mould (ca. 730; Jensen, 1990; Photo by Moesgaard Museum). (b) right: stone erected ca. 975 at Jelling (Jutland) with runic inscriptions and Jesus without the cross, but with Viking style ornaments (Jensen, 2004; Photo by J. Jedbo, Danish National Museum).*

Denmark was united as a kingdom around 700, but little is known about the formation process. The early historian Saxo (ca. 1200) claims that "Danmark" is named after the first king, Dan, and that his brother Angul became king of and named England ("Angleterre"), in what appears as more a tale of legends than facts. He describes how the most powerful and conservative local magnates assembled at Isøre (by the Northern inlet of Roskilde Fiord on Sealand, today Rørvig) to elect each new Danish king and make him promise to adhere to existing law, a story corroborated by the English writer Niger (ca. 1160). During the late $19^{th}$ century, some Danish historians interpreted the Isøre councils as indicating an early form of democracy, but as already noted by Olrik (1899) this is hardly the case. Before 700, there is not much point in distinguishing between Danes and the inhabitants of northern Germany, Sweden or Norway. Østergaard (1994) constructs a list of Danish kings, starting by 720 with names found in annals from other parts of Europe. Significant early monarchs are Godfred the Great (804–810), Horik I (819–827), and, after Swedish occupation 895–917, Knud I (from the

Danish settlement in France; see below), who in 918 regained power and was followed by Gorm the Old (948–958), Harald III Blåtand (Bluetooth; 958–987) and Sven I Tveskæg (Swein Forkbeard; 987–1014).

Two features of the development during the Iron and Viking ages are important for estimating changes in energy use. One is the progress in technology, such as the highly increased sophistication of metallurgic processes (for example, the shiny patterns in the buckle shown in Fig. 5.4b are made by a technique called *niello*, based on applying an alloy containing sulphur, copper, silver and lead). The other is the population size, which is surprisingly poorly known for the time period in question. The beginning of registration of property such as farms and other buildings (for reasons of taxation, or counting the family plots given by the king to each of his warriors) gives one indication. Another is the number of churches. Both are useful only at the end of the Viking period, suggesting a population size of some 300 000 in Denmark proper (from the southwestern rampart "Danevirke", built around 700* west of Hedeby to defend the southern border, to the east border of the trans-Øresund areas currently occupied by Sweden). A simple derivation of this population estimate would take the church number of about 2000 (using data from the Knýtlinge Saga by Þórðarson (1259) for year 1200, as interpreted by Scharling, 1889) and multiply it by an average congregation size of 150. Only the first two cities (Hedeby and Ribe) would have larger congregations around 1000, but they would add at most some 10 000–20 000 to the total population.

The population model used in the preceding chapters is now reconsidered, taking into account likely changes in birth and death rates, and specific migration events such as the Cimbrian campaign described above. In this way one can hope to get a basic level of consistency in the demographic development. Figs. 5.7 and 5.9 give the results using the following assumptions. The female fertility case *a* of Fig. 4.20 is used though the Bronze and Iron ages, as there is no concrete evidence supporting the decline of Fig. 4.20 case *b*. The number of people involved in the Cimbrian migration would suggest that the total Danish population was high enough to have over 200 000 people living in the part of west Jutland with conditions (perhaps emphasized by an extreme flooding situation) causing people to move away. Case *b* does not lead to sufficient subsequent population levels. The even more substantial migration during the Viking period (see below) would indicate even higher populations, and perhaps signs of overpopulation (relative to available

* Jensen (2004) describes several stages over several centuries of the Danevirke construction and other major canal and dam constructions in the 8th century.

food supply). However, this is not necessarily the case. The Viking activities all over Europe could be spurred by aspirations for adventure or for territorial power gains. Some authors suggest a very high level of fertility during the Viking age (invoking high levels of promiscuity as a reason; Brøndsted, 1960). The demographic model used here adopts a 2% increase in fertility after year 780, primarily because it makes it easier to reproduce the population counts at the end of the Viking period (Scharling, 1885) without having to assume very dramatic decreases in death rates (cf. Wittwer-Backofen *et al.*, 2008).

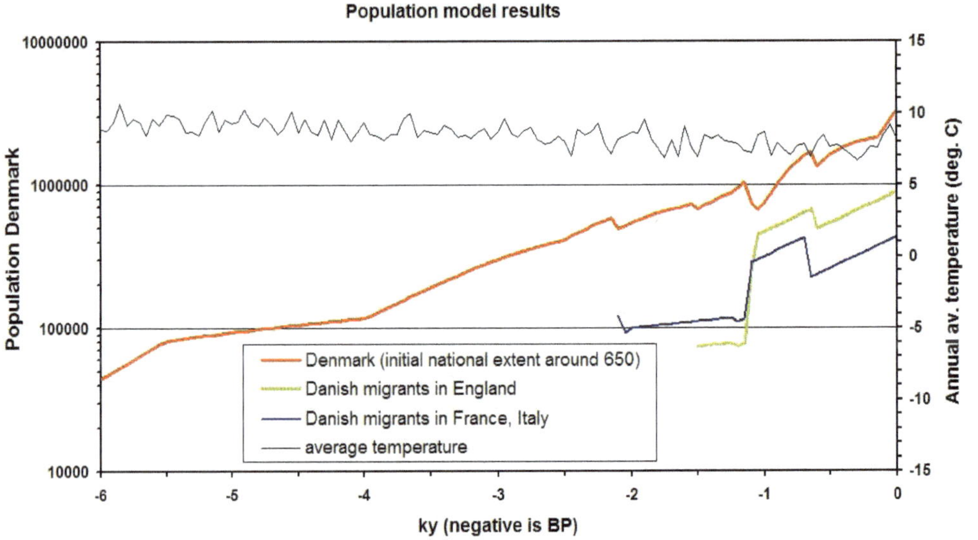


*Figure 5.7. Population model results taking into account migration and conquests and their impacts on the population development in Denmark proper. Choices of birth and death rates are discussed in the text. As before, the continuation to the current time only indicates the effect of keeping parameters unchanged after 1050, the end of the Viking Age.*

An alternative explanation of the migrational activities invokes the Danish inheritance laws, which until the introduction of Christianity stated that all property be left to the eldest son. Thus the younger sons had no alternative other than to try their luck elsewhere. The validity of this as the main explanation for the Viking expeditions is questioned already by Velschow (1843), and Christensen (1983) suggests that an extended family structure developed during the late Iron Age, which would make the inheritance laws compatible with a stable structure not forcing some family members to migrate. Thus the explanation given above in terms of foreign expeditions as an expression of a desire for adventure and the increased power associated with territorial expansion may be the more important factors determining Viking migration.

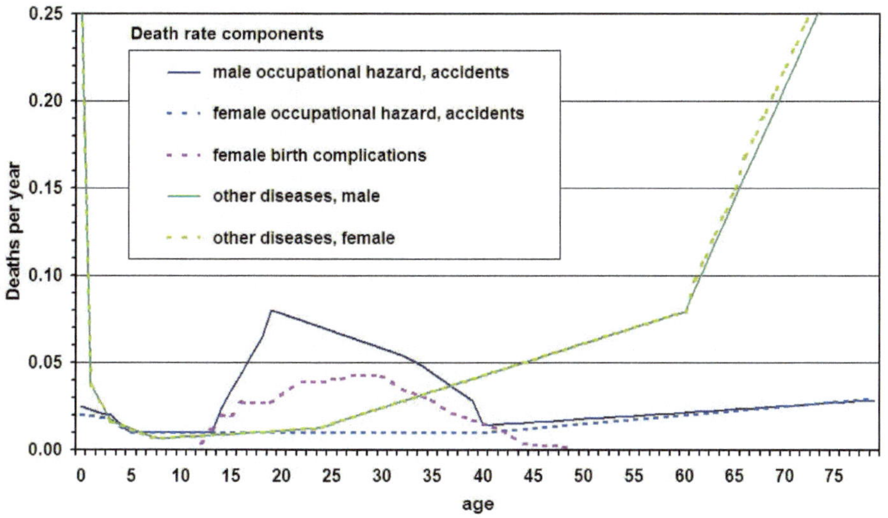


*Figure 5.8. Death rate components assumed for Danish warrior migrants invading other parts of Europe and bringing their families.*

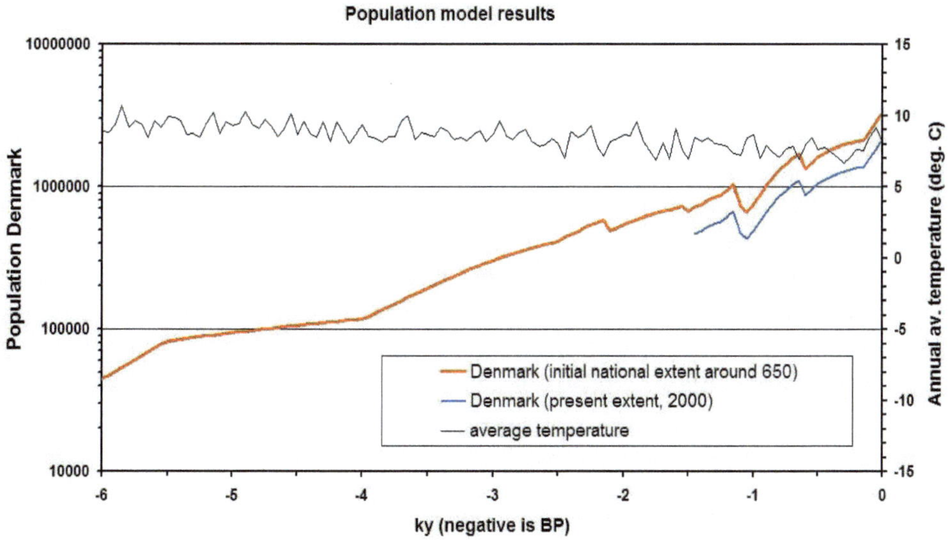


*Figure 5.9. Total Danish population development using parameters depicting changes up to the Middle Age period (as in Fig. 5.7), and the part of population (about 65%) living in current Danish territory.*

Death rates from disease seem to be declining systematically from Bronze to Iron and Viking ages. The 1% increase at the beginning of the

Bronze Age, due to a less varied diet, is considered replaced by a 2% decrease at the entrance to the Iron Age (relative to the time before agriculture, or 3% relative to the increased value during the Bronze Age), and further by another 3% decrease (5% relative to pre-agricultural times) during the Viking age. These changes would be due to having achieved adaptation to the new diet from plants and domesticated animals, as well as to improved medical insights (Grön, 1908).

On the other hand, the increased aggression caused increases in death rates among males in the approximate age range 14–39, assumed to be modest within Denmark (1% in Bronze Age, 2% in Iron Age due to power struggles associated with forming the nation, and back to 1% during the Viking age), but large among men involved in the foreign raids and conquests (100%, i.e. a doubling within the warrior age bracket, with a peak value at age 20 changing from 0.04, cf. Fig. 4.16, to 0.08 per year, as shown in Fig. 5.8).

The Cimbrian expedition involved only people from west Jutland and has been taken to constitute a loss of 20% of the overall Danish population by 2120 y BP. The population in Denmark is restored within two centuries and continues to grow. About 450 (1550 y BP), the Angles (presumed to come from Schleswig and surroundings) invade England, together with or preceded by the Saxons (from Holstein and other parts of northern Germany), taking advantage of Roman occupants leaving England after the collapse of the Roman Empire (Fig. 5.10). The Danish contribution to this conquest is taken as 10% of the Danish population. The Anglo-Saxons successfully settled in England, but not without fights with the original Britons (populating England in the Stone Age and probably coming from Portugal or west France, contributing the original Neolithic culture with hallmarks such as Stonehenge, and later Celts often assumed to come from the Munich region). Remarkably few Romans seem to have remained in England, probably due to the Roman perception of being "superior" to the local people, a trait characterizing all Roman occupation in Europe and the Middle East.

The Viking period sees a succession of raids, conquests and more or less permanent settling of Danes in several parts of Europe. Lindisfarne on the English east coast is raided in 793, Friesland in 834. Its important sea trade centre Dorestad is destroyed by a tsunami in 864. The Danish Vikings enter the river Thames and launch an attack on London in 835. Their King Horik I, after having devastated Hamburg and had his warlord Ragnar Lodbrog conquer Paris in 845, loses a battle against the Anglo-Saxon king Æthelwolf in 851. His successor, Horik II, again attacks the English, now in East Anglia, and the Danes settle in York.

Horik II is followed by King Halfdan, who conquers London in 872 and starts a general colonization. The Anglo-Saxons retain sovereignty over Wessex, and Alfred the Great in 878 defeats the Danes at Edington, leading to an agreed division of England. Danes now govern north of the Thames (Adam Bremensis, ca. 1075; Clarke, 1985; Østergaard, 1994).

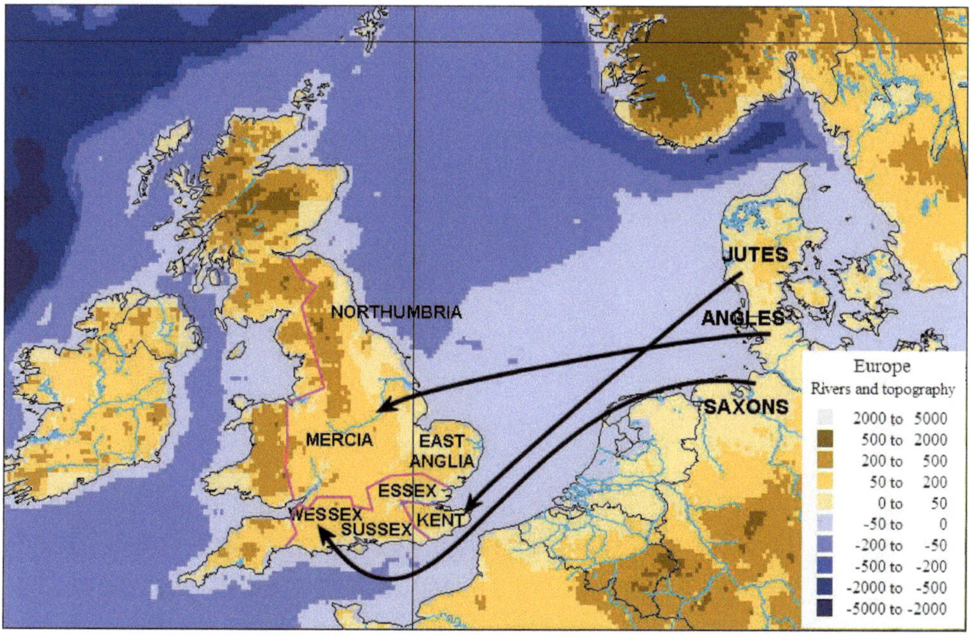


*Figure 5.10. English campaigns by the Saxons (400), and by the Jutes and Angles (450). The Britons retreated to the western and northern provinces, and to Bretagne in France (cf. Kinder and Hilgemann, 1964; Bede, ca. 720; Anglo-Saxon Chronicle, 787-1001).*

At the same time, a campaign on the continent at Gent was meant to strengthen the occupation of a part of France, where Rouen, Paris, Chartres and Tour had been Danish since 845, and where large sums of money had been paid by the local administrators to avoid the destruction of their cities. Despite occasional defeats, the Vikings were so feared that the continental ruler Carolus the Fat in 882 paid a huge sum of money to get the Viking to stop their raids. A decade later, many Danes (often called Normans and not distinguished from other Scandinavians) left France after outbreak of a serious epidemic disease. They landed in Kent to join the Danes in England. In 900, the Vikings returned and began a systematic colonization of France, led by the Danish/Norman Duke Rollo.

The colonization of (parts of) England is stepped up during the last part of the 9th century by the Danish king Gudrum, after having lost London in 886 and agreed with the Anglo-Saxons on a more permanent division of the country. The Danish territory, "Danelagen" (i.e. where the Danish law prevails), comprises the regions depicted in Fig. 5.11. When Canute the Great becomes king of all England (see below), the administration of the country is changed. Shires are formed, mostly around castles such as the five castles erected under Danelagen, and an earl is appointed for each shire, called the "shire-gerefa" (from the Danish word for earl, "greve"), or just sheriff.

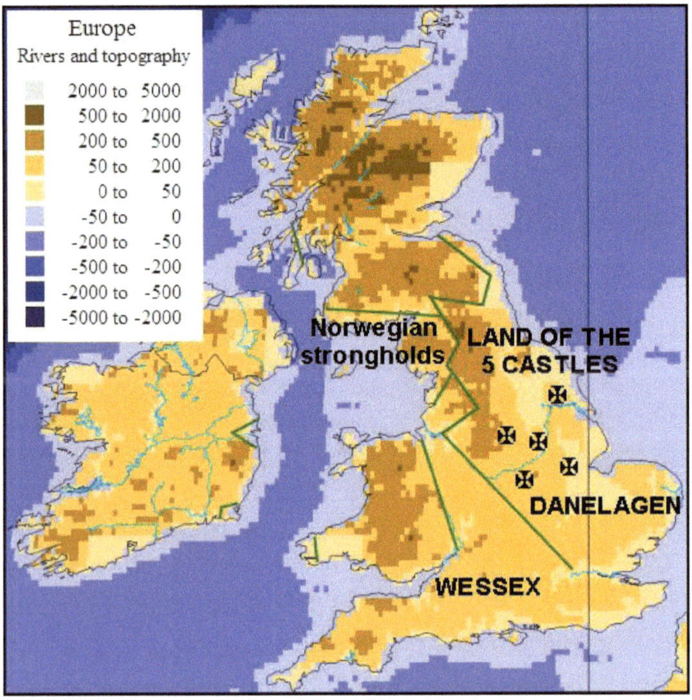


*Figure 5.11. Extent of Danelagen, with surrounding areas occupied by Anglo-Saxons or Norwegians (cf. Kinder and Hilgemann, 1964).*

The Swedish and Norwegian Vikings make similar conquests, the Swedes in eastern Europe, along the Russian rivers, and the Norwegians (formed formally as a nation by Harald Hårfager (Fairhair) in 872) concentrating on Scotland (from 839) and Ireland, where the Irish in 851 ask the Danes for help against the Norwegians. Although the Scandinavians lose Ireland in 901, both Danes and particularly the Norwegians already settled stay and contribute to the development of the country. Danes also reach Spain in 845 and from 860 Italy (Pisa and on the Adriatic side Bari), while the Norwegians reach Iceland, Greenland (890) and North America, probably Newfoundland (Leif Eriksen, 892;

Brøndsted, 1960). An overview of the Viking expeditions is given in Fig. 5.12.

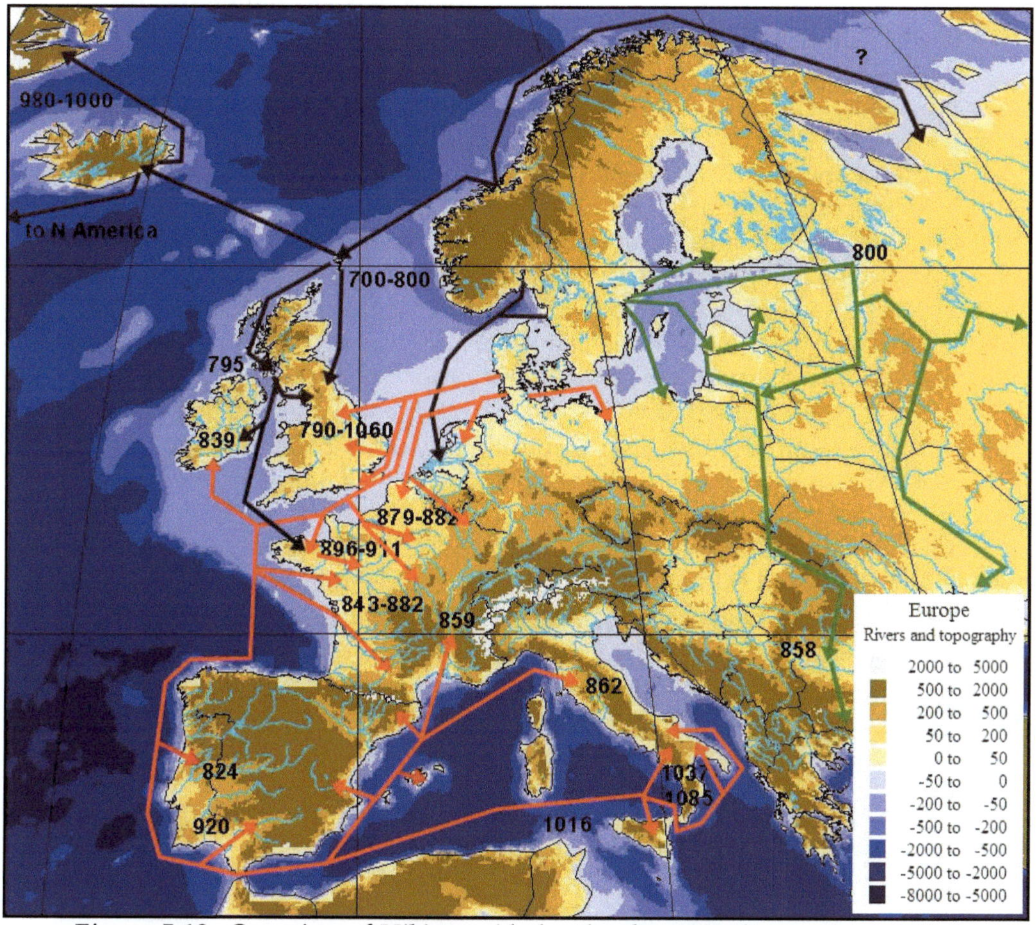


*Figure 5.12. Overview of Viking raids by the three Nordic people, and their settlements abroad (cf. Kinder and Hilgemann, 1964).*

It is evident that the conquests and migrations, particularly to France and England, must have involved a substantial part of the Danish population. The demographic model assumes that in the late 9th century, 20% of the Danish population migrates to England, as warriors or settlers, and another 20% to France and other continental destinations. During the mid-10th century, Denmark and Norway are at war in both Jutland and Scania, but the Danish King Harald III Blåtand wins the decisive battle at Hardaland in 960, after which he divides Norway in such a way that the southern part is controlled by him.

A new migrational move happens during the late 10th century, after a war with German emperor Otto II over Hedeby in Schleswig (south Jutland), successfully ended by Danish king Sven Tveskæg (Forkbeard). In 983, he again turns to England and, after a decade of attempts, finally in 1013 takes London. Unfortunately, he dies the year after, and his successor, Knud II (Canute) the Great must fight the Anglo-Saxon King Edmund a couple of years, before a peace agreement dividing the country is agreed upon. When Edmund dies, Canute becomes king of all England and the Danelag subsequently accepts a new wave of migrant settlers from Denmark, assumed in the demographic model to constitute 20% of the Danish population. The successful reign of Canute also leads to the conquest of Norway, where Canute inserts Haakon as king. This lasts to 1035, when the local warlord Magnus recaptures Norway. Support for the view that the Danish armies must have been of substantial size, and the settlers following even more, comes from observations made of the number of ships used in each campaign during the two centuries from 850 onwards, amounting to some 200–300 ships in each instance, implying the involvement of some 10 000 people in each voyage to a particular destination. Canute IV the Holy at the end of the Viking Age (1085) sent as many as 1000 ships to England, and planned to borrow an additional 600 ships from friends in Flanders, but he was then killed by Danes angry with his excessive levels of taxation (cf. Østergaard, 1994).

*Figure 5.13. Reconstructions of Viking warship (a, left: Havhingsten-Skuldelev 2, 1040, cf. Fig. 5.14) and merchant ship (b, right: Roald Ege, 1030; Roskilde Viking Ship Museum; photos by author).*

As noted earlier, late Iron Age expeditions used rowed ships, but sails were used during the Viking period (cf. Fig. 5.14; the introduction of sails happened very late relative to Mediterranean history, cf. Chapter 4 and Digby, 1954). However, sails may well have constituted a necessary condition for the campaigns to reach the volume they did, in terms of the quite sizeable numbers of Danes migrating to England and some of the other places visited. The reason is that ships with sails after transporting warriors or goods, could return with a small crew to collect relatives and peasant families, whereas the rowed ships required such a large crew (being the same as the warriors) that there would have been little room for more persons or cargo. These arguments counter the claim made by Sawyer (1962) that the Vikings invading England could have been only a few handfuls of people, but in any case, the total number of Danes estimated to have emigrated is highly uncertain.

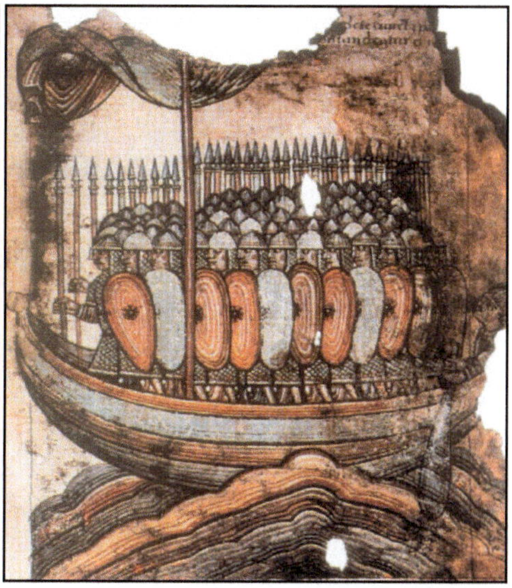

*Figure 5.14. Top left: Viking warship found at Gokstad (895; Oslo Vikingeskipsmuseum; photo by author). Right: picture from Angers Monastery (ca. 1100, depicting 9th century Viking approach; photo by Biblioteque Nationale de France, Paris). Left: scene from the Bayeux tapestry (ca: 1066; photo by Greenhalgh, 2003).*

The Vikings seem to have been careful not to steal from existing settlements in England. Cameron (1965) reaches this conclusion by analysing Danish settlements, and he finds that they are preferably made on marginal land not previously cultivated. Danish Vikings and the farmers they brought along were very considerate people! Successful blending into the existing English society is underlined by language studies. The English language owes its richness of expression to being not just a mixture of Celtic, German, Danish and French, but rather a union of all these languages, with the original several terms for a given thing acquiring with time different shades of meaning.

Figure 5.12 shows that the Vikings undertook substantial operations in the Mediterranean region, including the formation of a settlement in southern Italy and Sicily in the first part of the 11th century. That, too, soon merged with the local population. The 9th and 10th century stops in Spain may have involved trade with the Arab emirate/caliphate formed after the invasion of Spain in 711 (and "assistance" in some of the following power struggles between Arab fractions). Arab coins are found in Scandinavia, but whether the trade was via the ocean routes or through the Russian rivers and Kiev remains uncertain (Skovgaard-Petersen, 1971). Generally, Hedeby and Dorestad are considered the hubs of Atlantic coastal trade, while Birka (near present Stockholm) took care of the eastern trade routes. For example, Denmark is believed to have imported considerable amounts of rye during the Iron and Viking ages, possibly from the Kiev region (Helbæk, 1971).

As seen in Fig. 5.7, the loss of 400 000 people migrating away from Denmark between 840 and 1030 was made up for in less than 200 years. The Danish populations in England and France (plus Italy) were about 400 000 and 300 000, respectively, during the Viking period, and increased by another 400 000 people during the next three centuries. Fig. 5.7 includes the effect of the Bubonic plague epidemic spreading to northern and western Europe around 1350. In Italy, nearly 70% of the population was lost, in France some 50%, and in England and Denmark around 30% (Scharling, 1885; Ulsig, 1991). The continuation of the curves after these events, with unchanged parameters, indicates that there could currently be some 400 000 Danes in France/Italy, and some 850 000 in England. Because the Danes mixed and integrated well into these societies, in reality these genes would be spread over much larger fractions of the respective populations.

The size of Denmark proper has changed with time, and Fig. 5.9 gives the part of the population in the Viking age territory of Denmark that lived within the current territory (missing the provinces Skåne/Scania, Halland and Blekinge in east Denmark and the southern half of south Jutland/Schleswig in south Denmark). In between, Denmark was even larger, regaining in 1459 both Holstein and Schleswig, the latter having in the 12th century been transformed into an independent Duchy. Norway adopted a Swedish king in 1315, and then a Danish child king,

Olav, in 1380, but having Margrethe (Margaret) I (1352–1412) as regent, formally elected so at Olav's death in 1387 and soon becoming recognized as regent in both Denmark, Norway and Sweden (see Chapter 6). She manoeuvred the three Scandinavian countries into a formal union (the Kalmar Union) in 1397, with herself as the common regent, acting on behalf of another child king. Sweden temporarily needed Danish help due to trade competition from the Hanseatic League (north Germany). The relation between Norway and Denmark had the character of an alliance between two independent nations, finding it convenient to share a monarch, as well as the administration, and certain cultural and educational institutions. Sweden felt growing disagreements and finally left the union in 1523. The joint administration of Norway and Denmark (seated in Denmark!) lasted until the Napoleonic Wars, which Denmark–Norway lost, while Sweden was on the winning side and was given Norway by the 1814 Treaty of Kiel. The Norwegians did not accept this solution, but they were then invaded by Sweden, forcing a union on Norway that lasted to 1905.

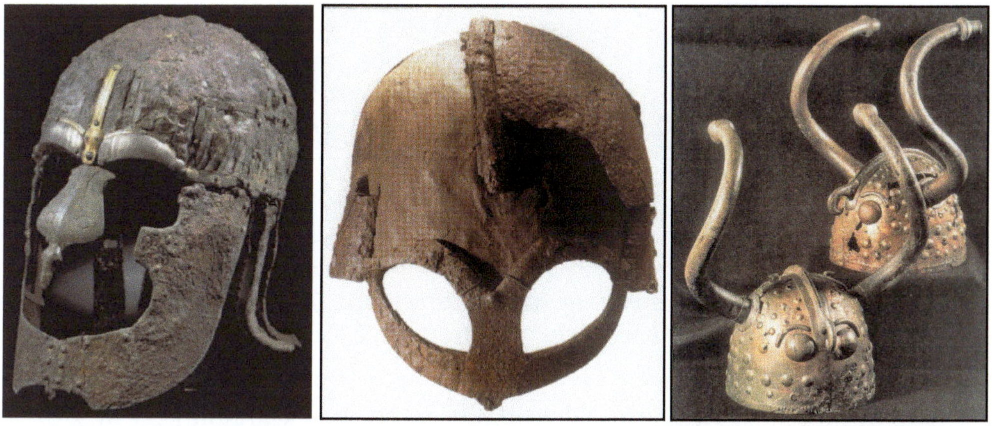

*Figure 5.15. Left to right: helmet from Vendel Church, Uppland (ca. 600; photo © S. Hallgren, Swedish Historical Museum, Inventory No. 9785:XII, 120458); Viking age helmet from Gjermundbu (Norway; Grieg, 1947; photo by Universitetets Oldsagssamling, Oslo); and two helmets from the Bronze Age (Veksø Bog, ca. 2800 y BP; Jensen, 1946; photo by Danish National Museum).*

It is rather amusing to note that the importance as a major European power which Denmark enjoyed during the Viking period is a rather common feature: nearly all European countries, small or large, have had such a period lasting a limited length of time! Danish migrants stood out by having little interest in staying "racially pure". They mixed freely with any existing population, and after a few generations were no longer seen as a particular group in their new society. Rulers with Danish ancestry in England and France, or with Swedish ancestry in Russia or Ukraine, were considered no more alien than the nobility from Holstein or Mecklenburg–made Danish monarchs during the later centuries. It should also be mentioned that several of the warlords leading Norwegian and Danish campaigns abroad (e.g. Rollo) felt little affinity with the motherland. One suggestion is that their adventures abroad were spurred by not wanting to submit to the sovereignty of a king. After all, these were the early days of Scandinavian nation and kingdom formation (cf. the discussion in Skovgaard-Petersen, 1971). The designs of helmets depicted in Fig. 5.15 underline the nature of later Iron Age and Viking warfare as man-to-man fighting (mostly with swords). The Bronze Age helmets shown for comparison are clearly more meant to impress the enemy than to protect the head. Surprisingly few Viking age military artefacts have survived.

*Figure 5.16. Christian baptizing scene (Østergaard, 1994; drawing dated to 814 from manuscript in Bayerisches Staatsbibliotek, Munich). Copy at Danish National Museum.*

The formal subscription to Christianity in Denmark announced in 975 by Harald III Bluetooth followed two centuries of Danish resistance to approaches from the Roman Church, but was actually preceded by several of the Danish rulers abroad adopting this new religion. In Denmark Horik I (813–854) and particularly Horik II (854–ca. 870) accepted both missionary activities (e.g. by the monk Ansgar) and also the building of churches, but did not themselves engage in Christian activities. Christened abroad were Rollo in France (founding Normandy in conquered areas by a peace settlement with Charles the Simple-minded, also marrying his daughter around 915) and Guthrum (christened name Æthelstan, died 890) in England, baptized by his Saxon opponent Alfred the Great (871–899). The written sources of condescending descriptions of the Viking rule are primarily by members of the Catholic Church and may be coloured by evaluating foreign people entirely on their willingness to convert to the Christian religion. Later Danish kings may have thought that the well-organized Catholic Church could help them subdue the many local warlords challenging the idea of a common nation under a king. In reality, they bought themselves a greater nuisance, having to render much tax money and influence to the Church (see Chapter 6). Fig. 5.16 shows a baptizing scene from Germany. In Denmark, the Church often had to bribe the victims to engage in the cere-ceremony.

*Figure 5.17. Foundation elements of the defence wall at Danevirke, laid 737 (Jensen, 2003; photo by H. Andersen, Danish National Museum).*

Building technology advanced considerably during the Viking age, concentrating on wood treatment (bending, pre-stressing, joint-making, impregnating, waterproofing and painting) and carpentry skills. The defensive wall structure at Danevirke mentioned above, just south of Hedeby, had a foundation with wooden structures preventing the sand–clay earth material from sliding down from the walls (see Fig. 5.17). Cities were formed with segments for various kinds of trade and workshops, as well as for living. Harbour areas allowed sheltering stocks of merchandise and easy access to ships (Fig. 5.18). All were made from wooden planks and structures, more often interlaced than nailed together. Dwellings were also made of wood, in a style that is still in widespread use (Fig. 5.19). The lengthwise walls of the house are slightly bent (bulging at the middle), a characteristic style demanding skilled craftsmen such as carpenters. This curved type of house is also found in cities, such as a 48–m–long house for which soil imprints of the posts have been found in Lejre (Sealand), and in the military buildings at Trelleborg (Fig. 5.21 below). One should note the professional wood-shingle tiled roof construction. The first Christian church in Denmark was small and not too different in architecture from the dwelling-style buildings (Fig. 5.20). The decorated piece found (from the band under the roof, shown to the left in Fig. 5.20) is in domestic Viking style, so the Roman Catholic art style seems not to have been imported along with the religion. Impressive stone churches in Danish architecture came later, during the 12th century (Chapter 6).

*Figure 5.18. Possible reconstructions of the city of Ribe marketplace around 725 (left; Jensen, 1997) and the harbour of Hedeby in the Viking age (right; Schietzel, Crumlin-Pedersen, 1980). Drawings by F. Bau, Danish National Museum.*

*Figure 5.19. Reconstruction of a Viking age village dwelling (ca. 1000; Jensen, 2004). "Salshuset", Vikingelandsbyen, Albertslund.*

Military buildings like Trelleborg (980–981) in Denmark had a more permanent character than the camps made during campaigns abroad (Fig. 5.21). All buildings in the Viking age were made of wood, which is perhaps natural considering that most of Denmark does not possess rocky underground, and at first there would have been plenty of wood. Later, when ships, cities and building constructions required so much wood that it could have become scarce, there would at least be import options from regions fairly nearby (such as Sweden). In Southern Europe, stone as a building material already had a long history. Stone became introduced in Denmark only after the Viking age, but because the ice ages had left large numbers of stones in the Danish soil stone materials must have been available earlier (a ploughing farmer would hit stones of varying size quite frequently), so the likely explanation is still that no shortage of wood was perceived until the medieval period.

*Figure 5.20. Reconstruction of the earliest church building (b, right) for which remains (a, left) and foundation post-holes have been found, at Hørning in Jutland, ca. 1065 (Jensen, 2004; photos by F. Bau and L. Larsen, Danish National Museum).*

The military installation at Trelleborg shown in Fig. 5.21 is one of several such structures (Fyrkat, Aggersborg, Nonnebakken). They are all characterized by the basic elements of four curved-façade houses arranged in a square. The length of each house is 100 Roman feet (29.33 m), and the whole layout has a precision of better than 0.2% (Christensen, 1966). Encircling the buildings is in most cases a circular wall, divided into four quadrants, each with a unit of four houses. The large circular wall in Aggersborg holds four segments, each with three squares of four houses. All is built with great accuracy, as inferred from the modest remains constituted mostly by the underground parts of the structures. The circular walls had complex inner wooden support structures like Danevirke (Fig. 5.17).

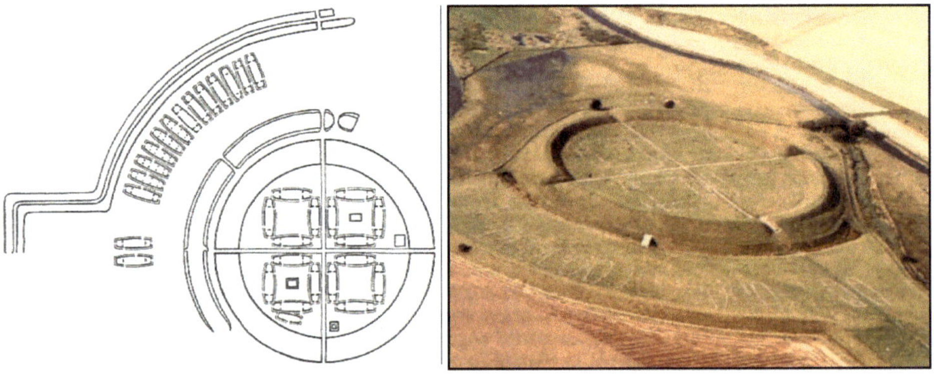

*Figure 5.21. The Trelleborg military installation built 980–981 near Slagelse (Sealand). Left: plan (Christensen, 1966; Danish Royal Library). Right: aerial view (Rieck et al., 1993; Danish National Museum).*

Estimates of the number of people living in the fortresses suggest about 300 in each square of four houses. Burial grounds at some of the sites suggest that the inhabitants included women and children. The precise purpose of the several fortresses, believed to have all built around the established date of Trelleborg (980–981) and used only for a relatively short period (no sign of maintenance, a small number of graves), remains unclear. Were they places where warriors were trained for action abroad, or did they serve domestic purposes, such as Harald III Bluetooth's fighting the opposition to his enforced christening of the Danes (Jensen, 2004)? The structure certainly made them easy to defend, considering the kinds of weapons available to attackers at the time, at least if sieges were short enough not to exhaust food and water supplies.

The per capita energy demand is explored in Tables 5.1 to 5.5, continuing where Table 3.1 left off, with delivered energy estimates for the time periods around 5, 3 and 2 ky BP, followed by years 600 and 900. The end-use estimates (similarly continued from Table 3.2) are in Tables 5.6 to 5.10. The pictorial rendering of the main results for usage and energy quality categories, last made in Fig. 2.16, is expanded to year 900 in Fig. 5.22 for delivered energy and Fig. 5.23 for end-use energy.

Tables 5.1 to 5.5, as well as Tables 2.1, 2.2 and 3.1, display delivered energy. By "delivered" is meant delivered to the end user, and this energy differs from primary energy, if conversions have been made between the acquiring of the energy and its being supplied to the end user. This is an important distinction in fossil energy systems, where refining and conversion (e.g. to electricity) involves a considerable energy loss. It also has implications for use of renewable energy flows, if an initial step of feeding harvested biomass to animals before deriving food energy from it is involved, or if the biomass is converted to, say, liquid biofuels. For other renewable energy sources, such as hydro, wind or solar power, the device used to harvest the energy may already deliver it in the desired form. Still, the term primary energy does not count conversions happening before human "harvesting" of the energy (e.g. processes in the atmosphere converting solar energy to wind). It is to avoid these issues that the tables here provide delivered energy rather than primary energy, but there are cases where the losses before delivery are important to a society, as will be clear when the analysis is carried to the 20th century in Chapter 10.

*Table 5.1. Estimated energy use (delivered) per capita (adult person) in Denmark around 5 ky BP (late Stone Age).*

| *Denmark, 5 ky BP.* | *Human energy* | *Energy from fuels and external energy flows* | | | | | |
|---|---|---|---|---|---|---|---|
| *Rate of energy use (average W for an adult person)* | Food energy | Low-temperature heat | Medium-temperature heat | High-temperature heat | Mobile power (excl. food) | Mechanical power (stationary) | TOTAL |
| **Liveable surroundings** | | | | | | | **268** |
| Basic metabolism | 68 | | | | | | **68** |
| Heating | | 150 | | | | | **150** |
| Light | | | | 50 | | | **50** |
| **Food and water provision** | | | | | | | **102** |
| Hunting | 2 | | | | | | **2** |
| Gathering | 2 | | | | | | **2** |
| Fishing, aquaculture | 3 | | 3 | | | | **6** |
| Agriculture | 20 | | | | 15 | | **35** |
| Preparation, cooking | 5 | | 50 | | | | **55** |
| Water provision | 2 | | | | | | **2** |
| **Security, assault, expansion** | | | | | | | **5** |
| Weapon's use, armies | 3 | | | | | | **3** |
| Discouraging assaults | 1 | | | | | | **1** |
| Territorial transportation | 1 | | | | | | **1** |
| **Health-related efforts** | | | | | | | **20** |
| Food preservation | 2 | 5 | 10 | | | | **17** |
| Water quality management | 1 | 1 | | | | | **2** |
| Doctors, lay or professional | 1 | | | | | | **1** |
| **Relations and leisure** | | | | | | | **2** |
| Toys, artwork | 1 | | | | | | **1** |
| Social transportation | 1 | | | | | | **1** |
| **Occupational activities (non-agric.)** | | | | | | | **25** |
| Construction | 4 | | | | | 1 | **5** |
| Manufacture | 4 | | | | | | **4** |
| Distribution, trade and services | 2 | | | | | | **2** |
| Transportation of goods | 3 | | | | 10 | | **13** |
| Transportation for services, commuting | 1 | | | | | | **1** |
| **Raw materials procurement** | | | | | | | **17** |
| Mining | 2 | | | | | | **2** |
| Materials for construction and industry | 6 | | 2 | 1 | | | **9** |
| Energy provision | 6 | | | | | | **6** |
| TOTAL | **141** | **156** | **65** | **51** | **25** | **1** | **439** |

*Table 5.2. Estimated energy use (delivered) per capita (adult person) in Denmark around 3 ky BP (high Bronze Age).*

| *Denmark, 3 ky BP.* | *Human energy* | *Energy from fuels and external energy flows* | | | | | |
|---|---|---|---|---|---|---|---|
| *Rate of energy use (average W for an adult person)* | Food energy | Low-temperature heat | Medium-temperature heat | High-temperature heat | Mobile power (excl. food) | Mechanical power (stationary) | TOTAL |
| **Liveable surroundings** | | | | | | | **313** |
| Basic metabolism | 68 | | | | | | **68** |
| Heating | | 200 | | | | | **200** |
| Light (when not from heating fires) | | | | 45 | | | **45** |
| **Food and water provision** | | | | | | | **124** |
| Hunting | 2 | | | | | | **2** |
| Gathering | 2 | | | | | | **2** |
| Fishing, aquaculture | 5 | | 3 | | | | **8** |
| Agriculture | 25 | | | | 30 | | **55** |
| Preparation, cooking | 5 | | 50 | | | | **55** |
| Water provision | 2 | | | | | | **2** |
| **Security, assault, expansion** | | | | | | | **7** |
| Weapons, armies | 5 | | | | | 0 | **5** |
| Discouraging assaults | 1 | | | | | | **1** |
| Territorial transportation | 1 | | | | | | **1** |
| **Health-related efforts** | | | | | | | **27** |
| Food preservation | 2 | 2 | 20 | | | | **24** |
| Water quality management | 1 | 1 | | | | | **2** |
| Doctors, lay or professional | 1 | | | | | | **1** |
| **Relations and leisure** | | | | | | | **2.2** |
| Musical instruments | 0.1 | | | | | | **0.1** |
| Theater, dance | 0.1 | | | | | | **0.1** |
| Toys, artwork | 1 | | | | | | **1** |
| Social transportation | 1 | | | | | | **1** |
| **Occupational activities** | | | | | | | **36** |
| Construction | 5 | | | | | 1 | **6** |
| Manufacture | 4 | | 2 | 1 | | | **7** |
| Distribution, trade and services | 2 | | | | | | **2** |
| Transportation of goods | 4 | | | | 16 | | **20** |
| Transportation for services, commuting | 1 | | | | | | **1** |
| **Raw materials procurement** | | | | | | | **29** |
| Mining | 3 | | | | | | **3** |
| Materials for construction and industry | 6 | | 5 | 2 | | | **13** |
| Energy provision | 6 | | | | 7 | | **13** |
| TOTAL | **153.2** | **203** | **80** | **48** | **53** | **1** | **538.2** |

In Table 5.1, pertaining to the late Stone Age, the required space heating (i.e. delivery of low-temperature heat energy to dwellings) has been taken as 25 W per $m^2$ of floor area, on average over the year, as typical for an uninsulated structure. The floor space is taken on average as 6 $m^2$ per person. This space heating is provided by indoor fires that may also be delivering light. Energy for cooking and other food preparation is listed separately. It also contributes to space heating, but modestly, because cooking is not done in all the hours where space heating is needed, and may actually not contribute to comfort because ventilation of odour may carry away the additional heat. Further outdoor fires for light and security are added as medium-temperature heat, at the same level as in the older Stone Age. The medium-temperature heat used for fisheries is for train or fat oil (night fishing), and the non-human energy input to agriculture is from draft animals, assumed to exist at a quantity of one for every four humans and to be used 6 hours a day.

*Table 5.3. Estimated energy use (delivered) per capita (adult person) in Denmark around 2 ky BP (year 0, Iron Age).*

| *Denmark, 2 ky BP.* | *Human energy* | *Energy from fuels and external energy flows* | | | | | |
|---|---|---|---|---|---|---|---|
| *Rate of energy use (average W for an adult person)* | **Food energy** | **Low-temperature heat** | **Medium-temperature heat** | **High-temperature heat** | **Mobile power (excl. food)** | **Mechanical power (stationary)** | TOTAL |
| **Liveable surroundings** | | | | | | | **353** |
| Basic metabolism | 68 | | | | | | **68** |
| Shelter, buildings (reduced heat need) | | 5 | | | | | **5** |
| Heating | | 240 | | | | | **240** |
| Light | | | | 40 | | | **40** |
| **Food and water provision** | | | | | | | **138** |
| Hunting | 2 | | | | | | **2** |
| Gathering | 1 | | | | | | **1** |
| Fishing, aquaculture | 5 | | 3 | | | | **8** |
| Agriculture | 30 | | | | 40 | | **70** |
| Slaughtering, preparing, cooking | 5 | | 50 | | | | **55** |
| Water provision | 2 | | | | | | **2** |
| **Security, assault, expansion** | | | | | | | **13** |
| Weapons, armies | 5 | | | | | 0 | **5** |
| Discouraging assaults | 1 | | | | | | **1** |
| Territorial transportation | 2 | | | | 5 | | **7** |
| **Health-related efforts** | | | | | | | **30** |
| Food preservation | 5 | 7 | 15 | | | | **27** |
| Water quality management | 1 | 1 | | | | | **2** |
| Doctors, lay or professional | 1 | | | | | | **1** |
| **Relations and leisure** | | | | | | | **8.2** |
| Musical instruments | 0.2 | | | | | | **0.2** |
| Theater, dance | 1 | | | | | | **1** |
| Toys, artwork | 1 | | | | | | **1** |
| Social transportation | 2 | | | | 4 | | **6** |
| **Occupational activities** | | | | | | | **71** |
| Construction | 8 | | 2 | | 6 | 2 | **18** |
| Manufacture | 6 | | 5 | 2 | 4 | 0 | **17** |
| Distribution, trade and services | 4 | | | | 2 | | **6** |
| Transportation of goods | 4 | | | | 25 | | **29** |
| Transportation for services, commuting | 1 | | | | | | **1** |
| **Raw materials procurement** | | | | | | | **78** |
| Mining | 4 | | | | 8 | 1 | **13** |
| Materials for construction and industry | 8 | | 20 | 10 | 10 | 1 | **49** |
| Energy provision | 6 | | | | 10 | 0 | **16** |
| TOTAL | **173.2** | **253** | **95** | **52** | **114** | **4** | **691.2** |

The delivered energy use allowing agricultural labour to be performed, both for humans and animals, is the additional food required to perform the muscle work associated with tilling, ploughing, sowing and harvesting. Energy used for food preservation comprises drying at low temperature and cooking at medium temperature. The ratio of the two is tipped towards cooking, relative to the earlier values of Table 3.1. Occupational activities involve more manufacturing (carts, boats, ploughs, etc.) than in the previous period, and hence more energy use. Because manufacture is increasingly performed at specialized workshops, there is an energy use added for transportation of goods, which was not present at 8 ky BP. Trade at long distances is taken at the same values as in the preceding period. The energy used for procuring materials for construction and manufacture is lower than at 8 ky BP (due to higher efficiency), and starts to comprise medium–and (small amounts of) high-temperature heat, indicating the beginning of metalwork and sophistication beyond just collecting flint for manufacturing stone tools.

*Table 5.4. Estimated energy use (delivered) per capita (adult person) in Denmark around 600 (late Iron Age).*

| *Denmark, 600.* | *Human energy* | *Energy from fuels and external energy flows* | | | | | |
|---|---|---|---|---|---|---|---|
| *Rate of energy use (average W for an adult person)* | **Food energy** | **Low-temperature heat** | **Medium-temperature heat** | **High-temperature heat** | **Mobile power (excl. food)** | **Mechanical power (stationary)** | **TOTAL** |
| **Liveable surroundings** | | | | | | | **356** |
| Basic metabolism | 68 | | | | | | **68** |
| Shelter, buildings (reduced heat need) | | 8 | | | | | **8** |
| Heating | | 240 | | | | | **240** |
| Light | | | | 40 | | | **40** |
| **Food and water provision** | | | | | | | **149** |
| Hunting | 1 | | | | | | **1** |
| Gathering | 1 | | | | | | **1** |
| Fishing, aquaculture | 6 | | 3 | | | | **9** |
| Agriculture | 30 | | | | 50 | | **80** |
| Slaughtering, preparing, cooking | 5 | | 50 | | | | **55** |
| Water provision | 3 | | | | | | **3** |
| **Security, assault, expansion** | | | | | | | **21** |
| Weapons, armies | 7 | | | | | 1 | **8** |
| Discouraging assaults | 1 | | | | | | **1** |
| Territorial transportation | 2 | | | | 10 | | **12** |
| **Health-related efforts** | | | | | | | **29** |
| Food preservation | 4 | 2 | 20 | | | | **26** |
| Water quality management | 1 | 1 | | | | | **2** |
| Doctors, lay or professional | 1 | | | | | | **1** |
| **Relations and leisure** | | | | | | | **9.7** |
| Musical instruments | 0.3 | | | | | | **0.3** |
| Theater, dance | 0.3 | | | | | | **0.3** |
| Toys, artwork | 1 | | | | | | **1** |
| Written materials | 0.1 | | | | | | **0.1** |
| Social transportation | 2 | | | | 6 | | **8** |
| **Occupational activities** | | | | | | | **88** |
| Construction | 9 | | 3 | | 8 | 2 | **22** |
| Manufacture | 7 | | 6 | 5 | 5 | 0 | **23** |
| Distribution, trade and services | 5 | | | | 2 | | **7** |
| Transportation of goods | 5 | | | | 30 | | **35** |
| Transportation for services, commuting | 1 | | | | | | **1** |
| **Raw materials procurement** | | | | | | | **89** |
| Mining | 5 | | | | 9 | 1 | **15** |
| Materials for construction and industry | 9 | | 25 | 12 | 10 | 1 | **57** |
| Energy provision | 7 | | | | 10 | 0 | **17** |
| TOTAL | **181.7** | **251** | **107** | **57** | **140** | **5** | **741.7** |

The Bronze Age situation of Table 5.2 still assumes 25 W/m$^2$ for space heating, but now has an assumed floor area of 8 m$^2$/cap. Use of outdoor fires is diminishing. Use of draft animals increases, assuming now one such for each two humans. The development in livestock has been investigated in Sweden and Germany (Kristiansen, 1998), based on bones found. A somewhat inconclusive result (for Swedish locations) of possible increase in pig and sheep and decline in cows is unlikely to be transferable to Denmark, where the human population increases (Fig. 5.7, contrasting with the development in other parts of Europe), and where the height of people (Fig. 4.4) and increased soil treatment by draft-animal-driven ploughs suggest a high proportion of cattle. There may also have been more horses than Kristiansen finds for Sweden, although (as concluded in Chapter 4) much of the Danish cart-moving was done by oxen, leaving the use of horses to the richest people. The energy used for food preservation is going down, because of the introduction of (imported) salt, and more (but still very little) energy is used for cultural undertakings. The energy used for construction, manufacture and moving goods increases slowly, and draft animals are introduced in the raw materials sector (e.g. collecting wood for growing numbers of fireplaces, and transporting stone material for tool making). Ox carts are assumed to be used 4 hours/day/cap.

In Table 5.3, the situation in the middle of the Iron Age is analysed. The agricultural scene is changing (Näsman, 2009; Stenak *et al.*, 2009). Family holdings and small villages had (perhaps gradually) been replaced by larger units, with proprietors and landlords controlling a sizeable workforce, including a substantial number of slaves. It is not known whether it was economic factors that forced some Danes into slavery. The capture of foreign people made slaves, as it unfolded in the Roman empire, could not have had a counterpart in the relatively isolated north. Chapter 4 has argued that greater wealth and the introduction of concepts such as heritage and property ownership in itself created a layered society. This, however, needed not be as bad as accepting slavery. One line of reasoning could be that knowledge of the extended slavery in the Mediterranean and Middle East regions had come to the attention of the newly created upper class in Denmark, and that they just copied it. Lines of information flow had certainly already been established. Late Iron Age farms got fenced in; the use of earlier commons were reserved for certain property owners; and, towards the end of the Iron Age, the larger estates were able to maintain several levels of trade, nationally (e.g. to the newly established towns) and internationally. The fields were growing crops in rotation, including periodic use as grazing ranges with the benefits of fertilization from animal manure, which allows much higher subsequent crop yields than possible with just ashes from burned natural vegetation, as used in the Stone Age. Still, moving villages occasionally took place, when the yields diminished in a given location. The livestock was during the coldest winter months kept in stables or compounds in or near the settlement houses, and they were fed hay (harvested on meadows) and part of crops or crop residues less suited as human food. This procedure meant that livestock manure could be easily collected and administered to fields selected for high-yield crops, creating (from the late Iron Age) a more permanent field–meadow division.

*Table 5.5. Estimated energy use (delivered) per capita (adult person) in Denmark around 900 (Viking period).*

| *Denmark, 900.* | *Human energy* | *Energy from fuels and external energy flows* | | | | | |
|---|---|---|---|---|---|---|---|
| *Rate of energy use (average W for an adult person)* | Food energy | Low-temperature heat | Medium-temperature heat | High-temperature heat | Mobile power (excl. food) | Mechanical power (stationary) | TOTAL |
| **Liveable surroundings** | | | | | | | **360** |
| Basic metabolism | 68 | | | | | | **68** |
| Shelter, buildings (reduced heat need) | | 10 | | | | | **10** |
| Insulation | | 2 | | | | | **2** |
| Heating | | 240 | | | | | **240** |
| Light | | | | 40 | | | **40** |
| **Food and water provision** | | | | | | | **144** |
| Hunting | 1 | | | | | | **1** |
| Gathering | 1 | | | | | | **1** |
| Fishing, aquaculture | 6 | | 3 | | | | **9** |
| Agriculture | 25 | | | | 50 | | **75** |
| Slaughtering, preparing, cooking | 5 | | 50 | | | | **55** |
| Water provision | 3 | | | | | | **3** |
| **Security, assault, expansion** | | | | | | | **36** |
| Weapons, armies | 10 | | | | | 1 | **11** |
| Discouraging assaults | 1 | | | | | | **1** |
| Territorial transportation, ships (sails) | 4 | | | | 20 | | **24** |
| **Health-related efforts** | | | | | | | **29** |
| Food preservation | 4 | 2 | 20 | | | | **26** |
| Water quality management | 1 | 1 | | | | | **2** |
| Doctors, lay or professional | 1 | | | | | | **1** |
| **Relations and leisure** | | | | | | | **14.9** |
| Musical instruments | 0.3 | | | | | | **0.3** |
| Theater, dance | 0.5 | | | | | | **0.5** |
| Toys, artwork | 1 | | | | | | **1** |
| Written materials | 0.1 | | | | | | **0.1** |
| Social transportation | 3 | | | | 10 | | **13** |
| **Occupational activities** | | | | | | | **106** |
| Construction (incl. Ships) | 12 | | 5 | | 10 | 3 | **30** |
| Manufacture | 8 | | 10 | 5 | 3 | 0 | **26** |
| Distribution, trade and services | 6 | | | | 2 | | **8** |
| Transportation of goods | 6 | | | | 35 | | **41** |
| Transportation for services, commuting | 1 | | | | | | **1** |
| **Raw materials procurement** | | | | | | | **106** |
| Mining | 6 | | | | 10 | 1 | **17** |
| Materials for construction and industry | 10 | | 30 | 20 | 10 | 1 | **71** |
| Energy provision | 8 | | | | 10 | 0 | **18** |
| TOTAL | **191.9** | **255** | **118** | **65** | **160** | **6** | **795.9** |

The growth in human shelter (and thus heated) area per person is during the Iron Age assumed to increase to 12 m$^2$. This and the increasing intensity of agriculture raised energy use, as did the growth in non-farming occupations such as traders, merchants, doctors, midwives, miners, blacksmiths, carpenters and, not least, soldiers. All contributed new activities requiring energy inputs, and particularly the large increase in domestic metal industries (bog iron extraction, high-temperature smelters, founding and many types of artisan metal processing). The energy accounts of Table 5.3 show new entries both in high-temperature heat usage and in using animal power not only for mobile applications, but also for stationary mechanical power in mines or manufacturing industries. Energy spent in connection with military activities is increasing, both for armoury production and for transportation, whether by horse, cart or ship. The Jutes and Angles boarding for England is one example, although happening between the dates depicted in Tables 5.3 and 5.4.

Table 5.4 shows that the upward trend in energy use continues towards the end of the Iron Age, with expenditures related to military operations showing considerable growth. Dwelling space (13 $m^2$/cap.) is now flattening out, but provision for soldiers and the hardware they use increases substantially. The establishment of the kingdom also entails new activities (e.g. in the building and construction sector). In both agriculture and transport of goods, increased efficiencies (better roads, better carts, etc.) make the delivered energy stay constant of go down, despite higher levels of activity.

In the Viking age, Table 5.5 shows that military requirements are now a significant cause of energy expenditure, in manufacture of weapons and equipment, in building ships and in providing other means of transportation. Also, non-military transportation increases, for reasons of trade and of social interaction (confined to the upper stratum of society). Wood is becoming scarce, at least locally, so more effort is required to secure supplies, including import from more distant regions (within or outside the county). Apart from the military emphasis, energy use is growing less rapidly than during the Iron Age; dwelling size stabilizes at about 15 $m^2$/cap.; leisure activities increase, but, as is their nature, using very little energy. The 15 $m^2$/cap. floor space estimate could be on the high side, e.g. compared with the estimate above of the number of people occupying, for example, the military establishments at Trelleborg, but all such estimates are subjective, as no tangible evidence has been uncovered.

Tables 5.6 to 5.10 give the end-use energy supply corresponding to Tables 5.1 to 5.5, using estimates of the efficiency of each type of energy conversion process.

In Fig. 5.22a, the results of Tables 5.1 to 5.5 are summarized over the entire period of investigation, up to year 900, divided on energy usage categories. In Fig. 5.22b, a similar summary is made, but divided on energy qualities. Both trends are shown on a per capita basis. It is seen that both quantities have been growing from 8 ky BP, where a minimum occurred because the efficiency of wood burning increased more than the per capita demand.

*Table 5.6. Estimated end-use energy use per capita (adult) in Denmark around 5 ky BP (late Stone Age).*

| *Denmark, 5 ky BP.* | *Human energy* | *Energy from fuels and external energy flows* | | | | | |
|---|---|---|---|---|---|---|---|
| *Rate of energy end-use (average W for an adult person)* | Food energy | Low-temperature heat | Medium-temperature heat | High-temperature heat | Mobile (excl. food) | Mech. power (stat.) | TOTAL |
| **Liveable surroundings** | | | | | | | **91** |
| Basic metabolism | 68 | | | | | | **68** |
| Heating | | 20 | | | | | **20** |
| Light | | | | 3 | | | **3** |
| **Food and water provision** | | | | | | | **21.6** |
| Hunting | 0.6 | | | | | | **0.6** |
| Gathering | 0.6 | | | | | | **0.6** |
| Fishing, aquaculture | 1 | | 1 | | | | **2** |
| Agriculture | 6 | | | | 5 | | **11** |
| Preparation, cooking | 1.7 | | 5 | | | | **6.7** |
| Water provision | 0.7 | | | | | | **0.7** |
| **Security, assault, expansion** | | | | | | | **1.6** |
| Weapons, armies | 1 | | | | | | **1** |
| Discouraging assaults | 0.3 | | | | | | **0.3** |
| Territorial transportation | 0.3 | | | | | | **0.3** |
| **Health-related efforts** | | | | | | | **6.7** |
| Food preservation | 0.7 | 2 | 3 | | | | **5.7** |
| Water quality management | 0.3 | 0.4 | | | | | **0.7** |
| Doctors, lay or professional | 0.3 | | | | | | **0.3** |
| **Relations and leisure** | | | | | | | **0.6** |
| Toys, artwork | 0.3 | | | | | | **0.3** |
| Social transportation | 0.3 | | | | | | **0.3** |
| **Occupational activities** | | | | | | | **5.9** |
| Construction | 1.3 | | | | | 0.3 | **1.6** |
| Manufacture | 1.3 | | | | | | **1.3** |
| Distribution, trade and services | 0.7 | | | | | | **0.7** |
| Transportation of goods | 1 | | | | 1 | | **2** |
| Transportation for services, commuting | 0.3 | | | | | | **0.3** |
| **Raw materials procurement** | | | | | | | **6.2** |
| flint | 0.7 | | | | | | **0.7** |
| Materials for construction and industry | 2 | | 1 | 0.5 | | | **3.5** |
| Energy provision | 2 | | | | | | **2** |
| TOTAL | **91.4** | **22.4** | **10** | **3.5** | **6** | **0.3** | **133.6** |

Figure 5.22a and Table 5.1 show that the energy delivered to "liveable surroundings" reaches a level of 353 W/cap. by 2 ky BP and stays at that level to 900. At the same time, the energy used for "food and water provision", having risen after the introduction of agriculture, also reaches a plateau. The growth is in "occupational activities" and in the "raw material procurement" required to carry out these. New emerging

activities related to health, security and relations appear on the chart, but consume only small amounts of energy.

*Table 5.7. Estimated end-use energy use per capita (adult) in Denmark around 3 ky BP (high Bronze Age).*

| *Denmark, 3 ky BP.* | *Human energy* | *Energy from fuels and external energy flows* | | | | | |
|---|---|---|---|---|---|---|---|
| *Rate of energy end-use (average W for an adult person)* | Food energy | Low-temperature heat | Medium-temperature heat | High-temperature heat | Mobile (excl. food) | Mech. power (stat.) | TOTAL |
| **Liveable surroundings** | | | | | | | **96** |
| Basic metabolism | 68 | | | | | | **68** |
| Heating | | 25 | | | | | **25** |
| Light | | | | 3 | | | **3** |
| **Food and water provision** | | | | | | | **29.1** |
| Hunting | 0.6 | | | | | | **0.6** |
| Gathering | 0.6 | | | | | | **0.6** |
| Fishing, aquaculture | 1.5 | | 1 | | | | **2.5** |
| Agriculture | 8 | | | | 10 | | **18** |
| Preparation, cooking | 1.7 | | 5 | | | | **6.7** |
| Water provision | 0.7 | | | | | | **0.7** |
| **Security, assault, expansion** | | | | | | | **2.3** |
| Weapons, armies | 1.7 | | | | | 0 | **1.7** |
| Discouraging assaults | 0.3 | | | | | | **0.3** |
| Territorial transportation | 0.3 | | | | | | **0.3** |
| **Health-related efforts** | | | | | | | **7.7** |
| Food preservation | 0.7 | 1 | 5 | | | | **6.7** |
| Water quality management | 0.3 | 0.4 | | | | | **0.7** |
| Doctors, lay or professional | 0.3 | | | | | | **0.3** |
| **Relations and leisure** | | | | | | | **0.7** |
| Musical instruments | 0.03 | | | | | | **0.03** |
| Theater, dance | 0.03 | | | | | | **0.03** |
| Toys, artwork | 0.3 | | | | | | **0.3** |
| Social transportation | 0.3 | | | | | | **0.3** |
| **Occupational activities** | | | | | | | **10.1** |
| Construction | 1.7 | | | | | 0.3 | **2** |
| Manufacture | 1.3 | | 1 | 0.5 | | | **2.8** |
| Distribution, trade and services | 0.7 | | | | | | **0.7** |
| Transportation of goods | 1.3 | | | | 3 | | **4.3** |
| Transportation for services, commuting | 0.3 | | | | | | **0.3** |
| **Raw materials procurement** | | | | | | | **10** |
| flint | 1 | | | | | | **1** |
| Materials for construction and industry | 2 | | 2 | 1 | | | **5** |
| Energy provision | 2 | | | | 2 | | **4** |
| TOTAL | **95.7** | **26.4** | **14** | **4.5** | **15** | **0.3** | **155.9** |

Looking at the distribution on energy qualities, food energy is seen to grow from 140 to 286 W/cap. at the end of the period looked at. This is due to increased hard labour for people working in agriculture (slaves or free persons) and in the military, with rowing warships as an obvious example. Whether the upper class could avoid increased labour is not clear. Written accounts indicate that kings and commanders fought and were often among the killed in military battles. Physical work did not seem to have been disdained. Energy-wise, to create 100 W of work requires 300 W of additional food energy (Sørensen, 2004). The use of low-temperature heat increases, due to larger indoor areas and also because such heat was used for an increasing number of tasks in kitchen and industry. The same is true for use of medium-temperature heat and high-temperature heat, becoming important around 2 ky BP for the metal industries (iron works, coke-burning kilns, etc.). Mobile power is growing quickly from the introduction of oxen, and again when ships start to use sails.

*Table 5.8. Estimated end-use energy use per capita (adult) in Denmark around 2 ky BP (year 0, Iron Age).*

| ***Denmark, 2 ky BP.*** | ***Human energy*** | ***Energy from fuels and external energy flows*** | | | | | |
|---|---|---|---|---|---|---|---|
| *Rate of energy end-use (average W for an adult person)* | **Food energy** | **Low-temperature heat** | **Medium-temperature heat** | **High-temperature heat** | **Mobile (excl. food)** | **Mech. power (stat.)** | **TOTAL** |
| **Liveable surroundings** | | | | | | | **106** |
| Basic metabolism | 68 | | | | | | **68** |
| Shelter, buildings | | 5 | | | | | **5** |
| Heating | | 30 | | | | | **30** |
| Light | | | | 3 | | | **3** |
| **Food and water provision** | | | | | | | **36.9** |
| Hunting | 0.6 | | | | | | **0.6** |
| Gathering | 0.3 | | | | | | **0.3** |
| Fishing, aquaculture | 1.7 | | 1 | | | | **2.7** |
| Agriculture | 10 | | | | 13 | | **23** |
| Preparation, cooking | 1.7 | | 8 | | | | **9.7** |
| Water provision | 0.6 | | | | | | **0.6** |
| **Security, assault, expansion** | | | | | | | **4.1** |
| Weapons, armies | 1.7 | | | | | 0 | **1.7** |
| Discouraging assaults | 0.3 | | | | | | **0.3** |
| Territorial transportation | 0.6 | | | | 1.5 | | **2.1** |
| **Health-related efforts** | | | | | | | **10.7** |
| Food preservation | 1.7 | 3 | 5 | | | | **9.7** |
| Water quality management | 0.3 | 0.4 | | | | | **0.7** |
| Doctors, lay or professional | 0.3 | | | | | | **0.3** |
| **Relations and leisure** | | | | | | | **2.9** |
| Musical instruments | 0.07 | | | | | | **0.07** |
| Theater, dance | 0.3 | | | | | | **0.3** |
| Toys, artwork | 0.3 | | | | | | **0.3** |
| Social transportation | 0.7 | | | | 1.5 | | **2.2** |
| **Occupational activities** | | | | | | | **26.4** |
| Construction | 3.3 | | 1 | | 2 | 0.7 | **7** |
| Manufacture | 2 | | 2 | 1 | 1.3 | 1.5 | **7.8** |
| Distribution, trade and services | 1.3 | | | | 0.7 | | **2** |
| Transportation of goods | 1.3 | | | | 8 | | **9.3** |
| Transportation for services, commuting | 0.3 | | | | | | **0.3** |
| **Raw materials procurement** | | | | | | | **29.4** |
| flint | 1.3 | | | | 2 | 0.3 | **3.6** |
| Materials for construction and industry | 2.5 | | 10 | 5 | 3 | 0.3 | **20.8** |
| Energy provision | 2 | | | | 3 | 0 | **5** |
| TOTAL | **103.2** | **38.4** | **27** | **9** | **36** | **2.8** | **216.4** |

*Table 5.9. Estimated end-use energy use per capita (adult) in Denmark around 600 (late Iron Age).*

| *Denmark, 600.* | *Human energy* | *Energy from fuels and external energy flows* | | | | | |
|---|---|---|---|---|---|---|---|
| *Rate of energy end-use (average W for an adult person)* | **Food energy** | **Low-temperature heat** | **Medium-temperature heat** | **High-temperature heat** | **Mobile (excl. food)** | **Mech. power (stat.)** | TOTAL |
| **Liveable surroundings** | | | | | | | **109** |
| Basic metabolism | 68 | | | | | | **68** |
| Shelter, buildings | | 8 | | | | | **8** |
| Heating | | 30 | | | | | **30** |
| Light | | | | 3 | | | **3** |
| **Food and water provision** | | | | | | | **41.3** |
| Hunting | 0.3 | | | | | | **0.3** |
| Gathering | 0.3 | | | | | | **0.3** |
| Fishing, aquaculture | 2 | | 1 | | | | **3** |
| Agriculture | 10 | | | | 17 | | **27** |
| Preparation, cooking | 1.7 | | 8 | | | | **9.7** |
| Water provision | 1 | | | | | | **1** |
| **Security, assault, expansion** | | | | | | | **7.2** |
| Weapons, armies | 2.3 | | | | | 0.3 | **2.6** |
| Discouraging assaults | 0.3 | | | | | | **0.3** |
| Territorial transportation | 1.3 | | | | 3 | | **4.3** |
| **Health-related efforts** | | | | | | | **8.3** |
| Food preservation | 1.3 | 1 | 5 | | | | **7.3** |
| Water quality management | 0.3 | 0.4 | | | | | **0.7** |
| Doctors, lay or professional | 0.3 | | | | | | **0.3** |
| **Relations and leisure** | | | | | | | **3.2** |
| Musical instruments | 0.1 | | | | | | **0.1** |
| Theater, dance | 0.1 | | | | | | **0.1** |
| Toys, artwork | 0.3 | | | | | | **0.3** |
| Written materials | 0.03 | | | | | | **0.03** |
| Social transportation | 0.7 | | | | 2 | | **2.7** |
| **Occupational activities** | | | | | | | **31.8** |
| Construction | 3 | | 1 | | 2.7 | 0.7 | **7.4** |
| Manufacture | 2.3 | | 4 | 2 | 1.7 | 0 | **10** |
| Distribution, trade and services | 1.7 | | | | 0.7 | | **2.4** |
| Transportation of goods | 1.7 | | | | 10 | | **11.7** |
| Transportation for services, commuting | 0.3 | | | | | | **0.3** |
| **Raw materials procurement** | | | | | | | **37.6** |
| flint | 1.7 | | | | 3 | 0.3 | **5** |
| Materials for construction and industry | 3 | | 13 | 8 | 3 | 0.3 | **27.3** |
| Energy provision | 2.3 | | | | 3 | 0 | **5.3** |
| TOTAL | **106.3** | **39.4** | **32** | **13** | **46.1** | **1.6** | **238.4** |

Looking at the development in end-use energy, in Fig. 5.23a, the picture is nearly the same, with some differences in efficiency, favouring particularly energy used in manufacture and raw material extraction. The differences are much more pronounced when looking at the energy qualities in Fig. 5.23b. Efficiencies are comparatively high for use of mechanical energy in stationary or mobile applications, and for process heat of higher temperatures. At the middle of the Viking Age, the total per capita energy use is three (end-use) or four (delivered energy) times higher than the food energy used for human metabolism, ratios often used to characterize the development of societies from rudimentary to more advanced stages.

*Table 5.10. Estimated end-use energy use per capita in Denmark around 900 (Viking period).*

| *Denmark, 900.* | *Human energy* | *Energy from fuels and external energy flows* | | | | | |
|---|---|---|---|---|---|---|---|
| *Rate of energy end-use (average W for an adult person)* | **Food energy** | **Low-temperature heat** | **Medium-temperature heat** | **High-temperature heat** | **Mobile (excl. food)** | **Mech. power (stat.)** | TOTAL |
| **Liveable surroundings** | | | | | | | **116** |
| Basic metabolism | 68 | | | | | | **68** |
| Shelter, buildings | | 10 | | | | | **10** |
| Insulation | | 2 | | | | | **2** |
| Heating | | 33 | | | | | **33** |
| Light | | | | 3 | | | **3** |
| **Food and water provision** | | | | | | | **39.6** |
| Hunting | 0.3 | | | | | | **0.3** |
| Gathering | 0.3 | | | | | | **0.3** |
| Fishing, aquaculture | 2 | | 1 | | | | **3** |
| Agriculture | 8.3 | | | | 17 | | **25.3** |
| Preparation, cooking | 1.7 | | 8 | | | | **9.7** |
| Water provision | 1 | | | | | | **1** |
| **Security, assault, expansion** | | | | | | | **12.6** |
| Weapons, armies | 3.3 | | | | | 0.3 | **3.6** |
| Discouraging assaults | 0.3 | | | | | | **0.3** |
| Territorial transportation | 1.7 | | | | 7 | | **8.7** |
| **Health-related efforts** | | | | | | | **8.3** |
| Food preservation | 1.3 | 1 | 5 | | | | **7.3** |
| Water quality management | 0.3 | 0.4 | | | | | **0.7** |
| Doctors, lay or professional | 0.3 | | | | | | **0.3** |
| **Relations and leisure** | | | | | | | **5.5** |
| Musical instruments | 0.1 | | | | | | **0.1** |
| Theater, dance | 0.1 | | | | | | **0.1** |
| Toys, artwork | 0.3 | | | | | | **0.3** |
| Written materials | 0.03 | | | | | | **0.03** |
| Social transportation | 1 | | | | 4 | | **5** |
| **Occupational activities** | | | | | | | **34.6** |
| Construction | 4 | | 2 | | 3 | 1 | **10** |
| Manufacture | 2.6 | | 4 | 2 | 1 | 0 | **9.6** |
| Distribution, trade and services | 2 | | | | 0.7 | | **2.7** |
| Transportation of goods | 2 | | | | 10 | | **12** |
| Transportation for services, commuting | 0.3 | | | | | | **0.3** |
| **Raw materials procurement** | | | | | | | **47.6** |
| Mining | 2 | | | | 3 | 0.3 | **5.3** |
| Materials for construction and industry | 3.3 | | 16 | 14 | 3 | 0.3 | **36.6** |
| Energy provision | 2.7 | | | | 3 | 0 | **5.7** |
| TOTAL | **109.2** | **46.4** | **36** | **19** | **51.7** | **1.9** | **264.2** |

The energy use in agriculture (Tables 5.1 to 5.5) goes along with an intensified cultivation in larger units and on new land, and the villages also increased in size during the Viking age (Stenak *et al.*, 2009). This led to increased harvest output and to increasing herds of livestock. Not only was it possible to feed a growing population (cf. the discussion above, dismissing food shortage as a reason for the Viking age expeditions abroad) but there was an increasing surplus that could be exported. That this did happen follows from records of taxes and duties levied on agricultural products by proprietors and by the upcoming state representatives of the king.

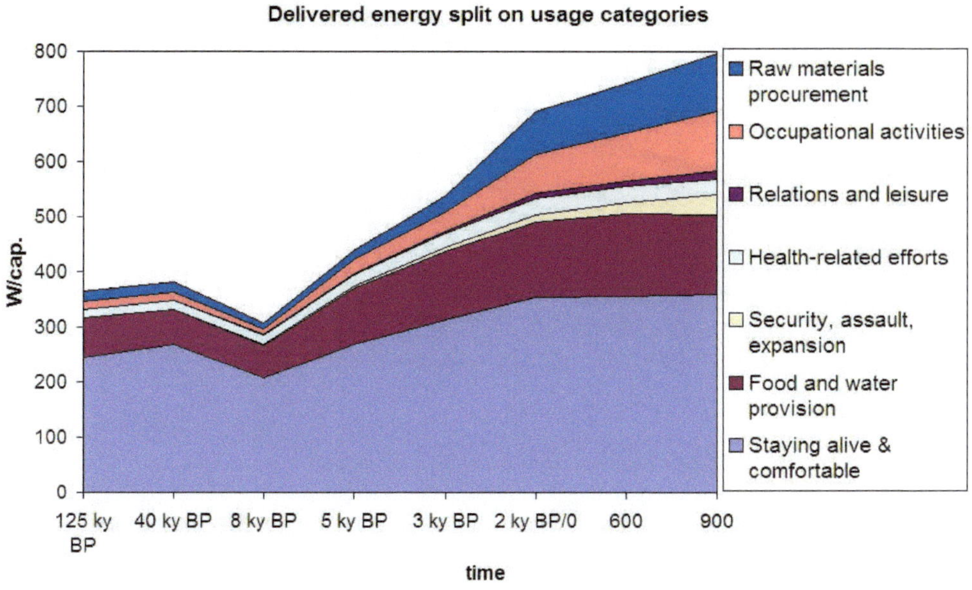


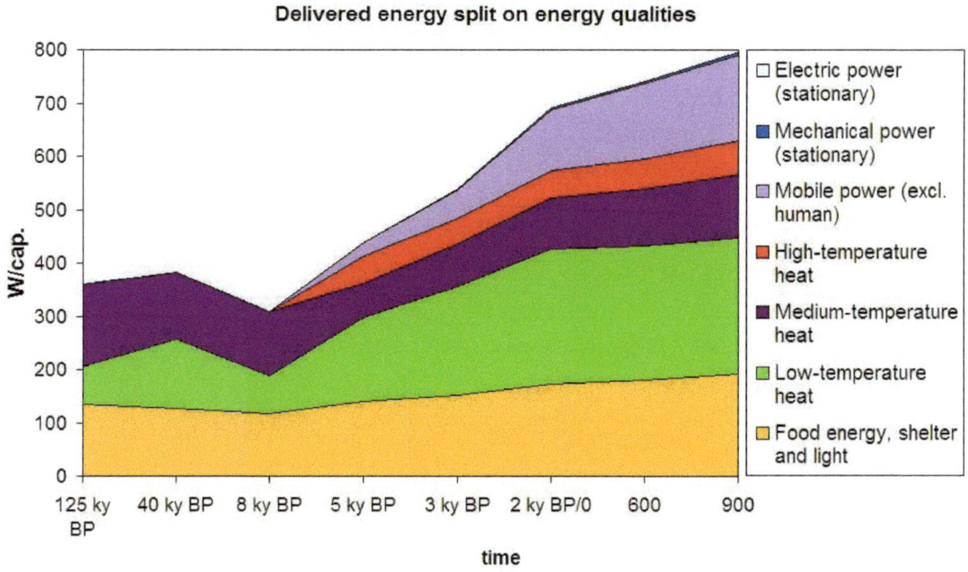


*Figure 5.22. Estimated energy use 125 to 1.1 ky BP, distributed on usage categories (a: upper panel) or on type of energy (b: lower panel).*

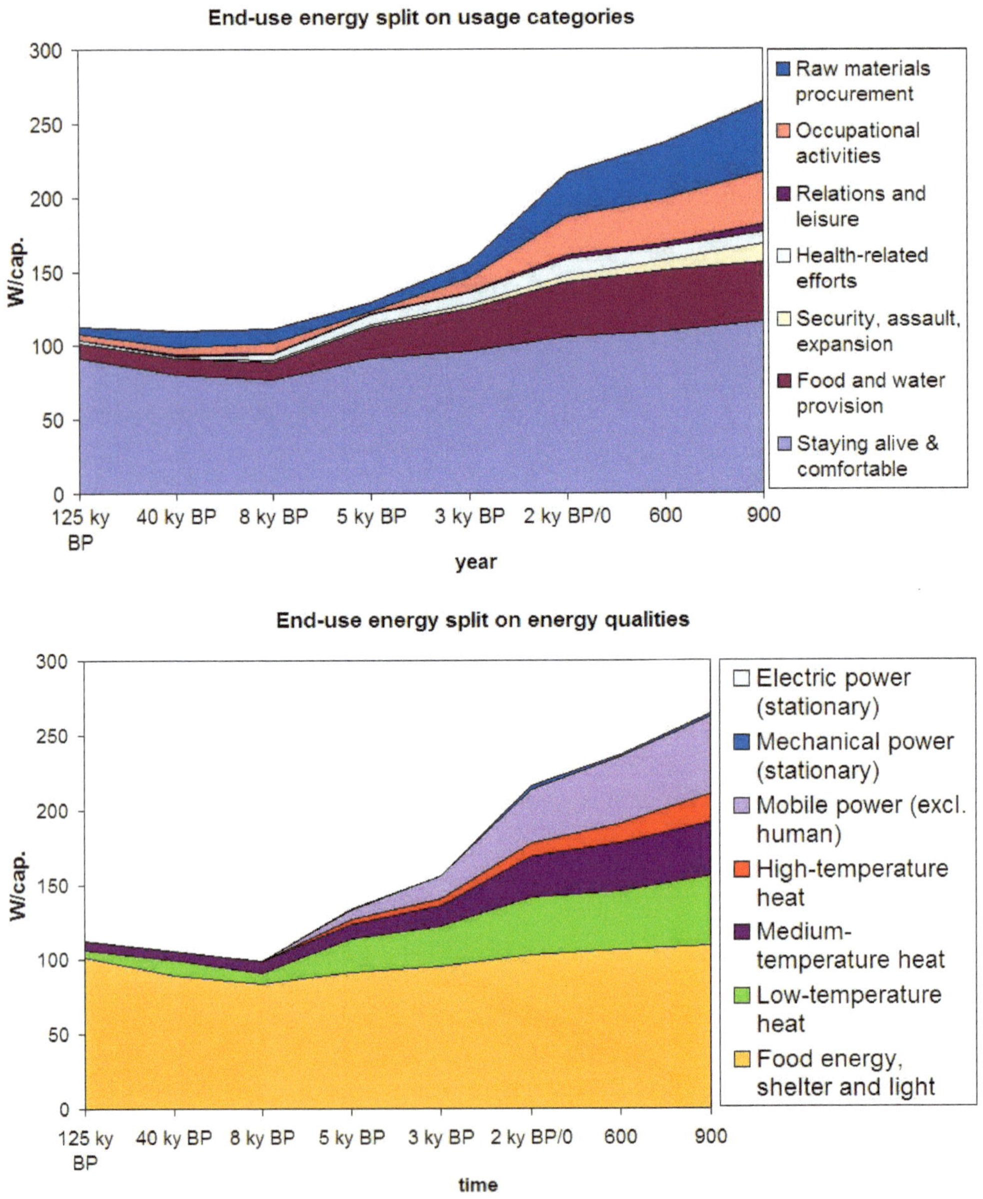


*Figure 5.23. Estimated end-use energy 125 to 1.1 ky BP, distributed on usage categories (a: upper panel) or on types of energy (b: lower panel).*

In Fig. 5.24, the per capita delivered and end-use energy figures have been multiplied by the population size, taken from Fig. 5.9, using the curve for the current Danish territory. At the late Iron Age or the Viking age, the total energy use is similar to that generated by a large contemporary coal-fired power station. Because the territory of Denmark or the territory occupied by Danes has changed very often with time, long-term development charts such as Fig. 5.24 use a fixed geographical territory taken as the present one.

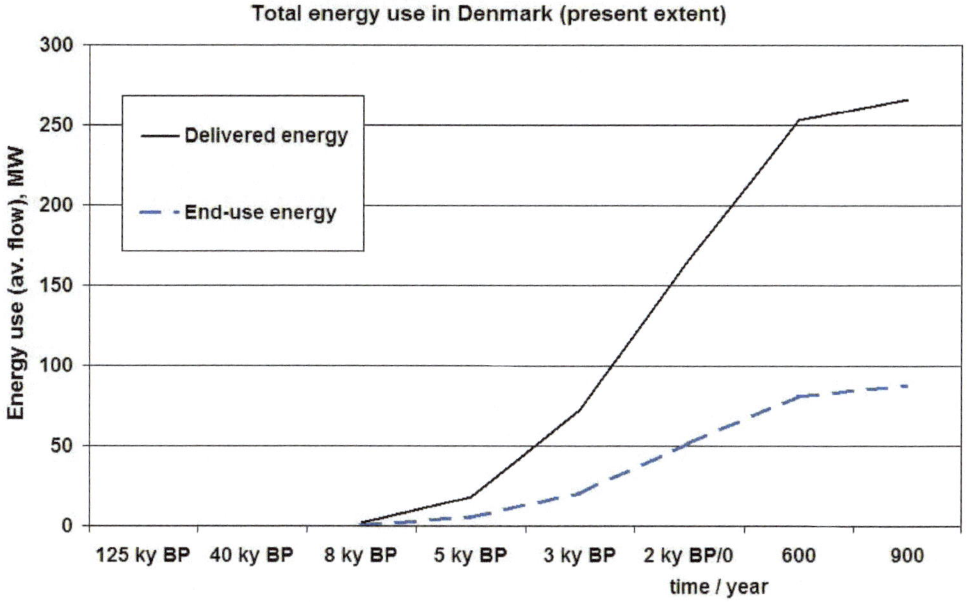


*Figure 5.24. Total Danish energy use, based on the population of Fig. 5.9, current territory and the delivered or end-use energy of an adult person (Figs. 5.22 and 5.23), but corrected for the actual age distribution of the population by assuming children of age 0–4 y to use 25% of the adult energy use, those of age 5–9 y 50%, and those of age 10–14 y 75% of an adult. Because energy is estimated only at specific times, the perturbations of population size caused by migration, for example, do not show.*

### *References:*

Adam Bremensis, ca. 1075. *Gesta Hammaburgensis Ecclesiae Pontificum*. Annotated 3rd edition by B. Schmeidler, 1917, Hahnsche Buchhandlung, Hannover and Leipzig.

Anglo-Saxon Chronicle, 787—1001. Edition by G. Garmonsway (ed.), London, 1972, and *Two of the Saxon Chronicles, parallel.* J. Earl edition, revised by C. Plummer, Clarendon Press, Oxford, 1892.

Bede, ca. 720. *Historia Ecclesiastica Gentis Anglorum* [*A History of the English Church and People*]. L. Sherley-Price, annotator, Harmondsworth, 1955.

Bibby, G., 1980. *Spadens vidnesbyrd*. Wormanium, Højbjerg.

Bondevik, S., Mangerud, J., Dawson, S., Dawson, A., Lohne, Ø., 2005. Evidence for three North Sea tsunamis at the Shetland Islands between 8000 and 1500 years ago. *Quarternary Sci. Rev.* **24**, 1757–1775.

Brøndsted, J., 1960. *Vikingerne*. Gyldendal, Copenhagen.

Bugge, A., 1904. Bidrag til det sidste afsnit af nordboernes historie i Irland. *Aarbøger for Nordisk Oldkyndighed og Historie.* **19**, 248–315.

Cameron, K., 1965. *Scandinavian settlement in the territory of the five boroughs: The place-name evidence.* Inaugural lecture, reprinted 1987 by the English Place-Name Society. The territory is normally recognized as the land of the 5 castles, and Cameron may have wrongly translated the Danish term ("5 borge").

Christensen, A., 1966. Mellem Vikingetid og Valdemarstid: Et forsøg paa en syntese. *Historisk Tidsskrift*, **12** (2), 31–56.

Christensen, C., 1983. Begrebet bol – et vidnesbyrd om Vikingetidens storbondesamfund. *Historisk Tidsskrift* **14** (4), 1–34.

Clarke, H., 1985. The North Sea: A highway of invasions, immigration and trade (5th to 9th centuries AD). *The North Sea* (A. Bang-Andersen, B. Greenhill, E. Grude, eds), pp. 39–47. Norwegian University Press, Stavanger.

Danish Royal Library, 2010. Web exhibition "Danmark på verdenskortet/de ældste tider", on website http://www.kb.dk

Digby, S., 1954. Boats and ships. *A History of Technology* (C. Singer, E. Holmyard, A. Hull, eds), p. 735. Clarendon Press, Oxford.

Fuglesang, S., 1990. Kvinnemasker. Figure 68 in *Faces of the Past* (P. Kjærum, R. Olsen, eds). Kongelige Nordiske Oldskriftselskab & Jysk Arkæologisk Selskab, Copenhagen.

Greenhalgh, M., 2003. Artserve. Australian National University. http://rubens.anu.edu.au

Grieg, S., 1947. Germundbufunnet: En høvdingegrav fra 900-årene fra Ringerike. *Norske Oldfunn VIII*, Oslo.

Grön, F., 1908. *Altnordische Heilkunde*. Erven Bohn, Harlem, The Netherlands.

Hebbeln, D., *et al.*, 2006. Late Holocene coastal hydrographic and climate changes in the eastern North Sea. *The Holocene* **16**, 987–1001.

Helbæk, H., 1971. Da rugen kom til Danmark. *Kuml*, 279-296.

Jensen, J., 1922. *Den lange rejse*, vol. 4 in the complete 1938 edition: *Cimbrenes tog*. Gyldendal, Copenhagen. Johannes Jensen received the Nobel-prize in literature 1944.

Jensen, J., 1946. Hjelmene fra Veksø mose. *Acta Archaeologica* **17**, 99–115.

Jensen, J., 2003. *Danmarks oldtid. Ældre Jernalder.* Gyldendal, Copenhagen.

Jensen, J., 2004. *Danmarks oldtid. Yngre Jernalder og Vikingetid.* Gyldendal, Copenhagen.

Jensen, S., 1990. Odin fra Ribe. Figure 71 in *Faces of the Past* (P. Kjærum, R. Olsen, eds). Kongelige Nordiske Oldskriftselskab & Jysk Arkæologisk Selskab, Copenhagen.

Jensen, S., 1997. Bydannelser. *Da klinger i muld* (S. Hvass, B. Storgaard, eds), pp. 202-205. Kongelige Nordiske Oldskriftselskab & Jysk Arkæologisk Selskab, Aarhus Universitetsforlag.

Jørgensen, L., 1990. Bornholmske masker. Figure 64 in *Faces of the Past* (P. Kjærum, R. Olsen, eds). Kongelige Nordiske Oldskriftselskab & Jysk Arkæologisk Selskab, Copenhagen.

Kinder, H., Hilgemann, V., 1964. *Atlas zur Weltgeschichte*, vol. 1. Deutscher Taschenbuchverlag, Köln. Revised Danish edition 1991, Munksgaard, Copenhagen.

Kristiansen, K., 1998. *Europe before history.* Cambridge University Press, Cambridge.

Lang, A., 1955. Traces of lost North European Sea Charts of the 15th century. *Imago Mundi* **12**, 31–44.

Livius, T., -1. *History of Rome from its foundation, Books 65, 67–68* (translated from the 4th century Latin excerpt, *Periochae,* by J. Lendering, who also estimates the publication date on http://www.livius.org/li-In/livy/livy.htm

Näsman, U., 2009. Jernalderens driftsformer i arkæologisk belysning. *Danske landbrugslandskaber gennem 2000 år* (Odgaard, B., Rømer, J. (eds), 99–116. Aarhus Universitetsforlag, Aarhus.

Niger, R., ca. 1160. Book reprinted as No. 19 in *Nachrichten der Königlischen Wissenschaftsgesellschaft zu Göttingen* (R. Pauli, ed., 1880). The part concerned with Denmark is repeated in extenso by Olrik (1899).

Olrik, H., 1891. En dansk pilgrimsfærd fra begyndelsen af 12. århundrede. *Historisk Tidsskrift* **6**, 232–235.

Olrik, J., 1899. Harald Héns love. *Historisk Tidsskrift* **7** (2. series), 177–212.

Østergaard, B., 1994. Sven Estridsens danmarkshistorie. Danmarks politiske historie ca. 890–965. *Historie/Jyske Samlinger*, 242–248. http://tidsskrift.dk, ID=75452.

Plutarch, M., ~100. *The Parallel Lives. Vol. 9: The Life of Marius.* English translation: the Loeb Classical Library edition, 1920. http://penelope.uchicago.edu

Poulsen, J., 1980. Offerfundet fra Vognserup Enge. No. 63 in *Danefæ* (P. Glob, ed.). Det kongelige Nordiske Oldskriftselskab & Jysk Arkæologisk Selskab. Danish National Museum, Copenhagen.

Ptolemaïos *alias* Claudius (after becoming a Roman citizen), ~150. *Geographia, Vol. 2, Oikoumené: Germania.* Redrawn and published 1482 in Ulm, "cleaned" version made from different sources by Schütte, 1915.

Rasmussen, F., 2005. Mid- to late-Holocene land-use change and lake development at Dallund Sø, Denmark: vegetation and land-use history inferred from pollen data. *The Holocene* **15**, 1116–1129.

Rieck, F., Andersen, S., Roesdahl, E., 1993. Landets forsvar. *Da klinger i muld* (S. Hvass, B. Storgaard, eds), pp. 210–214. Kongelige Nordiske Oldskriftselskab & Jysk Arkæologisk Selskab, Aarhus Universitetsforlag.

Sarnthein, M., Kreveld, S., Erlenkeuser, H., Grootes, P., Kucera, M., Pflaumenn, U. Schulz, M., 2003. Centennial-to-millennial-scale periodicities of Holocene climate and sediment injections off the western Barents shelf, 75°N. *Boreas* **32**, 447–461.

Sawyer, P., 1962. *The Age of the Vikings.* Edward Arnold, London.

Saxo Grammaticus, ca. 1200. *Gesta Danorum.* First Danish translation by A. Vedel: *Den danske krønicke.* Støckelman & Gutteruitz, 1595 (Royal Danish Library; ProQuest Early European Books website). English translation of the first 10 Chapters by O. Elton (New York, 1905), available at http://www.gutenberg.org

Scharling, W., 1885. Befolkningens størrelse i ældre og nyere tid. Tilvækstforhold. In *Danmarks Statistik* (V. Falbe-Hansen and W. Scharling, eds.), afsnit "Befolkningen", pp. 321-429. Forlagsbureauet. Copenhagen.

Scharling, W., 1889. Kirketallet og Folketallet i Danmark i det 13. Aarhundrede. *Historisk Tidsskrift* **6** (2), 264–318.

Schietzel, K., Crumlin-Pedersen, O., 1980. Havnen i Hedeby. *Skalk* (3).

Schmitt, L., Larsson, S., Burdukiewicz, J., Ziker, J., Svedhage, K., Zamon, J., Steffen, H., 2009. Chronological insights, cultural change, and resource exploitation on the west coast of Sweden during the late paleolithic/early mesolithic transition. *Oxford J. Archaeology* **28**, 1–27.

Schütte, G., 1915. Nord- og Mellemevropa efter den rensede Ptolemaios. *Geografisk Tidskrift* **23**, 257–268.

Skovgaard-Petersen, I., 1971. Oversigt: Vikingerne i den nyere forskning. *Historisk Tidsskrift* **12** (5), 3–75.

Sørensen, B., 2004. *Renewable Energy.* 3rd edition, Elsevier Academic Press, Burlington.

Stenak, M., Rømer, J., Näsman, U., Odgaard, B., 2009. Landskabsforandringer gennem 2000 år. *Danske landbrugslandskaber gennem 2000 år* (Odgaard, B.,

Rømer, J., eds), pp. 283–302. Aarhus Universitetsforlag, Aarhus.

Strabo, ~20. *Geography*, section 7.2.1 (English translation by H, Hamilton and W Falkoner). Georg Bell & Sons, London, 1903. http://old.perseus.tufts.edu

Thomas, D., ~75. *Gospel of Thomas.* Part of Nag Hammadi scrolls from 4th century, translations by T. Lambdin (http://home.sol.no/~noetic/nagham/gthlamb.html) or by S. Patterson and M. Meyer (http://www.gnosis.org/naghamm/gosthom.html). Gnostic Society Virtual Library at http://home.sn.no/home/noetic/libe.htm

þórðarson, Ó., 1259. Knýtlinge Saga (saga of the descendants of the Danish king Canute). In *Jómsvíkíngasaga ok Knytlínga með tilheyrandi þáttum*, pp. 209–402, vol. 11 in "Fornmanna Sögur, eftir Gömlum handritum", Norræna Fornfræða Fèlags, Copenhagen, 1928 (B. Kolbeinsson, ed.), Danish translation by C. Rafn in *Jomsvikinga Saga og Knutlinga tilligemed Sagabrudstykker og Fortællinger vedkommende Danmark*, pp. 159–357, vol. 11 in "Oldnordiske sagaer", Det Kongelige Nordiske Oldskrift-Selskab, Copenhagen, 1929. Both available at http://books.google.com based on copies in the library of Harvard University, Boston.

Ulsig, E., 1991. Pest og befolkningsnedgang I Danmark i det 14. århundrede. *Historisk Tidsskrift* **15** (6), 20–43.

Velschow, H., 1843. Om Folkemængden i Danmark i Midten af det trettende Aarhundrede. *Historisk Tidsskrift* **1** (4) 1–52.

Wittwer-Backofen, U., *et al.*, 2008. Basics in Paleodemography: A comparison of age indicators applied to the early Medieval skeleton sample of Lauchheim. *American Journal of Physical Anthropology* **137**, 384–396.

# 6

# Medieval society: Church building and a little dissent

The Roman Church brought stone materials into Danish architecture. During the 12th century, 2358 stone or brick churches were erected all over Denmark, followed by a few hundred more during the 13th century (Scharling, 1889; based on the Knytlinge saga mentioned in Chapter 5). Farmers and poor people continued to live in wooden houses, but soon the richer people wanted to copy the new style, and particularly in cities, streets became lined with impressive brick façades. The habit spread to the countryside, first with estate residences for the big landowners, later extending brick usage to many standard farmhouses.

Where did the money come from to pay architects, builders and raw material providers? The answer is to be sought in the continuing transformation of society. From providing basic food and shelter needs, the Stone and metal age societies had moved on to provide jewellery, art and weapons, and by the Viking age construction, including the means of transportation (ships, horses, carriages, roads), had reached a high level, thereby creating professions such as carpenters, artisans, merchants and warriors, all earning salaries beyond the self-contained economy of farmers chiefly providing for the family or their immediate community. By the medieval period, which in Danish context may be defined as 1050 to 1536 (ending at the formal adoption of Protestantism), a large number of activities had created a complex and diversified society. The largest percentage of the population was still employed in agriculture, but the city professions had come to control a major part of the money flow. Political power was divided between king, nobility and church, sometimes collaborating and sometimes opposing each other. On a slightly more basic level, one may distinguish between three kinds of power:

- Economic power
- Military power

• Spiritual power

Economic power is traditionally associated with those able to trade goods. It enabled the large farmers to turn the peasants into slaves. Similar control over people could be obtained by military threat. That approach was used by warlords and kings to expand their territories and get the conquered territories to pay for survival, through ransom or, more conveniently (for the sovereign), by recurrent taxation. The Roman Church had realized that spiritual power could supplement the other forms, by imposing beliefs (such as a promised afterlife, to be granted on condition that certain services and recognition were accorded to the Church). This did not mean that the Church shied away from acquiring and using military or economic power. The claim for 10% of all revenues in society to be paid to the Church was universal in all regions infiltrated by the Roman Church (it seems to have been invented in France; see Szomlaiski, 1975).

The Medieval period may have been a "Dark Age" in respect to questions of human rights and equity, but as regards technical and economic development it was not nearly as much so as is sometimes suggested (although there were certainly restrictions and censorship on art, literature and science that did not depict an Earth-centred cosmology or a conformist view of Catholic morals). However, engineering progressed, and the increased emphasis on law and regulation also had positive sides in allowing the smooth functioning of more complex societies based upon ordered relations between people.

While the money to build churches in cities and villages came from the 10% taxation and from generous gifts from the king and from rich citizens (in exchange for a prominent grave location and often after a measure of coercion exerted by the Church), the physical resources required would demand a significant development of raw materials and construction industries. Below, an attempt is made to estimate the amount of energy needed for the surge of church–building, directly and indirectly. This will enter the total energy balances, trying to encompass also the energy associated with other construction work in society, notably the foundation or expansion of cities, as well as energy connected with existing and novel end-use activities.

In the minds of most ordinary people, the (Roman Catholic) Christian Church became synonymous with brutal taxation, often involving armed tax collectors torturing simple people to give up parts of their grain stores, livestock and property in order to support (at least so it would seem) the extravagant castles and churches allowing bishops and other high-ranking Church representatives to live in luxury. This explains many of the difficulties that the new religion experienced in Denmark during the centuries following its introduction. Danes did not like taxes. They still do not. Even in recent years, incompetent governments have been elected just on the basis of promises to cut (or not to raise) taxes.

The Roman Church may initially have been called in by the Danish king in the hope that it could help him subdue obstinate local chieftains, but the outcome can be seen as the creation of a new power struggle between the king and the Church, one that may be seen as finally won by the king at the Reformation in the 16th century, but also one that created prospects for the local chieftains to influence things. They were sometimes leaning towards the king, acquiring titles of nobility from him, and at other times fighting or assassinating kings, in order to get complacent rulers who accepted their roles in society.

*Figure 6.1. Esbønderup church (north Sealand), its main part from 1130, tower from 1270 and extensions from 1735 (photo by author).*

The construction of churches took place both in cities and in smaller communities such as the villages formed during the preceding centuries, when farmers felt a need to move together in order to be better able to defend themselves against possible raids. The 10–11th century wooden churches have left only traces (see Fig. 5.20 and Randsborg,

2008), and they were replaced by new and larger structures complying with shifting architectural fashions, including additions (towers, chorus, apses, sacristies and chapels), but in a fashion that often left the original architecture intact (Exner and Ebert, 1968). Figs. 6.1 to 6.3 show some typical styles of 12th century village churches.

*Figure 6.2. Jørlunde church (Sealand). Originally built 1185, with tower and chorus added later. Like many other village churches, a white chalk layer cove over the bricks has been added (photo by author).*

*Figure 6.3. Østerlars church (Bornholm Island in the Baltic Sea). This round church, built in 1120 (with support pillars added about 1600), is one of a handful with round rather than square geometry (photo by author).*

The building materials preferred during the early period were travertine (mainly bicarbonate, $Ca(HCO_3)_2$), sandstone or glacially deposited rock material such as granite. Later, bricks made from extensively available clay deposits found in Denmark were used. The Danish clay deposits first used provided bright red bricks, used for nearly all buildings between 1200 and 1950, when these clay deposits were finally exhausted and other deposits delivering less stunning yellow bricks took over (currently, artificial colouring is in widespread use). The brick size has diminished with time, starting around $10 \times 14 \times 30$ $cm^3$ and diminishing to about $6 \times 10 \times 20$ cm$^3$.

Round churches such as the Østerlars church (diameter 17 m) depicted in Fig. 6.3 were serving as defence retreats during times of attack (note gun slits in second–floor façade).

*Figure 6.4. Mural frescoes in Nørre Herlev church (Sealand), depicting (left) a lady in a barrow, being rolled into the flames expelled by a fierce dragon, and (below) a male (possibly St. Laurentius) is roasted over a blown fire-bed (photo by author).*

The Esbønderup church depicted in Fig. 6.1 shows how the basic stone church has been expanded using bricks, both upwards with vaults and sidewards with tower and apse. These features are found in the history of most 12th–century Danish churches.

During the 14th and 15th centuries, many of the village churches were ornamented with mural frescoes. Because there were so many churches, these were made by artisan ateliers rather than by individual artists. The work of these artisan groups, such as the Isefjord workshop responsible for the frescoes shown in Fig. 6.4, shows humour and appreciation of the diversity in human life, as well as an enjoyment in painting devils torturing fair ladies or adventurous males. Is the torturing of St. Laurentius meant to pity the monk (Fig. 6.4b), or does the placement among the last judgement pictures suggest a place for monks in hell? Irreverent suggestions are common in carved ornaments on church chairs and pulpits from the same period. One church sports a folk-dance fresco without the slightest religious connotation (Fig. 6.5). The focus on hell (in Fig. 6.4) was obviously selected in an effort to scare people into listening to the doctrines of the Church. The institution had also identified sexuality as an important driver of human activities and put great effort into trying to control sexual behaviour (defying the tolerance exhibited by the historical Jesus).

*Figure 6.5. Folk dance scene from fresco in Ørslev Church, Funen (engraving reproduced from Erslev, 1898–1905, made after a drawing by J. Kornerup now in the Danish National Museum). The original frescoes in Danish churches are showing signs of decay, due to ammonia pollution from current Danish chemical agriculture (Skytte, 2010).*

An estimate of the energy used in constructing a medium-size (10 × 20 × 30 m$^3$) basic church (no bell tower or other additions) in the 12th century is made in Table 6.1 and scaled up to the actual building activity of an average of 20 churches per year. The basis for the investigation is an inventory of the materials used for churches built during the 12th century. Most of the early ones had foundations and walls made from stones found on fields and coasts nearby (Fig. 6.6). The bedrock in Denmark is covered by 2 to 4 km of clay and sand layers, so stone material was restricted to what was brought along by the glaciation and deposited when the ice melted. As indicated in Table 6.1, no quarry-style mining was required, only digging it up. However, some stones had to be split or adjusted by cutting in order to fit as building blocks for church walls. According to the estimate made in Table 6.1, this took more energy than digging up the stones. For other stone materials, the embedded energy was less. The ones available in Denmark were travertine from chalk deposits and occasional sandstone. Digging is assumed to have involved human labour, and the supplied-energy use is thus taken as the additional food energy to be taken in, equal to three times the muscle work performed (cf. Chapter 1).

*Table 6.1. Estimated energy expenditure for 12th century church construction, considering only the core building, not later additions.*

| ***Church building 12th century*** | ***MJ/kg*** | ***kg/m$^3$*** | *GJ/y* | ***Annual average energy usage flow (kW for 20 churches per year)*** | | | | |
|---|---|---|---|---|---|---|---|---|
| *Activity (20 churches constructed each year, with an average core volume of 10x20x30 m$^3$)* | *Specific energy content and density of materials* | | **End-use energy** | **Extra food energy** | **Low-temperature heat** | **High-temperature heat** | **Mobile power (excl. food)** | TOTAL |
| **Materials** | | | | | | | | |
| Raw materials | | | | | | | | |
| Glacial origin granite (30%) dug up | 0.8 | 2800 | **11424** | 1087 | | | | **1087** |
| Travertine (10%) cut out | 0.5 | 1200 | **1020** | 97 | | | | **97** |
| Sandstone (10%) cut out | 0.7 | 2100 | **2499** | 238 | | | | **238** |
| Adobe (earth/straw), for annexes | 0.1 | 1600 | | | | | | **0** |
| Gravel/sand (5% of wall volume) | 0.1 | 1600 | **136** | 13 | | | | **13** |
| (used for mortar and foundation) | | | | | | | | **0** |
| Clay for bricks, tiles (50%) | 0.1 | 2000 | **1700** | 162 | | | | **162** |
| Processing and unit sizing | | | | | | | | |
| Cutting granite | 1 | | **14280** | 1367 | | | | **1367** |
| Soft stone cutting | 0.5 | | **2805** | 269 | | | | **269** |
| Travertine drying | 0.3 | | **612** | 2 | 58 | | | **60** |
| Adobe mixing and drying | 0.5 | | **solar** | | | | | |
| Mortar manufacture | 2 | | **2720** | 259 | | | | **259** |
| Brick, tiles burning | 2.5 | 1800 | **38250** | 39 | | 3600 | | **3639** |
| Iron and steel | | | | | | | | |
| Nails, bars, ornaments (500 kg) | 32 | 7200 | **320** | 2 | | 30 | | **32** |
| Wood | | | | | | | | |
| Supports, rafters (100 m$^3$) | 0.5 | 700 | **700** | 67 | | | | **67** |
| Doors, seats, divisions, etc. | 0.3 | 700 | **42** | 4 | | | | **4** |
| **Transportation** | | | | | | | | |
| Quarry to construction site | | | | | | | | |
| Local materials (20%, <10 km) | 2 | | **15000** | 2 | | | 474 | **476** |
| Non-local materials (80%, >50 km) | 10 | | **300000** | 20 | | | 9513 | **9533** |
| (cart with oxen on dirt road, ship) | | | | | | | | |
| **Construction** | | | | | | | | |
| Building craftsmen, workmen | | | **1260** | 40 | | | | **40** |
| Ornamentation artisans | | | **130** | 4 | | | | **4** |
| (time-use 20 man-years/church) | | | | | | | | |
| ***Total*** | | | ***392898*** | *3672* | *58* | *3630* | *9987* | ***17347*** |

*Danish materials: ashlar (granite, sandstone, travertine), bricks (percentages given are for 20 churches built per year)*
*Wall thickness = 0.4 m (e.g. granite ashlars or double layers of bricks), floors/foundation 0.4 m, roof thickness 0.2 m*
*Church core: average volume of materials taken as 850 m$^3$ (floor/foundation, roof and sides)*
*Input energy is calculated as end-use energy divided by a conversion efficiency (e.g. food to muscle work)*
*End-use energy is taken as time-independent, based on Baird et al. (1997) and Sørensen (2007)*

*Figure 6.6. Stones washed ashore on north Sealand beach, ready to be picked up and used as building materials (photo by author).*

Travertine was at a reasonable distance from a few church building sites, such as Roskilde, but in most other cases, transportation of the stone material from the location of deposits to the construction site was difficult and (in energy terms) expensive, according to the estimate made in Table 6.1. Available roads were poorly suited for transporting heavy stone material by carts driven by oxen, and this fact, as well as the more convenient craftsman work offered by bricks, made bricks replace most of the stone materials used for churches by the end of the 12th century. Brick-burning was already an established technology, and the church builders brought along standardization of brick sizes and wall erection procedures. In Table 6.1, the end-use energy associated with brick-making and with transportation of building materials is taken from current studies, but the delivered energy is estimated to be much larger than it would be today, due to low conversion efficiencies. For transportation, the high rolling resistance and inefficient power transmission system of heavily loaded carts are taken to imply a ten times lower overall efficiency than for a modern lorry. The rationale for this method of estimation is that end-use energy is not supposed to change as long as the general chain of actions used to satisfy a given demand is not altered. This is the case here, where both are road transport (or occasional sea transport). Still, the energy given in the right-hand side of Table 6.1 does not include the metabolism of draft animals, but only the muscle power they deliver. This is in contrast to muscle power delivered by humans, where the three times higher food energy is given, but that is just a convention.

The result of the investigation is that the actual construction work is a minor energy user, while the raw materials and their transportation to the building site are the over-riding components of energy use. In total, the estimated average rate of energy expenditure for church-building is about 17 MW throughout the 12th century, and this being only for the basic church structure. Adding apses, bell towers and other auxiliary structures would have increased this, and as the century progressed, other stone and brick buildings were constructed, at first by the king and the wealthy segment of society (a segment that widened with time). The energy for these buildings is likely to have been as large (or probably even larger than) that used for church-building, adding altogether close to 40% to the energy used a few hundred years earlier, when all structures were made from wood (cf. Table 5.5).

*Figure 6.7. Egeskov castle (Funen), built 1554 by Jakob Brokkenhus in a similar style to that of the 13th century Nyborg castle, also on Funen (photo by H. Sørensen, 1955).*

Over the following centuries, both the king and local squires built castles such as the one depicted in Fig. 6.7, and so did the bishops of the Church, who felt they had to be able to defend themselves against citizens angry over the taxes collected, or bandits just wanting to lute the property of the church officials. An early example is Søborg castle, possibly built by Archbishop Eskil (cf. Saxo Grammaticus, ca. 1200), later captured by the king to use as a prison (e.g. for Archbishop Jens Grand suspected of plotting against the state). Such castles became targets for later anger against king or church, and most of them were torn down during quarrels such as "the feud of the count" (1534; Fig. 6.8). Also, a number of convents were established by orders from southern Europe. Again, few have survived (Fig. 6.9).

*Figure 6.8. Ruin of Søborg Castle (north Sealand), allegedly fortified by Archbishop Eskil in the mid-12th century and destroyed in 1534, in connection with the attempted take-over of the country by Count Kristoffer of Oldenburg (Heise, 1899–1906; further stone removal was authorized by the king in 1577, for the building of Kronborg Castle; Fig. 7.5; photo by author).*

Stone and brick technologies allowed buildings to use new technologies (e.g. for space heating). The convent in Esrum (Fig. 6.9) was equipped with hypocausts leading warm air from a basement boiler-room to the upstairs rooms to be heated, through channels in the floor with sliding controls (Olsen, 1875; Kornerup, 1879). When firing wood in the boiler to heat a large stone heat-store present in the basement structure, a damper placed in the duct could close access to living rooms and send the exhaust to the chimney. The same damper, released by controls located in the upstairs rooms, would open for transfer of heat stored in the large heated stone mass in the basement to the upper room space, at the same time closing the path to the chimney (Hertz, 1997). Figure 6.10 illustrates the system. Such hypocausts were already common in the Mediterranean region a millenium earlier, during the Roman period (Sørensen, 2010); they later developed into the current central heating systems, using air from heat exchangers or hot water circulation.

*Figure 6.9. Remaining wing of Esrum Convent (north Sealand), built by the French-based Cisterciensis Order in mid-12th century (photo by author).*

*Figure 6.10. Modes of operating a hypocaust central heating system, such as the one proposed by Hertz (1997), on the basis of Roman counterparts (cf. Sørensen, 2010), to have been part of the Esrum convent construction. Valve-equipped tiles were found in 1875 near the site of the remaining building (Fig. 6.9), but they have later been lost (drawing by author).*

The medieval period also found new ways of satisfying the demand for high-quality energy, first by waterwheels and towards the end of the period by windmills. Waterwheels even then remained considerably more abundant than windmills, even in a country such as Denmark, with very modest elevations of water (the highest point in the country is some 170 m above sea level). The waterwheel (horizontal or vertical) technology developed through stages (illustrated for vertical wheels in Fig. 6.11).

- Undershot wheel (introduced about 2500 y BP in Roman Empire; about 900 in Denmark).
- Overshot wheel (~2200y BP by Romans, ~1100 in Denmark).
- Breastshot wheel (~300 in Roman Empire, ~1400 in Denmark).

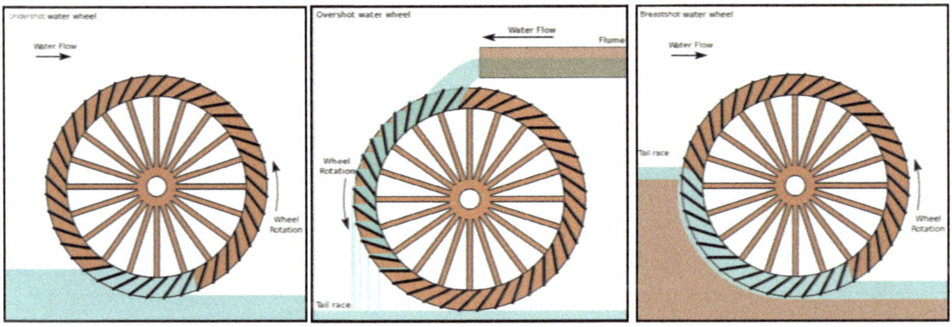


*Figure 6.11. Waterwheel types (from Wikipedia by Creative Common License, drawn by a person with chosen identity KVDP, 2009).*

Only traces of waterwheels from the early medieval period remain in Denmark (Steensberg, 1959; Fischer, 2004). The lifetime of the moving parts was less than a hundred years, and they were then replaced. Figure 6.12 shows a more recent example. The power gained was primarily used for grinding cereals, and occasionally industrial production (sawmills, hammer-mills). Typical heads were just a few metres, and later constructions were associated with natural or artificial reservoirs (stemming up water in lakes), in order to be independent of seasonal stream flows. The total number of watermills installed would by the late 12th century have been more than a hundred, but with total average power production of the order of only one megawatt, as judged from the physical size (assuming later replacements of at least the same size) and water heads available. The technology was truly reserved for high-quality purposes. During the following centuries, the number of watermills would have reached several hundred (Hjelholt, 1932; estimate based on water- and windmill registration for 17th century taxation purposes; Böcher, 1952, finds 1000 watermills present by 1838).

As mentioned in Chapter 5, the Danish settlers in France and England soon became integrated into the local populations, and when Hardeknud, the king following Canute the Great in England, died childless in 1042, there was no reason for England to look to Denmark for a new king. In Denmark, a range of monarchs followed, with varying skills and ability to deal with the problems that kept appearing in domestic as well as foreign policy.

*Figure 6.12. Undershot hammer-mill built 1765 (restored 1960–82) at Hellebæk for a north Sealand military arms factory. Earlier watermills on the site include a cereal-grinding mill from 1576 and a previous hammer-mill from the end of the 16th century, associated with a cannon foundry. A partly artificial reservoir system provides the low-head water (cf. Museum Council of Frederiksborg County, 2009; photo by author).*

Assaults from German sovereigns and rivalry with Norway were the persistent problems, soon followed by papal attempts to control Denmark through the German bishop in Hamburg, but repudiated by king Svend Estridsen (1047–1074). The Wends had been defeated in 1043, and peace with Norway was agreed after a battle in 1062 (see Adam Bremensis, ca. 1075). Disputes between the kings and influential citizens played an important role, and several Danish laws were common laws and regulations of royal succession stating limits to royal liberties of ruling (as described in the two earliest historical accounts not authored by foreign writers: Chronicon Roskildense, ca. 1140, and Saxo Grammaticus, ca. 1200). Saxo, of the noble Hvide family (Weibull,

1978), focuses his elegant tale on kings and victories, while the anonymous Chronicon writer, believed to have been associated with the cathedral at Roskilde, is more concerned with the fate of the losing parties. Both use Adam Bremensis (ca. 1075) as a source.

Bishop Eskil at Roskilde provides an example of the political role adopted by the Church. In 1136, Eskil instigated a revolt against king Erik Emune, involving a number of local squires, and in 1137, after the king had regained military superiority and subdued the rebels, the king was murdered (Saxo Grammaticus, ca. 1200). Eskil was at the side of the following king, Erik Lam, in defending the country against a Swedish invasion attempt (1142–43), and the reward allowed him to establish Esrum Convent (Fig. 6.9). However, king Lam abdicated and the following chaos (three king claimants) allowed German Friedrich I Barbarossa to interfere and insist on his right to decide who should rule Denmark. Order was restored by King Valdemar I (the "Great"), who by 1157 had defeated the rival who had just killed the second claimant, and he inserted his friend Absalon as bishop in Roskilde, thereby regaining a national unity that he hoped could resist the German threat. Absalon turned out to be a capable naval commander, who defeated the Wends and regained Danish supremacy over the western part of the Baltic Sea. The Danevirke defence wall towards Germany (Fig. 5.17) was fortified and the achievements of Absalon overshadowed those of the following king (Knud). After Knud, Valdemar II (1202–1241; nickname Sejr, or "victory") took over. He spent much time fighting in northern Germany and the east Baltic nations. In 1223 he was kidnapped by the German count Henrik of Schwerin. A ransom was paid and most Danish possessions in northern Germany had to be given up, and the Germans were accorded unlimited trade rights in Denmark.

Valdemar II appealed to the Pope, who relieved Denmark from all the commitments agreed to under pressure, and demanded that the ransom money was returned to Denmark. This did not happen, and Valdemar II invaded Germany, but was defeated at Bornhöved (Fig. 6.13), after which experience he kept a lower profile, although still opposing German trade campaigns in the Baltic region. After Valdemar II, weak kings continued to fight enemies within and outside the country. The period 1250–1286 saw three Danish kings being murdered, and the kings' latitude of power was further restricted in favour of squires and counts, through coronation charters. Unfortunately the charters did not restrict the financing of the Crown's activities by pawning the land. Creditors, mainly based in Holstein, tried to rule the country without a king from 1332, with little success. The main creditor, a Count Gerhard from Holstein, was murdered by a Danish squire in 1340, leading to the power being grabbed by a new king, Valdemar IV ("Atterdag", meaning "another dawn", 1340–1375), who repaid the loans through increased taxation and reconquered the pawned land if the Holstein mortgagees did not voluntarily release it (Erslev, 1898–1905; Szomlaiski, 1975; Skovgaard-Petersen *et al.*, 1977).

*Figure 6.13. 13th century illustration of the battle at Bornhöved. Redrawn in Steenstrup (1904) after a Saxon chronicle at the former Berlin Royal Library.*

The common people had less and less influence on their life conditions. Local councils and city councils were no longer listened to, and only the power struggles between king, landlords and clerical heads seemed to matter. To this came, as mentioned in Chapter 5, the Bubonic plague that is estimated to have cost some 30% of the population their lives during roughly a decade around 1350 (cf. Figs. 5.7 and 6.22 below). Property and taxation records show that many farms and villages had been emptied of inhabitants (Ulsig, 1983, 1991). The even stronger impact of the plague in other European countries made the demand for agricultural products fall and prices decline, to the disadvantage of exporting countries such as Denmark.* Valdemar IV added to the precarious situation by demanding that peasant slaves (for which euphemisms such as serf or adscript were used) should become available a certain fraction of their time for villeinage to the Crown. However, he got away with this, because many of the local landlords could see the macroeconomic necessity of his measures.

* The main Danish export articles were cattle, butter and fish (Kinder and Hilgemann, 1964).

The north German traders in the Hanseatic cities became stronger, and their organization in what was called "the German Order" (1226-1410) became a commercial (and later physical) threat to all of Scandinavia. The Order was a knight-type monk-state led by a "Grand Master". It had conquered the entire Baltic south coast in what was termed a crusade, and the "Hanse War" of 1361–1370 over the islands Öland and Gotland ended with a peace treaty signed at Stralsund, where Denmark had to accept the Hanseatic supremacy over the Baltic Region.

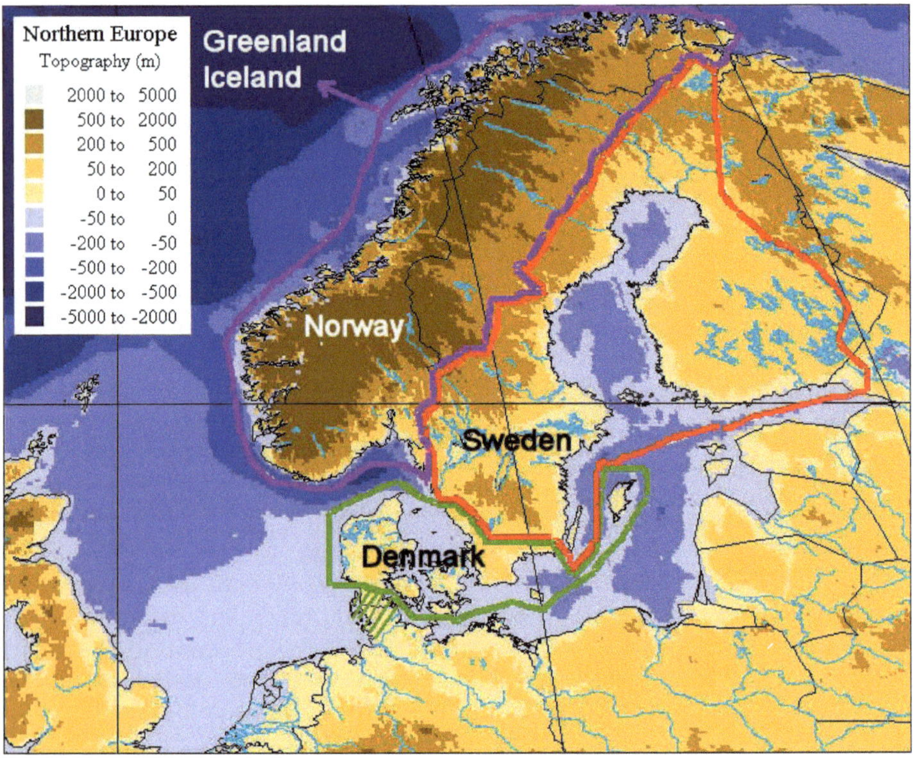


*Figure 6.14. Territory of the Nordic Union (1397–1523), with the subdivisions existing when the Kalmar treaty was drafted. In the northern part, the border between Sweden and Norway moved back and forth a number of times. The Danish Fief of Schleswig/Sønderjylland was under administration by the Holstein money-lending counts, a fact that regent Margrethe tacitly accepted while waiting for an opportunity to take it back (the opportunity to gain both Schleswig and Holstein came in 1460, after her death).*

Trading posts (called "factories") had been established all over the three Scandinavian countries, and the dissatisfaction with this German influence was the main reason that Sweden considered joining the already existing coalition between Norway and Denmark. Formally, a union was formed in 1397, after Margrethe I (1353–1412) in 1380 had become regent in both Denmark and Norway (see Chapter 5). It lasted until 1523, with Sweden being increasingly unhappy with the arrangement. The initial union territory is shown in Fig. 6.14. Little change occurred during the union period, but, both before and after, the northern Sweden–Norway border was moved back and forth. The document that formalized the Kalmar Union was less than a regular treaty, as it missed several necessary signatures (Fig. 6.15), but nobody questioned Margrethe I as the Nordic ruler. She was a cunning leader, renouncing formal approval if it was not forthcoming and satisfied with a *de facto* acceptance of her position as the ruler.

The reign of Margrethe I (formally representing as regent her son, the child king Olav, until his death in 1387, and later another minor, Erik, until her own death in 1412) coincided with improved economic conditions, and she managed to strengthen her position relative to the squires, who lost the right to fortify their estates, and had to give back the Crown property they had acquired during the reign of the preceding kings. Margrethe left several traditionally independent administrative positions unfilled, allowing her own offices to control military, legal, finance, tax and foreign affairs. Her sarcophagus stands as the centrepiece in Roskilde Cathedral (Figs. 6.16 and 6.17).

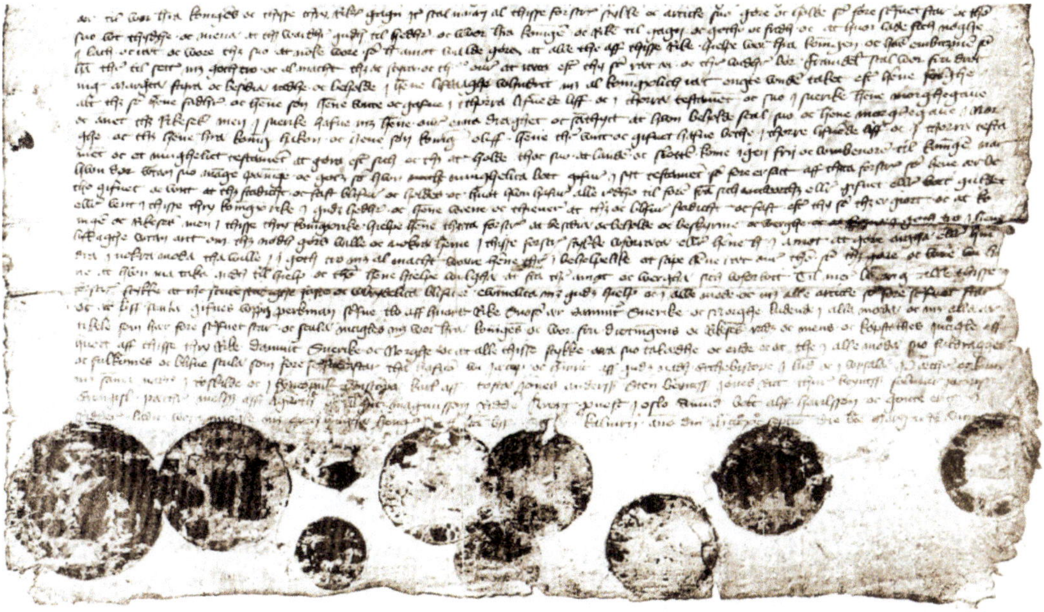

*Figure 6.15. Excerpt from the Kalmar Union Charter 1397, with the incomplete number of seals (Rigsarkivet/Nationalmuseet, Copenhagen; photo from Erslev, 1898–1905).*

*Figure 6.16 (top): the sarcophagus of Regent Margrethe I, finished in 1423, 11 years after her death, and placed in Roskilde Cathedral. The inset above details the head of the queen, a fine piece of alabaster work.*

*Figure 6.17 (left). Roskilde Cathedral (rebuilt several times, with the gothic spires added in 1635). (All photos by author.)*

The country squires placed increasing emphasis on cattle and sheep production, which was in higher demand for export than cereals, but which also required more primary land to be cultivated or used as rangeland, in order to provide fodder for the livestock animals. Small independent farmers disappeared, and the country nobility formally declared itself hereditary, thereby cementing the class division between the noble families and the farm-helping slaves. The squires enjoyed increasing relief of land taxes, but on the other hand, they had a commitment to provide manpower to the king's army. Acquiring new land was in competition with the Crown and the Church, of which the latter by the 15th century began to lose land ownership (Szomlaiski, 1975).

Before discussing the continuation of the Danish energy consumption curve, the demographic model must be updated. The population model that was used in the previous chapters was based on death rates with identifiable components that changed only when definite events demanded it, but with a birth rate distribution essentially unchanged since the Stone Age, only its overall amplitude being changed when data required it to be. The birth rate distribution last shown in Fig. 4.17 assumed female individuals to start giving birth soon after their bodies had matured to womanhood. There is ample evidence that this changed with time. Concrete data from 1901 gives the age distribution shown in Fig. 6.19, but that distribution, with a much later age of first becoming a mother, is known to have persisted for a while, as judged from the Church records on age of marriage and baptism of children (preserved in a few cases back to the early times of the Christian Church in Denmark; unfortunately, many churches have lost their records, typically in fires).

Except for the highest strata of the layered society (where other pressures prevented most girls from having early births), the morality imposed by the Church ensured a high age for first birth, combined with the fact that young farm-helping men did not have the economic means to marry and establish a family until rather late. There is no data to tell precisely when the change in age at first birth date took place, but a reasonable guess is that it happened in association with the adoption of the institutionalized Christian religion. Still, the change in habits could have happened earlier. The earlier fertility (number of offspring for a woman living through her entire fertile period) of 7–8 would imply an average distance between live births of 4 years. Starting as late as indicated in Fig. 6.19, a similar fertility would require reduction of the interval between births to about 2 years. This does describe the situation in the 19th century, but during the earlier periods, including the medieval ones, fertility was more likely to have been reduced to 4–5.

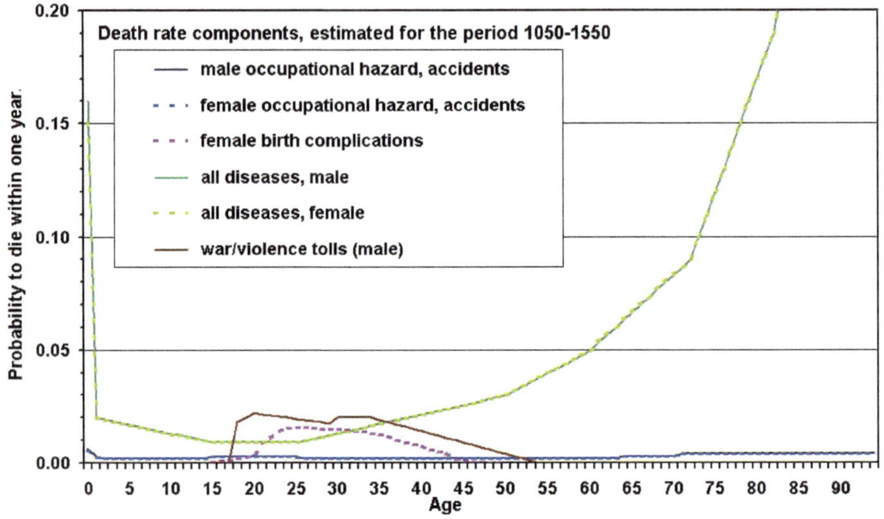


*Figure 6.18. Gender-specific mortality components for Danes used in the demographic model from 1050 (with later scaling adjustments).*

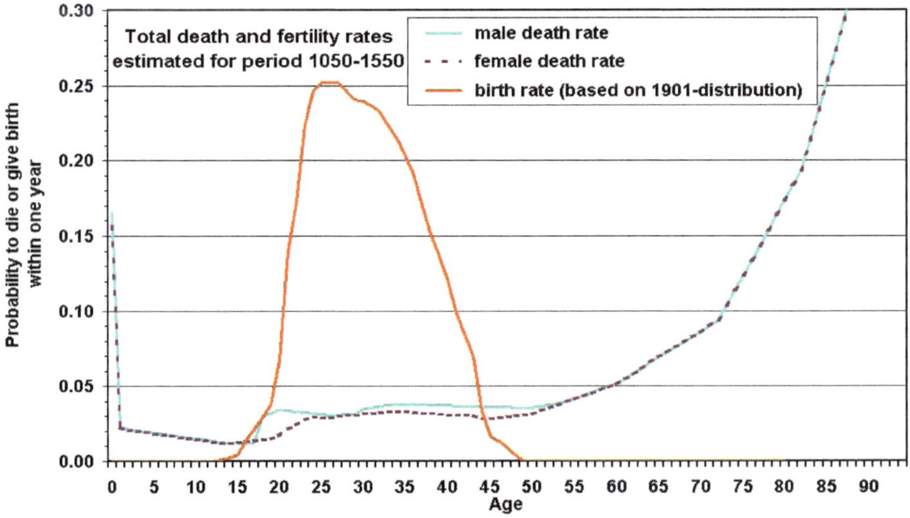


*Figure 6.19. Overall fertility and mortality rates for Danes used in the demographic model from 1050 (with later scaling adjustments).*

This is definitely the case from about 1800, where fertility statistics exist for all of Denmark (Mathiessen, 1976), or from 1650, where statistics exist for certain areas outside cities, such as Sealand (Lassen, 1965). The demographic model cannot tell which level of fertility is correct, because similar results may be obtained by adjusting both birth and death rates up or down in consort.

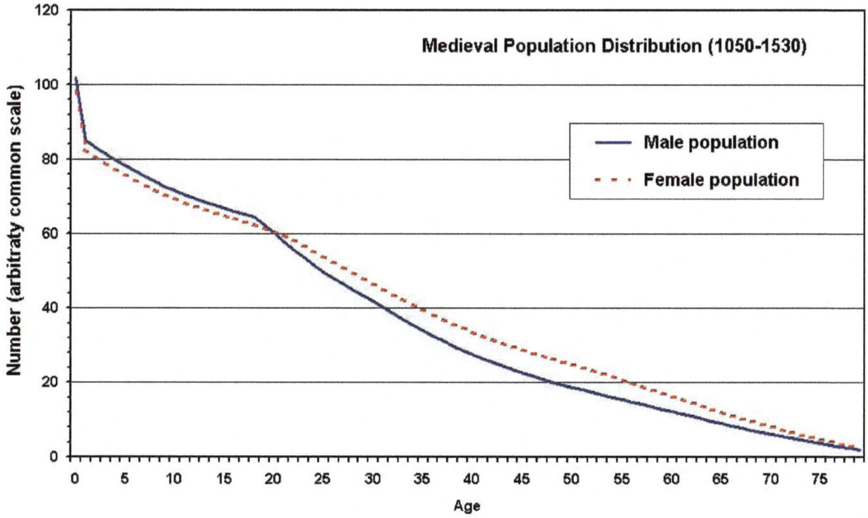


*Figure 6.20. Age distribution of Danish population in late medieval period (around 1350), derived from birth and death rates*

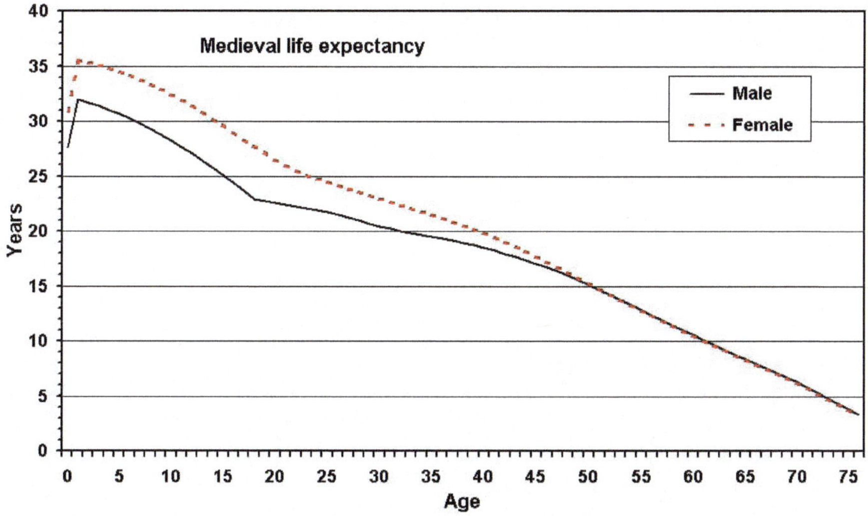


*Figure 6.21. Life expectancy of the Danish population in late medieval period (around 1350), derived from birth and death rates.*

The death rate components shown in Fig. 6.18 differ from the earlier ones shown in Fig. 4.16 in several aspects. While mortality from diseases has not changed dramatically, the death rates associated with accidents have greatly decreased, as people no longer had to confront large prey animals or risk breaking their legs far from the base camp. On the other hand, the male age group from 18 to 30s has death rates associated with armed conflicts and wars, which at times could be quite substantial. They have been singled out in Fig. 6.18, but still should be considered as averages, not going into the specific timing of assaults and battles (although this could have been done, as these were increasingly being recorded). The death rates of women in connection with childbirth are substantial in Fig. 4.16, and although they are still important in Fig. 6.18, they have been reduced due to better hygiene and midwifery knowledge.

The increase in overall violent death rate for young males (Fig. 6.18) makes the gender-specific age distributions change in favour of females, as seen in Fig. 6.20, and reflected in the life expectancy as a function of age achieved, shown in Fig. 6.21. The corresponding figures for the earlier times are Figs. 4.18 and 4.19, and the overall shape of both curves is seen to have changed little.

Specific incidents that have influenced birth or death rates are incorporated into the demographic model, for which the population estimates are given in Fig. 6.22 for the period ending 1770 (but with the curves continued to present time, still only as illustration and for comparison with the model changes described in the following chapters).

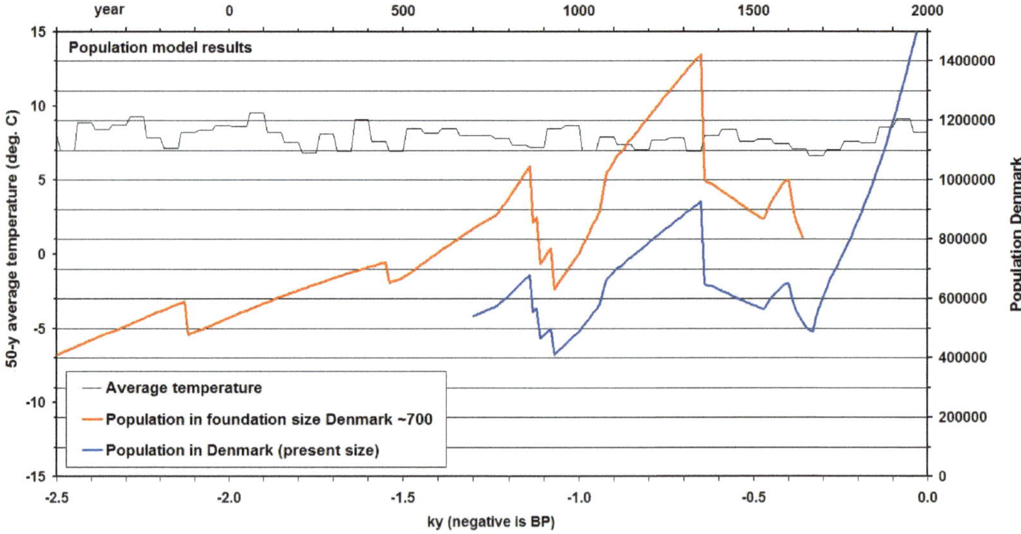


*Figure 6.22. Danish population development with incorporation of new demographic model components for the period 1050 to 1770. The model output is given both for the present (year 2000) area of Denmark and for the year 700 area at the formation of the kingdom.*

The early Danish medieval society saw a substantial growth in population. The loss by immigration, particularly to England and France, had already been made up for by the beginning of the medieval period, and although only a relatively small fraction of the land area was under cultivation, agricultural production enjoyed fertile soils. During the late medieval period, harvest yields per unit area were on average reduced by more than a factor of two (Stenak *et al.*, 2009), caused both by expansion of agriculture to more marginal land and by insufficient access to nutrients (despite rotation between crops, grass for grazing and fallow, only a fraction of the nutrients were returned to the fields through manure and through roots and stubbles ploughed in, and of course no chemical fertilizers were yet employed, as they were first introduced by the mid- to late 19th century).

The weakness of the Danish state, with rivalry between king, squires and nobility, as well as the Church, does not appear to have slowed population growth before 1350, not even when the state collapsed and Denmark was in the hands of foreign pawnbrokers. This is, at least, the result of the analysis carried out by Velschow (1843), estimating the 1250 population in current Denmark as 900 000 to 1 million and the population in the extended Denmark as around 1.5 million, on the basis of conscription records supplemented by estimates of farm numbers and farmland. Denmark was at the time divided into districts, each of which (one "herr", or "herred" in modern Danish) was to provide 120 warriors to the king. The districts were of different size, depending on fertility of the land used for farming, but each corresponding to 120 "bol", where a bol was the unit of land that would provide food for family and dependants of one warrior. Velschow reckoned that 1 bol originally (when introduced in 620) corresponded to 13 people (half of which would be slaves). Based on tax accounts from King Valdemar II (Royal Chancellery, ca. 1240, cf. Kroman, 1936; Ulsig and Sørensen, 1981), Denmark had some 200 districts by 1231. Velschow estimates that crop yields had then doubled, and possibly also the number of people per warrior (Crown- and Church-owned land had no conscription obligation). The Velschow estimates quoted above have been given support in a detailed critical analysis by the founder of the Danish statistical bureau (Scharling, 1885), and although one later study finds them 15% too high (Arup, 1925), they do make sense relative to estimates of the number of conscribed men (Saxo Grammaticus, ca. 1200).

The largest uncertainty is probably the number of people per bol, but the results of Velschow are similar to independent estimates based on taxation data and on number of churches times estimated size of congregations (Scharling, 1889; for other perspectives on the quantity "bol" see Aakjær, 1959; Christensen, 1983). If the population estimates of Velschow were substantially too high, it would be difficult to account for the subsequent effect of the plague and the increasingly well-documented later behaviour of the population curve in Figure 6.22.

The death toll of the Bubonic plague epidemic that hit Denmark around 1350 is, as mentioned, taken as 30% of the population. It could have been smaller, but hardly below 20%, as judged by the number of farms left void of inhabitants (Ulsig, 1991). The period following the epidemic also shows a declining population. The model uses a 9% decline in birth rate and an 18% increase in violent deaths, considered as primarily associated with the lost Hanse War. The model population towards the end of the medieval period is in agreement with a literature estimate of about 570 000 (Medieval Source Book, 2006).

The Reformation period from 1536, to be dealt with further in Chapter 7, followed a civil war won by the Protestant King Christian III (defeating the count Christoffer of Oldenburg, who attacked with a mercenary army hired in Lübeck; cf. Fode, 1977). Towards the end of the 16th century, climate cooling and altered conditions for fishery and agriculture set in and developed during the following century, where a series of bad wars, notably with Sweden, were fought on Danish soil, taking heavy tolls of human lives. This is included in Fig. 6.22, with birth rates of 1.09 from 1530, 1.01 from 1595 and 1.12 from 1660, relative to those shown in Fig. 6.18, and corresponding rates of violent death changing from 0.5 to 2.5 and then to 0.15, death rates from diseases and other causes moving from 0.95 to 1.05 and back to 0.90. The model calculation is considered valid until year 1770, from when the first real census data from all of Denmark exists (Danmarks Statistik, 2010). The severe model decline in population from early 17th century to the end of the war with Sweden and the strong increase afterwards are corroborated by several studies based on data covering parts of Denmark (Lassen, 1965; Fussing, 1944; Hammerich, 1860; Mackeprang, 1907; Johan-sen, 1977). These sources agree on the time variations, but, because of the partial data used, not on the absolute population, which vary by as much as 200 000. The present demographic model agrees with Lassen (1965) in the critical period, for the population in current Denmark. Figure 6.22 also gives the population in the area constituting Denmark at its formation in around 700. The 1660 peace agreement with Sweden caused Denmark to lose the three counties east of Øresund.

When the Christian religion was first introduced in Denmark in 975–1200, morality concepts substantially different from those of earlier beliefs were imposed and largely accepted. The preoccupation of the

Church with sexual behaviour, called "sin" if it was not conducted in the form approved by the institution, is a likely cause for the delay of marriage and childbirth to long after biological maturity, although the delay could, as mentioned above, have happened earlier, and could have been affected by other factors, such as slavery and economic incapacity. A perhaps unexpected implication of delaying marriage and childbirth was the better opportunities for education and learning available through the young ages of both genders. It is in this period that a considerable number of schools appeared. All this formed the basis for further human development, and ultimately for the present technological society, which is actually rather contrary to the intentions embedded in the new morality introduced by the Christian Church.

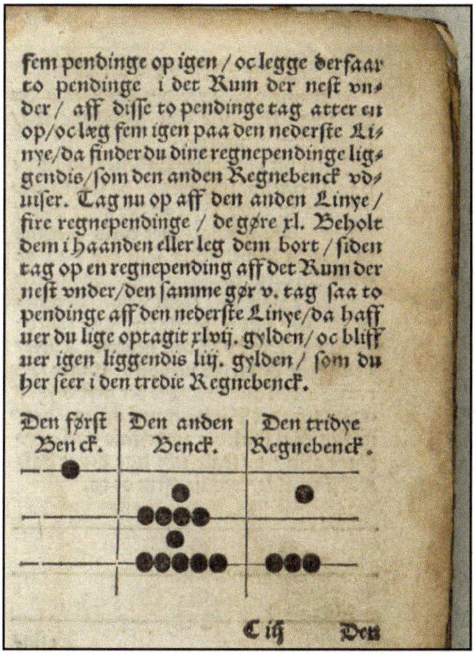
fem pendinge op igen / oc legge derfaar to pendinge i det Rum der neſt vnder / aff diſſe to pendinge tag atter en op/oc læg fem igen paa den nederſte Linye/da finder du dine regnependinge liggendis/ſom den anden Regnebenck vduiſer. Tag nu op aff den anden Linye / fire regnependinge / de gøre xl. Beholt dem i haanden eller leg dem bort / ſiden tag op en regnepending aff det Rum der neſt vnder/den ſamme gør v. tag ſaa to pendinge aff den nederſte Linye/da haffuer du lige optagit xlvij. gylden/ oc bliffuer igen liggendis liij. gylden / ſom du her ſeer i den tredie Regnebenck.



*Figure 6.23. Page from a Danish primer on calculus, presumably translated from German (Veyere, 1552). Manuscript held at the Royal Library, Copenhagen. The first Danish school was established at Hedeby by the missionary monk Ansgar in 828; religious schools were common in Danish cities from 1200, public schools not until the 17th century (Bredsdorff, 1969).*

An estimate of the development in energy use from 1200 to 1500 is made in Tables 6.2 to 6.4. As in the ones in the preceding chapters, there is a considerable uncertainty that will remain in these estimates until the appearance of measured data at the end of the 19th century. On the other hand, the estimates are certainly not grossly incorrect, partly because they take advantage of the continuity to constrain the development to one of internal consistency, and partly because a bottom-up approach has been used, identifying concrete activities requiring energy for each period of time. The energy requirement of each concrete activity has an uncertainty, and furthermore, activities needing energy could have been overlooked. Still, the overall development for the entire country (shown incrementally for each period) must show a long-range behaviour with only discontinuities that have identified reasons (such as migration or epidemic episodes), and otherwise continuous transition into the period of available statistical data.

The year 1200 charts (Table 6.2 and the corresponding end-use estimate in Table 6.5) take into account the energy use associated with church-building, estimated in Table 6.1. During the following centuries, the brick houses penetrate other types of building, as mentioned above, and continue to add to the energy use (Tables 6.3 and 6.6 for year 1350, and Tables 6.4 and 6.7 for year 1500). The raw materials extraction plays an important role, but because mining in Danish clay/sand geology is much less energy-intensive than mining in rock environments, the energy use and particularly the end-use energy expenditure in this sector only increases modestly. A much more significant increase in energy usage takes place in the raw materials transportation (requiring ships or roads, and oxen or horses to pull the carts) and processing sectors (e.g. for brick manufacture). This also implies more use of high-temperature heat (in kilns or industrial furnaces). The more sophisticated methods of manufacture give a premium for high-quality, easily regulated mechanical energy, which also appears in the medieval period.

With the Viking-period campaigns all over Europe ended, the total requirement for ships is somewhat reduced, representing now primarily merchant ships, and the wind has taken over as the primary energy supplier for propulsion at sea. Watermills are, as mentioned in connection with Fig. 6.12, already being built in substantial numbers before 1200, contributing to high-quality energy provision. The following centuries see a further expansion (Tables 6.3 and 6.4), and only by 1500 does the competition from windmills start. The transportation of raw materials for construction remains high in energy use. The earlier wooden buildings could be made from local forest resources, while the stone and brick materials had to be transported from locations favoured with suitable geological deposits. Brick-making becomes a major industry, with a relatively limited number of production centres, located near the clay deposits.

The end-use energy in some cases behaves different from the delivered energy, due to assumptions on energy conversion efficiency. In the case of using wind or elevated water instead of human muscle energy represented by the energy in the additional food intake required, efficiency appears to improve, as the primary input to wind or watermills is not listed, but only the output. Draft animals continue to play an important role for road transportation.

*Table 6.2. Estimated energy use (delivered) per capita (adult person) in Denmark around year 1200.*

| *Denmark, 1200.* | *Human energy* | *Energy from fuels and external energy flows* | | | | | |
|---|---|---|---|---|---|---|---|
| *Rate of energy use (average W for an adult person)* | Food energy | Low-temperature heat | Medium-temperature heat | High-temperature heat | Mobile power (ex. food) | Mechanical power (stationary) | TOTAL |
| **Liveable surroundings** | | | | | | | **345** |
| Basic metabolism | 68 | | | | | | **68** |
| Shelter, buildings (reduced heat need) | | 10 | | | | | **10** |
| Insulation | | 2 | | | | | **2** |
| Heating | | 240 | | | | | **240** |
| Light | | | | 25 | | | **25** |
| **Food and water provision** | | | | | | | **162** |
| Hunting | 1 | | | | | | **1** |
| Gathering | 1 | | | | | | **1** |
| Fishing, aquaculture | 5 | | 3 | | | | **8** |
| Agriculture, grinding flour | 25 | | | | 50 | 1 | **76** |
| Slaughtering, preparing, cooking | 5 | | 60 | | | | **65** |
| Water provision | 5 | | | | 6 | | **11** |
| **Security, assault, expansion** | | | | | | | **41** |
| Weapons, armies | 8 | | | | | 1 | **9** |
| Discouraging assaults | 1 | | | | | | **1** |
| Territorial transportation, ships | 4 | | | | 27 | | **31** |
| **Health-related efforts** | | | | | | | **29** |
| Food preservation | 4 | 7 | 15 | | | | **26** |
| Water quality management | 1 | 1 | | | | | **2** |
| Doctors, lay or professional | 1 | | | | | | **1** |
| **Relations and leisure** | | | | | | | **16** |
| Music | 0.3 | | | | | | **0.3** |
| Theater, dance | 0.5 | | | | | | **0.5** |
| Toys, artwork | 1 | | | | | | **1** |
| Written materials | 0.2 | | | | | | **0.2** |
| Social transportation | 4 | | | | 10 | | **14** |
| **Occupational activities** | | | | | | | **152** |
| Construction (incl. Ships) | 14 | | 8 | | 12 | 3 | **37** |
| Manufacture | 10 | 5 | 10 | 10 | 6 | | **41** |
| Distribution, trade and services | 8 | | | | 4 | | **12** |
| Transportation of goods | 15 | | | | 45 | | **60** |
| Transportation for services, commuting | 2 | | | | | | **2** |
| **Raw materials procurement** | | | | | | | **124** |
| Mining | 8 | | | | 12 | 2 | **22** |
| Materials for construction and industry | 12 | | 30 | 25 | 12 | 1 | **80** |
| Energy provision | 10 | | | | 12 | | **22** |
| TOTAL | **214** | **265** | **126** | **60** | **196** | **8** | **869** |

Figures 6.24 and 6.25 sum up the information in the tables, and place it on a timescale continued from Figs. 5.22 and 5.23 in the preceding Chapter. Figure 6.24a shows the increasing importance of the raw materials and manufacturing sectors. The human food energy converted into muscle power in 1200 and 1500 is on average about 140 W for an adult person (Fig. 6.24b), corresponding to about 47 W of mechanical power. Assuming a maximum of 12 working hours, the 94 W during this period is about the maximum a person can sustain, and possibly an overestimate, at least for women. The general trend can be explained like

this: from 6 ky BP the curve starts to grow, because the transition from hunting and fishing to agriculture entails more labour to be provided. The increase accelerates during the Viking period, because of an increased demand for agricultural products for warriors and for export, and because farm labourers are being treated increasingly as slaves, explaining the extended working hours. The overseas explorations and fighting additionally make demands on rowing manpower, all requiring increased food intake.

*Table 6.3. Estimated energy use (delivered) per capita (adult person) in Denmark around year 1350 (entries of "0" mean smaller than a half unit).*

| *Denmark, 1350.* | *Human energy* | *Energy from fuels and external energy flows* | | | | | |
|---|---|---|---|---|---|---|---|
| *Rate of energy use (average W for an adult person)* | Food energy | Low-temperature heat | Medium-temperature heat | High-temperature heat | Mobile power (ex. food) | Mechanical power (stationary) | TOTAL |
| **Liveable surroundings** | | | | | | | **345** |
| Basic metabolism | 68 | | | | | | **68** |
| Shelter, buildings (reduced heat need) | | 10 | | | | | **10** |
| Insulation | | 2 | | | | | **2** |
| Heating | | 240 | | | | | **240** |
| Light | | | | 25 | | | **25** |
| **Food and water provision** | | | | | | | **143** |
| Hunting | 0.5 | | | | | | **0.5** |
| Gathering | 0.5 | | | | | | **0.5** |
| Fishing, aquaculture | 4 | | 3 | | | | **7** |
| Agriculture, grinding flour | 20 | | | | 50 | 1 | **71** |
| Slaughtering, preparing, cooking | 4 | | 50 | | | | **54** |
| Water provision | 4 | | | | 6 | | **10** |
| **Security, assault, expansion** | | | | | | | **33** |
| Weapons, armies | 8 | | | | | 1 | **9** |
| Discouraging assaults | 1 | | | | | | **1** |
| Territorial transportation, ships | 3 | | | | 20 | | **23** |
| **Health-related efforts** | | | | | | | **29** |
| Food preservation | 4 | 7 | 15 | | | | **26** |
| Water quality management | 1 | 1 | | | | | **2** |
| Doctors, lay or professional | 1 | | | | | | **1** |
| **Relations and leisure** | | | | | | | **16.2** |
| Musical instruments | 0.3 | | | | | | **0.3** |
| Theater, dance | 0.5 | | | | | | **0.5** |
| Toys, artwork | 1 | | | | | | **1** |
| Written materials | 0.4 | | | | | | **0.4** |
| Social transportation | 4 | | | | 10 | | **14** |
| **Occupational activities** | | | | | | | **127** |
| Construction (incl. Ships) | 13 | | 6 | | 8 | 2 | **29** |
| Manufacture | 13 | 4 | 8 | 8 | 5 | 0 | **38** |
| Distribution, trade and services | 7 | | | | 3 | | **10** |
| Transportation of goods | 13 | | | | 35 | | **48** |
| Transportation for services, commuting | 2 | | | | | | **2** |
| **Raw materials procurement** | | | | | | | **115** |
| Mining | 7 | | | | 10 | 2 | **19** |
| Materials for construction and industry | 11 | | 30 | 25 | 10 | 1 | **77** |
| Energy provision | 9 | | | | 10 | 0 | **19** |
| TOTAL | **200.2** | **264** | **112** | **58** | **167** | **7** | **808.2** |

*Table 6.4. Estimated energy use (delivered) per capita (adult person) in Denmark around year 1500 (totals rounded).*

| *Denmark, 1500.* | *Human energy* | *Energy from fuels and external energy flows* | | | | | |
|---|---|---|---|---|---|---|---|
| *Rate of energy use (average W for an adult person)* | **Food energy** | **Low-temperature heat** | **Medium-temperature heat** | **High-temperature heat** | **Mobile power (ex. food)** | **Mechanical power (stationary)** | TOTAL |
| **Liveable surroundings** | | | | | | | **345** |
| Basic metabolism | 68 | | | | | | **68** |
| Shelter, buildings (reduced heat need) | | 10 | | | | | **10** |
| Insulation | | 2 | | | | | **2** |
| Heating | | 240 | | | | | **240** |
| Light | | | | 25 | | | **25** |
| **Food and water provision** | | | | | | | **159** |
| Hunting | 0.5 | | | | | | **0.5** |
| Gathering | 0.5 | | | | | | **0.5** |
| Fishing, aquaculture | 5 | | 3 | | | | **8** |
| Agriculture, grinding flour | 25 | | | | 50 | 6 | **81** |
| Slaughtering, preparing, cooking | 4 | | 50 | | | | **54** |
| Hot and cold water provision | 4 | 5 | | | 4 | 2 | **15** |
| **Security, assault, expansion** | | | | | | | **20** |
| Weapons, armies | 8 | | | | 1 | 1 | **10** |
| Discouraging assaults | 1 | | | | | | **1** |
| Territorial transportation, ships | 3 | | | | 6 | | **9** |
| **Health-related efforts** | | | | | | | **29.1** |
| Food preservation | 4 | 7 | 15 | | | | **26** |
| Water quality management | 1 | 1 | | | | | **2** |
| Doctors, lay or professional | 1 | | | | | | **1** |
| Hospitals | 0.1 | | | | | | **0.1** |
| **Relations and leisure** | | | | | | | **16.2** |
| Musical instruments | 0.3 | | | | | | **0.3** |
| Theater, dance | 0.5 | | | | | | **0.5** |
| Toys, artwork | 1 | | | | | | **1** |
| Written materials | 0.4 | | | | | | **0.4** |
| Social transportation | 4 | | | | 10 | | **14** |
| **Occupational activities** | | | | | | | **172** |
| Construction (incl. Ships) | 13 | | 9 | | 12 | 3 | **37** |
| Manufacture | 12 | 13 | 13 | 13 | 8 | 1 | **60** |
| Distribution, trade and services | 8 | | | | 3 | | **11** |
| Transportation of goods | 14 | | | | 45 | | **59** |
| Transportation for services, commuting | 3 | | | | 2 | | **5** |
| **Raw materials procurement** | | | | | | | **134** |
| Mining | 6 | | | | 15 | 3 | **24** |
| Materials for construction and industry | 12 | | 25 | 30 | 15 | 3 | **85** |
| Energy provision | 10 | | | | 15 | 0 | **25** |
| TOTAL | **209** | **278** | **115** | **68** | **186** | **19** | **875** |

During the medieval period, construction and manufacture increase dramatically, but this primarily transfers labour from agriculture to raw materials extraction, transportation of goods and to the building or production industries. The altered ownership of agricultural assets developing during the medieval period points in the direction of further transformation of farm work into slave-like arrangements. Thus, if there is an overestimate of human labour, it is hardly more than 25%. The fact that children contribute less muscle power, although they certainly do work, is taken into account in the totals shown in Fig. 6.26.

*Table 6.5. Estimated end-use energy per capita in Denmark around year 1200.*

| ***Denmark, 1200.*** | ***Human energy*** | ***Energy from fuels and external energy flows*** | | | | | |
|---|---|---|---|---|---|---|---|
| *Rate of energy end-use (average W for an adult person)* | **Food energy** | **Low-temperature heat** | **Medium-temperature heat** | **High-temperature heat** | **Mobile power (ex. food)** | **Mechanical power (stationary)** | **TOTAL** |
| **Liveable surroundings** | | | | | | | **116** |
| Basic metabolism | 68 | | | | | | **68** |
| Shelter, buildings (reduced heat need) | | 10 | | | | | **10** |
| Insulation | | 2 | | | | | **2** |
| Heating | | 33 | | | | | **33** |
| Light | | | | 3 | | | **3** |
| **Food and water provision** | | | | | | | **39.1** |
| Hunting | 0.3 | | | | | | **0.3** |
| Gathering | 0.3 | | | | | | **0.3** |
| Fishing, aquaculture | 1.7 | | 1 | | | | **2.7** |
| Agriculture, grinding | 8.3 | | | | 12 | 0.3 | **20.6** |
| Slaughtering, preparing, cooking | 1.7 | | 10 | | | | **11.7** |
| Water provision | 1.5 | | | | 2 | | **3.5** |
| **Security, assault, expansion** | | | | | | | **13.5** |
| Weapons, armies | 2.6 | | | | | 0.3 | **2.9** |
| Discouraging assaults | 0.3 | | | | | | **0.3** |
| Territorial transportation, ships | 1.3 | | | | 9 | | **10.3** |
| **Health-related efforts** | | | | | | | **9.3** |
| Food preservation | 1.3 | 2 | 5 | | | | **8.3** |
| Water quality management | 0.3 | 0.4 | | | | | **0.7** |
| Doctors, lay or professional | 0.3 | | | | | | **0.3** |
| **Relations and leisure** | | | | | | | **5.87** |
| Musical instruments | 0.1 | | | | | | **0.1** |
| Theater, dance | 0.1 | | | | | | **0.1** |
| Toys, artwork | 0.3 | | | | | | **0.3** |
| Written materials | 0.07 | | | | | | **0.07** |
| Social transportation | 1.3 | | | | 4 | | **5.3** |
| **Occupational activities** | | | | | | | **51.5** |
| Construction (incl. Ships) | 4.5 | | 2.6 | | 4 | 1 | **12.1** |
| Manufacture | 3.3 | 2 | 4 | 4 | 2 | 0 | **15.3** |
| Distribution, trade and services | 2.7 | | | | 0.7 | | **3.4** |
| Transportation of goods | 5 | | | | 15 | | **20** |
| Transportation for services, commuting | 0.7 | | | | | | **0.7** |
| **Raw materials procurement** | | | | | | | **58.9** |
| Mining | 2.6 | | | | 4 | 0.7 | **7.3** |
| Materials for construction and industry | 4 | | 18 | 18 | 4 | 0.3 | **44.3** |
| Energy provision | 3.3 | | | | 4 | | **7.3** |
| TOTAL | **116** | **49** | **41** | **25** | **61** | **3** | **294** |

The end-use energy of Figs. 6.25a and 6.25b, corresponding to the delivered energy of Fig. 6.24a and 6.24b, again shows a higher increase in the manufacturing and transportation sectors than in the raw material extraction sector. The decline in mobile power exhibited in Fig. 6.25b signals the end of the travel-intensive Viking period, which is not compensated by increased transportation of raw materials and goods (the latter was already substantial in the Viking period, with its widespread international trade).

Finally, the overall drop in energy use between 1200 and 1350 is at the end influenced by the Bubonic plague epidemic (cf. Chapter 5, Figs. 5.9

and 6.22), not only in total energy (Fig. 6.26) but also in per capita values, due to the decline in the European market for trading agricultural products. Several authors speak of a crisis for Danish agriculture (Ulsig, 1983; Enemark, 1991), not only due to the many deserted farms left by tenants succumbing to the plague, but also because of reduced demands for agricultural export to countries south of Denmark, where the toll of the plague was considerably higher than the 30% population loss estimated for Denmark.

*Table 6.6. Estimated end-use energy per capita in Denmark around year 1350.*

| ***Denmark, 1350.*** | ***Human energy*** | ***Energy from fuels and external energy flows*** | | | | | |
|---|---|---|---|---|---|---|---|
| *Rate of energy end-use (average W for an adult person)* | **Food energy** | **Low-temperature heat** | **Medium-temperature heat** | **High-temperature heat** | **Mobile power (ex. food)** | **Mechanical power (stationary)** | **TOTAL** |
| **Liveable surroundings** | | | | | | | **116** |
| Basic metabolism | 68 | | | | | | **68** |
| Shelter, buildings (reduced heat need) | | 10 | | | | | **10** |
| Insulation | | 2 | | | | | **2** |
| Heating | | 33 | | | | | **33** |
| Light | | | | 3 | | | **3** |
| **Food and water provision** | | | | | | | **36.2** |
| Hunting | 0.2 | | | | | | **0.2** |
| Gathering | 0.2 | | | | | | **0.2** |
| Fishing, aquaculture | 1.3 | | 1 | | | | **2.3** |
| Agriculture, grinding | 6.7 | | | | 12 | 0.3 | **19** |
| Slaughtering, preparing, cooking | 1.3 | | 10 | | | | **11.3** |
| Water provision | 1.3 | | | | 2 | | **3.3** |
| **Security, assault, expansion** | | | | | | | **11.3** |
| Weapons, armies | 2.7 | | | | | 0.3 | **3** |
| Discouraging assaults | 0.3 | | | | | | **0.3** |
| Territorial transportation, ships | 1 | | | | 7 | | **8** |
| **Health-related efforts** | | | | | | | **9.3** |
| Food preservation | 1.3 | 2 | 5 | | | | **8.3** |
| Water quality management | 0.3 | 0.4 | | | | | **0.7** |
| Doctors, lay or professional | 0.3 | | | | | | **0.3** |
| **Relations and leisure** | | | | | | | **5.9** |
| Musical instruments | 0.1 | | | | | | **0.1** |
| Theater, dance | 0.1 | | | | | | **0.1** |
| Toys, artwork | 0.3 | | | | | | **0.3** |
| Written materials | 0.1 | | | | | | **0.1** |
| Social transportation | 1.3 | | | | 4 | | **5.3** |
| **Occupational activities** | | | | | | | **49.4** |
| Construction (incl. Ships) | 4.3 | | 2 | | 2.7 | 0.7 | **9.7** |
| Manufacture | 4.3 | 2 | 4 | 4 | 1.7 | 0 | **16** |
| Distribution, trade and services | 2.7 | | | | 1 | | **3.7** |
| Transportation of goods | 4.3 | | | | 15 | | **19.3** |
| Transportation for services, commuting | 0.7 | | | | | | **0.7** |
| **Raw materials procurement** | | | | | | | **52** |
| Mining | 2.3 | | | | 3 | 0.7 | **6** |
| Materials for construction and industry | 3.7 | | 18 | 15 | 3 | 0.3 | **40** |
| Energy provision | 3 | | | | 3 | 0 | **6** |
| TOTAL | **112** | **49** | **40** | **22** | **54** | **2** | **280** |

When the per capita energy use is multiplied by the population size at each instance in time, a more dramatic picture of Fig. 6.26 emerges. The effect of the 1347 plague is seen to have a considerably more sizeable impact on the total energy use than the massive migration of people out of the present Danish territory during the earlier period 600–900.

*Table 6.7. Estimated end-use energy per capita in Denmark around year 1500.*

| *Denmark, 1500.* | *Human energy* | *Energy from fuels and external energy flows* | | | | | |
|---|---|---|---|---|---|---|---|
| *Rate of energy end-use (average W for an adult person)* | **Food energy** | **Low-temperature heat** | **Medium-temperature heat** | **High-temperature heat** | **Mobile power (ex. food)** | **Mechanical power (stationary)** | **TOTAL** |
| **Liveable surroundings** | | | | | | | **121** |
| Basic metabolism | 68 | | | | | | **68** |
| Shelter, buildings (reduced heat need) | | 10 | | | | | **10** |
| Insulation | | 2 | | | | | **2** |
| Heating | | 38 | | | | | **38** |
| Light | | | | 3 | | | **3** |
| **Food and water provision** | | | | | | | **46.1** |
| Hunting | 0.2 | | | | | | **0.2** |
| Gathering | 0.2 | | | | | | **0.2** |
| Fishing, aquaculture | 1.7 | | 1 | | | | **2.7** |
| Agriculture, grinding | 7.5 | | | | 15 | 3 | **25.5** |
| Slaughtering, preparing, cooking | 1.3 | | 10 | | | | **11.3** |
| Hot and cold water provision | 1.3 | 2 | | | 2 | 1 | **6.3** |
| **Security, assault, expansion** | | | | | | | **6.6** |
| Weapons, armies | 2.7 | | | | 0.3 | 0.3 | **3.3** |
| Discouraging assaults | 0.3 | | | | | | **0.3** |
| Territorial transportation, ships | 1 | | | | 2 | | **3** |
| **Health-related efforts** | | | | | | | **9.3** |
| Food preservation | 1.3 | 2 | 5 | | | | **8.3** |
| Water quality management | 0.3 | 0.4 | | | | | **0.7** |
| Doctors, lay or professional | 0.3 | | | | | | **0.3** |
| Hospitals | 0.03 | | | | | | **0.03** |
| **Relations and leisure** | | | | | | | **6.0** |
| Musical instruments | 0.1 | | | | | | **0.1** |
| Theater, dance | 0.13 | | | | | | **0.13** |
| Toys, artwork | 0.3 | | | | | | **0.3** |
| Written materials | 0.13 | | | | | | **0.13** |
| Social transportation | 1.3 | | | | 4 | | **5.3** |
| **Occupational activities** | | | | | | | **61.7** |
| Construction (incl. Ships) | 4.3 | | 3 | | 4 | 1 | **12.3** |
| Manufacture | 4 | 4 | 6 | 7 | 2.7 | 0.3 | **24** |
| Distribution, trade and services | 2.7 | | | | 1 | | **3.7** |
| Transportation of goods | 4.7 | | | | 15 | | **19.7** |
| Transportation for services, commuting | 1 | | | | 1 | | **2** |
| **Raw materials procurement** | | | | | | | **61.3** |
| Mining | 2 | | | | 5 | 1 | **8** |
| Materials for construction and industry | 4 | | 15 | 20 | 5 | 1 | **45** |
| Energy provision | 3.3 | | | | 5 | 0 | **8.3** |
| **TOTAL** | **114** | **58** | **40** | **30** | **62** | **8** | **312** |

Recent studies have tried to follow the degree and nature of cultivation through farming during the period from 1000 to the present (Stenak *et al.*, 2009), as a particular issue in the study of changing landscapes. The area distribution of tree species and forest extent, of natural vegetation such as heather, of grass (natural or cultivated) and of cereal crops can be followed through pollen analysis, as done for particular sites with time-stratified soil layers (Odgaard and Nielsen, 2009). Figs. 6.27a and 6.27b give two examples, for landscapes originally dominated by forest or by meadow.

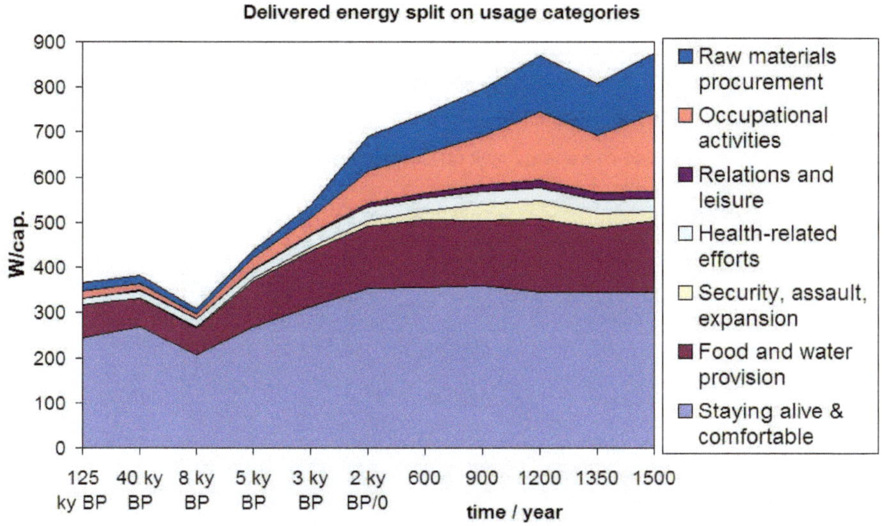


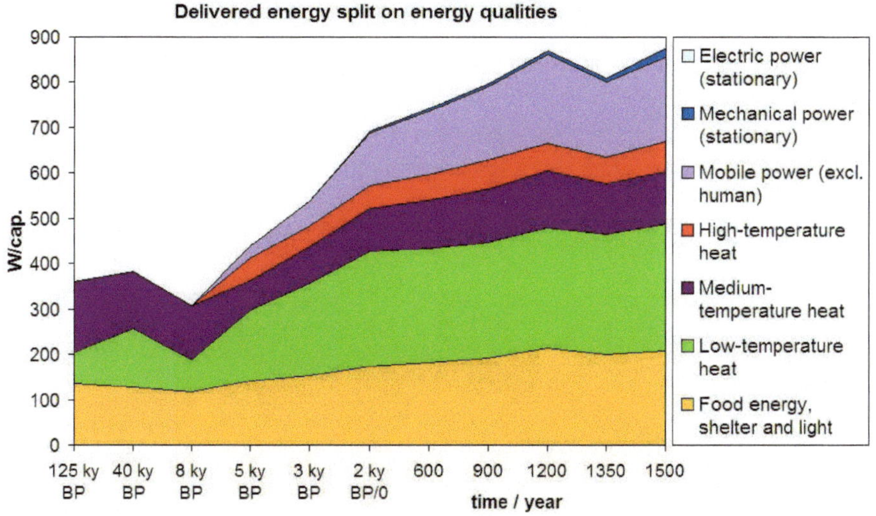


*Figure 6.24. Estimated energy use per capita in Denmark from 125 ky BP to 1500, distributed on usage categories (a, above) or on type of energy (b, below).*

In Fig. 6.28, the percentage of Danish land used for cultivation of cereal or grass (typically in three shifts: cereal cultivation, fallow, livestock grazing) is estimated as it develops with time. Agricultural export had already started about 500 (Velschow, 1843), and the 6–8 times cultivated-area increase between 1000 and 1350 shown in Fig. 6.28 is much higher than the 2.5 times increase in population (according to Fig. 6.22), so a combination of increasing export or declining yields must be invoked. Probably both, unless the specific locations used by Stenak *et al.* (2009) are not representative for the entire country. Velschow (1843) estimated a doubling of yields between 500 and 1250, implying more than a factor of three increase in exports.

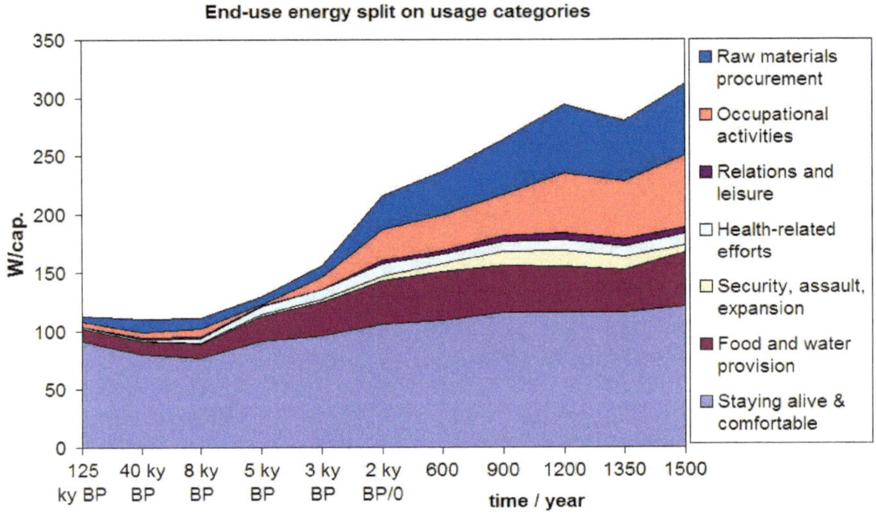


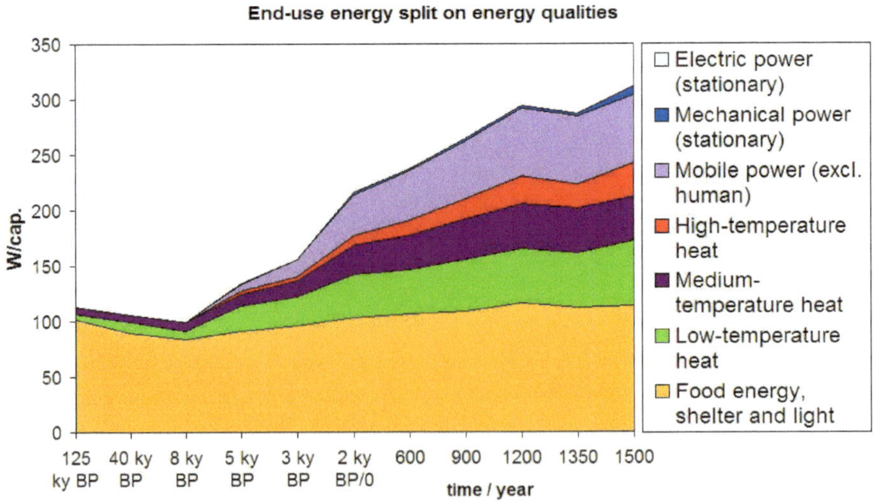


*Figure 6.25. Estimated end-use energy per capita in Denmark from 125 ky BP to 1500, distributed on usage categories (a, above) or on type of energy (b, below).*

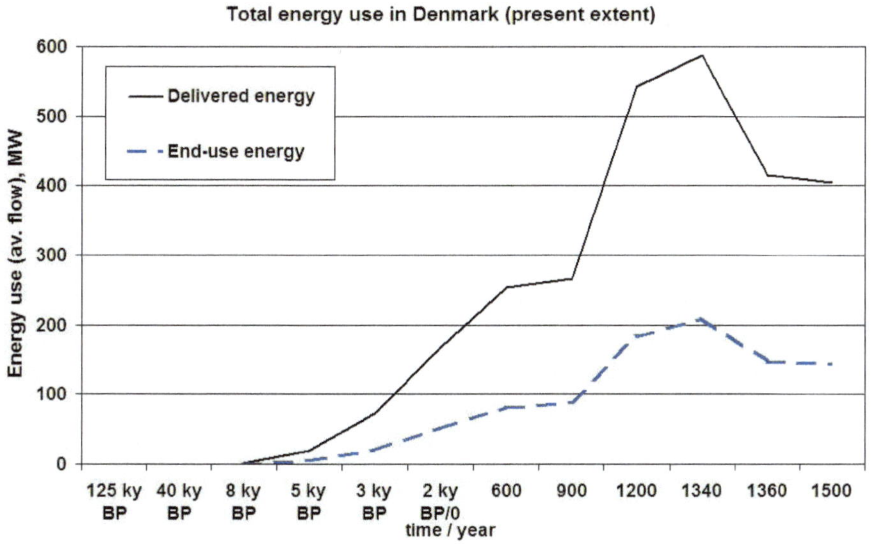


*Figure 6.26. Total Danish energy use (delivered and end-use), for the population that at a given time occupied the current territory (based on Fig. 6.22, and Figs. 6.24–6.25). Note that the timescale is not equidistant. The lower energy use of children, relative to the values given in Figs. 6.24–6.25 for adults, has been taken into account by a linear decline at ages below 15, subtracted on the basis of the age profile of the population for each period of time.*

Fig. 6.28 further shows a modest reversal in cultivated area development by the time of the Bubonic plague, followed by a higher growth rate (the area increasing by close to a factor of 3 from 1350 to 1850). Again, this could be the result of increasing exports, but is probably dominated by declining yields. The population growth during the period 1350–1850 is near zero (according to Fig. 6.22, a decline followed by a return to the population 500 years earlier). Average yields may have declined due to extending cultivation to poorer soils, but on the other hand, ploughs and tilling equipment improved, and use of animal manure and rotation would have acted in the opposite direction. Stenak *et al.* (2009) suggest that yields could have been reduced to below 50% of the early ones between 1250 and 1800, based on the assumption that population remained unchanged during this period, but they also find that yields after 1750 exhibited a persistent increase, a statement now backed up by actual measurements. The locations behind Fig. 6.28 fall into two groups: the fertile soils of mid-Sealand and Funen, and the less fertile sandy or heather-grass land of north Sealand and north and west Jutland.

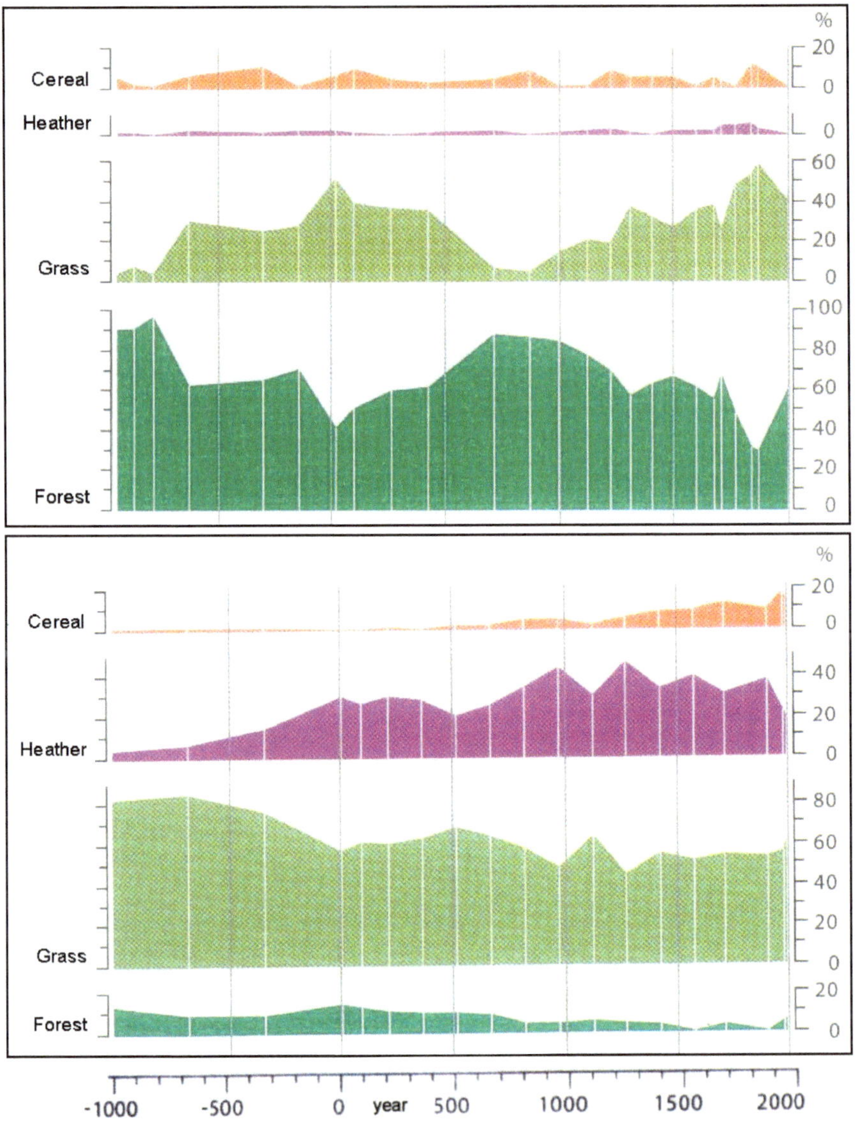


*Figure 6.27. Reconstructed vegetation development during the last 3 ky, for the forest area around Store Gribsø (Sealand, at top), and the grass/heather area near Navnsø (north Jutland, at bottom). From Odgaard and Nielsen, 2009 (courtesy of Aarhus University Press).*

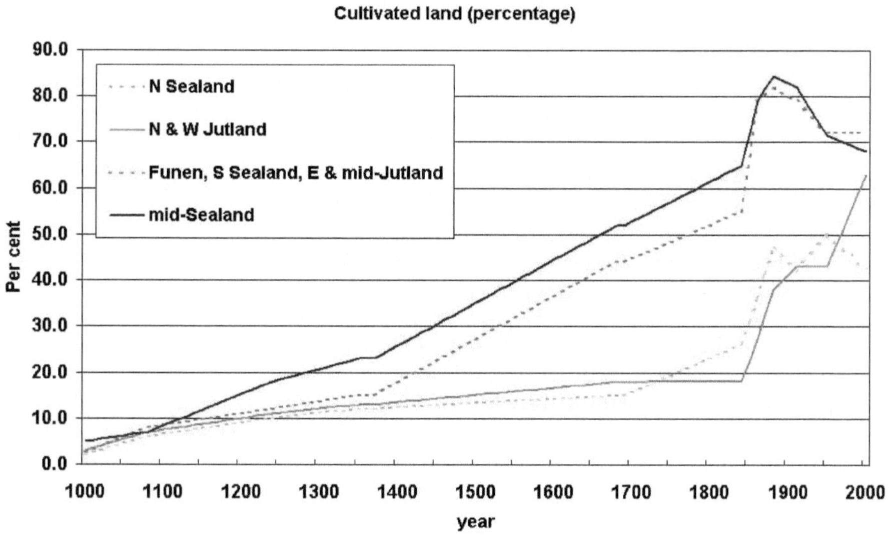


*Figure 6.28. Degree of cultivation in specific areas of different parts of Denmark as it develops in time (based on Stenak et al., 2009).*

## *References:*


Aakjær, S., 1959. Måleenheder i landmål og jordvurdering i Middelalderens Danmark. *Historie/Jyske Samlinger* **5**, 175–303.

Adam Bremensis, ca. 1075. *Gesta Hammaburgensis Ecclesiae Pontificum*. Annotated 3rd ed. by B. Schmeidler, 1917, Hahnsche Buchhandlung, Hannover & Leipzig.

Arup, E., 1925. *Danmarks Historie*, vol. 1. Hagerup, Copenhagen.

Baird, G., Alcorn, A., Haslam, P., 1997. The energy embodied in building materials – updated New Zealand coefficients and their significance. *IPENZ Transactions* **24**, 46–54.

Böcher, S., 1952. Vandkraften som initial lokaliseringsfaktor for dansk industri. *Geografisk Tidsskrift* **52**, 33–49.

Bredsdorff, M., 1969. *Læsestykker til den danske folkeskoles historie.* Gyldendal's Pedagogic Library, Copenhagen.

Christensen, C., 1983. Begrebet bol. *Historisk Tidsskrift* **14**, 1–34.

Chronicon Roskildense, ca. 1140. Anonymous writer of first Danish history account by a Dane. Earliest known copy (13th century) at Kiel University Library. Danish translations by M. Bredsdorff (1883) and as "Roskildekrøniken" by J. Olrik, Selskabet for historiske kildeskrifters oversættelse, 1898.

Danmarks Statistik, 2010. *Statistikbank.* Befolkning. http://www.statistikbanken.dk, Danish Department of Statistics, Copenhagen.

Enemark, P., 1991. Vesteurope, Lybæk og dansk handel i senmiddelalderen. *Historisk Tidsskrift* **15**, 361–401. Available at http://tidsskrift.dk

Erslev, K., 1898–1905. Den senere middelalder. (1241–1481). Books 1–4 in vol. 2 of Steenstrup *et al.*, 1897–1906. Det Nordiske Forlag, Copenhagen.

Exner, J., Ebert, J., 1968. *400 danske landsbykirker.* Gyldendal Publ., Copenhagen. Details of individual churches can be found in the tourist pamphlets issued by parochial councils and available at the churches, as well as on the websites of individual churches.

Fischer, C., 2004. *Tidlige danske vandmøller.* Silkeborg Museum/Jysk Arkæologisk Selskab. Århus University Press.

Fode, H., 1977. *Fra reformation til folkestyre, 1523–1849.* Gyldendal Publ., Copenhagen.

Fussing, H., 1944. Øernes befolkning under Karl Gustavkrigene. *Historisk Tidsskrift* **11** (1), 287–333.

Hammerich, F., 1860. Bidrag til belysning af den danske stats folkemængde, næringsveie og finanser i midten af det syttende Aarhundrede. *Historisk Tidsskrift* **3** (2), 28–106.

Heise, A., 1899–1906. Reformationenens Tidsalder (1520–1536). Book 2, part 3 in vol. 3 of Steenstrup *et al.*, 1897-1906. Det Nordiske Forlag, Copenhagen.

Hertz, J., 1997. Esrum Kloster – arkæologisk belyst. In *Bogen om Esrum Kloster* (S. Frandsen, J. Jørgensen, C. Tortzen, eds), pp. 43–71. Frederiksborg Amt Publ., Hillerød.

Hjelholt, H., 1932. Ansættelsen til mølleskyld i Christian V's matrikel. *Historisk Tidsskrift* **10** (2), 36–58.

Johansen, H., 1977. Danmarks folketal omkring år 1700. *Historie/Jyske Samlinger* **12**, 360–374.

Kinder, H., Hilgemann, V., 1964. *Atlas zur Weltgeschichte*, Vol. 1. Deutscher Taschenbuchverlag, Köln. Revised Danish edition 1991, Munksgaard, Copenhagen.

Kornerup, J., 1879. Minder om Cistercienserklostret i Esrum, om dets stifter og dets forbindelser med Clairvaux. *Aarbøger for Nordisk Oldkyndighed og Historie*, 1–18.

Kroman, E., 1936. *Kong Valdemars Jordebog*. Et haandskrifts historie. Danish National Archive, Copenhagen; and in German: *Acta philologica Scandinavica* (11).

Lassen, A., 1965. *Fald eller fremgang*. Jysk Selskab for Historie, Sprog og Litteratur, Århus University Press.

Mackeprang, E., 1907. Et brudstykke af en folketælling fra 1645. *Nationaløkonomisk Tidsskrift* **3** (15), 248–270.

Matthiessen, P., 1978. *Befolkningens vækst*. Berlingske Leksikon Bibliotek, Copenhagen.

Medieval Source Book, 2006 (P. Halsall, ed.). Fordham University, http://www.fordham.edu/halsall/sbook.html

Museum Council of Frederiksborg County, 2009. Webpage: Industrihistoriens Danmarkskort: Hellebæk. http://www.kulturarv.dk

Odgaard, B., Nielsen, A., 2009. Udvikling i arealdækning i Danmark i perioden 0–1850. *Danske landbrugsland-skaber gennem 2000 år* (Odgaard, B., Rømer, J., eds), pp. 41–58. Aarhus Universitetsforlag, Aarhus.

Olsen, M., 1875. Indberetning vedr. Esrum Kloster udgravning. Danish National Museum Archive of Correspondence.

Randsborg, K., 2008. Kings' Jelling: Gorm & Thyra's palace, Harald's monument & grave, Svend's cathedral. *Acta Archaeologica* **79**, 1–23.

Royal Chancellery, ca. 1240. Taxation basis and revenues under king Valdemar II (in a series of document copies made by J. Jutæ and K. Jwl at Sorø's cisterciencis convent, the tax sheets being borrowed from the Chancellery by abbey head Hemming, who was a member of the State Council). Parts of the collection of unrelated documents were in the 18th century published in Sweden under names such as "Kong Valdemars Jordebog" and "Liber census Daniæ". A facsimile with Danish translation (vol. I.2) and commentary volumes may be found in S. Aakjær (1926–45): *Kong Valdemars Jordebog*. Akademisk Forlag, Copenhagen, 1980.

Saxo Grammaticus, ca. 1200. *Gesta Danorum*. First Danish translation is by A. Vedel: "Den danske krønicke". Støckelman & Gutteruitz, 1595 (Royal Danish Library; ProQuest Early European Books website). An English translation by O. Elton (New York, 1905) is available at http://www.gutenberg.org

Scharling, W., 1885. Befolkningens størrelse i ældre og nyere tid. Tilvækstforhold. *Danmarks Statistik* (V. Falbe-Hansen and W. Scharling, eds), section "Befolkningen", pp. 321-429. Forlagsbureauet. Copenhagen.

Scharling, W., 1889. Kirketallet og Folketallet i Danmark i det 13. århundrede. *Historisk Tidsskrift* **6** (2) 264–318.

Skovgaard-Petersen, I., Christensen, A., Paludan, H., 1977. Tiden indtil 1340. *Danmarks historie*, vol. 1 (A. Christensen, H. Clausen, S. Ellehøj, S. Mørch, eds), Gyldendal Publ., Copenhagen.

Skytte, L., 2010. Effekten af luftbåren ammoniak på nedbrydningen af den danske kulturarv. PhD. Thesis, Syddansk Universitet, Odense.

Sørensen, B., 2007. Assessing current vehicle performance and simulating the performance of hydrogen and hybrid cars. *Int. J. Hydrogen Energy* **32**, 1597–1604.

Sørensen, B., 2010. Renewable Energy Origins and Flows. Introduction to vol. 1 of *Renewable Energy. Earthscan Reference Collections* (B. Sørensen, ed.), pp. 1-20. Earthscan, London.

Steensberg, A., 1959. En skvatmølle i Ljørring. *Kuml*, 130-145.

Steenstrup, J., 1904. Valdemarenes tidalder (1157–1241). Book 4 in vol. 1 of Steenstrup *et al.*, 1897-1906. Det Nordiske Forlag, Copenhagen.

Steenstrup. J., Erslev, K., Heise, A., Mollerup, V., Fridericia, J., Holm, E., Jørgensen, A., 1897–1906. *Danmarks Riges Historie.* Det Nordiske Forlag, Copenhagen.

Stenak, M., Rømer, J., Näsman, U., Odgaard, B., 2009. Landskabsforandringer gennem 2000 år. *Danske landbrugslandskaber gennem 2000 år* (Odgaard, B., Rømer, J., eds), pp. 283–302. Aarhus Universitetsforlag, Aarhus.

Szomlaiski, L., 1975. *De nordiske lande bliver til. Tiden før 1523.* Vol. 1 in "Hovedlinjer i Nordens Historie" (R Thomsen, ed.). Gyldendal Publ., Copenhagen.

Ulsig, E., 1983. Den senmiddelalderlige agrarkrise. *Historie/Jyske Samlinger* **15**, 219–233. Available at http://tidsskrift.dk

Ulsig, E., 1991. Pest og befolkningsnedgang i Danmark i det 14. århundrede. *Historisk Tidsskrift* **15** (6), 20–43. Available at http://tidsskrift.dk

Ulsig, E., Sørensen, A., 1981. Studier i Kong Valdemars Jordebog – Plovtalsliste og Møntskat. *Historisk Tidskrift* **14** (2), 1–26.

Velschow, H., 1843. Om Folkemængden i Danmark i Midten af det trettende Aarhundrede. *Historisk Tidsskrift* **1** (4) 1-52.

Veyere, H. (translator; author not known), 1552. *Regne Bog*. Copenhagen.

Weibull, C., 1978. Vem var Saxo? *Historisk Tidsskrift* **13** (5) 87–96.

# 7

# Renaissance: The missed opportunities

## 7.1 International trade

Denmark is well suited for agriculture. Solar radiation energy is about half of that in Sahara and with strong seasonality, but the days are longer during the growing season, thus as a whole providing a perfect incidence of light for photosynthesis without scorching the plants. Water supply is ample, generally without the flood-causing torrents or the long dry periods common in Southern Europe, and finally, the nutrient availability is excellent. Danish soils are deep glacial clay and sand layers, which after millennia of plant growing provides a thick humus layer, not prone to being washed away during rain or eroded and carried away by wind in dry periods. The initial felling of natural vegetation followed by burning tree stubs in most cases did not lead to significant nutrient loss, because cultivated plants and animals were introduced at about the same time, so that the animals could return manure and its nutrients in at least a partial recycling process (cf. Chapter 4). In coastal areas, further fertilization was achieved by people spreading seaweed or kelp (ashes from combusted seaweeds). The introduction of shifts, working the land-plots in rotation for cereals, cattle grazing and fallow, ensured continuing productivity.

Today, Danish agriculture could feed some 12 times the current Danish human population, if vegetable-derived food were the only source. In practice it feeds less, because a part of the food is going through the animal-raising cycle with its pronounced losses (first of all having to use food of vegetable origin to sustain the life processes of the livestock animals). The situation 500 years ago was somewhat similar, because less intensive agricultural practices using fewer artificial fertilizers were compensated for by a smaller Danish population. What happened during the 16th century was that more areas were set aside for agriculture, at a pace determined by several factors (discussed below). There are no precise statistics on the overall development of exports, but the cultivation of new land is fairly well documented through property inventories made for purposes of taxation.

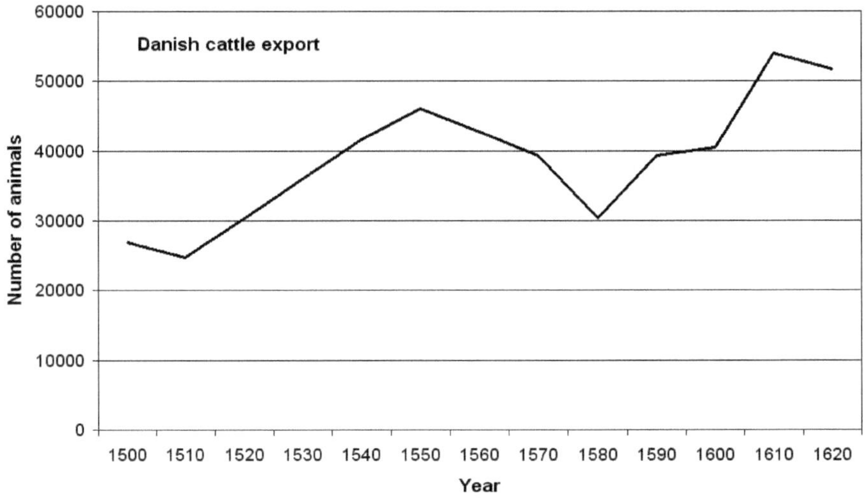


*Figure 7.1. Estimated cattle export from Denmark. The 1580 dip is partly due to competition from transient overproduction in Hungary and Poland (based on Blanchard, 1986).*

Several sources identify cattle (and mainly steers) as the main agricultural export article in the 16th century and onwards. During the summer half of the year, cattle were grazing pastures, and during the winter they were kept in barns and given mainly hay to eat.

During the particular periods up to an imminent sale campaign, additional food was provided for fattening, presumably including cereals (Bang, 1897; Willerslev, 1952). Main recipient regions were northern Germany and Holland, and substantial amounts of cattle was marched south over land, arriving somewhat emaciated because there were few feeding options along the roads and trails. Therefore, a new fattening process had to be initiated at the destination point, at least for steers arriving during the fall period, before selling at the markets. After 1615, the timing changed so that most exports followed winter fattening in barns and took place in spring, where more roadside foraging was available (Frandsen, 1988).

The number of animals exported from Denmark has been estimated as rising from a few thousand in 1423 to 45 000 in the late 1560s, then a decline to 31 000 by 1580 and a new peak at 54 000 in 1610 (Blanchard, 1986). These numbers are consistent with data from the individual Danish cattle-shipping locations such as Ribe or Kolding, mentioned in Danish sources, with the reservation that only export by ship is accurate as the cattle driven to Schleswig-Holstein over land was mostly not recorded (Bang, 1897). The source of data is the revenue from customs duties recorded at individual customs houses (Venge, 1987; Frandsen, 1988; Baden, 1806; general taxation described in Berntsen, 1655, Fig. 7.2). Baden adds remarks on earlier trade (1200–1513), chiefly based on Huitfeldt (1599a, 1599b, 1600). Also a number of horses were exported, as recorded by the customs office at Gottorp (in Schleswig; see Hybel and Poulsen, 2007). Derived animal products for export included meat not only from cattle but also from pigs, such as bacon, and other products including fat, butter, tallow and hides.

*Figure 7.2. Frontispiece from Arent Berntsen's 1655 account of taxes and revenues in Denmark and Norway (from Kildeskriftselskabet, Copenhagen).*

*Figure 7.3. Street in the south Jutland city of Ribe, one of the main centres for hand-ling export shipments to Germany (photo by author).*

Cereals were occasionally imported or exported, but on a highly variable basis, depending on annual harvest yields in Denmark and on regulatory restrictions. A potential export market was present, due to generally increasing prices on the European markets, particularly during the period 1565–1580 (Gamrath and Petersen, 1980; Frandsen, 1988). Subsequently, prices declined, but not below the pre-1565 level.

Forestry products constituted an important commodity, especially timber for the shipbuilding industry, producing vessels both for trade and for war. Fuelwood was equally important for use in fireplaces, furnaces for cooking, and ovens for heating water and indoor spaces. While fuelwood was used locally, timber export was extensive in Europe because of the lack of resource-correlation with demand. Norway developed an extensive export of timber because of its large forestry assets and because of the convenience of shipping by sea, from port cities such as Bergen. Danish timber was roughly sufficient for national shipbuilding, but there was no surplus for export (Hybel and Poulsen, 2007).

Fish ranked at a level comparable to cattle as a shipping commodity, both from Norway and from several coastal cities in Denmark. A traditional catch in Denmark had, since 1200 (and presumably earlier) been herring, abundant in the Øresund Strait. The Hanseatic League had since the 14th century been in charge of fish trade in the Baltic and beyond. Weibull (1922) estimates, supported by Christensen (1972), that by 1400, they would have bought some 25 000 ton of fish from the Danish markets (of which a third went through Lübeck markets), and sold them on to European destinations. One of the products imported by Denmark through Lübeck was salt used for preservation of fish and meat (Hybel and Poulsen, 2007). The catches fell during the 15th century and by 1500 were only between a third and a quarter of the amount a century earlier. During the 16th century, catches regained a sizeable magnitude, but again declined in the 17th century, despite added fleets operating from the north of Jutland and extending the catch to cod, haddock, flounder and plaice (Holm, 1996). The Hanseatic towns declined in importance, and Danish ports themselves became significant shipping points.

The trade with the countries south of Denmark involved imports of a range of goods, not least for the wealthier people in Denmark. From before 1500, items such as a range of metals, lime (for mortar), stone building materials (such as granite or marble), beer, wine, cloth (e.g. silk and velvet) and clothes, ammunition and jewellery had been imported, and also occasionally exported (beer, clothes). Internal trade included hobs and malt for beer-brewing (Unger, 2004), dyes for cloth colouring, flour, butter, milk and honey, construction materials such as timber, stones, gravel, sand and bricks, and so on (Baden, 1806). In the 16th century, items such as spices (other than those previously imported from the Middle East) and tobacco were added, having been made possible by European colonization of foreign continents (Hybel and Poulsen, 2007). With the invention of printing, books became a European trade commodity of considerable importance (Fudge, 2007).

Conducting trade was not exactly easy for those having food or other merchandise to sell. There was a web of legislation preventing a lot of obvious trade opportunities. A set of privileged cities had obtained monopoly on trade, both within the country and by import or export. The king wanted a countable number of merchants that he could control and tax whenever his war expenses required so. Farmers driving their cattle to Germany or fishermen sailing a load of fish to Sweden were considered a nuisance (and probably with some cause, as regards their desire to avoid taxes and customs duties). However, it is difficult to avoid entrepreneurship, as witnessed by a stream of laws and regulatory documents issued by the Crown, followed by further documents where the king complains that his laws are still not followed and issues diffuse threats to the offenders (e.g. the 1582 document by King Frederik II, 1559-1588; Fig. 7.4). Legislation required that the peasants drew excess cattle to the estates of the aristocracy, where it would be fattened and sold through the concessioned city merchants. King Christian II (1520–1521) in 1521 outlawed all export of grazing cattle (i.e. that of the peasants), and restricted the right to fatten cattle by purchased fodder to the Crown and nobility (and before the Reformation also bishops), making only home-grown fodder available to farmers. The kind interpretation of this regulation is that the king wanted to prevent the peasant population from starving if their food were exported, but the fact was that export could have earned the farmers much more than selling to the nobility or Crown. Meat export was similarly forbidden. To this came the restrictions from the fact that most peasants were not independent but worked as slaves to an estate owner (cf. Section 7.2), who would also control the fate of their harvest and animal production (Frandsen, 1988). However, some trade activities by ordinary people seem to have been tacitly accepted, such as the small ships carrying goods between north Jutland and Norway (Hvidtfeldt, 1935), but other sources say that such shipments were also clandestine (e.g. hiding meat for export under a thin layer of herring; Bang, 1897).

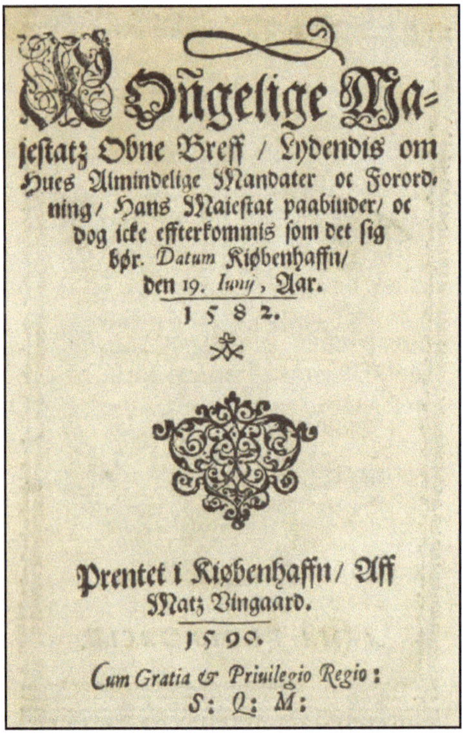
Kongelige Maiestatz Obne Breff / Lydendis om Hues Almindelige Mandater oc Forordning / Hans Maiestat paabiuder / oc dog icke efterkommis som det sig bør. Datum Kiøbenhaffn / den 19. Iunij, Aar. 1582.

Prentet i Kiøbenhaffn / Aff Matz Vingaard. 1590.

Cum Gratia & Priuilegio Regio: S: Q: M:

*Figure 7.4. Open letter issued 1582 by King Frederik II, complaining that his restrictions (e.g. on export of agricultural products) are not followed (Frederik II, 1590). Danish Royal Library, Copenhagen, item LN 1377 4°, digitized by Proquest Europeam Early Books project.*

On top of the limitations posed by trade restrictions, the repeated taxation by the king had a very negative effect on the economic activity outside the privileged nobility clan that served as partners to the king ("feudalization"; cf. Lauritsen, 1987). The largest amounts of customs revenues were already collected at Elsinore from 1429 (Fig. 7.5; fortifications on both sides of the sound, Øresund, ensured compliance by passing ships of any nationality), but, as mentioned, there were additional customs houses (e.g. at Gottorp, Ribe and Kolding; Venge, 1987). Several European protests were overheard, and historians like Huitfeldt

(1599a) draw up defensive statements on these "internal Danish waters that had never in history been open to foreigners"! Interestingly, the Great Belt passage to and from the Baltic remained free to pass through, as it was too wide to allow effective patrolling, but on the other hand, it could be difficult for ships to manoeuvre through it due to moving sand banks. The Øresund customs duty was eventually accepted by the Netherlands and the Hanseatic League by treaties in 1544 and 1560. During the several north European wars in the 16th century, Denmark on several occasions tried to help its allies by closing the Øresund passage (Venge, 1987).

*Figure 7.5. Customs collection venue at the sound between Elsinore and Elsingborg employed fortifications on both sides (Kronborg Castle and Kärnan tower; Braun and Hogenberg, 1585; map from Danish Royal Library, 2010a).*

The Danish kings spent money at a high rate, and when increased taxation of the incomes of the Danish nobility was insufficient, an additional cargo tax was introduced in 1567 on all Øresund shipping, to be paid in Copenhagen. This policy (accompanied by even more increases in spending) was continued by King Christian IV (1588–1648), who also at times forbade imports, and at other times placed high taxes on several imported goods. Fig. 7.6 shows the ordinance forbidding import of textiles in 1623. The Danish kings foolhardily took part in any European war (and there were many) under the illusion that Denmark was a superpower, as it perhaps briefly had been during Viking times.

Forordning
At intet vden Rigit giort / farffued eller
bered Klæde / Kirsey / Foderting / Bay / Sarge eller
noget saadant maa vdi Seeland indføris.
WJ Christian den Fierde / med Guds Naade / Danmarckis / Norgis / Wendis oc Gottis Konning / Hertug vdi Slesuig / Holsten / Stormarn / oc Ditmersken / Greffue vdi Oldenborg oc Delmenhorst / etc. Giøre alle witterligt / at efftersom Wi Naadigst / Kiøbmends handelen oc Handverckerer her vdi Vort Rige Danmarck at befordre oc forsette / haffuer for raadsomt oc nødvendigt anseet oc eractet / med Voris Elskelige Danmarckis Rigis Raads Raad oc samtycke at lade Vor Naadigst Forbud oc Forordning vdgaa / at

*Figure 7.6. Import prohibition on textiles and other items, issued by King Christian IV in 1623 (Customs and Revenue Museum; Venge, 1987).*

The ruler's addiction to war was a common European problem, but realizing the economic possibilities of opening economies to import and export was more successfully converted into policy by several countries in Europe other than Denmark. The new goods brought back from colonized regions outside Europe were traded and incorporated into European lifestyles, leading to substantial heightening of living standards in southern and western Europe.

The extravagant lifestyles were soon copied by Danish kings, as witnessed by a very large number of castles and other impressive buildings being erected, including Kronborg Castle in Elsinore (Helsingør, built by Frederik II on top of previous buildings still visible in the basement) and culminating during the reign of Christian IV. His showpieces include the Stock Exchange and Rosenborg Castle in Copenhagen, and Frederiksborg Castle in Hillerød, built in the architectural style called "Dutch renaissance" (Fig. 7.7). Today we are happy with these fine buildings, but it threw Denmark into bottomless debts at the time and, along with the costs of war, greatly damaged conditions for the common man. Records show that taxes could in many cases not be collected simply because the person owing had no possibility of ever earning the amount of cash demanded (Venge, 1987). This was true even if the work of the person was generating reasonable revenue, because only a small fraction of work compensations were in the form of cash. Most of the Danish economy was still based on payment in natural products such as grain, meat or return services, and, in consequence, a person could be doing fairly well and still be unable to come up with the cash demanded by the king's tax collectors. The lords were better off, being allowed to trade in cattle and participate in the royal projects, and many stately country houses were constructed, notably during the early 16th and 17th centuries.

*Figure 7.7. Frederiksborg Castle in north Sealand, built in 1600–1620 by King Christian IV, after tearing down most of an earlier castle from 1560, and faithfully rebuilt after a fire in 1859 (photo by author).*

One outcome of the debt situation prevailing for both the Crown and the citizens of Denmark was a lack of capital for investment in the new technologies becoming available in Europe, or generally for embarking on new projects, notably in agriculture or manufacture. This is one reason that farming yields continued to decline: without population growth, no increases in food production seemed justifiable (Fig. 6.22). Another reason, as mentioned earlier, was that little grain export was allowed and only temporary surges of cattle export, so the steady increase in actually cultivated area (Fig. 6.28) must indicate a declining area yield, estimated to have decreased from about eight-fold around 1050 down towards two-fold around 1650 (Nielsen and Dalgaard, 2009; Frandsen, 1988). The "fold" is defined as the harvest yield needed for next year's seed, so "two-fold" is only twice the amount to set aside for continuity.

The list of 16th and 17th century wars is long. King Christian IV attacked Sweden in the Kalmar War (1611–1613) and marched into the 30-year War (1614–1648, Denmark joining 1625), his army being crushed in 1626 by the Catholic army at Lutter in Germany. Sweden sought revenge by invading Jutland 1643 and Skåne 1644. In 1645 a peace treaty forced Denmark–Norway to yield Halland, Gotland, the Danish part of Estonia, Härjedalen and Jämtland to Sweden. In 1657, the next Danish king, Frederik III, in turn attacked Sweden through Germany, but completely lost, with Sweden ending up occupying nearly all of Denmark, and through a treaty by early 1658 took over Skåne, Blekinge and Bornholm in eastern Denmark, and Bohuslän and Trøndelag (near Trondheim) from Norway. Only the city of Copenhagen held out, and in the subsequent peace settlement in late 1658, Denmark–Norway regained jurisdiction over the island of Bornholm and Trøndelag. This was not the end of the hostilities. Denmark tried in the Scanian War (1675–1679) to regain Skåne, but failed.

The Danish kings did not behave differently from other European monarchs, always looking for options for territorial gain (sometimes disguised as wars over religious differences), but the small size of the country combined with the inflexible structure of labour organisation in the dominating agricultural sector made the war losses and property damage in occupied parts of the country have a larger impact on serious investments for establishing new types of production or business.

The Danish kings actually tried to copy some actions that had worked well in other countries, such as acquiring colonies in the Asia–Pacific region, but succeeded only in occupying a small area called Trankebar on the Indian east coast, despite efforts to also conquer Sri Lanka. Trankebar was made the supplier to the newly formed Danish Ostindian Trade Company, and, failing to become a more substantial colonial power (later a few acquisitions were made in the Caribbean, see Fig. 8.19, and African Goldcoast areas, but subsequently sold to the US and England, respectively), the Danish king Christian IV turned to Greenland and Iceland, originally explored by Norway but thereafter mainly being serviced and (in the case of Greenland) populated by

Danes (maintaining, it seems, reasonable relations with the indigenous Inuit population). The king now formed trade companies with monopolistic rights to import and export, causing the establishment of a very unfair price structure, making the population of these northern parts of the kingdom suffer unduly over the next several centuries (Wikipedia, 2010).

The nearly uninterrupted state of war made the king establish a number of soldiers' garrisons in different parts of Denmark, and notably in the Sealand region near Copenhagen. The king had scientific advisers informing him of the dietary requirements for making fierce soldiers, and as they identified fish as an important kind of food for producing strong men, the king ensured the establishment of a number of new fishing villages with contracts to furnish certain volumes of fish to the garrisons. One example is Gyldeleye (now Gilleleje), placed at the northern tip of Sealand (Fig. 7.8), 4 km north of Søborg, the small city (with full trade privileges) formed during the 12th century around the archbishop's castle (see Fig. 6.7). In the beginning of the 16th century, the king built houses along the beach for a group of fishermen (coming from Søborg, from where they had previously conducted a bit of summer seasonal fishing from the shores now becoming Gyldeleye, staying the nights in sheds, and bringing the fish back to Søborg, stored in salty brine for winter consumption). In return, the fishermen would now deliver a fixed amount of fish, mainly cod, each year to the Crown, presumably for the soldiers in the closest garrison near Elsinore. Any surplus that the fishers might catch could be sold at market prices (Bang, 1917; Terslin, 1933). Because both land and buildings were owned by the king, the concepts of rent and tax may have been simplified. Before the fishing villages here and in other parts of the country were established, the king's tax collectors had followed the fishermen to their temporary stalls, and people of other professions were strictly forbidden to go fishing, indicating that this was a widespread practice for an extra profit or a free meal for the family. Fishermen without a special contract with the king were required to pay several taxes: a sand tax for use of the beach, an oar tax for using their boats and a booth tax if they sold fish from a stall (Steenstrup, 1906).

The Gyldeleye fishers initially had golden days and sold the surplus catches illegally (fig. 7.9), because, of the two legal trade cities, the trip to Elsinore was considered too long, and the prices obtained in the decaying town of Søborg (today a small village) too low. The Danish kings generally had a fatherly attitude towards "their" fishermen, and when the catch declined dramatically towards the end of the 16th century, Frederik II gave each family a strip of land to grow vegetables and berries, and keep a cow or pig or some chickens, to make up for the missing income from fishing. Later, when the sand drift (see Chapter 8) destroyed the fishermen's houses, the king helped them financially to rebuild their homes. The cause of the disappearance of the herring from the seas around Sealand is not known. The "little ice age" lowering of temperature (Fig. 6.22) appears too small to have this effect. In following centuries, fish catches again increased, but with other species than the previously dominant herring, and new periods of disappearing stocks set in at intervals.

*Figure 7.8. North Sealand (Blaeu, 1662; at the Danish Royal Library, 2010a).*

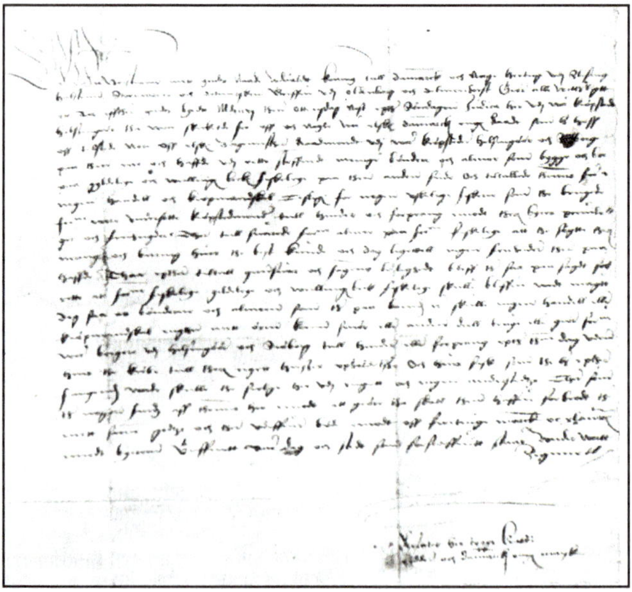

*Figure 7.9. Letter from King Christian III's marschal T. Krabbe, 1536, to the Gyldeleye fishing community, informing them that selling fish bypassing the market towns of Elsinore and Søborg will not be tolerated (Terslin, 1933; Elsinore city archives, letters on paper, vol. IV, 1509–1584, the document is now at the Danish State Archives).*

Foundation stones of the houses built in Gyldeleye for the fishermen in the 16th century have been excavated (see Fig. 8.5), but the buildings themselves are lost. The population in the late 16th century reached some 70 fishermen, plus their families and a priest employed in their self-erected church. Court records tell of many incidences of violence between the men, and of attempts at cheating in property deals (Frandsen, 1987; based on Royal Chancellery letter of 24th December 1580; document in Danish State Archives). The Crown's civil servants had an eye for details of what went on in every corner of the country.

## 7.2 Social stratification and incompetent rulers

The lot of the peasants, still constituting some 90% of the Danish population, was largely deplorable. They were no longer called slaves (this name was abandoned after the Viking age), but rather adscripts and copyholders. The 16th century social hierarchy had servants and lodgers at the bottom (both lodging at somebody else's dwelling against labour services), then a trifle higher the copyholders bound to a farm owned by someone else, and finally the much smaller groups of professionals (such as blacksmiths and grocers) and of land or other property owners (from a few simple independent farmers to merchants, landlords and aristocracy).

The adscription-like arrangements had slightly different conditions in different parts of Denmark, but generally involved a copyholder's contract (Fig. 7.11) providing the peasant with a piece of land (and normally a farmhouse, some cattle and seeds for the following year, if this was a novel arrangement and the current year's harvest had already been completed). In return, the peasant should pay the landlord an annual amount (usually in the form of grain, milk and meat, as cash was rare and this was in addition to taxes paid to church or king), and should deliver a certain number of hours of labour weekly (villeinage) to whatever task the landlord wanted performed. The contract was for life, and in some regions (e.g. Sealand) also involved the family of the farmer, allowing the landlord to fulfil his conscription requirement to the king's army by picking any suitable sons of his peasants. On average, 25% of the men at farms were conscribed (cities had no conscription obligations). The peasant was not allowed to leave the estate where he was born but could be forced to shift to another holding of the landlord, such as a neglected or deserted farm. In some periods, the landlord could also sell his peasants (Frandsen, 1988). Such arrangements were introduced in the 14th century and were first called "vornedskab" and from 1733 "adscription".

By the end of the 18th century, adscription became formally abandoned, but it took even longer to lift the spatial entrapment in practice, because the copyholders rarely had alternative options. However, a gradual improvement did materialize, whereby the peasants little by little became capable of purchasing the farms they had cultivated, albeit in most cases with a mortgage deed replacing the copyholder contract. However, the mortgaging was a purely economic arrangement. As for army duty, the conscription management was in 1788 moved from the landlords to regional recruiting offices, furthering the process of removing the power of the landowners over peasants (Lind, 2008). General conscription was not introduced until 1848.

*High and low – a genealogic digression*

The stratification of society during the 16th and 17th centuries may be illustrated by two people selected from the direct line of ancestors to the author. My family tree has been carried 11 generations back, on both my mother's and my father's side. From a genetic point of view, this gives over 2000 individuals of interest, but not all have been identified, due to records lost (mostly in fires). Sources of information comprise church entries at baptism, marriage and funeral, land registry, taxation and property ownership records, deeds and mortgage papers, census and soldier recruitment notes, any criminal records, succession and inheritance specifications, and finally, private or official letters and other material that may have survived.

Nearly all my father's ancestors were peasants and never ventured outside a $20 \times 20$ km$^2$ area in northwest Jutland, while my mother's ancestors comprise a variety of professions, including before the 20th century mayors, priests or vicars, undertakers, bellringers, teachers, school masters, land agents, market gardeners, postmasters, judges, aldermen, public trustees, district bailiffs, ropemakers, fire chiefs, train station masters, police officers, coppersmiths, retailers, tailors and sailors. Many stayed on the island of Funen, but there were also several settling in other parts of Denmark or overseas (mostly in Europe or Canada).

In may father's line, members of the earliest five identified generations (born 1610 to 1740) were all copyholders near the Simested Stream in Rind District, northwest Jutland. Then came a very industrious Jens Pedersen Støttrup (1765–1845), who with his wife Johanne Thomasdatter worked his little (3500 m$^2$) copyholder farm on the heathland so well that the proprietor decided in 1807 to move him to a larger (21 000 m$^2$) but decayed and not very fertile piece of land on the other side of the Stream, swapping with its incompetent copyholder, who then got Jens's well-kept farm. Jens and Johanne managed to also make the new farm produce a surplus over the dues to the squire, and in 1827 they were able to buy the farm for 325 dollars ("rigsdaler", the Danish currency at the time), with 125 dollars down payment and the rest mortgaged. The transactions are documented in public records (see Krogh, 1954; Nordisk Slægtsforskning, 1990), but there is additional evidence told orally, first by Jens to his son Jens, who told the stories on to his son Jens, who had a brother with two children becoming successful writers, and both very interested in Danish history: Marie "Thit" Jensen (1876–1957) and Johannes V. Jensen (1873–1950; Nobel Prize in literature, 1944). Johannes interviewed his uncle and wrote the story about Jens Pedersen Støttrup's two farms as the last short story in his book on the land of his roots (Jensen, 1923).

*Figure 7.10. One of the few remaining heath areas in Denmark. They were common in the 16th century northwest Jutland region (photo by author).*

As an impression of the type of soil the early family copyholders had to put up with, Fig. 7.10 shows one of the few heath or moor areas remaining today. The army recruiting office reports Jens Pedersen Støttrup's height to be 1.62 m, in line with the declining height of Danes up to the end of the 19th century (cf. Fig. 4.4). Since milk intake is an important determinant for human height, and the deed protocols for the farms where Jens spent his childhood indicate presence of a livestock holding, one would conclude that most of the milk produced had to be delivered to the owner's estate (Lille Restrup). The sale that made Jens rise to independent farmer was of course made possible by the consent of the landlord, which he did not have to give, but did, in line with the more democratic sentiment emerging at that time. Jens and Johanne passed the farm to one of their sons, in return for being cared for to their death, which occurred some 15 years later. Around the time of retiring the remaining 200 dollars of debt had been fully repaid.

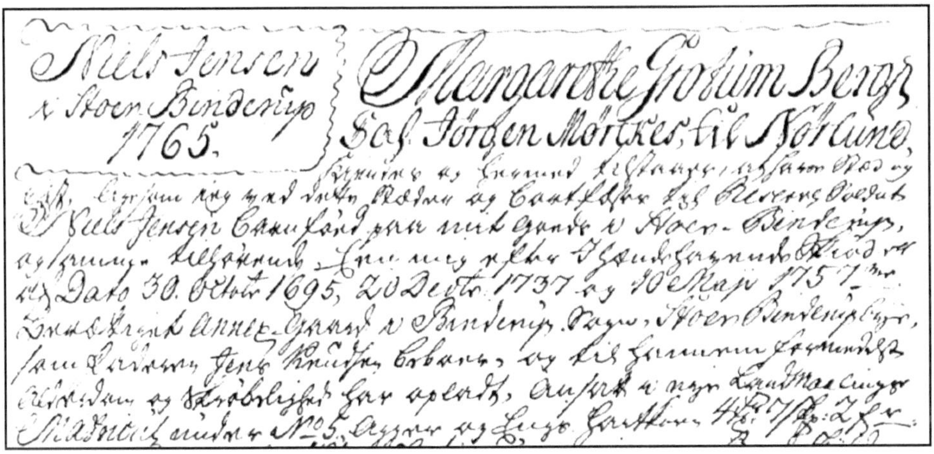
Niels Jensen
i Hoer Bindrup
1765.
Margarethe Grotum Berg
Sal. Jörgen Mörkes til Nörlund

*Figure 7.11. Copyholder's contract, 1765, issued to the brother of Karin Jensdatter, the mother of Johanne, wife of Jens Pedersen Støttrup, discussed in the text (Copyholding Protocol Office at Nørlund, Jutland Archive; now at Danish State Archives).*

The family generations before Jens Pedersen Støttrup had far fewer assets, except for one who was not a peasant but a shopkeeper. Figure 7.11 shows part of a typical copyholder contract (that of Johanne's grandfather). At the death of the last of Jens's grandparents, the stocktaking of assets to divide between heirs amounted to the following: the clothes being worn (plus some torn ones), two sheep, a kettle, two dishes and a broken plate, a cupboard and a bed, two candlesticks, and a defective stove. The earliest identified ancestor of Jens on the male side, Jens Amdisen (about 1630–1685), also died in poverty. He was copyholder not of a farm but of a windmill in Lørup (owned by a Viborg Chapter), but the mill burned down in 1681 and his face was badly burnt trying to fight the fire. As the charity of the Church was not extended to him, he had to live as a beggar during his last years (according to a legal statement letter concerning the fire, reproduced in Nordisk Slægtsforskning, 1990).

On my mother's line of ancestors I will focus on Hans Jørgensen (1612–1681), from 1655 mayor of Frederikodde Fortress (east Jutland, now the city Fredericia) and from 1660 mayor of Bogense town on Funen. He was married to the daughter, Kristine Steen, of the former Bogense mayor, Jens Madsen. Frederikodde was conquered by the Swedish army in 1657, where it lost over 1000 lives, but was liberated the following year. In 1658, Hans Jørgensen travelled in Skåne (Scania), where he tried to raise Danish-disposed citizens and the city of Malmø's non-Swedish commandant to break away from the Swedish suzerainty. That would also have saved a farm he owned at Aagerup north of Ringsjön, but the revolt failed and the rebels were jailed (Drysenius, ca. 1690). Hans escaped because he was in Copenhagen at the time.

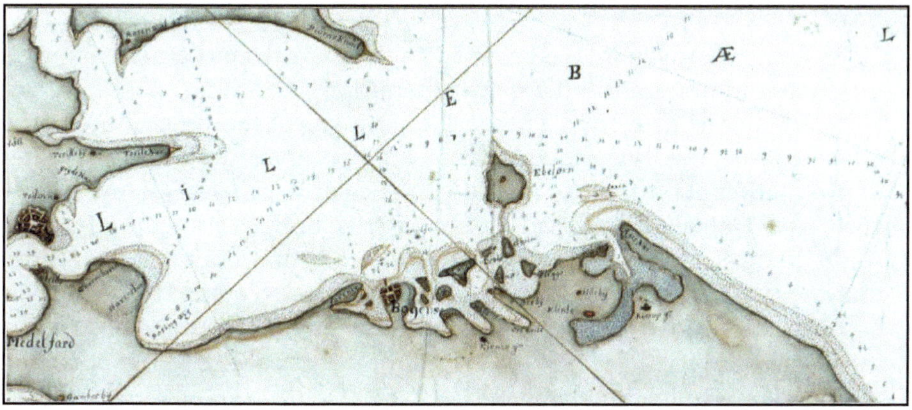

*Figure 7.12. Æbelø at Bogense (J. Sørensen, naval map, 1700; Danish Royal Library in Copenhagen, 2010a).*

*Figure 7.13. Bogense church and first town hall (J. Brun, 1754; Danish Royal Library in Copenhagen, 2010b).*

His new mayoral post was very challenging, as the Bogense town was very poor and deserted after the war (50% of its houses unoccupied; Søkilde, 1892). Hans's own farm near Frederiksodde was also destroyed, but he managed to bring Bogense around to a reasonable condition after only a few years. King Frederik III rewarded Hans Jørgensen for this and for his heroic wartime conduct by giving him the island of Æbelø to keep as long as he lived, in addition to other estates near Bogense that the king traditionally entrusted a mayor, who was not otherwise salaried (these are specified in the crown records later published by West, 1908, as having a value over $ 20 000 – compare with the price of $ 325 given above for the farm bought by Jens Pedersen Støttrup in Jutland). Æbelø is connected to Funen just north of Bogense by a causeway (Fig. 7.12). It offered fine forestry and a bit of farming, but it is today most known for its fauna, including the stag-beetle, not present anywhere else in Denmark (Friis, 1981). Hans Jørgensen's quarters at the Bogense town hall are shown in Fig. 7.13.

Hans Jørgensen was in 1660 elected to the Danish parliament as one of the first commoners to take a seat there (Barfod, 1895). Most members were aristocrats or clergy. The parliament was at the time limited to an advisory role vis-à-vis the king. It could be seen as replacing the things of pre-medieval period (e.g. at Isøre, N. Sealand, cf. Ch. 5). However, by 1660, the Danish kings had declared themselves autocratic, causing some distress among the aristocracy, who felt that they lost power and that the parliament was only for show. This lasted until 1848, when the absolute monarchy was given up, presumably influenced by fear that the French Revolution (1789–1799) might be spreading to the north and might endanger the kings' necks.

*Figure 7.14. Mayor Hans Jørgensen (1612– 1681). Contemporary painting in family possession (Nielsen, 1947).*

Hans Jørgensen provided generous work for the Danish courts, from lower ones to the Supreme Court, with over 30 claims by him or against him regarding land payments, rent, compensation for property lost in the Swedish War, and releases from payments for provisions destroyed during his wartime assignment as manager of Frederiksodde provisions and building activities. He initially won the cases at court, but those concerning the Frederiksodde period were later taken up again by the King Christian V (1670–1699) following Frederik III, along with many similar cases against other wealthy members of the upper classes, using petty arguments of missing vouchers to demand large repayments with interest to the king. Hans Jørgensen, of course, could not produce the more than 20-year-old receipts lost in the Swedish assault and wrote two petitions to the king, mentioning his highly appreciated services to the previous king and asking for mercy to "a poor man". But Christian V stands firm and Hans Jørgensen dies (from grief?) a few weeks later. His widow (not the same as Peter Madsen's daughter) writes a third petition and although the king now annuls the war transaction claims, only a small house is left her after other claims have been settled by execution levied against the real estate holdings. Yet she retains parts of Æbelø, as the Crown lawyers forget that is was given only for a finite period (Barfod, 1895).

Forordning
Om
Consumptions Middel / Qvæg Tiende Skat / Item Told oc Accises Forhøyelse oc seiglet Papir.

Kiøbenhaffn /
Hos Peder Morsing / Kongl. M. oc Univ. Bogtr.
ANNO 1657.

*Figure 7.15. King Frederik III's ordering of a general consumption tax, 1657 (P. Morsing Print, Copenhagen; in Customs and Revenue Museum, Copenhagen).*

The king tries to fill the hole created in the state finances after the several lost wars with Sweden, not just by court actions against landowners, but also by adding to the several existing taxes a general consumption tax levied onto all merchandise sold in Denmark, including food

(Venge, 1987; Fig. 7.15). The level of taxation peaked in 1678 (Feldbæk, 1993; Fig. 7.16).

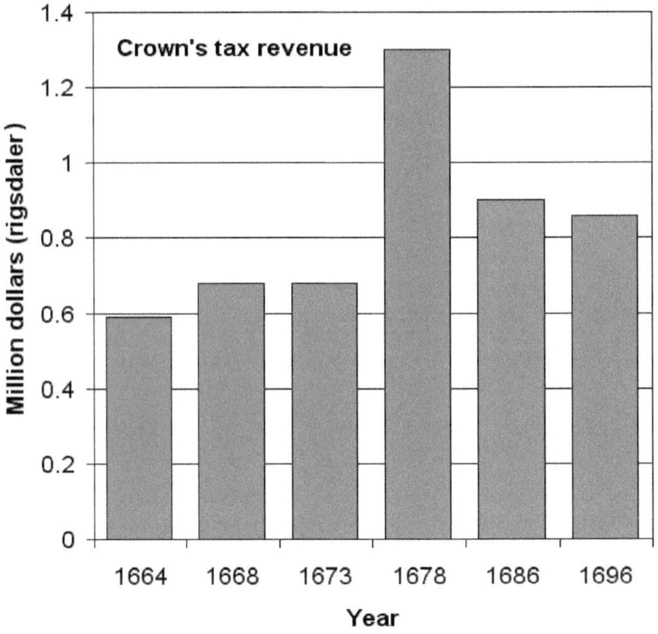


*Figure 7.16. Variations in the Danish crown's revenue from taxation in the period around the Scania war (based on data from Jespersen, 1989).*

Tax calculation was made easier by a new type of land registration implemented from 1682, to be discussed in Section 7.3.

Perhaps the European wars during the medieval and Renaissance period can be explained in terms of a process of establishing and delineating the new top-level structures called "nations". However, wars kept breaking out. They often appear as a game played between the kings of European countries–kings that were mostly part of a heavily interbred common family. Ordinary people felt little difference in belonging to one or another territorial entity, or being ruled by one or another member of the European royal families, but they surely felt the negative effects of killings and shortage of food and other commodities each time the rulers clashed. Some of the wars were connected to religious differences among Christian fractions showing little Christian tolerance towards each other, a feature that seems common to all institutionalized religions. The price was paid in terms of absent social improvements.

## 7.3 Energy in agricultural production

Christian V reformed the basis for taxation of the farmers by an ambitious project of land registration. Surveyors were in 1681 sent out to measure the areas of fields for each farm in the country, and assessors subsequently evaluated the quality of each plot, estimating the average harvest yield over good and bad years. The annual tax was then set as a fixed amount for each area, which was recorded in a register together with tariffs for hay from pastures, grazing cattle or pig's consumption of mast in forests. After rounds of hearing complaints by farmers, the registry was finalized in 1688 (Pedersen, 1928). The annual average cereal production at a given location is estimated to have varied between 6 and 16 times the fixed tax specified as barrels of grain (Dam, 2009). The data have been digitized (Dam, 2004) and forms the basis for the interpretation made in the following figures. Since the data are available at the village level, the survey starts in a specific region, taken as the neighbourhood of Bogense town discussed in Section 7.2.

Figure 7.17 shows the administrative divisions on districts and parishes, the position of villages, and how many farms there are at each village. In Fig. 7.18*a*, the cultivated area for each village is given, converting the original numbers of "barrels of land" into $km^2$ by multiplying by a factor of 0.0055. In Fig. 7.18*b*, the harvest yields in an average year are estimated from the taxation amounts, assuming that the variation in quality of the soil has been correctly assessed by the Christian V teams, so that a typical ratio of actual production and the amounts to be paid as tax can be taken as 12, the mean value for the range given above. The tax yields exempt fallow fields (resulting from rotation) and are originally given in barrels of "hartkorn" (i.e. grains of barley or rye, and one barrel of hartkorn is assumed to weigh 100 kg). The energy content of cereal grains is in the range of 14–16 kJ/kg (FAO, 2003), and 14.7 kJ/kg has been used to obtain the estimate shown in Fig. 7.18*b*.

It is seen that the distribution of cultivated area, of number of farms and of harvest yields are roughly similar in this limited area of Funen. This is not quite the case if all of Denmark is considered. Figure 7.19 shows the distribution of farms on parishes for the whole of Denmark, as it looked after the losses in the wars with Sweden. Figure 7.20 shows the distribution of cultivated land on parishes and Fig. 7.21 the distribution of harvest yields in energy units, calculated in the way described above. Now the differences in quality of soil between eastern and western Denmark stand out clearly: the yields are significantly lower in the western part of the country. One might thus expect a larger area to be required for feeding the family of each farmer, but such a difference is not evident, implying that the west Jutland farmers were not only poorer, but also closer to starvation than their counterparts on the islands or the eastern coast of Jutland.

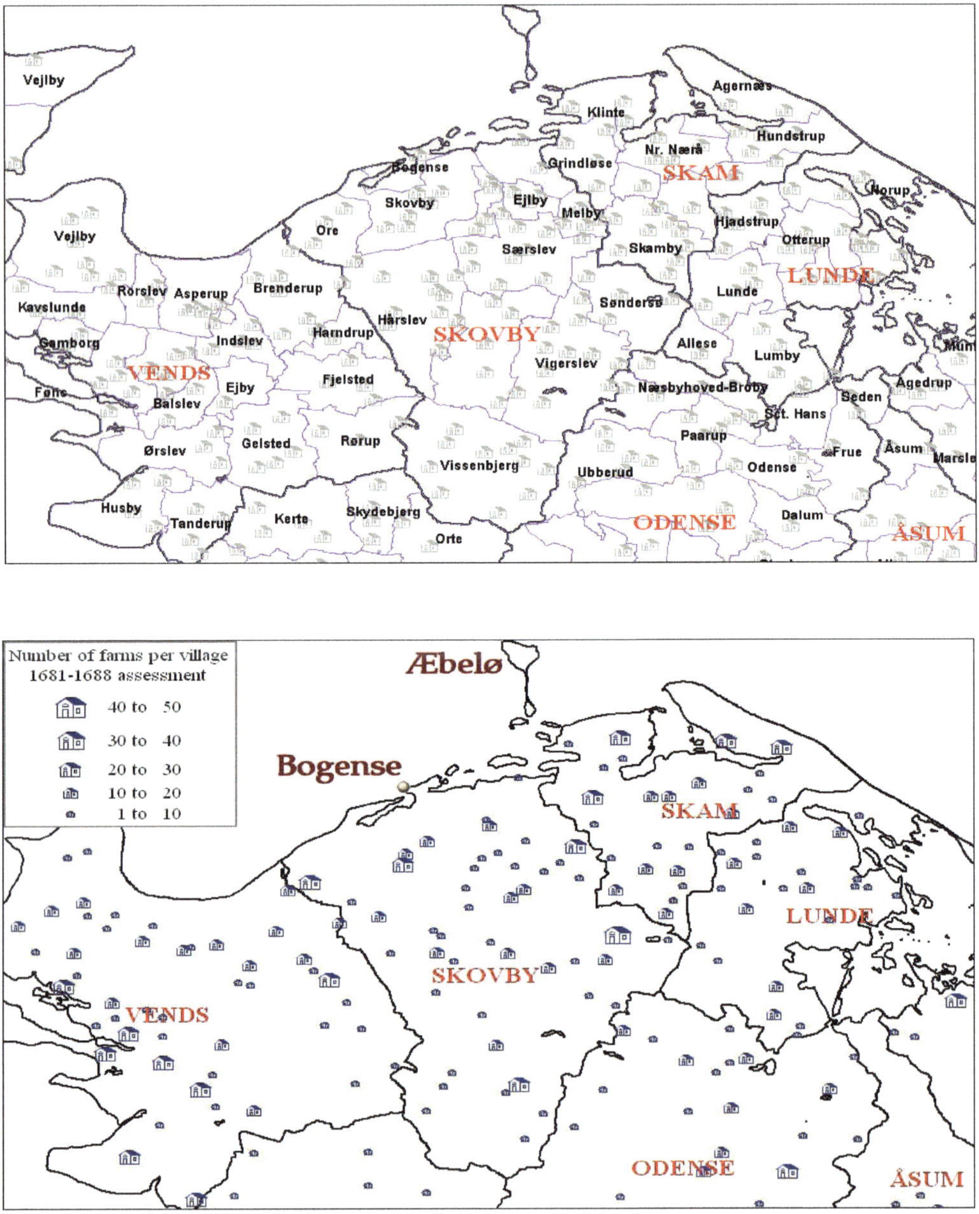


*Figure 7.17. Villages in northwest Funen with administrative divisions, 1688 (a, top). Number of farms in each village (b, below; based on the Registry of King Christian V and Dam, 2004).*

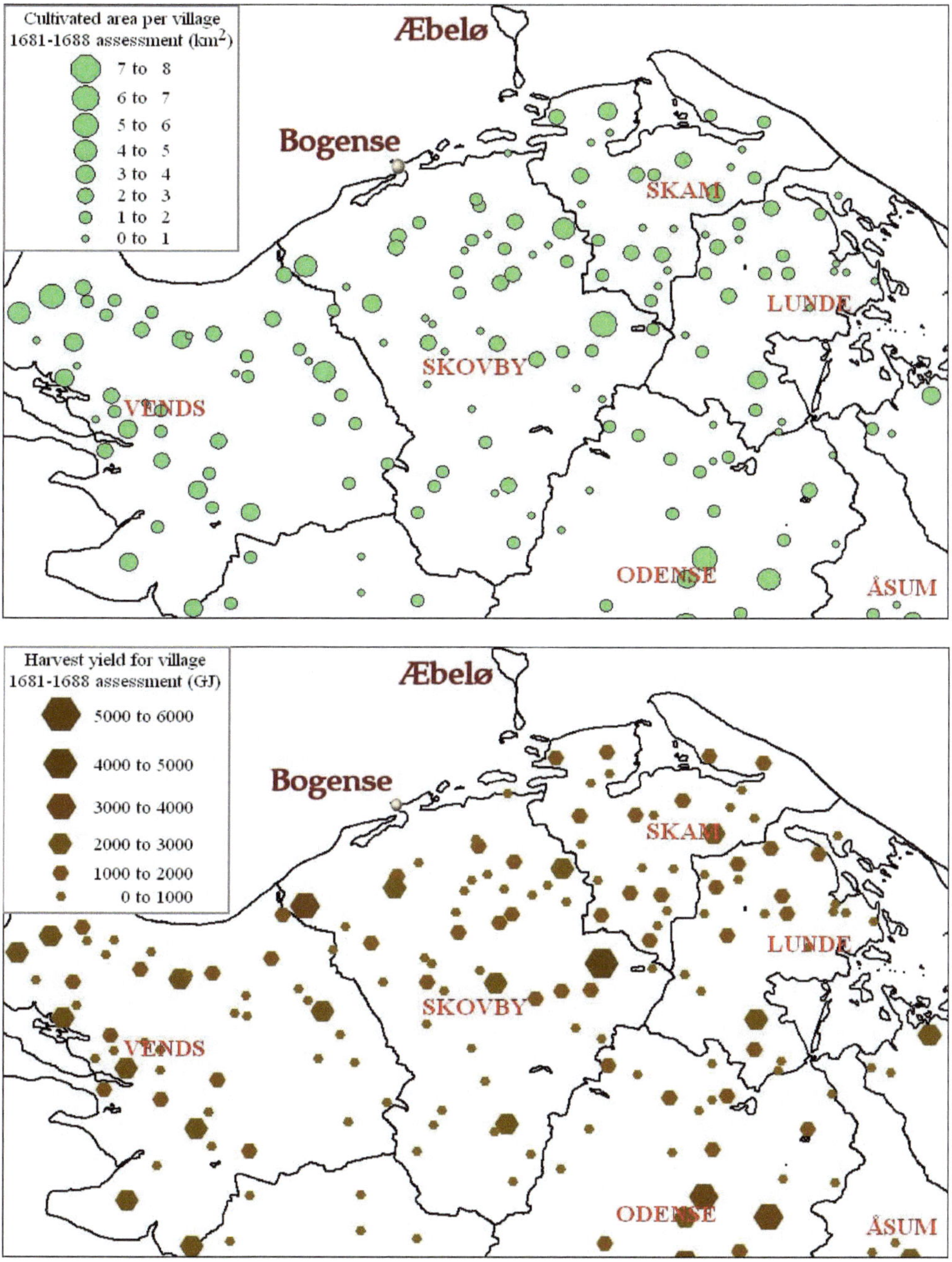


*Figure 7.18. Cultivated areas for villages in northwest Funen, 1688 (a, top). Average harvest yields of each village, expressed as the energy content of the harvest of a typical year (b, below; calculation based on data in the Registry of King Christian V and Dam, 2004).*

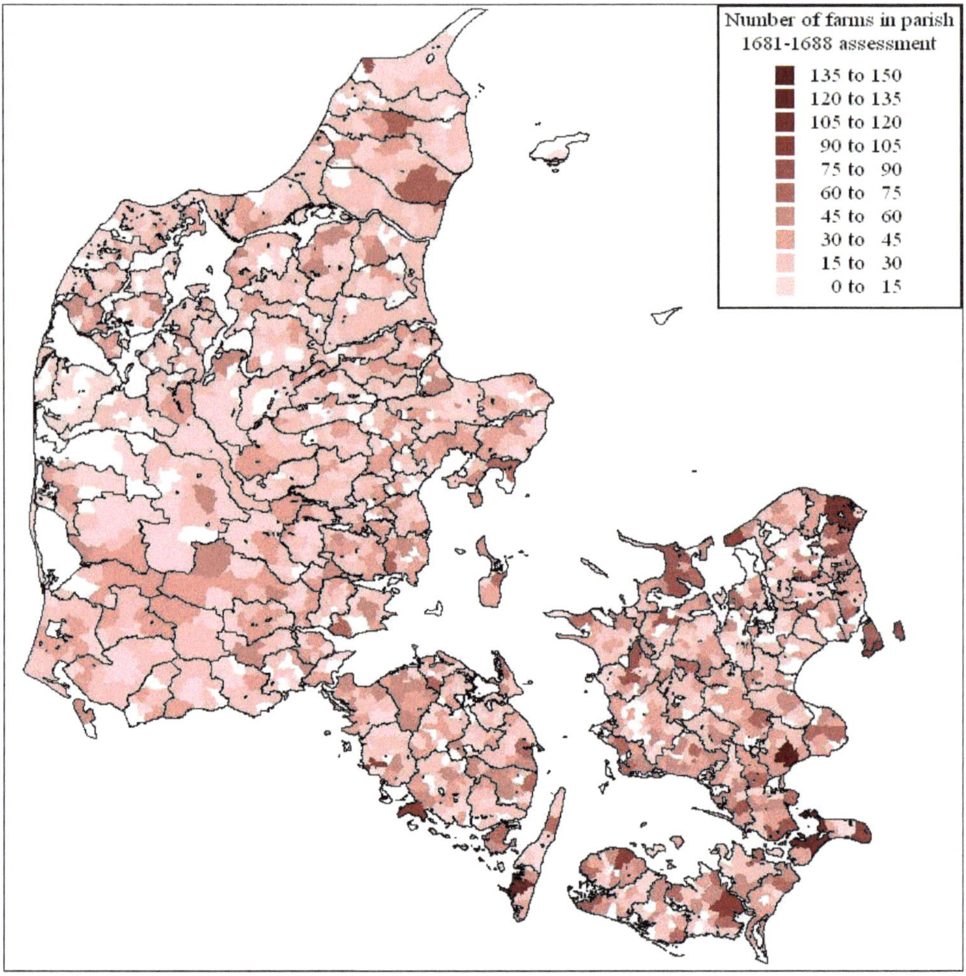


*Figure 7.19. Number of farms per parish in 1688 (based on the Registry of King Christian V and Dam, 2004).*

The energy yields in Fig. 7.21 allow a check to be made on the population model presented in Fig. 6.22. According to this, Denmark would have a population of around 600 000 towards the end of the 17th century, and increasing. The energy contents in the average annual harvest figures estimated in Fig. 7.21 add up to 3.37 PJ. Assuming that 25% is lost in storage (going mouldy or eaten by rodents during storage), the remaining 2.53 PJ $y^{-1}$ could feed 670 000 people consuming food at an average rate of 120 W (cf. Sørensen, 2010), in agreement with the estimate behind Fig. 6.22. Further food (e.g. from livestock and fish), would have been available, indicating that, in an average year, Denmark was able to maintain self-sufficiency in food, while allowing for possible food losses above 25%. The 120 W/cap. includes provision for an amount of muscle work to be performed. It is seen from Fig. 7.20 that the four areas delivering data to Fig. 6.28 are not always representative for the entire region (particularly not for north Sealand).

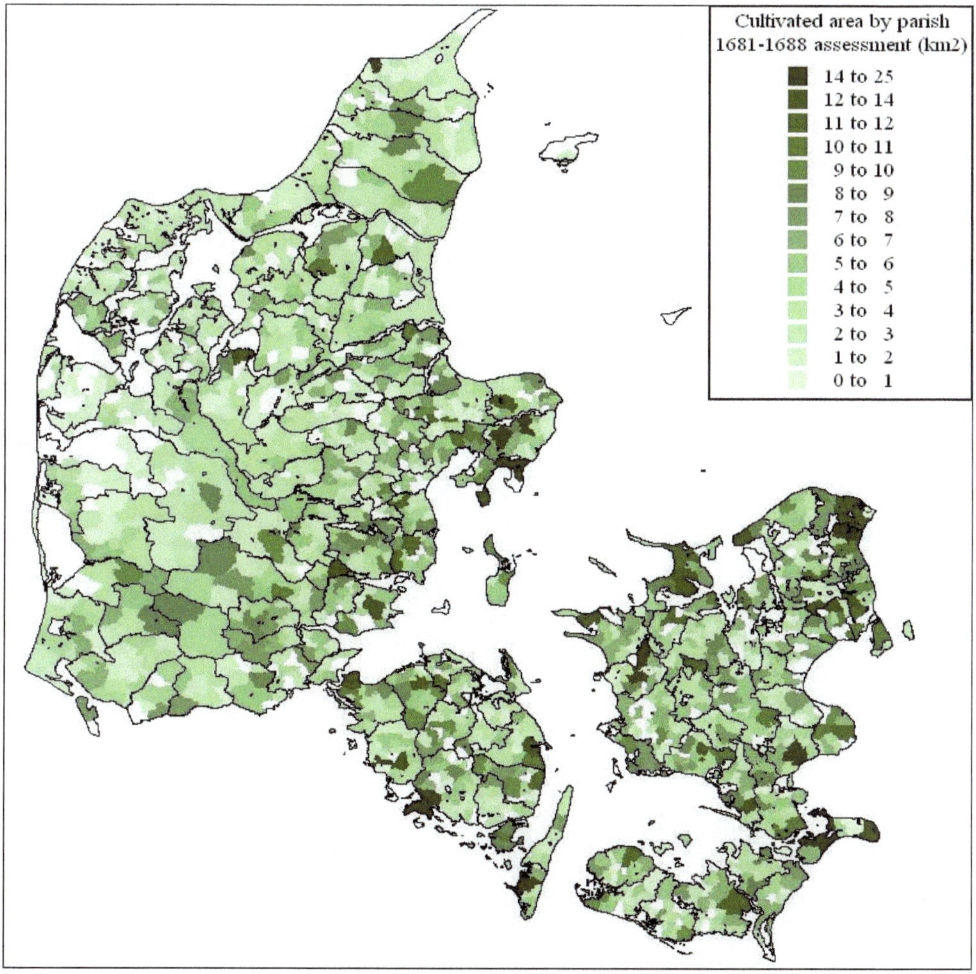


*Figure 7.20. Cultivated area (km$^2$) in each parish by 1688 (based on the Registry of King Christian V and Dam, 2004).*

The population living in cities was still small by the late 17th century. In 1672, an estimated 41 000 people lived in Copenhagen, and a similar number in all the other market towns together, the largest being Aalborg with 4200 inhabitants (Degn, 1987). Census-taking did not materialize until a century later.

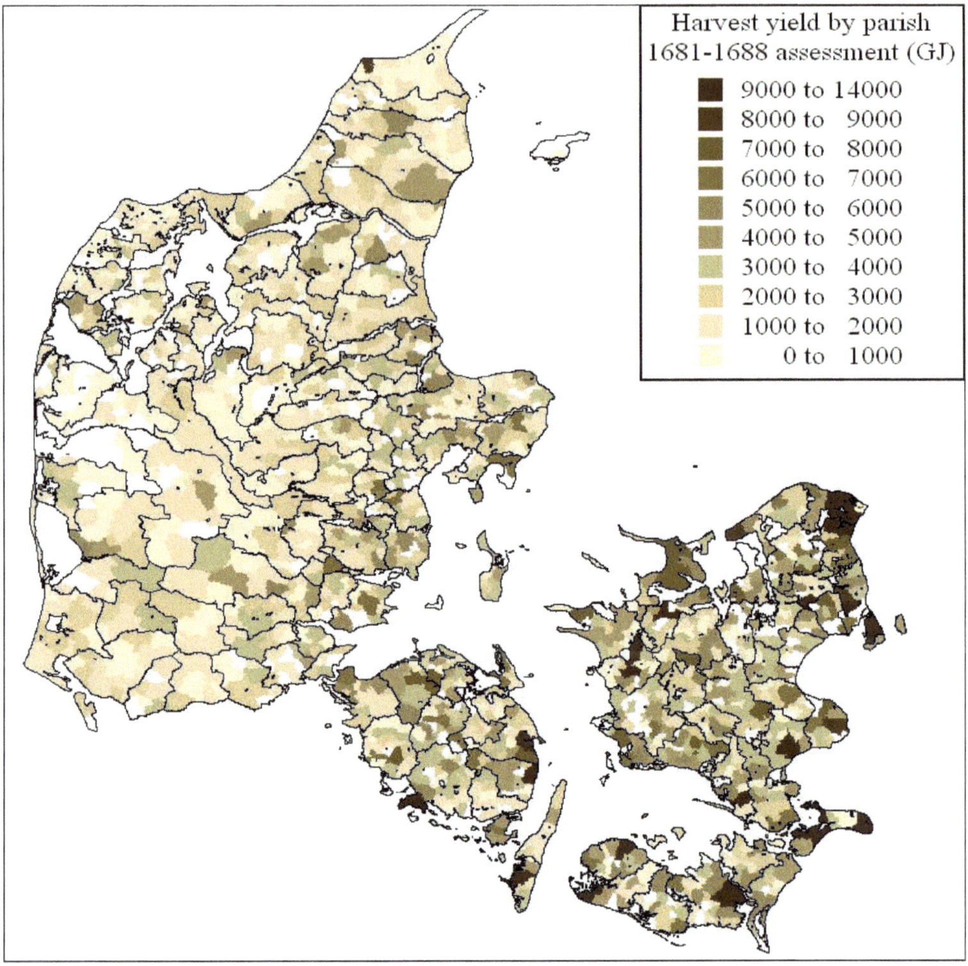


*Figure 7.21. Typical annual harvest yields by parish, expressed in energy units (GJ $y^{-1}$; calculation based on the 1688 Registry of King Christian V and Dam, 2004).*

## 7.4 Church, knowledge and education

The transition from Catholicism to Protestantism had involved some initial fighting, intertwined with questions of royal succession, but once in place it went fairly smoothly, with existing papal priests and monks in convents allowed to function until their death, during a gradual transfer of the Church's property to the king.

Protestantism was more moral-based and ascetic than most Catholic teaching. The historical Jesus stressed that moral imperatives could not be boiled down to obeying a list of commandments (cf. Chapter 5), but this was exactly what the Lutheran Church proposed, by analogy with the Jewish Moses Laws.

*Figure 7.22. A gospel writer helped by a divine prompter and sitting in psychedelic surroundings. The picture is in the introduction to the Danish translation of the New Testament (published 1524). The book is in the Danish Royal Library, Copenhagen, LN 270 8°, digitized by the Proquest "Early European Books" project.*

According to the philosophy of Luther's Protestantism, the New Testament containing the foundation scriptures of the Christian faith had to be translated to local languages so that everyone could read them without an intermediary priest. A Danish translation appeared in 1524 (Fig. 7.22), including an "Introduction to the reader", unheard of in the Catholic Church. The presentation of Luther's thoughts to the Danish public was most persuasively made by a Johannit cleric, Hans Tawssen (Tausen) from Viborg (Tawssen, 1528). Books arguing for and against the cause soon appeared from the Catholic supporters and from those demanding change (Bishops' reply, 1533, to Luther, 1533; Anonymous, 1535). Once the transition had been carried through, the rituals of the church ceremonies were gradually adapted to the new philosophy, keeping those things that seemed to have worked well. For example, the community spirit had traditionally been promoted during masses by music, and this was preserved, except that the Latin (Gregorian, see

Fig. 7.23) song tradition was gradually replaced by national hymns, written byDanish songwriters and composers. Among the composers were the blind Mogns Pedersøn (ca. 1560–1623), from whom a number of polyphonic hymns, mostly for five voices, are preserved (a single voice excerpt is shown in Fig. 7.24). Similar efforts had been carried out in other countries, such as by John Taverner (1490–1545) in England.

*Figure 7.23. Score of Gregorian chant from a widely disseminated compendium suggesting to Catholic priests or monks which songs to sing on each day or special occasion through the year (Venice, ca. 1575; in author's collection).*

*Figure 7.24. First song in Mogns Pedersøn's* Praetum Spirituale *(1620), published in Denmark in contrast to his earlier madrigals (1609); from the historic sheet music collection of the Danish Royal Library (2010c).*

*Figure 7.25. Frontispiece of the first Danish cookbook (Anonymous, 1625). The picture gives an impression of the energy use in kitchens. In Danish Royal Library, Copenhagen (2010c).*

Both painting and music were important ingredients of Danish renaissance life, with many works imported from southern Europe. The works of Danish artists are mostly very traditional and not incorporating the new developments of the Italian Renaissance, but probably this says more about the kings and aristocracy who paid for the artwork than on the virtues of the artists. Sophistication made its entry in many areas of social life. In 1616, the first Danish cookbook was published (Fig. 7.25, from second edition).

Historical interest grew. In addition to history writers such as Saxo Grammaticus (Chapter 6) and Huitfeldt (1600), medieval folksongs were collected and eventually printed (Muus, 1555; Vedel, 1591). Among writers, several focused on writing text to hymns for use in church (Kingo, 1689), but some wrote novels on real or fictitious, often heroic characters (Anomymous, 1577, 1695).

Regarding the continuation of the demographic development shown in Fig. 6.22, based on mortality and fertility assumptions in Figs. 6.18 and 6.19, the 16th and 17th centuries brought about changes associated with the further stratification of society, as well as others associated with epidemic outbreaks, notably during the periods 1662–1669 and in 1711. Mortality was also strongly affected by war casualties. Inspection of the genealogical data (such as was discussed in Section 7.2) shows that the new, Protestant vicars topped the fertility list, with often 16–18 children per male priest (typically wearing out two wives), although childless couples appear with considerable frequency in the family accounts, presumably signalling infertility of one member of the couple (Melbye, 1783; Nielsen, 1947). The population model will be continued in the following chapters.

Another growing interest in Renaissance/Reformation Denmark was science. A tradition had been established already during the 13th century, with the botanist Henrik Harpestreng (main opus *Liber herbarium* on herbs) and the philosopher Boethius de Dacia (1270), influenced by Aristotle and Averroes. The University of Copenhagen was founded in 1479, probably as a reaction to the establishments of universities by the Hanseatic League in Rostock (1419) and by Sweden in Uppsala (1477). In the 16th century, observations by the astronomer Tycho Brahe (1546–1601) found new stars (Brahe, 1573) and provided the data that were used by Kepler to establish the elliptic orbits of planets around the Sun. Brahe's *Uranienborg-Stjerneborg* (Fig. 7.26) observatory on the island of Hven in the Øresund Strait was funded by over 1% of the state's total income at the time (Kragh, 2005). Brahe did not use telescopes as they were first invented in Holland and Italy 1608–1609, but instead he viewed the sky with instruments mounted on basement foundations and using cylindrical tubes for calibration to known stars so that accurate coordinates could be determined, even during daytime (e.g. the mural sextant shown in Fig. 7.27).

*Figure 7.26. Uranienborg (upper panel), the Tycho Brahe Hven castle, with living quarters, printing offices and library. Stjerneborg (lower panel), the observatory annex with underground sites for the instruments used (Brahe, 1598). In Danish Royal Library (2010d).*

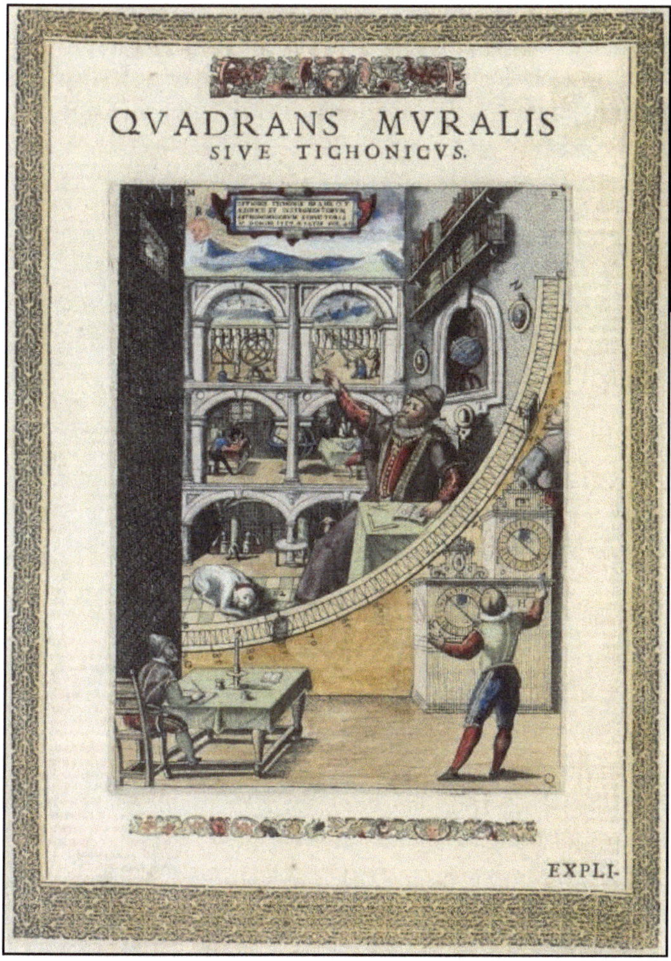


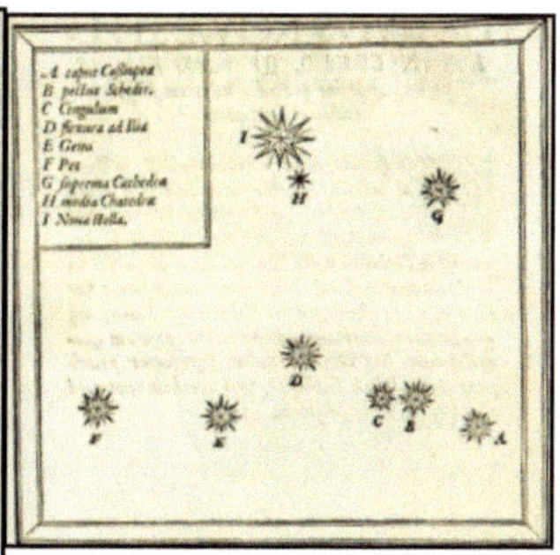

*Figure 7.27. Left: mural sextant built by Tycho Brahe (seen in middle of picture; Brahe, 1598). Above: reporting a new star: a supernova observation (Brahe, 1573). Both in Danish Royal Library (2010d, e).*

Ole Worm (1588–1654) started the first Danish museum, inspired by similar collections he had seen while travelling abroad (Shackelford, 1999). He was a teacher in medicine at the Copenhagen University, but in his own work he failed to understand the breakthrough in understanding blood circulation by William Harvey in England, in contrast to his nephew Thomas Bartholin (1616-1680), who pursued the work and discovered the lymph circulation (Bartolini, 1653; Fig. 7.28*a*). A prolific scientist was Niels Stensen (Nicolas Steno; 1638–1686), who made major contributions to anatomy, geology (stratification of sediments) and solid-state physics (Stenonis, 1667, 1669; see Fig. 7.28*b*).

Ole Rømer (1644–1710) made significant contributions to physics, determining the finite speed of light (Rømer, 1676; Fig. 7.29), but also had interests in various fields of engineering (just as had Leonardo da Vinci). He spent considerable time at the French Academy of Sciences in Paris and constructed the mechanism for the fountains in the Versailles gardens, among several other technical installations in France and Denmark, chiefly of hydraulic nature (Frederichsen and Tortzen, 2001).

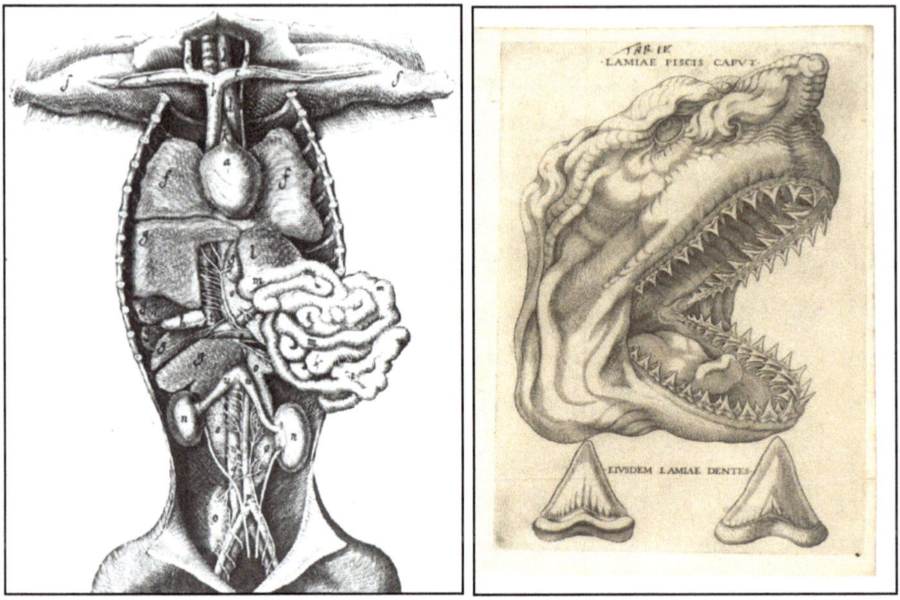


*Figure 7.28. Lymph glands illustration from Thomas Bartholin's 1653 book (a, left). Frontispiece of Niels Stensen's 1667 work on shark anatomy (b, right). Works held in Danish Royal Library (2010e).*

Later Danish scientists include Ole Borch (1626-1690), the first professor of chemistry (although the subject had been pursued long before his time). He is mostly known for his preparation of a wine for the king, made from crushed and distilled gold sheets, contrasting with the elixirs claimed to transform other metals into gold by the alchemists (Fink-Jensen, 2001). The scientific tradition in Denmark was carried on in the 19th and 20th centuries by Hans Ørsted's finding of electromagnetic induction and Niels Bohr's theory of the atom, based on the philosophical view that irreconcilable ("complementary") opposites might after all coexist peacefully.

The emphasis on knowledge was a central part of the thinking behind the Reformation period, and not only was high-level teaching carried out at the new university, but also primary and secondary schools gradually transformed the process of learning to become a basic part of young people's lives. General school attendance was not yet compulsory, but those sent to school were little by little offered more than the religious education characterizing the pre-reformation school. A Church Ordinance in 1539 still defined the purpose of the school as religious, but because everyone now was supposed to personally read the bible, skills in reading became essential (see Fig. 7.30), and skills in writing were often added. The children of peasants were sent to school, at least when there was no work in the fields (even today, schools close during summer and during the potato-harvesting period in October). At the same time, commercial entrepreneurs needed people with training in calculus and persuaded cities to establish schools with a curriculum of calculus, accounting and writing as major subjects (Fig. 6.23; Bredsdorff, 1969). Trade with other countries, as well as front-end science, required knowledge of foreign languages, which started to be taught (in addition to the required Latin), and dictionary aids became available (Fig. 7.31).

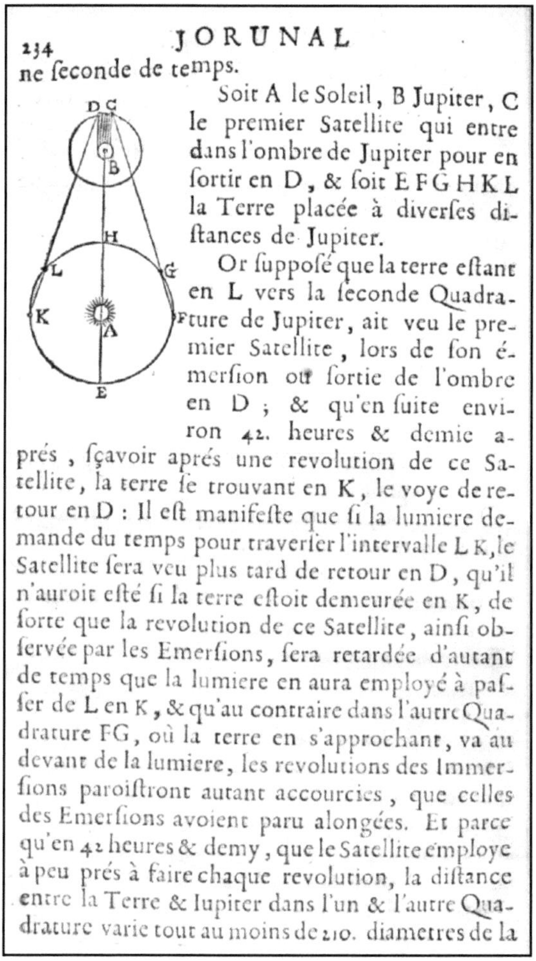


234 JORUNAL

ne ſeconde de temps.

Soit A le Soleil, B Jupiter, C le premier Satellite qui entre dans l'ombre de Jupiter pour en ſortir en D, & ſoit E F G H K L la Terre placée à diverſes diſtances de Jupiter.

Or ſuppoſé que la terre eſtant en L vers la ſeconde Quadrature de Jupiter, ait veu le premier Satellite, lors de ſon émerſion ou ſortie de l'ombre en D; & qu'en ſuite environ 42. heures & demie aprés, ſçavoir aprés une revolution de ce Satellite, la terre ſe trouvant en K, le voye de retour en D: Il eſt manifeſte que ſi la lumiere demande du temps pour traverſer l'intervalle L K, le Satellite ſera veu plus tard de retour en D, qu'il n'auroit eſté ſi la terre eſtoit demeurée en K, de ſorte que la revolution de ce Satellite, ainſi obſervée par les Emerſions, ſera retardée d'autant de temps que la lumiere en aura employé à paſſer de L en K, & qu'au contraire dans l'autre Quadrature FG, où la terre en s'approchant, va au devant de la lumiere, les revolutions des Immerſions paroiſtront autant accourcies, que celles des Emerſions avoient paru alongées. Et parce qu'en 42 heures & demy, que le Satellite employe à peu prés à faire chaque revolution, la diſtance entre la Terre & Iupiter dans l'un & l'autre Quadrature varie tout au moins de 210. diametres de la

*Figure 7.29. From Ole Rømer's article on the hesitation of light (Rømer, 1676; in the Sorbonne Library, Paris; photo by A. Bülow-Jacobsen). During the years after its first publication, other versions appeared in different journals, with different misprints (such as seen in the header of this original version; Frederichsen and Tortzen, 2001).*

*Figure 7.30. First Danish ABC (Anonymous, 1634; the edition shown is from 1649). Both in Danish Royal Library (2010c).*

True secular and common learning was not introduced until the early 19th century, upon recommendation from a progressive group of squires (Reventlow, 2010), inspired by the ideas of Rousseau in France as transmitted through squire von Rochow in Mecklenburg (Meyer, 2003).

### 7.5 Mills and energy use

Energy use increased during the renaissance period, due to new types of activities that people engaged in. Chores of the agriculture and food sector were mechanized (e.g. milling, which could take advantage of water- or windmills). Traction power in ploughing and hauling harvested crops was increasingly using horses rather than bullocks, as was road transportation. Sea transport of goods required an increasing number of ships powered by the wind, as did the naval fleet. Metal casting and manufacture required energy, and so did space heating and hot water. It is still difficult to obtain quantitative estimates of the energy use in each of these categories, but at least the land registry effort mentioned above did in principle include windmills, because there was a mill tax in addition to taxation of whatever farming was attached to the windmill property.

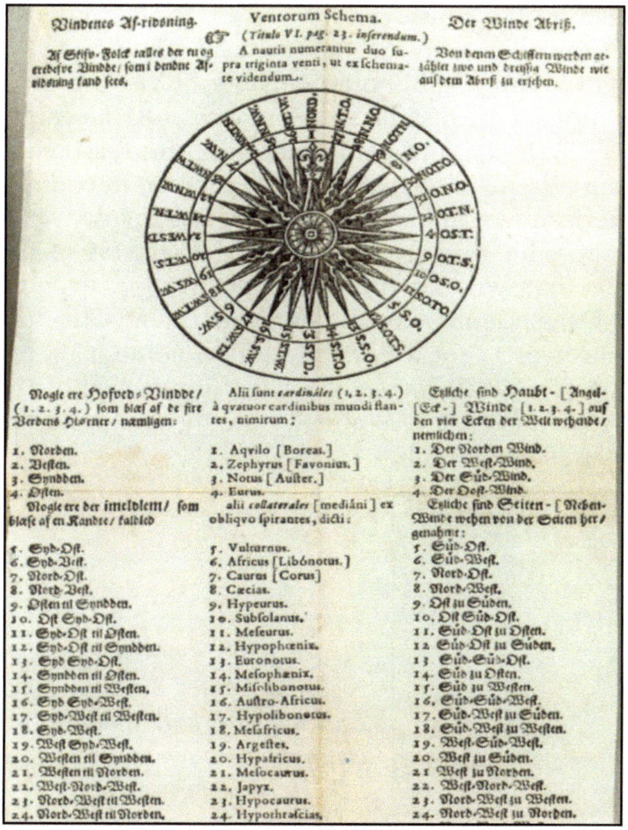

| Vindenes Af-ridsning. | Ventorum Schema. (Titulo VI. pag. 23. inserendum.) | Der Winde Abriß. |
|---|---|---|
| Af Skibs-Folck talles der tu og tredesve Vindde/ som i denne Afridsning sand sees. | A nautis numerantur duo supra triginta venti, ut ex schemate videndum. | Von denen Schiffern werden gezählet zwo und dreyssig Winde wie auf dem Abriß zu ersehen. |
| Nogle ere Hofved-Vindde/ (1. 2. 3. 4.) som blæs af de fire Verdens Hiørner/ nemligen: | Alii sunt cardinales (1, 2. 3. 4.) à qvatuor cardinibus mundi flantes, nimirum; | Etliche sind Haubt- [Angel- [Eck-] Winde [1. 2. 3. 4.] auf den vier Ecken der Welt wehende/ nemlichen: |
| 1. Norden. | 1. Aqvilo [Boreas.] | 1. Der Norden Wind. |
| 2. Vesten. | 2. Zephyrus [Favonius.] | 2. Der West-Wind. |
| 3. Synden. | 3. Notus [Auster.] | 3. Der Süd-Wind. |
| 4. Osten. | 4. Eurus. | 4. Der Ost-Wind. |
| Nogle ere der imellem/ som blæse af en Kandte/ kalded | alii collaterales [mediani] ex obliqvo spirantes, dicti: | Etliche sind Seiten- [Neben-Winde wehen von der Seiten her/ genahmt: |
| 5. Syd-Ost. | 5. Vulturnus. | 5. Süd-Ost. |
| 6. Syd-Vest. | 6. Africus [Libónotus.] | 6. Süd-West. |
| 7. Nord-Ost. | 7. Caurus [Corus] | 7. Nord-Ost. |
| 8. Nord-Vest. | 8. Cæcias. | 8. Nord-West. |
| 9. Osten til Synden. | 9. Hypeurus. | 9. Ost zu Süden. |
| 10. Ost Syd-Ost. | 10. Subsolanus. | 10. Ost Süd-Ost. |
| 11. Syd-Ost til Osten. | 11. Meseurus. | 11. Süd-Ost zu Osten. |
| 12. Syd-Ost til Synden. | 12. Hypophœnix. | 12 Süd-Ost zu Süden. |
| 13. Syd Syd-Ost. | 13. Euronotus. | 13 Süd-Süd-Ost. |
| 14. Synden til Osten. | 14. Mesophænix. | 14. Süd zu Osten. |
| 15. Synden til Vesten. | 15. Misolibonotus. | 15. Süd zu Westen. |
| 16. Syd Syd-West. | 16. Austro-Africus. | 16. Süd-Süd-West. |
| 17. Syd-West til Westen. | 17. Hypolibonotus. | 17. Süd-West zu Süden. |
| 18. Syd-West. | 18. Mesafricus. | 18. Süd-West zu Westen. |
| 19. West Syd-West. | 19. Argestes. | 19. West-Süd-West. |
| 20. Westen til Synden. | 20. Hypafricus. | 20. West zu Süden. |
| 21. Westen til Norden. | 21. Mesocaurus. | 21 West zu Norden. |
| 22. West-Nord-West. | 22. Japyx. | 22. West-Nord-West. |
| 23. Nord-West til Westen. | 23. Hypocaurus. | 23. Nord-West zu Westen. |
| 24. Nord-West til Norden. | 24. Hypothrascias. | 24. Nord-West zu Norden. |

*Figure 7.31. Page from a pictorial language book, giving concepts in three languages, Danish, Latin and German, here listing the names of winds coming from different directions (Comenius, 1672). In Danish Royal Library (2010c).*

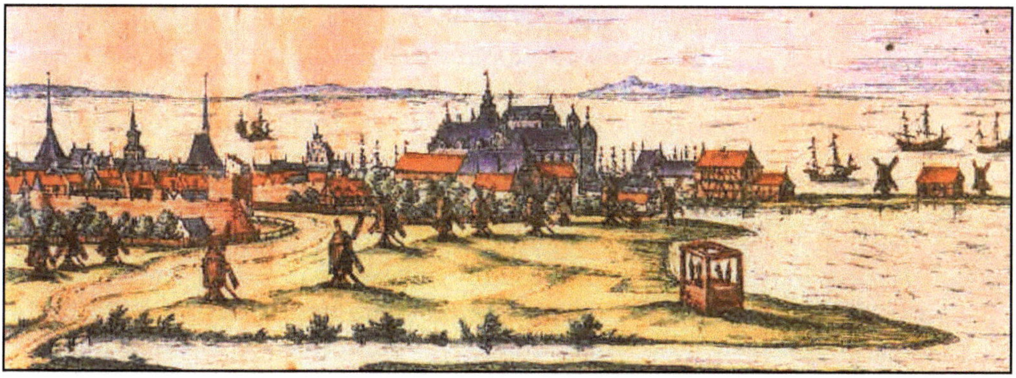

*Figure 7.32. Excerpt from a picture of Copenhagen city by Hogenberg (1587), showing more than a dozen windmills operating outside the city limits (along with the gallows of the hanging scaffold!). Many prints exist, some hand-coloured such as this one (Danish Mail and Tele Museum, Copenhagen).*

One sees by going through the registry books that at least one in three villages had a windmill, but there were additional windmills serving the cities. Figure 7.32 shows the considerable number of windmills already erected around Copenhagen in the 16th century. The windmills are of the post-mill type, with a mill-house capable of yawing on a wooden foundation mount, a type installed in Denmark until the end of the 18th century (Fig. 7.33). It is a four-bladed turbine with reasonably slow rotation, providing a large annual energy production, in contrast to the multibladed American windmill aimed at operating in low winds and as many hours as possible through the year (Sørensen, 1995). The reason this is possible is that wind conditions are fairly good, even on the eastern side of the Danish land area (the prevailing wind direction in Denmark is from the west). Yet a better technical solution is the Dutch type of windmill (Fig. 10.2), where only the upper part of the mill-house is yawing, and this type was already the most common in Denmark when the Pibemølle was built.

*Figure 7.33. Pibemølle, a late post-mill erected 1789 in north Sealand, and still in operating condition (photo by author).*

With the considerations made above, the total number of windmills in the middle of the 17th century would be about 4000. To this come a number of waterwheels (cf. Fig. 6.12), but there are few substantial height differences in Denmark, and the water streams suitable for extraction of waterpower are few. Yet the total number of watermills could have been quite considerable, as Böcher (1952) finds as many as 200–300 watermills operating in 1838.

In France, Italy and southern Germany, advances in use of watermills were substantial (Ramelli, 1588) and, thanks to improvements of technology (Agricola, 1556; Troitzsch, 1997), mining of minerals experienced a rapidly increasing volume, allowing the satisfaction of requirements also in regions such as Denmark without mining options. In addition, a wealth of new ideas were researched, some of them, like the steam engine (Porta, 1601; Caus, 1615; Branca, 1629; based upon an idea by Heron, 60), to be put in widespread use over the following centuries, while others like the solar engine (Fig. 7.34) are still today the subject of experimentation (e.g. the high-temperature metal melting installation at

Odeillo in southern France; Trombe and Vinh, 1973). It is a sign of the flowering intellectual environment of the Renaissance that Caus came up with the idea of the device shown in Fig. 7.34 only six years after the first use of lenses in telescopes, and the suggestions for use of steam engines eight years earlier. Caus was also a renowned organ-builder and wrote books on organ construction and harmonic theory in musical composition. Of course, the roots of concentrating sunlight and use of steam for mechanical work go back to Archimedes (ca. -210) and Heron (60) in Greek antiquity.

*Figure 7.34. Solar radiation concentrated through lenses to produce steam, used to raise water in a fountain (Caus, 1615; Conservatoire Numérique, Conservatoire National des Arts et Métiers, Paris, at http://cnum.cnam.fr).*

The per capita energy use and end-use energy in 1650 are estimated in Tables 7.1 and 7.2. The trends of Figs. 6.24 to 6.26 will be taken up and continued in Chapter 8.

The total amount of energy used in 1650 has not changed much relative to year 1500, but the efficiency of stoves and fireplaces for heating has improved, although it remains low, and that of transportation likewise; ships in particular are using wind resources more effectively, causing the required human energy inputs to decrease. In agriculture, any improved efficiency has been eaten up by the larger areas required to obtain the same harvest, and, finally, increases in energy input for the many new buildings and tools, equipment and fancy objects used in the cities add to energy use when they are manufactured domestically. Raw materials such as timber, clay and sand for the construction industry are used increasingly and add to Danish energy use, while metals do not as they are imported. See Chapters 11 and particularly 12 for a discussion of the energy embedded in imports, but not included in the national tables.

*Table 7.1. Estimated energy use (delivered) per capita (adult person) in Denmark around year 1650.*

| *Denmark, 1650* | *Human energy* | *Energy from fuels and external energy flows* | | | | | |
|---|---|---|---|---|---|---|---|
| *Rate of energy use (average W for an adult person)* | Food energy | Low-temperature heat | Medium-temperature heat | High-temperature heat | Mobile power (ex. food) | Mechanical power (stationary) | TOTAL |
| **Liveable surroundings** | | | | | | | **350** |
| Basic metabolism | 68 | | | | | | **68** |
| Effect of shelter, buildings | | 10 | | | | | **10** |
| Insulation | | 2 | | | | | **2** |
| Heating | | 250 | | | | | **250** |
| Light (including vax candles) | | | | 20 | | | **20** |
| **Food and water provision** | | | | | | | **172** |
| Hunting | 0.5 | | | | | | **0.5** |
| Gathering | 0.5 | | | | | | **0.5** |
| Fishing, aquaculture | 7 | | 3 | | | | **10** |
| Agriculture, grinding flour | 30 | | | | 60 | 7 | **97** |
| Slaughtering, preparing, cooking | 4 | | 40 | | | | **44** |
| Hot and cold water provision | 4 | 8 | | | 5 | 3 | **20** |
| **Security, assault, expansion** | | | | | | | **33** |
| Weapons, armies | 14 | | | | 1 | 1 | **16** |
| Police, property protection | 1 | | | | | | **1** |
| Territorial transportation, ships | 6 | | | | 10 | | **16** |
| **Health-related efforts** | | | | | | | **26.3** |
| Food preservation | 3 | 7 | 13 | | | | **23** |
| Water quality management | 1 | 1 | | | | | **2** |
| Doctors, lay or professional | 1 | | | | | | **1** |
| Hospitals | 0.3 | | | | | | **0.3** |
| **Relations and leisure** | | | | | | | **23.4** |
| Musical instruments | 0.3 | | | | | | **0.3** |
| Theater, dance | 0.6 | | | | | | **0.6** |
| Toys, artwork | 1 | | | | | | **1** |
| Written materials | 0.5 | | | | | | **0.5** |
| Social transportation | 6 | | | | 15 | | **21** |
| **Occupational activities** | | | | | | | **195** |
| Construction (incl. Ships) | 10 | | 12 | | 12 | 3 | **37** |
| Manufacture | 8 | 15 | 15 | 14 | 8 | 1 | **61** |
| Distribution, trade and services | 7 | | | | 10 | | **17** |
| Transportation of goods | 15 | | | | 60 | | **75** |
| Transportation for services, commuting | 3 | | | | 2 | | **5** |
| **Raw materials procurement** | | | | | | | **125** |
| Mining | 5 | | | | 15 | 3 | **23** |
| Materials for construction and industry | 11 | | 20 | 30 | 13 | 2 | **76** |
| Energy provision | 11 | | | | 15 | 0 | **26** |
| TOTAL | **219** | **293** | **103** | **64** | **226** | **20** | **925** |

*Table 7.2. Estimated end-use energy per capita (adult person) in Denmark around year 1650.*

| *Denmark, 1650* | *Human energy* | *Energy from fuels and external energy flows* | | | | | |
|---|---|---|---|---|---|---|---|
| *Rate of energy end-use (average W for an adult person)* | **Food energy** | **Low-temperature heat** | **Medium-temperature heat** | **High-temperature heat** | **Mobile power (ex. food)** | **Mechanical power (stationary)** | TOTAL |
| **Liveable surroundings** | | | | | | | **133** |
| Basic metabolism | 68 | | | | | | **68** |
| Shelter, buildings (reduced heat need) | | 10 | | | | | **10** |
| Insulation | | 2 | | | | | **2** |
| Heating | | 50 | | | | | **50** |
| Light (including vax candles) | | | | 3 | | | **3** |
| **Food and water provision** | | | | | | | **55.1** |
| Hunting | 0.2 | | | | | | **0.2** |
| Gathering | 0.2 | | | | | | **0.2** |
| Fishing, aquaculture | 2.2 | | 1 | | | | **3.2** |
| Agriculture, grinding | 10 | | | | 18 | 5 | **33** |
| Slaughtering, preparing, cooking | 1.3 | | 13 | | | | **14.3** |
| Hot and cold water provision | 1.3 | | | | 2 | 1 | **4.3** |
| **Security, assault, expansion** | | | | | | | **10.4** |
| Weapons, armies | 4.5 | | | | 0.3 | 0.3 | **5.1** |
| Police, property protection | 0.3 | | | | | | **0.3** |
| Territorial transportation, ships | 2 | | | | 3 | | **5** |
| **Health-related efforts** | | | | | | | **9** |
| Food preservation | 1 | 2 | 5 | | | | **8** |
| Water quality management | 0.3 | 0.4 | | | | | **0.7** |
| Doctors, lay or professional | 0.3 | | | | | | **0.3** |
| Hospitals | 0.1 | | | | | | **0.1** |
| **Relations and leisure** | | | | | | | **9.8** |
| Musical instruments | 0.1 | | | | | | **0.1** |
| Theater, dance | 0.2 | | | | | | **0.2** |
| Toys, artwork | 0.3 | | | | | | **0.3** |
| Written materials | 0.2 | | | | | | **0.2** |
| Social transportation | 2 | | | | 7 | | **9** |
| **Occupational activities** | | | | | | | **73.1** |
| Construction (incl. Ships) | 3.3 | | 6 | | 4 | 1 | **14.3** |
| Manufacture | 2.7 | 5 | 8 | 8 | 2.5 | 0.3 | **26.5** |
| Distribution, trade and services | 2.3 | | | | 3 | | **5.3** |
| Transportation of goods | 5 | | | | 20 | | **25** |
| Transportation for services, commuting | 1 | | | | 1 | | **2** |
| **Raw materials procurement** | | | | | | | **59** |
| Mining | 1.7 | | | | 5 | 1 | **7.7** |
| Materials for construction and industry | 3.7 | | 14 | 20 | 4.2 | 0.7 | **42.6** |
| Energy provision | 3.7 | | | | 5 | 0 | **8.7** |
| TOTAL | **118** | **69** | **47** | **31** | **75** | **9** | **350** |

War preparations and war participation mean that arms manufacture, including warships, rises again, although on a *per capita* basis not reaching the level of the Viking period. An energy industry has begun to evolve since the establishments of cities, for which fuelwood cannot be gathered in the nearby surroundings, and where trade from other regions is thus being organized (shipping methods and routes will be further discussed in Chapter 9). Also, water provision changes with the formation of cities, from use of simple wells to extended pipe systems connecting sources of water such as lakes and the cities (Fig. 7.35). Sewage treatment likewise requires more sophisticated engineering solutions, increasing with the size of the city. On the other hand, plague outbreaks continued to happen from time to time (e.g. in 1656 and 1711; Frandsen, 1988), presumably causing a reduction in activities and energy use. The energy use for the activities mentioned above are included in Tables 7.1 and 7.2, but still remain fairly modest, due to the still small fraction of the total population living in cities.

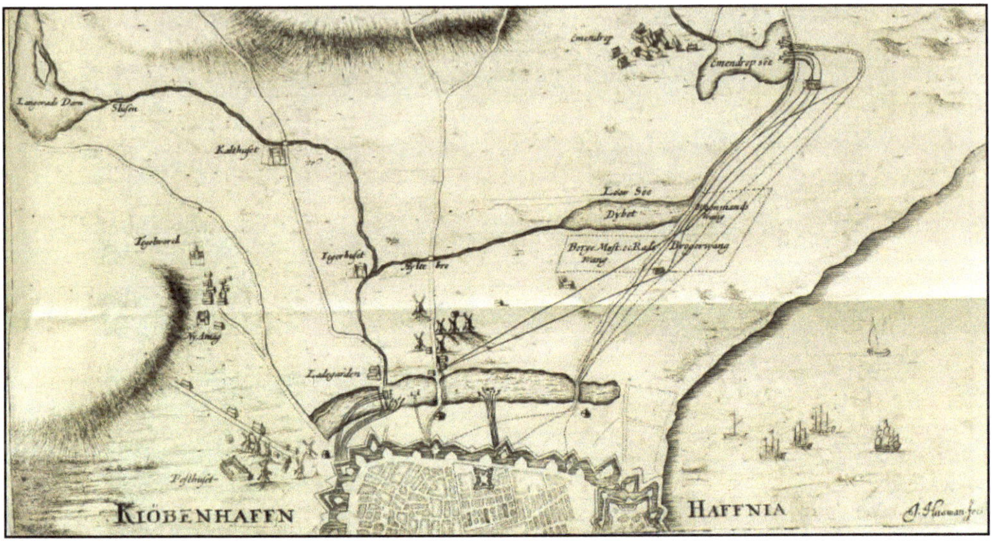


*Figure 7.35. Location of pipelines supplying Copenhagen city with fresh water from surrounding lakes. The nearest were used for sewage disposal. In Royal Danish Library, map-collection (Resen, 1677).*

*Figure 7.36. Heating stoves. Left: front plate of cast iron stove from about 1650 found at Loftsgård (Bornholms Museum). Right: Stove from around 1700, with multiple exhaust routes to increase heat liberation to the room (Troels-Lund, 1914).*

Medium-temperature heat for cooking was derived from stoves of the kind shown in Fig. 7.25, while space heating was mostly obtained with a tiled or cast-iron stove (Fig. 7.36). Tiled stoves were often two-sided, and hence capable of heating two adjacent rooms at the same time.

In transportation, the carts used on farms, pulled by oxen or horses, were not significantly better in the Renaissance than during the preceding several centuries (Fig. 7.37), but the carriages used in cities improved, with added springs for greater comfort to the passengers and metal bands around the wheels for durability, particularly when cobbles were used for street pavement (Fig. 7.38). Roads outside cities were still scarce, but with improved gravel surfaces, and a variety of different carriage types were developed for passenger and freight transport.

*Figure 7.37. Example of cart used by farmers in 17th to 19th centuries (Slesvig Carriage Museum, Haderslev; drawing by author).*

*Figure 7.38. Example of city carriage used in 18th to 19th centuries (Langelands Museum; drawing by author).*

*References:*

Agricola, G., 1556. *De re metallica.* Basileae. English translation, Dover Publ., New York, 1950.

Anonymous, 1535. *Om den gamle Gud oc om den ny Gud.* Malmø. Book LN 193 8° in Royal Library, Copenhagen.

Anonymous, 1577. *Tuenne sandrue, ubarmhjetelige og klagelige tidende oc beretning om det gruelige tirannj som den blodhund Moscouiteren haffuer bedreffuit i Liffland.* M. Vingaard Publ. Copenhagen.

Anonymous, 1625. *Koge Bog: Indeholdende et hundrede fornødene stycker.* 2nd edition (1st edition 1616). Salomone Sartorio Publ., Copenhagen.

Anonymous, 1634. *ABC.* S. Sartorius Publ., Copenhagen. The 1649 edition was published by M. Winkler, Copenhagen.

Anonymous, 1695. *Kong Olger Danskis Krønicke.* Copenhagen. At Danish Royal Library.

Archimedes, ca. –210. *On burning mirrors.* Lost book; existence known from mention in other sources.

Baden, G., 1806. *Udkast til en Historie af Danmarks og Norges Handel og Næringsveje fra Oldtiden til Nutiden.* J. Schubothe, Copenhagen.

Bang, N., 1897. Dansk Studehandel i 2den Halvdel af det 16de Aarhundrede. *Historisk Tidsskrift* **7** (1), 328-345. Author also known by maiden name Ellinger.

Bang, T., 1917. Bidrag til Gillelejes Historie. Yearbook, pp. 56–65. *Frederiksborg Amt Historical Society*, Elsinore.

Barfod, F., 1895. Borgerlige rigsdagsmænd fra 1660. *Personalhistorisk Tidsskrift*, 3rd Series, **4**, 206–243. O. Wroblewski, Copenhagen and M. Lund, Christiania. Google scanning at http://www.achive.org

Bartolini, T., 1653. *Vasa Lymphatica.* S. G. Holst/P. Hakii, Copenhagen.

Berntsen, A., 1655. *Danmarks oc Norgis Frugtbar Herlighed.* Bergen (reproduced by the Society for Publishing of Sources to Danish History, Copenhagen, 1971).

Bishops' reply (1533). *Menige Danmarkis rigis biscoppers och prelaters christelige oc retlindige geenswar till the Lwtheriansche artickle.* Book LN 72 4° in Royal Library, Copenhagen.

Blaeu, J., 1662. *Grand Atlas* (inspired by H. Lauremberg), Amsterdam.

Blanchard, I., 1986. The continental European cattle trade, 1400–1600. *Economic History Rev.* 2. Series, **39** (3), 427–460.

Böcher, S., 1952. Vandkraften som initial lokaliseringsfaktor for dansk industri.

*Geografisk Tidsskrift* **52**, 33–49.

Boethius de Dacia, 1270. *Opuscula de aeternitate mundi. De summo bono. De somniis* (N. Green-Pedersen, ed., Copenhagen 1976). English translation by J. Wippel: *On the supreme good. On the eternity of the world. On dreams.* Pontificial Institute of the Medieval, Toronto, 1987.

Brahe, T., 1573. *De nova stelle.* At the Danish Royal Library, Copenhagen. http://www.kb.dk

Brahe, T., 1598. *Astronomiæ Instauratæ Mechanica.* At the Danish Royal Library, Copenhagen. http://www.kb.dk

Branca, G., 1629. *Le machine.* Plate 25, Iacomo Mascardi Publ., Rome.

Braun, G., Hogenberg, F., 1585. *Civitates Orbis Terrarum*, Köln.

Bredsdorff, N., 1969. *Læsestykker til den danske folkeskoles historie.* Gyldendal's Pedagogical Library, Copenhagen.

Caus, S. de, 1615. *Les raisons des forces mouvantes.* J. Norton, Frankfurt. Available at http://cnum.cnam.fr

Christensen, A., 1972. Denmark. In *Det nordiske syn på forbindelsen mellem Hansestæderne og Norden.* Aarhus University.

Christian IV, 1623. Forordning at intet uden Rigit gjort ... maa udi Seland indføres. Toldmuseet (Customs Museum), Copenhagen.

Comenius, J., 1672. *Dend ganske verden fuld af de ting, som kan sees og sandses, afmaled.* Hendrick og Jørgen Gøede Publ., Copenhagen.

Dam, P., 2004. *Det GIS-baserede bebyggelses- og ejerlavskort 1681-1688.* At http://hiskis-net.cafezaki.com/forside/tabid/ctl/register/default.aspx (accessed February 2009).

Dam, P., 2009. *Det øvrige vi nyde: studier af landgilde og landbrugsproduktion i Danmark i 1600-tallet.* PhD. Dissertation, University of Copenhagen, 324 pp., and personal communication.

Danish Royal Library, 2010a. Danmarks kortlægning. Web exhibit divided into ten exhibition cases, http://www.kb.dk

Danish Royal Library, 2010b. Map collection: Frederik V's Atlas, vol. 35–38, at http://www.kb.dk

Danish Royal Library, 2010c. Fra støv til guld. Web exhibit divided into cases such as "Guds ord", "Praktiske bøger" and "Første alfabet", http://www.kb.dk

Danish Royal Library, 2010d. Tycho Brahe web-exhibit at http://www.kb.dk/da/nb/tema/webudstillinger/brahe_mechanica and other titles.

Danish Royal Library, 2010e. Texts by Brahe, Rømer, Bartholin and Stensen at http://www2.kb.dk/tekster/denovastella, steno_haj, etc.

Degn, O., 1987. Byer, byhierarkier og byudvikling i Danmark 1550–1700. *Historie/Jyske Samlinger*, New Series, **17**, 527–546.

Drysenius, E., ca. 1690. Om det förræderiske anslag på Malmøe. Swedish National Archives, as quoted by Barfod (1895).

FAO, 2003. *Food energy – methods of analysis and conversion factors.* Food and Nutrition Paper 77, Food and Agricultural Organization of the United Nation, Rome.

Feldbæk, O., 1993. *Danmarks økonomiske historie*, 1500–1840. Systime Publ., Herning.

Fink-Jensen, M., 2001. Ole Borch – en dansk kemiker i 1600-tallet. *Dansk Kemi* **82** (4), 38–40.

Frandsen, S., 1987. Fiskersamfundene ved Sjællands nordkyst. *Bygd* **18** (1), 2–31. Nordsjællands Fiskerlejer/Gilleleje Museum.

Frandsen, K-E., 1988. 1536–ca. 1720. In *Det danske landbrugs historie II. 1536–1810* (C. Bjørn, ed.), pp. 9–209. The Farmers Historical Society, Odense.

Frederichsen, P., Tortzen, C., 2001. *Ole Rømer. Korrespondance og afhandlinger samt et udvalg af dokumenter.* C. Reitzel Publ., Copenhagen.

Frederik II, 1590. Kongelige Majestatz obne Breff, 1582. M. Vingaard, Copenhagen.

Friis, A., 1981. *De danskes øer*, vol. 2, pp. 206–218. Grafisk Forlag, Copenhagen.

Fudge, J., 2007. *Commerce and print in the early Reformation.* Brill Academic Publ., Leiden.

Gamrath, H., Petersen, E., 1980. *Danmarks Historie*, Vol. 2, 1340–1648. Gyldendalske Boghandel, Copenhagen.

Heron of Alexandria, 60. The aeolipile, Section 50 in *Pneumatica.* English translation by B. Woodcroft (Taylor, Walton & Maberly, London, 1851). Available at http://www.his-tory.Rocheter.edu/steam/hero/index.html

Hogenberg, F., 1587. Hafnia vulgo Kopenhagen. In *Civitates Orbis Terrarum* (G. Braun and F. Hogenberg, eds). Malines. Reproduced in Mingroot and Ermen, 1987.

Holm, P., 1996. Catches and manpower in the Danish fisheries, c. 1200–1995. In *The North Atlantic Fischeries, 1100–1976* (P. Holm, D. Starkey, J. Thór, eds.). *Studia Atlantica*, **1**. Study Series No. 7 of the N. Atlantic Fishing & Shipping Museum, Esbjerg.

Huitfeldt, A., 1599a. *Kong Hansis Krønicke.* 1481–1513. H. Waldkirch, Copenhagen.

Huitfeldt, A., 1599b. *Historiske Bescriffuelse.* 1448–1481. H. Waldkirch, Copenhagen.

Huitfeldt, A., 1600. *En kaart Chronologia.* 1182–1448. H. Waldkirch, Copenhagen.

Hvidtfeldt, J., 1935. Skudehandelen i det 17. aarhundrede. *Historie/Jyske Samlinger* **5** (2), 29–79.

Hybel, N., Poulsen, B., 2007. *The Danish Resources c. 1000–1550.* Brill Academic Publ., Leiden.

Jensen, Johannes, 1923. Jens Jensen Væver. In *Himmerlandshistorier.* Selections published in Danish 1906–2009 by Gyldendal Publ., Copenhagen (this story is in the 13th edition, 1956).

Jespersen, K., 1989. *Danmarkshistorie,* vol. 3: 1648–1730. Gyldendal Publ., Copenhagen.

Kingo, T., 1689. *Vinterparten.* Collection of 290 (largely own) hymns, the publication of which was cancelled by the king. Many appeared in later hymnbooks.

Kragh, H., 2005. *Fra middelalderen til den nye videnskab 1000–1730.* Aarhus Universitetsforlag, Århus.

Krogh, J., 1954. *Slægten Jens Jensen,* 159 pp., Holms Bookprinters, Skive.

Lauritsen, J., 1987. *Marseliskonsortiet – Et studie over forholdet mellem handelskapital og kongemagt I 1600-tallets Danmark.* Jydsk Selskab for Historie, Århus.

Lind, G., 2008. 1720–1814. In *Danmarks krigshistorie 700–1814* (O. Frandsen, K. Jespersen, eds), pp. 370–457. Gad Publ., Copenhagen.

Luther, M., 1533. *Formaning.* Danish translation published in Malmø. Book LN 145 8° in Royal Library, Copenhagen. Title page missing.

Melbye, T., 1783. *Forsøg til en Fyensk præstehistorie.* Iversen Publ., Odense. Thor Melbye is a descendant of Hans Jørgensen, discussed in Section 7.2, four generations down.

Meyer, C., 2003. Friedrich Eberhard von Rochow's education of the children in rural communities and its impact on urban education reforms in the 18th century. *Pædagogica Historica: Int. J. of the History of Education* **39**, 19–35.

Mingroot, E. van, Ermen, E. van, 1987. *Scandinavia in old maps and prints.* Mappamundo, Knokke, Belgium.

Muus, A., 1555. Heart-shaped book. Handwritten folksong collection. In Danish Royal Library, Copenhagen.

Nielsen, H., Dalgaard, T., 2009. Middelalderens landbrug (ca. 1050–1536). In *Danske landbrugslandskaber gennem 2000 år* (Odgaard, B., Rømer, J. (eds.), 177–204. Aarhus Universitetsforlag, Aarhus.

Nielsen, P., 1947. *Stamtavle over en gammel Bogense-slægt,* 35 pp. Vejle Amts Folkeblads Publ., Vejle.

Nordisk Slægtsforskning, 1990. *En slægt fra Støttrup,* 730 pp., Skals.

Pedersen, H., 1928. *De danske landbrug – fremstillet paa grundlag af forarbejderne til Christian V.s matrikel 1688.* Published posthumously with introduction by S. Aakjær. Gyldendal, Copenhagen.

Pedersøn, M., 1609. *Madrigale a cinqve voci.* A. Gardano e Fratelli Publ., Venice.

Pedersøn, M., 1620. *Praetum Spirituale. Messer, psalmer, motetter som brugelige ere udi Danmark oc Norge.* H. Waldkirch Publ. Copenhagen.

Porta, G. della, 1601. *Pneumaticorum libri tres.* J. Carlinum Publ., Naples.

Ramelli, A., 1588. *Le diverse et artificiose machine.* Paris, published by the author. Available at http://cnum.cnam.fr

Resen, P. (1677). *Atlas Danicus.* Engraved copperplates published as the 1st volume in an uncompleted series on Danish topology. In Danish Royal Library map-collection, http://www.kb.dk

Reventlow, L., 2010. His life 1751–1801. In a family biography at http://www.reventlow.dk/jl_reventlow.php

Rømer, O., 1676. Demonstration tovchant le movement de la lumiere . *Journal des Sçavans,* 233–236. English translation: *Philosophical Transactions* **136,** 1677.

Søkilde, N., 1892. Synsvidner efter Svenskekrigen. Fyen og Laaland. Summary by H. Nørgaard at http://www.fynhistorie.dk/node/12236, and in Barfod, 1895.

Sørensen, B., 1995. History of, and recent progress in, wind–energy utilization. *Annual Rev. Energy & Environment* **20**, 387–424.

Sørensen, B., 2010. *Renewable Energy.* 4th edition, Academic Press-Elsevier, Burlington.

Steenstrup, J., 1906. Nogle træk af fiskerbefolkningens historie. *Historisk Tidsskrift* 7 (6), 141–171. Web: http://tidsskrift.dk

Stenonis, N., 1667. *Canis carchariæ dissectum caput.* Stellæ, Florentiæ.

Stenonis, N., 1669. *De solido intra solidum naturaliter contento dissertationis Prodomus.* Stellæ, Florentiæ.

Tawssen, H. (1528). *Her haffue wij bodhae edt ønclelight klawaemooll och edt wenlight tilbudh.* Wiborg. Book LN 154 4° in Royal Library, Copenhagen.

Terslin, H., 1933. Bidrag til Gillelejes historie i det 16. aarhundrede. *Fra det gamle Gilleleje,* 17–44. The Museum Association's Publishers, Gilleleje.

Troels-Lund, T., 1914. *Dagligt liv i Norden i det sekstende aarhundrede.* Gyldendal/Nordisk Forlag, available at http://runeberg.org/dagligt

Troitzsch, U., 1997. Technischer Wandel in Staat und Gesellschaft zwischen

1600 und 1750. In *Mechanisiering und Maschinenbau 1600–1840,* pp. 11–270. Propyläen Verlag, Berlin.

Trombe, F., Vinh, A., 1973. Thousand kW solar furnace, built by the National Center of Scientific Research, in Odeillo (France). *Solar Energy* **15**, 57–61.

Unger, R., 2004. *Beer in the Midle Ages and the Renaissance.* University of Pennsylvania Press, Pittsburgh.'

Vedel, A., 1591. *It hundrede vduaalde Danske viser.* Author's own printing, Ribe.

Venge, M., 1987. *Fra åretold til toldetat. Middelalderen indtil 1660.* Danish Customs Historical Society, Copenhagen.

Weibull, C., 1922. *Lübeck och Skånemarknaden.* Lund University.

West, F., 1908. *Kronens Skøder.* Vol. 2, 1648–1688. C. Reitzel, Copenhagen.

Wikipedia, 2010. "Denmark" and "Former Danish colonies" pages at http://en.wikipedia.org/wiki/ under *Denmark.*

Willerslev, R., 1952. Dansk Studehandel fra 1560 til 1660. *Historie/Jyske Samlinger.* New Series **2**, 153–181.

# 8

# Aftermath: Ecological disasters and countermeasures

## 8.1 Sand drift and erosion

Around 1560, severe sand drift problems hit the coastal areas of north Sealand and northwest and west Jutland. They persisted, and it took 150–300 years to overcome the negative effects (Terslin, 1940; Pragensis, 1989). The causes were intensely debated then but appear more evident today: the forests that had long existed along the coastlines in question were cut down in an effort to increase land for agricultural cultivation and animal husbandry. This changed the wind profile during the windy periods that are abundant under Danish conditions.

Figure 8.1 illustrates the transformations of a wind profile for a situation where the wind direction is roughly perpendicular to the coastline, for different types of vegetation from the coast into the land area. The precise form of the profile of increasing wind upwards depends on the stability of the air masses and the roughness of the underlying surface. A neutral atmosphere has been assumed, corresponding to no temperature-induced motion (cf. Sørensen, 2010). Near the ground, the wind speed has to approach zero, due to friction at the land surface, and it then increases with height, most rapidly over surfaces with low roughness, such as water. It is seen that a shoreline forest already lifts the profile before the wind reaches the shore. This means that wind speeds near the ground are insufficient for moving sand. The wind is in this case also reduced when it reaches the cereal plants growing further inland. When the forest is felled, there will be high wind speeds close to the ground, capable of moving the sand at the shore inwards, which is the cause of the sand drift. Some of the sand will be carried far inland, possibly causing damage to agricultural plants.

However, as far as agriculture is concerned, the altered wind profile has a further and more severe effect. Small droplets will be emitted to the atmosphere above the sea surface, particularly when the waves approach the shore and form whitecaps. Because the seawater is salty, these droplets contain salt, and the foam emitted is called sea salt spray. When there is a forest along the shore, the salt-containing water particles will mostly be stopped by the trees and will not move further inland. However, as illustrated in Fig. 8.2, removing the forest causes the sea salt spray to move further inland, where it is eventually deposited on the vegetation of the fields. Most types of crops that were farmed over the preceding several centuries were harmed by such exposure to salt, causing the harvest yields to decrease. This was observed after the sand drift started in 1560, along with the associated dispersal of salt particles inland.

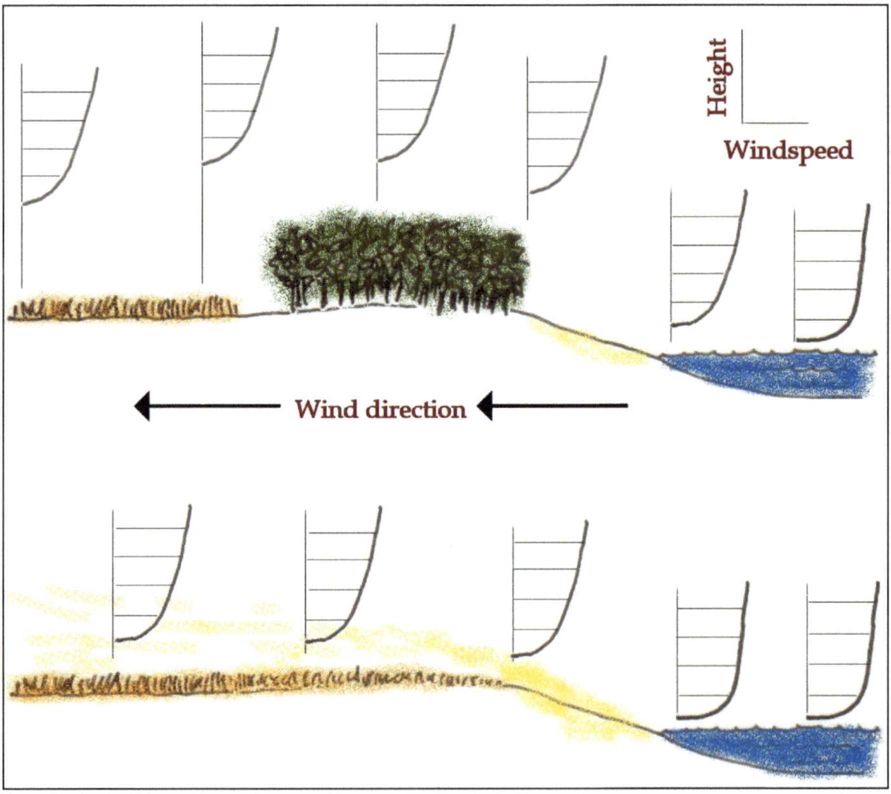


*Figure 8.1. Proposed mechanism of coastal sand drift. When forest is cleared and the succession of windspeed profiles changes as indicated, there will be transport of sand to locations inland, as indicated in the lower panel.*

The areas worst hit were the seashores with frequent onshore winds, which include the western coast of Jutland and the northern coast of Sealand. The latter is already accustomed to receiving large amounts of material in the form of stones, gravel and sand during storms from the north. The origin is rock eroded from mountains in Norway and transported through the ocean, ending up on Sealand's northern shores (Fig. 8.3). It may affect the Danish coast in different ways: soft cliffs may be eroded away, with loss of land entailed. On the other hand, the Norwegian material may also pile up on the shore, enlarging it and the land area. Material still in the water is moved around by currents in quite irregular patterns, making it difficult to construct harbours and also presenting danger for ships of going aground. The balance of all these effects was disturbed when the coastal forests were removed.

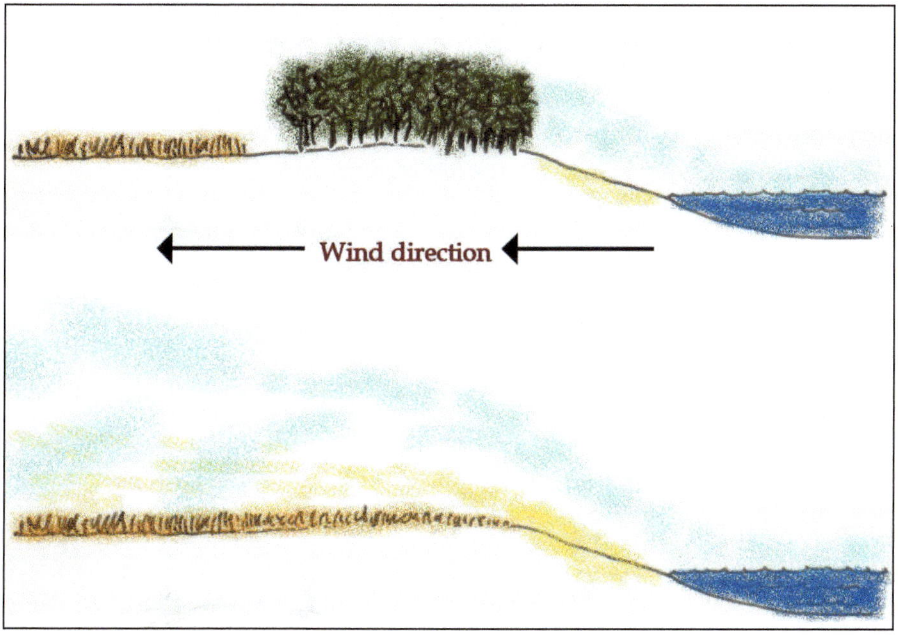


*Figure 8.2. The effect on sea salt spray dispersal when coastal forest is cleared. Rather than being stopped by the forest, the salt sprays can now be carried far inland, where they will be capable of depositing on and damaging crops.*

The effect of the sand drift was quite substantial, as documented by the 50–80% reduction in the yield expectations registered for taxation purposes. Farms in villages some 5–10 km inland were destroyed by sand coverage (e.g. Torup, the remains of which is shown in Fig. 8.4), and fishing villages such as Tisvilde, Gilleleje and Hornbæk on the Sealand north coast had by 1700 seen many houses made totally inhabitable by the sand (Fig. 8.5; Terslin, 1940). As some of the fishing communities were established by the king, the Crown replied favourably to an application for new houses built behind the sand dunes (Beyer *et al.*, 1989).

*Figure 8.3. Norwegian stones washed ashore by a winter storm on the north coast of Sealand, despite the presence of wave breakers (photo by author).*

*Figure 8.4. Recently excavated foundation of Torup village near Tisvilde. The village was destroyed by sand drift between 1560 and 1630 (photo by author).*

Figure 8.5 shows how an attempt was made to cover the shifting sand (the lower layer in the left panel) by mud or clay, only to see a second and even thicker sand layer accumulate above and cover the clay. The fishing communities suffered a succession of setbacks, from the plague of 1564 that claimed 21 lives in Gilleleje (some 6% of the population), the disappearance of fish from their waters by 1580 and the subsequent oscillation of catching opportunities (a census from 1645 finds about 150 persons present and the 1688 land registration by King Christian V mentioned in Chapter 7 consistently finds 30 houses inhabited), to the sand drift problems after 1700, which arrived later here than in the coastal villages further west (Frandsen, 1987; Beyer *et al.*, 1989). A reason for the late arrival of sand drift could be that several of the forests in the region were owned by the Crown, for which a practice was already in place to replant new trees when felling had taken place (necessary because of the king's huge wood requirements for ships and buildings, in addition to woodfuel presumably taken from residues).

*Figure 8.5. Excavation of house on Fabersvej 10 in Gilleleje, probably destroyed by sand drift as late as around 1770. Layers of shifting sand are visible in the picture above left (Beyer et al., 1989).*

*Figure 8.6. Farmers letting cattle graze on coastal land contribute to the advance of the sand. Note the stripe of bare sand-clay behind the cow (photo by author).*

Not only felling of forest, but also use of meadows near the sea for cattle to graze upon has been suggested as contributing to the creation of conditions for sand drift, because clear-grazing increases the risk of desertification, especially in combination with wind that can carry the top-soil humus layer away (Fig. 8.6). Once sand dunes were created, the wind would carry them around, causing accumulation in sheltered locations such as around buildings. The measured depth of sand deposits was two metres in some locations (Terslin, 1940). In the example shown in Fig. 8.5 (left), the sand layers reach about one metre.

22 3 Capitel.

3. Capitel.

Om Giødning i Særdelished eller ſærdeles Stykker om Giødning at ſamle og vide at bruge paa Ager og Eng.

Den Sædvane, ſom paa mange ſtender i Norge er, med at drive Fæet eller den ſtørſte deel af fra Gaarden og Hiem-Jorders Mark, langt af ſides udi Marcken og der heele Sommeren have gaaende i Græsgang med Dagen, og med Natten lade hvile udi et indhegnet ſted,

*Figure 8.7. Beginning of chapter on fertilizers in agriculture, from P. Juel's* A good farmer, *published 1721 (Hogben, 1940), original in Trondheim Library.*

There had been voices advocating a more efficient way of conducting agriculture, avoiding the need to expand the cultivated area. One such critic was Povel Juel (1673–1723), prefect in Norway and author of a book recommending expanded tillage and use of manure to fertilize the fields (Juel, 1721; Fig. 8.7). Juel was dismissed from his job after a quarrel with his superiors over his rough behaviour to foreign ship captains suspected of not paying the proper customs duties. He travelled to Copenhagen in the hope of support and offered to head a planned expedition to Greenland instead of Hans Egede (who subsequently carried out the expedition to the king's full satisfaction). Not receiving any support, Juel angrily approached the Swedish king and the Russian czar, suggesting that they conquered first Greenland and then, allegedly, Norway. Unfortunately, the Danish director general of the mail, who was also the king's security officer, Cristian Erlund, deciphered Juel's coded correspondence, and Juel was convicted and hanged (Pedersen, 2002). It is believed that the most important Nordic writer at the time, Ludvig

Holberg (1684–1754; born in Bergen, professor at Copenhagen University, and most known for his comedies in the style of Molière), may have used Juel to shape the character of the main person, Niels Klim, in a science fiction novel depicting an imaginary society with high levels of social equality, borrowing from Thomas More's *Utopia* (1516) (Andersen, 2010). Holberg was worried about being caught criticizing the king and the existing society, and he published the book in Latin through a German publisher (Holberg, 1741), a common way to circumvent censorship at the time.

## 8.2 Restoration of the ecological balance

The earliest attempt at stopping the sand before the land was transformed into desert was erection of wattle or stone fences. As late as the early 18th century, when King Frederik IV (1699–1730) had summoned expertise from Holland, only the same advice was repeatedly given: build more fences. The fences accumulated sand on the leeward side, and they did not have any appreciable slowing effect on the drift. However, in 1724 Johan Röhl from Oldenburg made an unsolicited visit to the Danish king, offering him a project to restore the sandy areas. The king and his advisers were now desperate and willing to try new ideas, so they accepted Röhl's offer. Röhl knew that re-establishing vegetation was the only way of stopping the sand drift, and the issue was how to get started, as few plants could fasten and grow on pure sand (Seeger, 1929; Kjærgaard, 1991). A natural vegetation that had already been tried in Jutland was lime and marram grasses, plants capable of fastening on sand but still open in above-ground structure, so that the wind would not be totally restrained from moving more sand. But Röhl was prepared to experiment with other plants, hoping to find a succession of plants that would take the restoration from lime and marram grasses to forest in steps. One idea emerging for an early intermediate kind of vegetation was wild roses. This turned out to be very successful, and even today much of the Sealand north shores have areas of wild roses, establishing themselves on sand dunes as easily as lime grass, but having a much denser growth and therefore offering better conditions for re-establishing a humus layer to the benefit of larger scrubs, and eventually trees (Fig. 8.8).

Because the sea salt spray damage to crops (Fig. 8.2) was felt far inland, part of the farming community eventually felt an interest in helping remedy the sand drift problem, and many peasants came to the coastal areas and helped plant new vegetation, during periods where there was no pressing work at the farms. Some expressed regrets that Röhl was always "ordering them about" (Holm, 1907).

*Figure 8.8. Wild roses establish themselves on top of sand dunes and offer better protection against drift than lime and marram grasses (photo by author).*

When King Christian VI (1730–1746) succeeded Frederik IV, the positive effects of Röhl's work was already clear, and Röhl was initially supported in his further work by the new king. However, the prefect of north Sealand, Gram (Fig. 8.9; no picture of Röhl has been found), who at first had supported Röhl by procuring the workforce and materials needed for the effort, increasingly became opposed, partly for reasons of jealousy and partly for lack of understanding of the nature of the sand drift. Röhl wanted to continue the gradual replacement of plants, until the original coastal forests had been restored, and had already made a demonstration pine plantation at Tibirke near Tisvilde, where the Sealand sand drift had started. He understood that only a strip of coastal forest could permanently solve the problem, but Gram did not believe that forests were necessary, and particularly not when they were exposed to coastal winds and thus might not provide maximum timber for the military. Gram used the military and the farmers as allies, as all of them believed they had reasons to fight the establishment of Röhl's proposed forests, and finally in 1738 they persuaded the king to dismiss Röhl (Seeger, 1931).

*Figure 8.9. Friedrich von Gram, prefect of Frederiksborg and Kronborg from 1718 to 1741. Painting at Boller Castle, Horsens (photo by V. Seeger, 1930).*

There were others who saw Röhl as being right, and by 1750 the royal gardener F. Beynall was allowed to establish a large pine plantation at Asserbo, between the coast and the Tibirke site. It was extended to Tisvilde and today is a beautiful forest, with trees becoming increasingly crooked as one gets nearer the shore, but performing their job of restraining the sand (Seeger, 1931, bases his account of the circumstances surrounding Röhl's activities on written correspondence between the parties, resting then in local, now in the national archives).

The delay in adequately dealing with the sand drift was presumably also influenced by the warfare, epidemics and associated poverty characterising Denmark from the Swedish wars in the 17th century and later. The occasional plague epidemics mentioned in Chapter 7 halted several activities. The 1711 outbreak started in Hornbæk on the Sealand north coast and soon spread to other parts, notably cities like Elsinore (Helsingør) and Copenhagen, helped by poor sanitary installations. This happened despite a harsh quarantine effort: houses where people were sick were sealed off, the military placed armed guards around major cities to prevent people entering the cities, and mail to and from Copenhagen was fumigated with juniper and sulphur at the checkpoints. The confined citizens were brought a limited and not particularly healthy selection of food (excluding pigs, milk, cheese, fruit and liquor!) through provisional government markets with fixed prices. The excess death toll estimates range from 21 000 (in Copenhagen; Kühle, 1993) to 25 000 (in entire country; Johansen, 2002). Some villages were hit hard, while others went free (Seeger, 1929).

*Figure 8.10. Faber's Grove aimed to stop cliff erosion at Nakkehoved. To the right is the the slope towards the beach (photo by author).*

The effects of the sand drift kept disrupting parts of the Danish society long after the restoration efforts had gained momentum. Vendsyssel, the northern part of Jutland, suffered a new setback in 1775, when the church in the town of Skagen at the northern tip was buried in sand and destroyed (Kjærgaard, 1991, with further references). The restoration efforts went better in northern Sealand, perhaps due to better funding from the (closer) king. In 1824, Christian Faber became lighthouse inspector for the two lighthouses at Nakkehoved on the northern tip of Sealand. He found the areas around the lighthouses barren as the grazing of cattle had bared the ground, and he embarked on a transformation of the land into a park-like forest grove of deciduous trees. The pine plantations undertaken to restore other sand-covered regions were at the time seen as less aesthetically attractive than deciduous forests, and Faber had the sentiments of the romantic period with him in suggesting this type of vegetation. On the other hand, trees like beech and elm were less likely to stick to the sandy soil, and many botanists advised against his plan. However, the king liked Faber and granted him the money. Faber had to plant over 100 000 trees on an area of 50 000

$m^2$, but succeeded in making a little less than a third establish roots, and he thereby created an interesting grove with several trees sliding down a 25 m high cliff sloping towards the beach, with the trees occasionally fastening halfway down (Terslin, 1940; Fig. 8.10). Experiments to find ways to avoid erosion of the shores were continued into the 20th century, with different types of wave breakers, parallel or perpendicular to the shoreline (Olsen, 1957; an example is seen in Fig. 8.3).

Incidentally, the lighthouses aimed at guiding seafarers were an institution started by primitive coastal bonfires in the 13th century. An early such lighthouse was erected in the south of the Sound (Øresund), on the eastern side at Falsterbo. By the mid-16th century, these had largely been replaced by parrot-type lighthouses (Fig. 8.11), and a century later by a pivot-type using a lever to lower the fire-basket to the ground rather than the hoist mechanism of the parrot lighthouses (Secher, 1900). Skagen, at the north tip of Jutland, received the first such lighthouse, and soon Nakkehoved followed. The next lighthouses on the Nakkehoved premises were built by Niels Tranbjerg and mounted on two towers in 1771, burning coal first in an open fireplace, but from 1800 in an enclosed house with lanterns (Friis, 1945; Fig. 8.12). Some 30 years later, the use of mirrors was used to direct the light beam (Terslin, 1957), and new lighthouse towers became taller. The lighthouses had a Crown concession to collect fees from passing ships, a fact that was not always appreciated by the shipmasters or the ship's owners.

*Figure 8.11. Parrot-type lighthouse. A reconstruction at Gilleleje Lighthouse Museum, Nakkehoved, of the lighthouses erected during the 1560s (photo by author).*

The energy use at the lighthouses was substantial, particularly for the early open constructions. At times, some of the lighthouses were closed down, because the fees paid by the passing ships could not pay for the fuelwood consumed, which seemingly cost more than the salary of the lighthouse master. This was initially private business, although an operating permission had to be obtained from the king, and the fees also went around the Crown (e.g. collected together with customs duties). The king had to deal with complaints from seafarers finding the flare lighted only a few hours each night, in order to save fuel (Secher, 1900).

*Figure 8.12. The smaller of two lighthouses operated from 1771 on coal at Nakkehoved on the north shore of Sealand. The choice of two close lights was based on wanting the signature to be different from that of the Kullen (Scania) lighthouse across the Sound (this was before intermittent, rotating lights allowed a simpler discrimination; photo by author).*

The "lamp money" eventually became an important part of the Crown's foreign policy, with increased revenues, but also increased responsibility for keeping the lighthouses running properly. The fuel used in the lighthouses was first wood, then train oil extracted from fish, and tallow, followed by coal and finally electricity. Today, all lighthouses have been closed or transformed into museums, as satellite-based general positioning systems (GPS) are furnishing the necessary information for sailors.

## 8.3 An energy supply crisis?

It has been suggested by Kjærgaard (1991) that not only did Denmark suffer from the ecological catastrophe caused by the sand drift in the 16th to 18th centuries, but also from an energy supply crisis, caused by increasing use of wood as timber for construction and as fuel for an expanding range of energy-related activities. While it is true that the demand for wood must have increased substantially, and that the expansion of land occupied by farmers did partially happen at the expense of Danish forests, the concrete arguments provided by Kjærgaard do not appear quite convincing. He suggests that the use of half-timbering for house-building is probably due to scarcity of wood. But the reality is that the half-timbering technique (Fig. 8.13) is simply a better technique than the previous all-wooden houses and considerably cheaper than the brick houses fashioned by the wealthier strata of the Danish population. The half-timbering frames provide the structural strength needed, and it is relatively unimportant what is put between the beams. Poor people used clay or mud, possibly with straw admixtures, in order to save money, and these houses dominate the rural areas as well as less affluent town areas. This could not have happened if wood was scarce, because then the price would have been unaffordable to the poor peasants.

*Figure 8.13. A 19th century half-timbered house in Gilleleje fishing village (Store Strandstræde; photo by author).*

The records show that Denmark solved its wood problem by import from neighbouring countries, notably Norway, but also sometimes Russia (Secher, 1900; Terslin, 1957). Sweden also has large forest surpluses and would have been closer, but the relations with Sweden were not the best. Wood prices must have been affordable, even to the peasants, since wooden structures were their first choice. It is therefore not warranted to talk about an energy crisis even though imports were required, but Kjærgaard may be right in noting that (imported) coal appeared on the scene exactly when the time was right for a new series of technology changes demanding increased energy inputs.

The advent of coal technology did not require much adaptation, as charcoal had already been used extensively. Producing charcoal from wood was an established profession from about 1500 (according to Chancellery letters kept at the Danish National Archive), and the charcoal burners were found abundantly in all parts of Denmark, again contradicting the claim that wood resources should have been near exhaustion. Charcoal was used by blacksmiths, by the Royal Mint, the Royal Pharmacy and the military powder factories. The price of charcoal subsequently declined and allowed common uses for cooking and heating. The charcoal burners were initially copyholder peasants forced to do the burning as part of their obligation to the squires, but from about 1600 coal-burning became a full-time profession (Schjøtt-Olsen, 1987). The following centuries saw several hundred charcoal-burning businesses of private or Crown ownership, and the production in charcoal stacks (Fig. 8.14) or kilns continued until the late 19th century where it was replaced by industrial retort furnaces. Charcoal burners such as the one reconstructed in Fig. 8.14 were carefully stacked heaps of fuelwood covered by spruce twigs and earth on top, ignited from the top and burning through to the bottom over a period of a couple of weeks, with careful venting and occasional topping up of wood fuel by the attendants (Oppermann, 1889; Petersen, 1987). The final product is more stable and burns more controllably compared with the wood source, and it has a higher volume energy density.

*Figure 8.14. Charcoal burning using the standing stack method in use from the 16th century in Sealand (photo from 1915 in Jagt- og Skovbrugsmuseet, also at Heritage Agency of Denmark, http://www.kulturarv.dk). In Jutland, the stacks had a central vent and a log foundation (Friederichsen, 1987).*

## 8.4 Energy usage 1650–1800

I have spent a fair amount of space describing the first well-documented ecological disaster in Denmark. Although the sand drift does not in itself influence energy, it certainly does indirectly, by changing the environmental surroundings, and thus the conditions for economic and energy-using activities. It does so in combination with other determining factors, such as the belligerent policy of the Danish kings and the limits on economic activity caused by the kind of social organization prevailing. The result is that the Danish population and energy developments are modest or negative during the period 1650–1800, despite the growth options furnished by the advances in technology taking place in several other European countries.

As regards population, the demographic model results shown in Fig. 6.18 can now be continued on the basis of concrete data, starting with the first Danish census in 1769. The results are shown in Fig. 8.15. The model results connect smoothly with the data, and, as discussed in Chapter 6, also with a number of studies based on 17th century pre-census data sources. It is seen that after the 300–400 years of mostly declining population, caused by successive wars and the aftermath of disease epidemics, there is a dramatic increase starting in 1680 after the devastating loss of lives and land in the Swedish war. Despite poverty remaining throughout the following century, the population continues to grow at an unprecedented rate, showing even further increases after 1800. The reason must be declining death rates, as fertility was already high. Identifying reasons for this and discussing the further changes in fertility and mortality will be undertaken in Chapter 9. Increased genetic diversity would have been predicted as an outcome of the abolishment of the adscription by the end of the 18th century (formally in 1788), when peasants were again allowed to move away from the location where they were born. However, this did only happened slowly, as the economic conditions did not allow many peasants to make use of their newly won freedom.

The Danish Statistical Office did not collect data relevant to energy statistics until after 1901, prompted by the increasing economic importance of energy imports, so before that it is still necessary to estimate energy use on the basis of activity accounts and indications from a variety of sources. Tables 8.1 and 8.2 gives the estimated energy use in 1800, before and after conversion taking place at the end-user, by analogy with the earlier tables, such as Tables 7.1 and 7.2 for the year 1650. Both demography and energy use will be taken up again in Chapter 9.

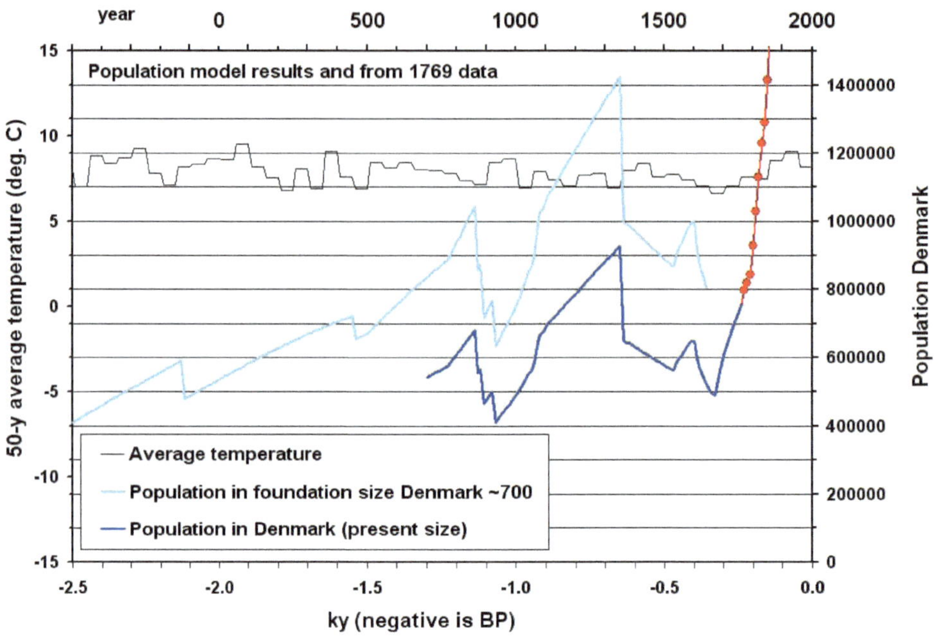


*Figure 8.15. Continuation of the results from the demographic modelling tool used in Chapters 2 to 6, also showing measured data from 1769 and onwards (in red) based on population data held at Danmarks Statistik (2010).*

The total per capita energy use shows modest change between 1650 and 1800, but the distribution is somewhat altered, with more emphasis on construction and service, manufacture and trade at the end of the period, in combination with higher conversion efficiencies. In agriculture, a substantial efficiency improvement was associated with the *enclosure legislation*, which in 1781 outlawed the partitioning of agricultural land ownership into tiny plots to be cultivated collectively by the villagers. The movement had started with prescient estate managers (count Christian Stolberg, inspector Torkel Baden) and landowners such as the influential count A. G. Moltke expressing the view that the Danish agricultural resources could provide much higher yields if they were cultivated appropriately. This involved getting rid of villeinage and adscription, as well as joining the strips of shared land into larger units that could be farmed effectively. These people carried though such reforms on their own estates and made the king set up commissions to provide recommendations on the matter (Dombernowsky, 1988). The final edict came after the king had tried by friendly persuasion to convince all landowners to carry through similar reforms. This had not been enough, due to resistance from conservative squires, who thought the reforms would negatively affect their hunting options!

*Table 8.1. Estimated energy use (delivered) per capita in Denmark around 1800, for an adult person.*

| *Denmark, 1800* | *Human energy* | *Energy from fuels and external energy flows* | | | | | |
|---|---|---|---|---|---|---|---|
| *Rate of energy use (average W for an adult person)* | Food energy | Low-temperature heat | Medium-temperature heat | High-temperature heat | Mobile power (ex. food) | Mechanical power (stationary) | TOTAL |
| **Liveable surroundings** | | | | | | | **387** |
| Basic metabolism | 70 | | | | | | **70** |
| Effect of shelter, buildings | | 12 | | | | | **12** |
| Insulation | | 5 | | | | | **5** |
| Heating | | 280 | | | | | **280** |
| Light (mostly vax candles) | | | | 20 | | | **20** |
| **Food and water provision** | | | | | | | **168** |
| Hunting | 0.5 | | | | | | **0.5** |
| Gathering | 0.5 | | | | | | **0.5** |
| Fishing, aquaculture | 7 | | 3 | | | | **10** |
| Agriculture, grinding flour | 25 | | | | 50 | 10 | **85** |
| Slaughtering, preparing, cooking | 4 | | 40 | | | | **44** |
| Hot and cold water provision | 4 | 15 | | | 5 | 4 | **28** |
| **Security, assault, expansion** | | | | | | | **29** |
| Weapons, armies | 10 | | | | 1 | 3 | **14** |
| Police, property protection | 2 | | | | | | **2** |
| Territorial transportation, ships | 5 | | | | 8 | | **13** |
| **Health-related efforts** | | | | | | | **28** |
| Food preservation | 2 | 7 | 13 | | | | **22** |
| Water quality management | 1 | 1 | | | | | **2** |
| Doctors, lay or professional | 1 | 2 | | | | | **3** |
| Hospitals | 0.5 | 0.5 | | | | | **1** |
| **Relations and leisure** | | | | | | | **18.8** |
| Musical instruments | 0.3 | | | | | | **0.3** |
| Theater, dance | 0.5 | | | | | | **0.5** |
| Toys, artwork | 1 | | | | | | **1** |
| Written materials | 1 | | | | | | **1** |
| Social transportation | 6 | | | | 10 | | **16** |
| **Occupational activities** | | | | | | | **220** |
| Construction (incl. Ships) | 8 | | 10 | | 10 | 5 | **33** |
| Manufacture | 10 | 15 | 25 | 15 | 10 | 10 | **85** |
| Distribution, trade and services | 9 | | | | 10 | | **19** |
| Transportation of goods | 17 | | | | 60 | | **77** |
| Transportation for services, commuting | 4 | | | | 2 | | **6** |
| **Raw materials procurement** | | | | | | | **101** |
| Mining | 3 | | | | 5 | 6 | **14** |
| Materials for construction and industry | 10 | | 10 | 20 | 10 | 10 | **60** |
| Energy provision | 10 | | | | 12 | 5 | **27** |
| TOTAL | **212** | **338** | **101** | **55** | **193** | **53** | **952** |

An example of the redistribution is shown in Fig. 8.16 (before) and 8.17 (after the reform). The gain in energy efficiency obtained by the enclosure movement may well have been small, but the reduction in work load meant that farmers could accomplish more, and, in the long run, it set the conditions necessary for improving and eventually mechanising the farming equipment. In some areas, the change was slowed by complaints from peasants that the distances to the new fields from their farmhouses were larger than to the old ones (eventually, farmhouses were moved from the village to the new land plots), plus endless debates over who should pay for and put fences around the new fields.

Some historians see the introduction and phase-out of adscription and later the altered structure of the cultivation arrangements as primarily determined by up- and downturns in cereal prices (Fode, 1977).

*Table 8.2 Estimated end-use energy per capita in Denmark around 1800. As in the previous "end-use" tables, this reflects the energy for the same service, had the best 2010 technology been used, not an absolute energy minimum.*

| *Denmark, 1800* | *Human energy* | *Energy from fuels and external energy flows* | | | | | |
|---|---|---|---|---|---|---|---|
| *Rate of energy end-use (average W for an adult person)* | **Food energy** | **Low-temperature heat** | **Medium-temperature heat** | **High-temperature heat** | **Mobile power (ex. food)** | **Mechanical power (stationary)** | **TOTAL** |
| **Liveable surroundings** | | | | | | | **153.5** |
| Basic metabolism | 68 | | | | | | **68** |
| Effects of shelter, buildings | | 12 | | | | | **12** |
| Insulation | | 5 | | | | | **5** |
| Heating | | 65 | | | | | **65** |
| Light (mostly vax candles) | | | | 3.5 | | | **3.5** |
| **Food and water provision** | | | | | | | **63.1** |
| Hunting | 0.17 | | | | | | **0.17** |
| Gathering | 0.17 | | | | | | **0.17** |
| Fishing, aquaculture | 2.2 | | 2 | | | | **4.2** |
| Agriculture, grinding | 8 | | | | 16 | 8 | **32** |
| Slaughtering, preparing, cooking | 1.3 | | 18 | | | | **19.3** |
| Hot and cold water provision | 1.3 | | | | 2 | 4 | **7.3** |
| **Security, assault, expansion** | | | | | | | **9.1** |
| Weapons, armies | 3 | | | | 0.3 | 1 | **4.3** |
| Police, property protection | 0.7 | | | | | | **0.7** |
| Territorial transportation, ships | 1.3 | | | | 2.8 | | **4.1** |
| **Health-related efforts** | | | | | | | **8.8** |
| Food preservation | 0.7 | 2 | 5 | | | | **7.7** |
| Water quality management | 0.3 | 0.4 | | | | | **0.7** |
| Doctors, lay or professional | 0.3 | | | | | | **0.3** |
| Hospitals | 0.1 | | | | | | **0.1** |
| **Relations and leisure** | | | | | | | **6.2** |
| Musical instruments | 0.1 | | | | | | **0.1** |
| Theater, dance | 0.2 | | | | | | **0.2** |
| Toys, artwork | 0.3 | | | | | | **0.3** |
| Written materials | 0.3 | | | | | | **0.3** |
| Social transportation | 2 | | | | 3.3 | | **5.3** |
| **Occupational activities** | | | | | | | **92.6** |
| Construction (incl. Ships) | 2.7 | | 3 | | 3.3 | 3 | **12** |
| Manufacture | 3.3 | 5 | 16 | 12 | 3.3 | 7 | **46.6** |
| Distribution, trade and services | 3 | | | | 3 | | **6** |
| Transportation of goods | 5.7 | | | | 20 | | **25.7** |
| Transportation for services, commuting | 1.3 | | | | 1 | | **2.3** |
| **Raw materials procurement** | | | | | | | **52.6** |
| Mining | 1 | | | | 1.7 | 4 | **6.7** |
| Materials for construction and industry | 3.3 | | 7 | 15 | 3.3 | 7 | **35.6** |
| Energy provision | 3.3 | | | | 4 | 3 | **10.3** |
| TOTAL | **114** | **89** | **51** | **31** | **64** | **37** | **386** |

The reduction in delivered energy from 1650 to 1800 is mainly caused by a decline in industrial activities (such as construction of castles and ships), related to the economic exhaustion following the lost Swedish wars. The end-use energy is only marginally changed, due to improvements in efficiency.

*Figure 8.16. Ottestrup village near Slagelse on Sealand in 1769, before the enclosure reform (© Kort & Matrikelstyrelsen, in Danish State Archives).*

As an example of the building standards prevailing among the poor (the vast majority) members of society around 1700, Fig. 8.18 shows the plan for a typical family house. The building walls and floors are essentially without insulation (heat loss *U*-value 1.85 W $m^{-2}$ $K^{-1}$), but the thatched roof provides considerable insulation (typically an *U*-value around 0.5 W $m^{-2}$ K, as extrapolated from Rockwool, 2010). K stands for degrees Kelvin or Celsius. This implies a total heat loss from the house of about

$$(112\ m^2 \times 1.85\ W\ m^{-2}\ K^{-1}) + (78\ m^2 \times 0.5\ W\ m^{-2}\ K^{-1}) = 246\ W\ K^{-1}.$$

Assuming the house to have been occupied by six people, and the lowest acceptable winter indoor temperature, on average over the entire house, as being 16 °C (which is 7 °C above the Danish annual average outdoor temperature), then an average heat supply of 287 W/cap. is required, in good agreement with the 270–290 W/cap. heat input implied by Tables 7.1 (1650) and 8.1 (1800) after correction for the age distribution in the population, derived from indoor fires, stoves and burners, including cooking and contributions of metabolic heat from people being in the building (~40% of time, during cold months).

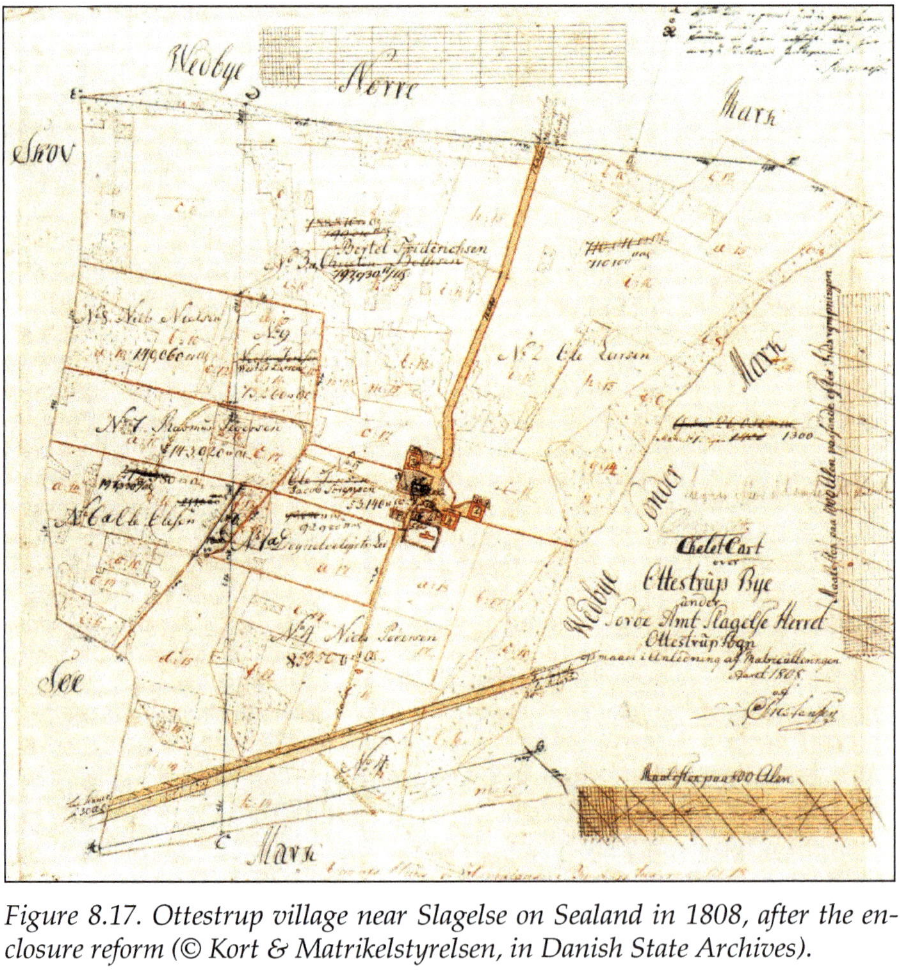


*Figure 8.17. Ottestrup village near Slagelse on Sealand in 1808, after the enclosure reform (© Kort & Matrikelstyrelsen, in Danish State Archives).*

On the energy supply side, small-scale hydro and wind energy continued to be the only sources of high-quality energy except for human or draft animal muscle work. An idea of the density of wind turbines considered necessary in new settlements is offered by the map excerpt of the colonial possessions at St. Croix in the Caribbean Sea shown in Fig. 8.19. Existing villages and city surroundings in Denmark generally had fewer wind turbines (but still a sizeable number), as demonstrated by the Land Registers for rural areas and in some cases the city archives, including pictures such as that of the Copenhagen surroundings (Fig. 7.32). Wind conditions are better in Denmark than in the Caribbean (Sørensen, 2010), and the technology used continued to be that of post mills, where the whole tower was yawing in order for the rotor blades to face the direction of wind. For a blade length of 5 metres, the annual average wind power intercepted at a fair inland location would be of the order of 20 kW, implying an average yield of a few kW, considering the estimated conversion efficiency of the post mills.

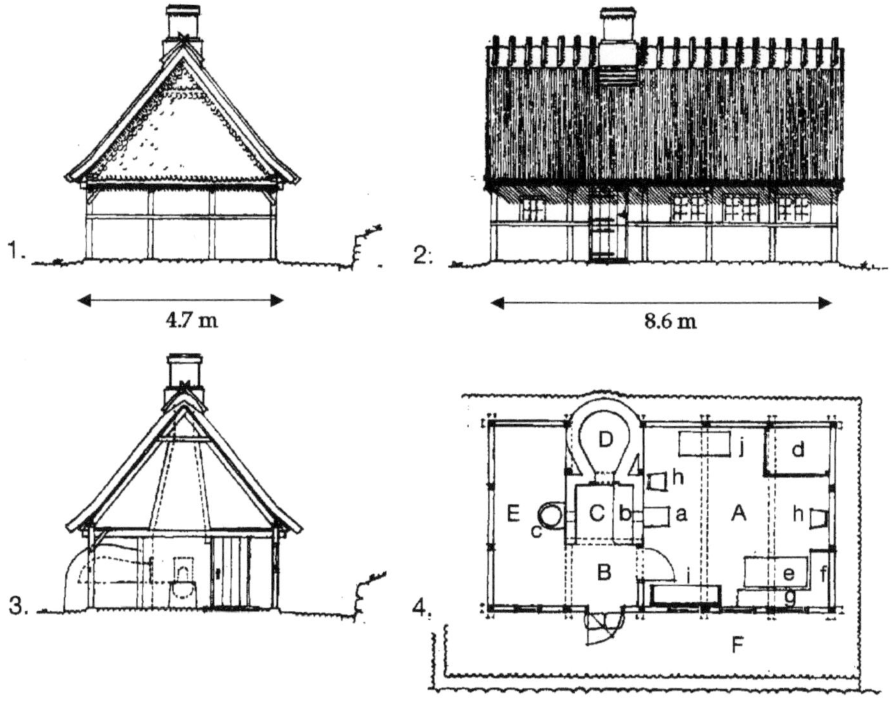


*Figure 8.18. Typical design and measures of the house of a poor family, around 1700, with thatched roof but otherwise poor insulation standards. D is owen (Handest, 1994).*

The windmills were originally used for grinding flour, but starting in the 18th century and continuing through the 19th century, other applications were introduced, such as running machinery for treshing and other mechanical tasks performed primarily in the farming communities where most wind turbines were erected, but also near cities, where demand for mechanical energy was increasing, but where wind conditions were generally less favourable than in the open country.

*Figure 8.19. Plots with windmills on the island of St. Croix in the Caribbean, a Danish colony later sold to the USA, in 1776 (from map drawing by I. Beck, at Danish Royal Library, Copenhagen; the city depicted is Christianstad).*

***References:***

Andersen, B., 2010. Paa tvende bene. Project described at the website http://holberg.nu/i-denne-jueltid.htm

Beyer, F., Egevang, R., Frandsen, S., Jarrum, E., 1989. *Historiske huse i Gilleleje fiskerleje.* Danish National Museum and Gilleleje Museum. Planning Agency of the Danish Department of the Environment.

Danmarks Statistik (2010). Statistikbanken – data og tal. Web resource at http://www.statistikbanken.dk (accessed Nov. 2010).

Dombernowsky, L., 1988. Ca. 1720–1810. In *Det danske landbrugs historie* (C. Bjørn, ed.), pp. 211–394. Farming History Society, Odense.

Fode, H., 1977. *Fra reformation til folkestyre 1523–1849.* In series "Hovedlinier i Nordens historie" (R. Thomsen, ed.), vol. 2, Gyldendal Publ., Copenhagen.

Frandsen, S., 1987. Fiskersamfundene ved Sjællands Nordkyst. *Bygd* **18** (1) 2–31.

Friederichsen, A., 1987. De nye kulsviere i Maarum. In *Fra Frederiksborg Amt Årbog* (P. Vestergaard, ed.), pp. 118–127. Kulsvierlandet–Frederiksborg Amts Historical Society.

Friis, N., 1945. Nakkehoved fyrs historie. In *Fra det gamle Gilleleje.* Yearbook, pp.

5–40. Gilleleje Museum Publishers.

Handest, B., 1994. Fiskerhusene i Raageleje – byggeskik og planløsning. *Gilleleje Museum* **31**, 3–30.

Hogben, L., 1940. *Videnskab for hvermand.* Vol. 2. Gyldendal, Copenhagen. Translation by I. Gudme with adaptations of *Science for the Citizen* (Unwin Bros., London, 1938).

Holberg, L., 1741. *Subterranean journey of Niels Klim, containing a new theory of the Earth and the unknown history of the fifth monarchy, extracted from the book collection of the blessed Abelin.* Latin version published in Leipzig. Danish translation *Niels Klims underjordiske rejse* by J. Baggesen, 1789, reissued by Gyldendal 1905, Copenhagen and Christiania.

Holm, E., 1907. Sandflugtens dæmpning i Tisvildeegnen. In *Fra Frederiksborg Amt Aarbog,* pp. 1–19. Frederiksborg Amts Historical Society, Helsingør.

Johansen, H., 2002. *Danish population history 1600–1939.* University Press of Southern Denmark, Odense.

Juel, P., 1721. *En god bonde, hans avl og biaering.* Crammius, Copenhagen.

Kjærgaard, T., 1991. *Den danske revolution 1500-1800. En økologisk tolkning.* Gyldendal, Copenhagen.

Kühle, E., 1993. *Danmark Historie Samfund.* Gyldendal, Copenhagen.

More, T., 1516. *Utopia.* Louvain. Available at http://www.Gutenberg.org/ebooks/2130

Olsen, H., 1957. Lidt havnehistorie 3. In *Fra det gamle Gilleleje.* Yearbook, pp. 33–48. Gilleleje Museum Publishers.

Oppermann, A., 1889. Dansk kulsvieri. In *Tidsskrift for skovvæsen* (J. Møller, ed.). Series B, vol. 1, pp. 37–128.

Pedersen, S., 2002. Postspionen. *Museumsposten* 2002-2. Danish Mail and Tele Museum, Copenhagen. http://www.teponia.dk

Petersen, B., 1987. Milebrændingens teknik. In *Fra Frederiksborg Amt Årbog* (P. Vestergaard, ed.), pp. 110–115. Kulsvierlandet–Frederiksborg Amts Historical Society.

Pragensis, J., 1989. Sandflugten og dens bekæmpelse i Gilleleje-området. In *Gilleleje Museum* **28**, 107–120.

Rockwool, 2010. Tag. http://guiden.rockwool.dk (accessed Dec. 2010).

Schjøtt-Olsen, C., 1987. Kulsvidning i 1500- og 1600-tallet. In *Fra Frederiksborg Amt Årbog* (P. Vestergaard, ed.), pp. 34–39. Kulsvierlandet–Frederiksborg Amts Historical Society.

Secher, V., 1900. Det danske fyrvæsens historie 1560–1660. *Samlinger til Jydsk Historie og Topografi,* Series 3, vol. 2, pp. 329–400. Jydsk Historisk-

Topografiske Selskab, at Tillges, Copenhagen. Google scanning at http://www.archive.org

Seeger, V., 1929. *Fra Kulsvierlandet I.* C. Nordlundes Publ., Hillerød.

Seeger, V., 1931. *Fra Kulsvierlandet II. Fra Frederiksborg og Kronborg Amter,* C. Nordlundes Publ., Hillerød.

Sørensen, B., 2010. *Renewable Energy,* 4th edition. Academic Press-Elsevier, Burlington.

Terslin, H., 1940. Sandflugten ved Nordsjællands kyst. In *Fra det gamle Gilleleje.* Yearbook, pp. 14–21. Gilleleje Museum Publishers.

Terslin, H., 1957. Belysningen af Øresund. In *Fra det gamle Gilleleje.* Yearbook, pp. 20–33. Gilleleje Museum Publishers.

# 9

# Industrialization and infrastructure development

Entering the 19th century, the industrialization era that had started in England reached the Danish shores. Industrial manufacture had of course taken place long before (shipbuilding from the Viking period, military clothes and gun factories from the 18th century, using hydropower for energy). The new concept had to do with a rapid change in the work arrangements, and with the massive use of machinery, notably powered by steam engines, having reached a much more mature form than the experimental Italian ones mentioned in Chapter 7.

However, the political environment continued to restrain progress, at least temporarily. The primary reason was the tension between Napoleon's imperialistic dreams in France and the existing British empire. Until his death in 1797, the Danish foreign minister A. Bernstorff had pursued a strict neutrality policy, but in 1800 the Danish King Christian VII (1766–1808) entered an alliance with Sweden, Prussia and Russia, aiming to defy the English blockade of France and its allies. England responded in 1801 by sending its navy, defeating the Danish navy and forcing Denmark to leave the alliance (Fig. 9.1). When Napoleon as a countermeasure tried to stop British trade by persuading neutral European countries to sever trade relations to England, the British in 1807 decided to attack Copenhagen and destroy or conquer the entire Danish fleet of military vessels (Fode, 1977). The Copenhagen attack was carried out from several sides, with indiscriminate carpet-bombing (Fig. 9.2). This is presumed to be the first instance in military history, where an assault has been directed entirely against the civilian population, aimed to terrorize rather than pursuing concrete military objectives. Such warfare became widespread during the 20th century. The 305 buildings burned down were still less than the 941 buildings that had been lost in an accidental 1795 Copenhagen fire starting in a fuelwood store during a storm (Fig. 9.3; Holm, 1904).

*Figure 9.1. Sea battle between English and Danish fleets outside Copenhagen, 1801 (from etching by C. Lorentsen made for Holm, 1903).*

The attack drove Denmark into an unfortunate alliance with Napoleon. Sweden, on the other hand, at first opposed Napoleon and had Finland taken from them by Russia. The Swedes then attempted to restore friendship with the French by electing the French marshal Bernadotte as king (under the name Karl Johan). However, Bernadotte preferred to make peace with Russia and join the fight against France, personally leading the defeat of Napoleon at Leipzig in 1813 and making way for the Kiel peace treaty in 1814, which gave Sweden sovereignty over Norway as compensation for renunciation of claims on Finland. Denmark had tried to avoid this by giving Norway a democratic constitution (at Eidsvoll, 1814) and by confirming its formal independence, but this was nullified after a short war, where Karl Johan physically invaded Norway and imposed a treaty, which inserted him as Norwegian king and in charge of foreign relations, but honoured the internal affair provisions of the Eidsvoll constitution. Denmark had by the Kiel treaty been accorded the former Norwegian colonies of Faroe Islands, Iceland and Greenland, obviously not considered valuable at the time.

Danish foreign trade, which in 1797 had been liberalized by new legislation on customs duty, stopped abruptly when Christian VII in 1807 announced the death penalty for trading with the enemies of Denmark. The restrictions were lifted in 1814 (Becker-Christensen, 1988), but by then had already forced the Danish state into bankruptcy (in 1813). By 1820, the economy and grain exports were again growing.

*Figure 9.2. The British siege of Copenhagen 1807. Filled red squares inside or outside the city walls are buildings destroyed during the attack, while letters, crosses and open squares indicate English gun positions and camps. Illustration in Holm (1903), based on map in the collection at Frederiksborg Museum.*

*Figure 9.3. The 1795 fire in Copenhagen (in its early stage). Drawing in Holm (1903), based on etching by Lahde. Copenhagen already had another bad fire in 1728 (Reiser, 1858).*

*Figure 9.4. First democratic constitution of Denmark, from 1849 (etching in Jørgensen, 1904; original in Danish State Archives). A parliament with two chambers was prescribed, and men over 30 could vote, provided that they had never received social poverty compensation and that they had their own household. This excluded 86% of the population (Dybdahl, 1975).*

A Danish National Bank was established in 1818, but it turned out to be incompetent, believing that the crisis could be dealt with by just printing less paper money. The resumed growth was due to loans accorded to Danish businesses by German banks in Hamburg, and as a consequence, this city took over the role Copenhagen had earlier as the trade and financing centre of northern Europe (Fode, 1977).

*Figure 9.5. National Assembly meeting of 1848 at Christiansborg (Copenhagen) for preparation of the new constitution (painting by Constantin Hansen, Frederiksborg Museum, Hillerød). The numbered persons are 1: A. Moltke (to become first prime minister), 2: D. Monrad, presumed to be main architect of constitution text, together with 3: O. Lehmann, and 4: C. Ploug (newspaper editor promoting the idea of incorporating Schleswig but not Holstein in Denmark). Public domain photo by P. Burton (Wikimedia Commons).*

### 9.1 Democracy and final defeat

A second international conflict had a profound influence on Denmark by the middle of the 19th century. Efforts to assemble the German provinces into a single country, severing the ties to the Austrian empire, involved (in order to create a balanced composition of the new federation) the incorporation of Schleswig and Holstein, two provinces (duchies) that at times had been under Danish jurisdiction and where the northern part of Schleswig (south Jutland) was Danish-speaking (as a contemporary German tourist guide puts it: "any person wearing wooden shoes will speak only Danish"; Essen, 1800). Holstein was already a member of the German Federation and had a democratic constitution, while the kings still were absolute rulers in Denmark. A democratic constitution such as the one instituted in Norway was not introduced in Denmark itself until finally by King Frederik VII in 1849 (called the "June-constitution"; see Figs. 9.4 and 9.5), and a three-year Danish–German war 1848–1850 had not brought any useful clarification on the future affiliation of the southern provinces.

In 1863, the Danish government led by C. Hall (following several other governments since the formal introduction of democracy in 1849) foolhardily declared Schleswig part of Denmark (but not Holstein, as some Danish hardliners would have liked). In 1864, this brought about a joint attack by Prussia and Austria, whereby they annexed both Holstein and Schleswig, later to become part of Prussia after the two German powers fell out. The Danish army was deployed at Dannevirke (Fig. 5.17) and destroyed in a battle at nearby Dybbøl. By then, the German army had also made a surprise landing at the unprepared peninsula of Als (just as attempted in the Iron Age; cf. the discussion of Fig. 4.38) and from there occupied half of Jutland, to the city of Århus. According to the annals of the British mediators having tried to get the parties to agree to a peaceful settlement, the German leader O. von Bismarck had been willing to offer Denmark to keep the northern part of Schleswig (as in the final settlement carried through in 1920 after World War I and a referendum), but the Danish government (led by D. Monrad, following Hall in December 1863) insisted on a confrontation and unrealistically thought that it could gain both duchies, or at least all of Schleswig (Møller, 1958; Buk-Swienty, 2009, 2010). The replacement of degenerate and incompetent kings by elected political parties and lay prime ministers had not significantly raised the quality of strategic decision-making. The usual consolation is "we lost all the wars but have all the songs" (Lehrer, 1965). The national Danish Songbook (1895–1942) contains dozens of songs mourning the heroes fallen at Dybbøl.

## 9.2 Enlightenment and learning

Despite the problems of foreign policy, industrious people made sure that a modernization of Danish society took place. This development was due not only to technicians and businessmen, but also quite substantially to humanists and other scientists, bringing about a learning and spiritual environment of unprecedented scope. This interest in bringing intellectual insights to the disposal of many ordinary people without formal higher education originated partly from the establishment of public media for mass consumption, such as newspapers and magazines, and partly from initiatives on general or continuing education, in Danish called "folkeoplysning", a term that more adequately expresses the information qualities as distinct from traditional education with tests and diplomas.

Central to this movement was the Danish Folk High School system, founded on the basis of ideas by the poet and priest Nikolaj Grundtvig (1783–1872). Grundtvig had found inspiration for these ideas during a stay in England 1829–1831 and wanted to create a "school for life", as opposed to the "black school" based on discipline and rote learning. As the Folk High School concept became developed, it offered courses of varying duration (a few weeks to a year) on any subject, including practical skills (handling new farming equipment, building windmills, welding, etc.) and arts (e.g. learning to paint or compose music), theoretical topics (from arithmetic skills to foreign languages) and discussion issues (e.g. based on philosophy, existential, historical, religious or political questions). There are no exams and no diplomas given – the benefit is entirely what the course participant is able and willing to learn. The first Folk High School was established in 1844 at Rødding in Schleswig, but in 1865 moved north of the border to Askov, after Schleswig had been ceded to Germany. Several more such schools followed (see Simon, 1989, an expanded version of Simon's thesis of 1963).

*Figure 9.6. Lecture (by L. Schrøder) at Askov Folk High School, painted 1903 by Erik Henningsen, known for drawing the "thirsty man" advertisement for Tuborg beer (© Ribe Kunstmuseum, id: m0072; photo permission bought).*

Folk High Schools became a vehicle for the following century of agricultural reforms, as well as for a number of practical and mindset-building activities, probably serving as a driving force behind the successes of the political debates leading to the formation of the welfare economy in Denmark during the following century (Fig. 9.6). Until about 1975, it was a widespread practice for young people leaving school to spend some time at a Folk High School before deciding on a profession or further education. In today's world of commercialized efficacy, such behaviour of learning without obtaining diploma-type proof is pejoratively termed "a fooling around year". Likewise, people divorcing or shifting jobs would often at the time take a Folk High School course in order to "clear their minds" before the next phase of life.

Discussion of different methods of learning and improved school structures took place in many countries during the 19th century, continuing the efforts of Geneva-born Jean-Jacques Rousseau during the French Enlightenment Period and the practical school work that in the 1770s was carried out by Eberhard von Rochow in Prussia (cf. the discussion of farming reforms in Chapter 7 and Alexander, 1918). Subsequent to the expansion of Folk High Schools in Denmark that was directed primarily at adult education of those above 18 years, there were around 1900 primary school reforms suggested by Maria Montessori in Italy, based on self-managed, project-oriented group work (Montessori, 1912). Today, many Montessori schools exist, particularly in the USA. The same pedagogical ideas became tested at university level from about 1970, notably in Germany and Denmark.

Many Danish schools in villages and cities were established after the reform thoughts of the Reventlow Commissions (cf. Chapter 7), including some of quite decent quality. The words of Grundtvig, that children below age 18 were best suited for physical activity seasoned with a little theoretical learning (Norrild, 1956), were honoured by establishment of gymnastics classes in the school schedule (Fig. 9.7). In 1814, a proper School Law was passed, with seven years of compulsory school attendance at the primary school (Bredsdorff, 1969). The law contained nice words on the importance of understanding the meaning of what was being read, and yet critics pointed out the continued emphasis on rote learning of what might as well have been dog Latin (Müller, 1849). Later school discussions have focused on whether to keep the students together in one line or divide them according to skills (requiring exams) or interests (producing natural scientists ignorant about humanistic sciences and *vice versa*). In current practice, the tendency is for politicians to regulate the curriculum content at a highly detailed level, referring to but largely missing the lessons from the fundamental learning methodology issues that were the focus of earlier debates.

*Figure 9.7. Gymnastic exercises in schoolyard at Gerrild School, 1824 (drawing made by a vicar, Kampf, at Statens Pædagogiske Studiesamling).*

### 9.3 Liberalism and enterprise

England was the first European country to combine the assets obtained by colonizing foreign continents with technological progress and thereby create an early industrial growth with substantial economic benefits, primarily to be reaped by the factory owners. Most other colonial powers in Europe, such as Spain and Portugal, had been happy to carry the South American gold home and exhibit the splendour derived from it, as well as to enjoy the new spices and tobacco brought back. As mentioned in Chapter 7, many technical inventions were made on the European continent, such as the steam engine explorations in Italy and France, but it was only with the practical design of James Watt (1736–1819; a Scottish instrument maker associated with University of Glasgow (who in 1769 obtained a patent on methods he had invented for improving the efficiency of steam engines), that the use of the technology took off, starting in 1776 when the first machines were installed in industrial settings.

The second foundation pillar for the industrial revolution was the newly formulated theoretical economic framework for financing and distributing wealth (Adam Smith, 1776, at Glasgow University). England and Scotland made use of this capitalistic market theory to multiply the wealth gathered from the colonies, by using it to finance the establishment of industrial factories for mass production. England treated its colonies in a differentiated way, recognizing that highly cultured old civilization like those of India or China should be approached differently from the less advanced indigenous populations of, say, North America or Australia. However, the reverence did not go as far as limiting the economic exploitation of the Asian conquests. The less developed colonies, such as Australia, were used for getting rid of poorly adapted British citizens, such as political fundamentalists, murderers and people having stolen bread to give to their starving children. Not surprisingly, after the United States in North America had become independent in 1776 and dominated by immigrants from the British Isles, it was the first to copy the industrial development of England and Scotland, and eventually surpassed these countries in turning industrial mass manufacture into wealth.

The main feature characterizing industry as opposed to earlier ways of manufacturing goods is the increased use of machinery that forces the people associated with the process to work at a pace determined by the technology rather than that of organization, in line with human needs for pauses and control of work sequences. Time schedules had to be followed more rigorously than before, and this required profound changes in the attitudes of the personnel, now called workers. This was achieved in England and later in other industrialising nations with the help of the Christian Church, and particularly the Protestant persuasions embracing the notion that "idleness is a sin" (Anthony, 1977). The typical 12-hour working day (Christensen, 1996) may not have been much different from that of previous times, but the change was in the intensity, removing options for occasional breathing. Only later were pauses in the work schedule introduced, because productivity studies revealed that continuous work did not optimize output.

Danish industrialization was not really taking off until the last decades of the 19th century, although more technology was appearing earlier in connection with the activities of skilled craftsmen. In fact, industry is not listed as a separate entity in Danish statistics for most of the 19th century, but lumped together with trade and crafts (Boje, 1976). Even more recently, it is difficult to follow the development due to the changing conventions used in statistical material, where sometimes the division between industry and craft is set at 6 employees, at other times at 20 employees, and where some industry employment data count only workers, but most also list office staff and other work categories. Also on a more fundamental note, the distinction between craft and industry is blurred. One may ask why the factory work of the late 19th century deserves the name "industry" more than the effort in previous centuries to build armadas of large war and merchant ships. The craftsmen who constructed the ships used available energy sources, such as animal power for hauling and wood burning for processing, and worked in large teams, very much as the factory workers did, albeit now with access to steam power as the most apparent difference. Attempts to use a distinction between employment of either tools or machines to define the difference are dubious, as industries even today use manual labour with simple tools rather than available robotic machinery to produce and assemble many products, whenever it appears less expensive.

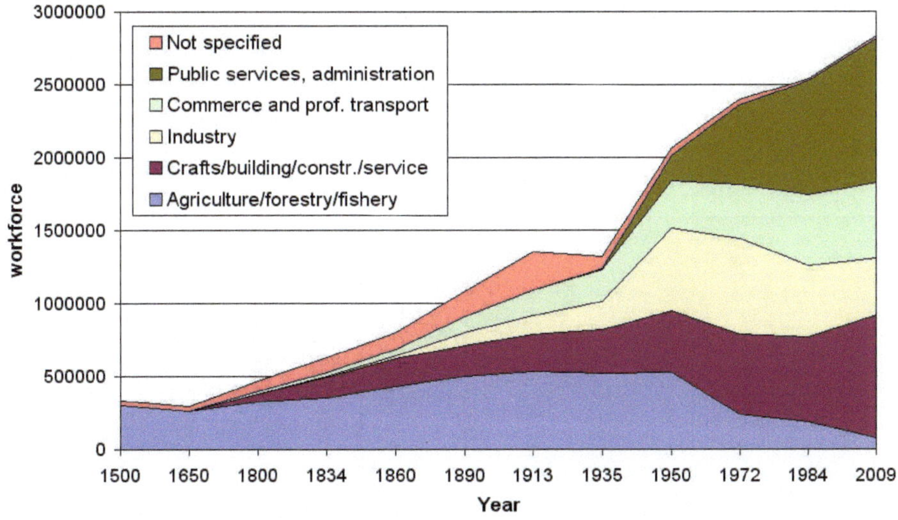


*Figure 9.8. Estimated development of Danish workforce from 1500 to the present. Based on Hansen, 1976–1977; Fode, 1977; Dybdahl, 1975; Kühle, 1993; Johansen, 1988; Hastrup, 1979; Danmarks Statistik, 1972–1989, 2010. See text for further explanations.*

Figure 9.8 shows an estimate of the variations in the distribution of the Danish workforce for different types of occupations, such as agriculture, skilled crafts, construction, manufacture, service or administration. For the recent decennium, statistical data about employment are used directly, while for the earlier periods some estimation has had to be exercised. The aim is basically to include the part of the population engaged in a trade, independently of whether the work is directly paid. That would include female work on farms and as maids and servants, but not the subsistence work within a family. One important factor is the age distribution of the population, which varies substantially between the medieval (Figure 6.20) and modern times, passing through the 19th century period of families with a large number of children, of which some age groups are also to be counted in the labour force. It has conservatively been assumed that the workforce is constituted by no more than 60% of the population, declining to the 50% claimed in statistical surveys around 1890 (Hansen, 1976–1977).

Around 1650, the negative impact of the Swedish wars is seen, and during the early 1930s, the effect of the first major global breakdown of the market economy is seen, starting in 1929 in the USA and then spreading to the rest of the world. For the rest of the period, the total workforce is increasing, including the recent decades with no Danish population growth, but with more women joining the paid labour market, a development starting at the beginning of the 20th century. The sizeable "not specified" category before 1935 would include women working as maids and other types of domestics, such as gardeners, cooks and servants. After the economic crisis during the 1930s, such employment rapidly vanished, and after World War II was replaced by non-domestic employment.

*Figure 9.9. The children working in the tobacco company C. W. Obel, assembled for the photo in 1897 (Aalborg City Archives).*

The percentage of the population counted in the workforce statistics for the period after 1950 actually counts only self-employed or salaried work, excluding unemployed persons (e.g. in 1984, the unemployment rate was around 10%, while in 1972 and 2007 it was close to 1%). In 1890, the workforce included in the survey constituted 50% of the total population, excluding pensioners, children and women not assigned a trade. Some children were part of the workforce, the fraction dropping from 10% to 5% of the workforce between 1875 and 1900 (Christensen, 1998), and being largest in cities (Fig. 9.9). From 1500 to 1860, the size of the workforce has been estimated on the following basis: the age distributions from the demographic model employed in Chapter 6 and below are used to obtain the fraction of the population aged over 16 years. The counting as part of the workforce is then extrapolated back from the late 19th century, being cautious not to allow change in the total population fraction at work that would increase it to more than 60%. The rationale for this is that although the fraction of the population contributing work of one or another kind was possibly higher in earlier times (considering gender-shared farm work), the smooth matching to recent data that excludes household subsistence work demands that a similar exclusion is attempted earlier.

The total number of people working in agriculture (plus forestry and fisheries) does not start to decline until as late as 1950, but then dwindles rapidly. However, from roughly 1800 to 1950, a constant agricultural labour force is capable of feeding a population growing by a factor of four, and in the 20th century even creates a growing surplus for export. During the 19th century, the reasons are partly better mechanical tools and partly more dedication of the farmers, now not being treated as slaves. Use of chemical fertilisers did not begin until the early 20$^{th}$ century. After 1950, more significant changes take place, first replacing imported high-protein fodder fed to the livestock animals by a second crop (such as lucerne/alfalfa), which in Denmark can be grown before or after the main cereal crop, and more recently by introducing an ecological ("organic") style of production, where pesticides are abandoned and fertilizer is derived from manure and residues, preferably after conditioning in a biogas plant (Sørensen, 2010, 2011).

## 9.4 Communication and transportation

Among the novel applications of Watt's steam engine were uses for propulsion in trains and ships. Early uses were in freight ships, initially with a reduced amount of sails for back-up if the engines should fail. The British tonnage of steam ships increased from 638 (8 ships) by 1815 to over 10 million (12 000 ships) by 1910 (Pearsall, 1985). The first steam ship in Denmark was the *Caledonia,* built in England and operated from 1819 by the Danish Mail Services between Copenhagen and Kiel in Holstein (Fig. 9.10). Danish sailors protested against the use of steam ships, which they felt required skills that they did not possess, and succeeded in forcing passengers to Caledonia to have to be rowed in small boats out to an anchoring position in the outer Copenhagen harbour, due to the claimed high danger of the steam engine exploding (Petersen, 1983;

Davidsen, 1910; indeed, infrequent explosion accidents with steam engines had occurred abroad, but claimed far fewer casualties than accidents while rigging or performing other work onboard sail ships).

*Figure 9.10. The paddle steam ship* Caledonia *inaugurated in 1819 for Danish mail runs between Copenhagen and Kiel, demonstrating its capability to move against the wind (Danish Mail & Telegraph Museum).*

*Figure 9.11. Steam-powered lorry constructed 1669–1671 by N-J. Cugnot on a grant from the French military. Although the functioning of this front-wheel-driven car was demonstrated, the idea was not pursued beyond the prototypes, probably due to difficulties in manoeuvring the front wheel with the heavy load on it. The drawing shows the second prototype in 1670 (Schieldrop, 1934). The third prototype is on display at the Musée des Arts et Métiers, Paris.*

Steam cars for the road had been investigated (e.g. by N-J. Cugnot in France; Fig. 9.11), but it was the railway systems that captured the market, with steam-powered trains for long distances and the horse-driven streetcars in cities (Fig. 9.12), in most places being operated until electrification in the late 19th century. Rail systems had also been in use in areas with mining, for transport of mined commodities since the 18$^{th}$ century (Paulinyi and Troitzsch, 1997). Early train examples are those of Richard Trevithick, moving on a circular track at circus exhibitions in London (1808–1809) and from 1814 the George Stephenson locomotives for serious transportation of people or coal (Thestrup, 1997). Rail systems were a natural choice, as they offered low friction, were cheaper than the roads that would have allowed similar speeds, and avoided complex vehicle steering mechanisms. Early wooden or cast-iron rails were soon replaced by less brittle wrought-iron rails capable of sustaining the high weight of trains and cargo.

*Figure 9.12. Horse-driven tram at Lake Square in Copenhagen, 1875 (photo in Danish Royal Library photo collection).*

Figure 9.13 shows the inauguration pamphlet for the first Danish railway, after some debate established in 1844 in Holstein as a carrot in the on-going debate on incorporating the duchies in Denmark. This and the subsequent construction of train services for passengers and goods in Denmark proper were mostly done by what became the state railway corporation, but with shares offered to private and municipal investors (Thestrup, 1997). A few minor lines were established by purely private companies, and around the end of the 19th century most major Danish destinations could be reached by train. Figure 9.14 shows the state rail network and the passenger and freight traffic on it by 1889.

*Figure 9.13. Prospect on the first Danish railway line, inaugurated 1844 between Altona and Kiel in Holstein (for political reasons; picture permission bought from Danish Railway Museum, Odense).*

While the first locomotives were imported, Danish constructions appeared from 1871, at the initiative of the Burmeister & Wain company (Fig. 9.15). Later came Danish-built diesel trains, but today, most trains are bought in Italy, despite recurring delivery delays. The Danish State Railways also operated the necessary ferries over the straits separating the Danish isles and Jutland from each other, but soon started to replace the ferries by bridges, starting with the narrowest distances. Figure 9.16 shows the construction in 1877 of the bridge across the Limfjord at Aalborg. The final step of Danish interconnection was taken in 1998, when the 25 km Great Belt bridge and tunnel connection was opened.

Steam engines required coal fuel, as did many stationary installations for industrial heat or mechanical power. For this reason, import of coal, a commodity not found in the glacial deposits forming the upper kilometres of Danish underground, increased substantially during the 19th and 20th century (Fig. 9.17). Other fuel imports were still insignificant during the 19th century, but kerosene started to get used in lamps. The net import data (import minus export) behind Fig. 9.17 are presented in Table 9.1, and discussion of the more recent part is continued in Chapter 11. The coal import data up to 1870 are derived from British customs duty records, given in terms of volume (barrels: 1 bbl = 0.17 $m^3$). The derivation of energy content has an uncertainty connected with the void fraction that again depends on the size distribution of the coal pieces in the barrel. A value of 136 kg per barrel is assumed, based on Møller-Holst (1877–1883), leading to 3386.4 MJ/kg, in good agreement with current values. The evaluation by Kjærgaard (1991) further makes use of the first volumes of Statistics Denmark (Falbe-Hansen and Schar-

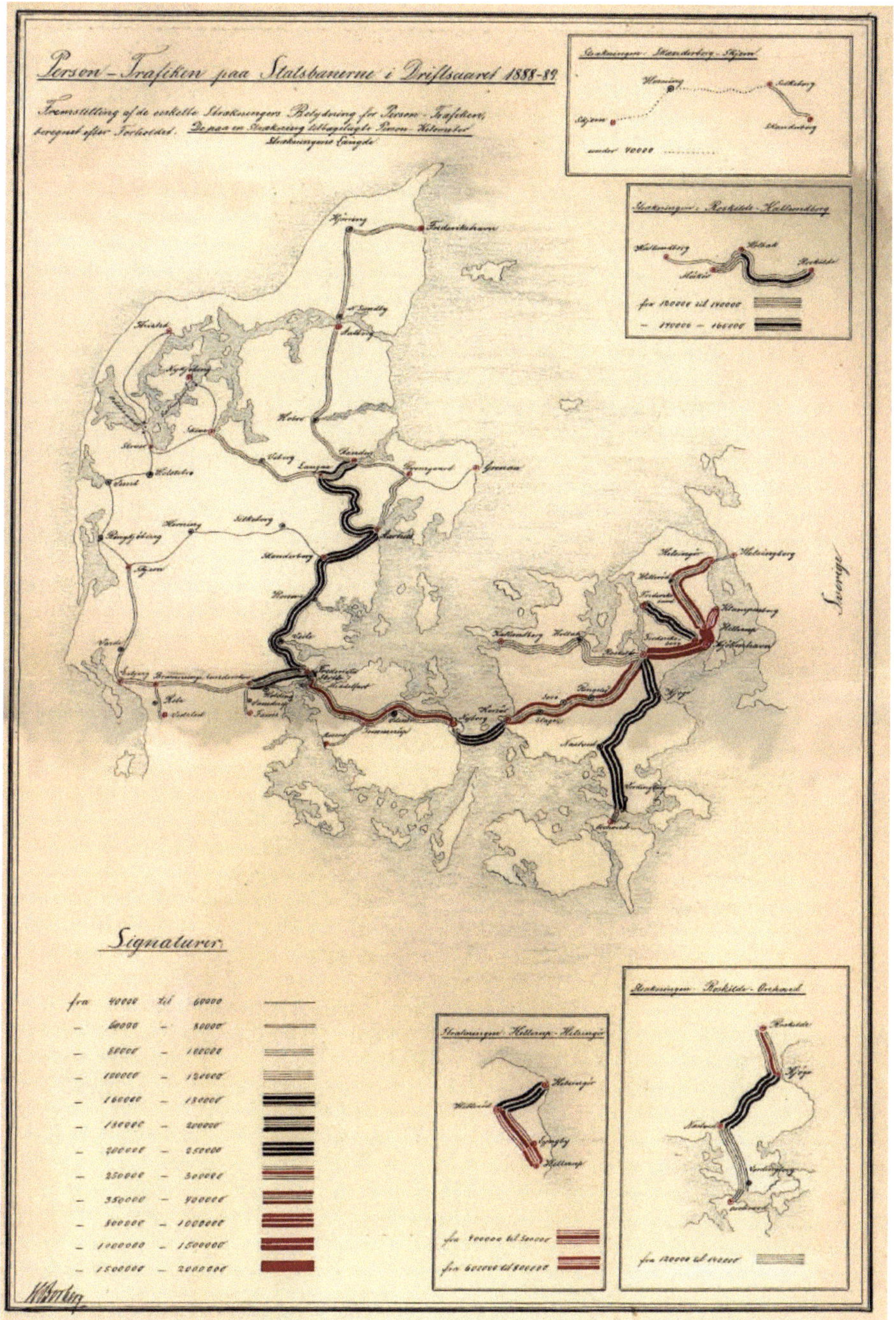

*Figure 9.14a, b. Passenger (this page) and freight traffic density (overleaf, including cattle hauls) on the Danish state railway lines by 1889 (picture permissions bought from Danish Railway Museum, Odense).*

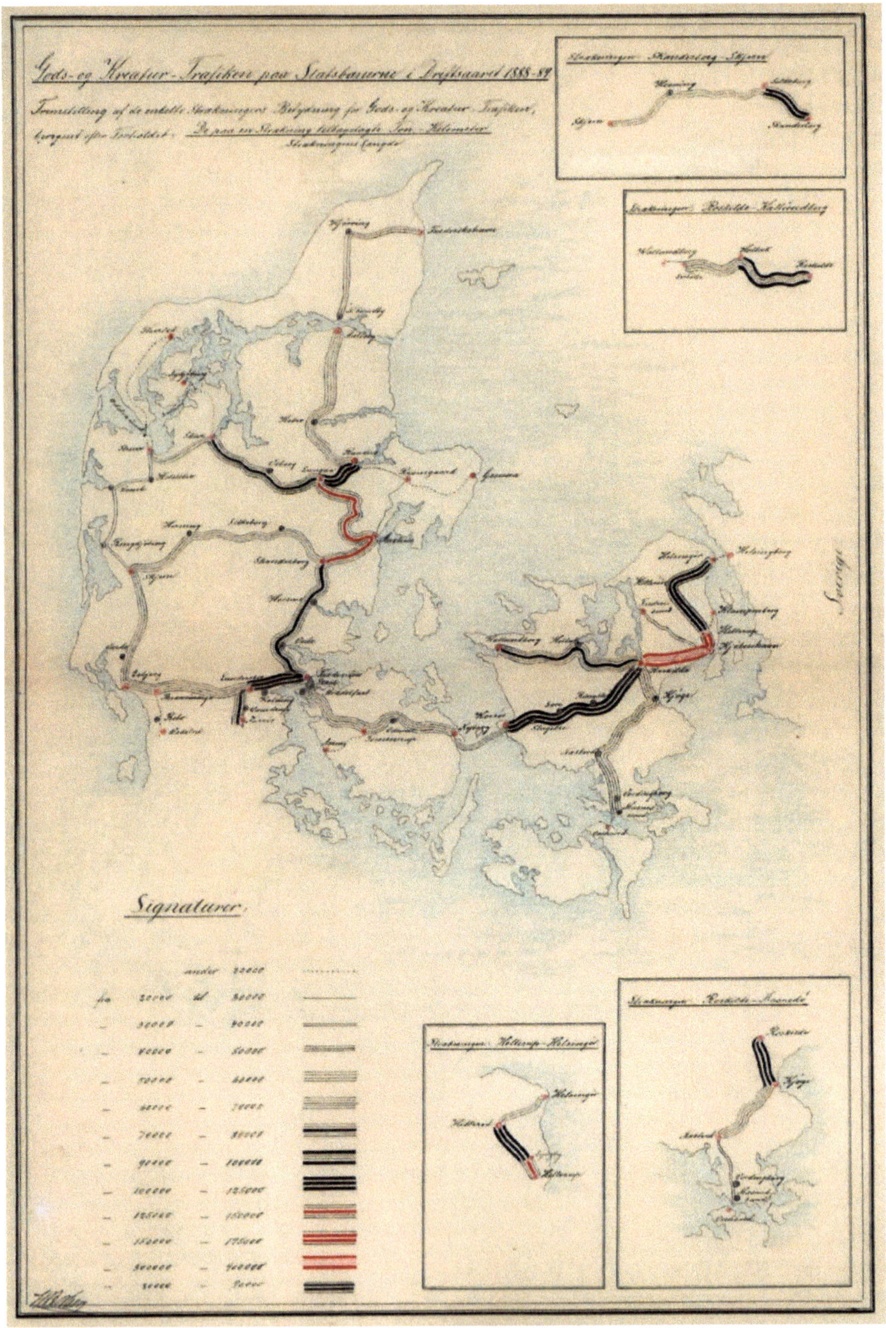


ling, 1885, and following years). The more recent periods are more systematically covered by data from Danmarks Statistik, although several methodology changes have come into effect.

*Figure 9.15. First Danish-built steam locomotive, 1871 (photo W. Bendtsen; picture permission bought from Danish Railway Museum, Odense).*

*Figure 9.16. Railway bridge over the Limfjord Sound in northern Jutland, during construction 1871 (photo W. Bendtsen; picture permission bought from Danish Railway Museum, Odense).*

As use of steam turbines increased, so did the efficiency of converting coal into mechanical power (Fig. 9.18). The efficiency of thermodynamic machines is determined by the second law of thermodynamics (Sørensen, 2010), stating that there is a maximum efficiency possible for transforming heat into mechanical work. This maximum efficiency depends on temperature, which is why the steam engine efficiency increased as it became possible to generate steam at higher temperatures. However, the improvement in efficiency was not able to follow the pace of the increase in usage, so fuel demand increased.

*Table 9.1. Annual average net Danish energy import flows (import minus export, in MW). Trade of energy-containing commodities such as food and timber is not included. Data sources: British customs records (Kjærgaard, (1991), Henriksen and Ølgaard (1960), Danmarks Statistik (1959, 1967) and the* Danish Energy Accounts *spreadsheet at Danmarks Statistik (2010).*

| Year | Coal | Oil | Natural gas | Fuelwood etc | Electric power |
|---|---|---|---|---|---|
| *1723* | 1 | | | | |
| *1753* | 3 | | | | |
| *1763* | 5 | | | | |
| *1773* | 6 | | | | |
| *1783* | 10 | 1 | | | |
| *1793* | 11 | 1 | | | |
| *1811* | 17 | 1 | | | |
| *1843* | 43 | 1 | | | |
| *1870* | 483 | 5 | | | |
| *1875* | 521 | 11 | | | |
| *1880* | 695 | 17 | | | |
| *1885* | 908 | 26 | | | |
| *1890* | 1034 | 36 | | | |
| *1895* | 1303 | 50 | | | |
| *1900* | 1544 | 58 | | | |
| *1905* | 1803 | 73 | | 24 | |
| *1910* | 2223 | 106 | | 20 | |
| *1915* | 3146 | 144 | | 12 | |
| *1920* | 2201 | 196 | | 16 | |
| *1925* | 3279 | 374 | | 4 | |
| *1930* | 4122 | 655 | | 4 | |
| *1935* | 4466 | 741 | | 0 | |
| *1940* | 3424 | 203 | | 2 | |
| *1945* | 1189 | 199 | | 0 | 0 |
| *1950* | 5371 | 2319 | | 0 | 21 |
| *1955* | 6141 | 4093 | | 0 | 18 |
| *1960* | 4674 | 6237 | | 0 | 43 |
| *1966* | 3809 | 16996 | | -7 | 70 |
| *1970* | 3012 | 27160 | | -12 | -483 |
| *1975* | 3379 | 23302 | | -2 | 103 |
| *1980* | 8014 | 19570 | | -2 | -141 |
| *1985* | 10303 | 11953 | -527 | 0 | 53 |
| *1990* | 8038 | 5913 | -1243 | 8 | 805 |
| *1995* | 10277 | 3011 | -2026 | 8 | -91 |
| *2000* | 5135 | -4476 | -3800 | 881 | 76 |
| *2005* | 4641 | -658 | -5276 | 422 | -328 |
| *2008* | 5966 | 15844 | -6774 | 655 | 166 |
| *2009* | 3594 | 8389 | -4750 | 692 | 38 |

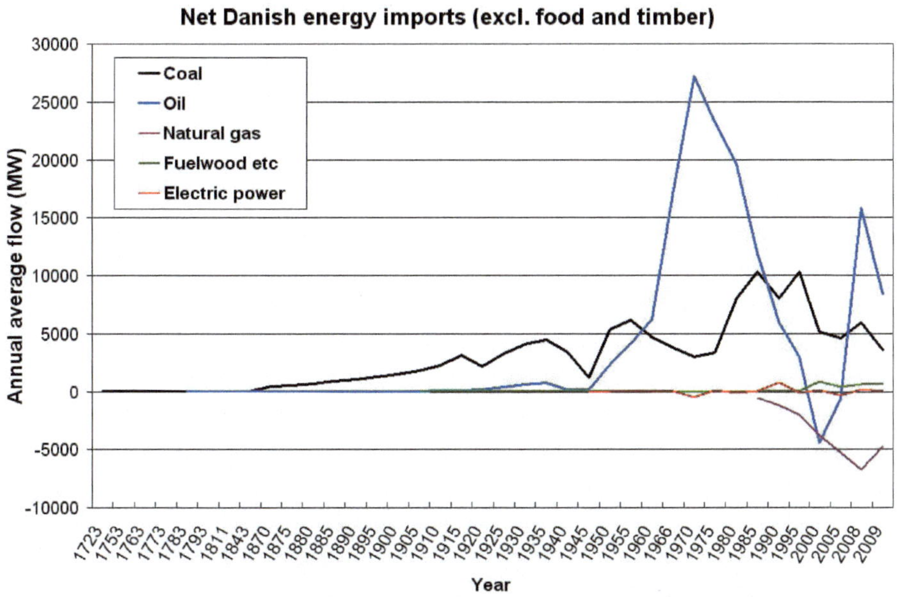


*Figure 9.17. Danish net import of energy in the form of fuels or electricity. The conventional economic boundaries are used (i.e. counting all Danish activities included in the national economy, such as worldwide shipping and transportation by Danish companies). See Table 9.1 and text for data sources.*

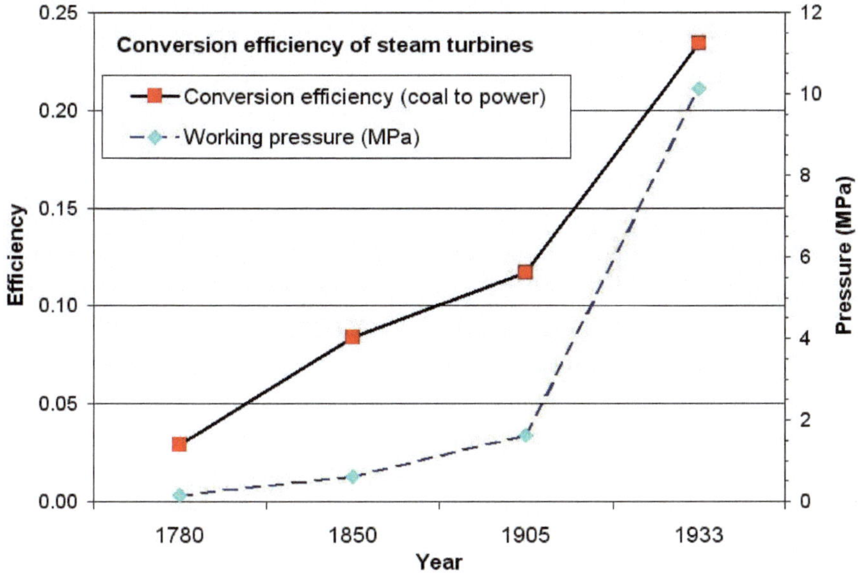


*Figure 9.18. Development in the efficiency of steam turbines and the pressure levels at which they are operating (based on Vol. 2, Part 3 in Hogben, 1938).*

The telephone, invented by the Scottish-American A. Graham Bell in 1876, soon became a preferred technology for communication. The first telephone company in Denmark was Copenhagen Telephone Inc., formed as a subsidiary of Bell's company in 1881. It soon received competition from a number of other private telephone companies, using telephones from L. M. Ericsson in Stockholm, and from 1885, Danish-built telephones from F. E. Sørensen in Copenhagen or E. Møller in Horsens. The picture of the 1898 Hjallese telephone central in Fig. 9.19 shows a Møller telephone. The other parts of the picture show an 1893 phone from the Kristiania company Elektrisk Bureau and the 1898 answering machine by Valdemar Poulsen, the Danish inventor of the tape recorder. Electric battery-based telegraph services started in 1854 and soon replaced the earlier optical semaphore-based telegraph systems.

*Figure 9.19a, b, c (clockwise). Kristiania phone (1893), Hjallese telephone exchange (Funen, 1898) and Valdemar Poulsen's tape-recorder based answering machine, invented in 1898 (Danish Mail & Tele Museum, Copenhagen).*

The regular Danish Mail Service also underwent several enhancements. Originally established in 1624, mailmen travelled on horses between intermediate stations that kept exchange horses, and sounded their post horns upon arrival. By 1815, the service was introducing spherical mail wagons (Fig. 9.20), for protecting the letters and preventing the mailmen from slowing the mail by taking passengers for an extra fee. The first stamp was issued in 1851, as a further streamlining of the mail processing. Within cities, uniformed mailmen delivered the mail after having sorted it in the mail offices. Mail officers were highly trusted civil servants, occasionally used by the crown to carry out espionage (such as in the affair of Juel mentioned at the end of section 8.1).

*Figure 9.20. Left: The spherical mail wagon, "invented" by the Danish Mail Services in 1815. Above: First Danish stamp, 1851 (Danish Mail & Tele Museum, Copenhagen).*

## 9.5 Gas comes to town

Gasification of organic material was first performed by the Belgian scientist Jean van Helmont (1577–1644). The retort process (heating without presence of air) was over the following centuries carried out on coal as demonstration for students at Belgian and German universities, but did not meet with practical use until the Scot William Murdoch in 1792 used piped coal gas to supply light to his house, and his apprentice Samual Clegg constructed a practical gas plant with chalk-based sulphur cleaning. The lighting system based on gas was demonstrated in London in 1807 and a commercial supply network was introduced in 1814. In Denmark, gaslights were first demonstrated in 1813 by J. Irgens, and the first city gas plant was built in 1857, in the western outskirt of Copenhagen. Its manager, J. Howitz, discovered that bog iron ore available in Denmark was a much cheaper alternative to chalk for sulphur removal (Edelberg *et al.*, 1932). Figure 9.21 shows the process of filling the retort oven by coal, Fig. 9.22 the lighting of a gas lamp in front of the Copenhagen gas plant, and Fig. 9.23 the Copenhagen gas grid as it was a year after its establishment. The circular signatures refer to large stores of gas. Several other Danish cities also established gas networks over the following decades.

*Figure 9.21 (above). Photograph of workers filling a retort oven with coal (Copenhagen Gasworks; Edelberg et al., 1932).*

*Figure 9.22 (left). Manual lighting-up of the gaslights in front of the Copenhagen gas plant (from 1895 drawing in Edelberg et al., 1932).*

Oil-based streetlights had existed before gas became available, at least in the larger cities. However, pedestrians often had to bring their own lights if venturing into the streets at night, such as shielded wax candles or torches. From the beginning of the 19th century, kerosene lamps were used inside the houses of wealthier citizens, while less affluent people continued to use candlelight.

*Figure 9.23. Extent of the Copenhagen gas grid in 1858. The gas-producing plant is in the lower left corner of the map (Edelberg et al., 1932).*

## 9.6 Electricity makes its entry

Electricity generated by rubbing certain materials against each other was known by Greek scholars more than 2000 years ago, and both theory and practical demonstrations had developed from the 18th century, based on work by people such as Galvani, Volta, Ørsted, Ampère and Faraday. Producing electric current by rotating magnets took off when it was realized (by the Danish engineer Søren Hjorth in 1851) that the magnetizing could be done by the generated current itself (Andersen *et al.*, 1942). Such electromagnets were the basic idea behind the dynamos being put into production in several European countries. The first use of this form of electricity was for carbon-arc light, based on the 1802 observation, described by Vasily Petrov and Humphry Davy, that a spark was generated when an electric circuit was switched off.

In Copenhagen, a demonstration of electric light had taken place in 1857, but it took a long time before the city decided to follow the several other European cities embarking on a process of replacing gas light by electric light. Figure 9.24 is from a second demonstration of arc-light in Fenruary 1871, illuminating a frozen lake in the city used for skating. Arc-light had high intensity and was suited for lighting night work, but less so for home use, although arc-lamps were also constructed. The next step was the incandescent light proposed in 1838–1840 by Jobart and Grove, but made practical only by 1879, when Edison replaced the large-size evacuated glass dome of the glowing metal filament in the early lamps by a bulb suited for mass production (Fig. 9.25).

*Figure 9.24. Demonstration of arc-light for night-skating in 1871 on a Copenhagen lake (Illustreret Tidende **12**, No. 594, February 1871, page 183; Andersen et al., 1942; also at Royal Library, http://www.illustrerettidende.dk).*

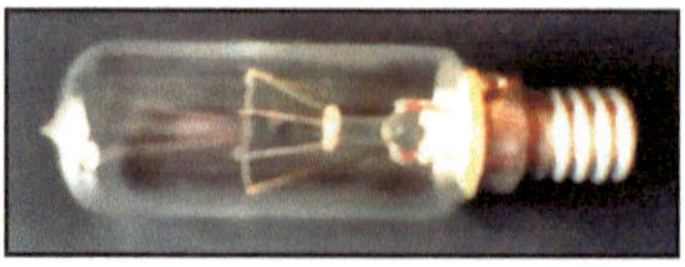

*Figure 9.25. Example of incandescent light bulb in use around 1900.*

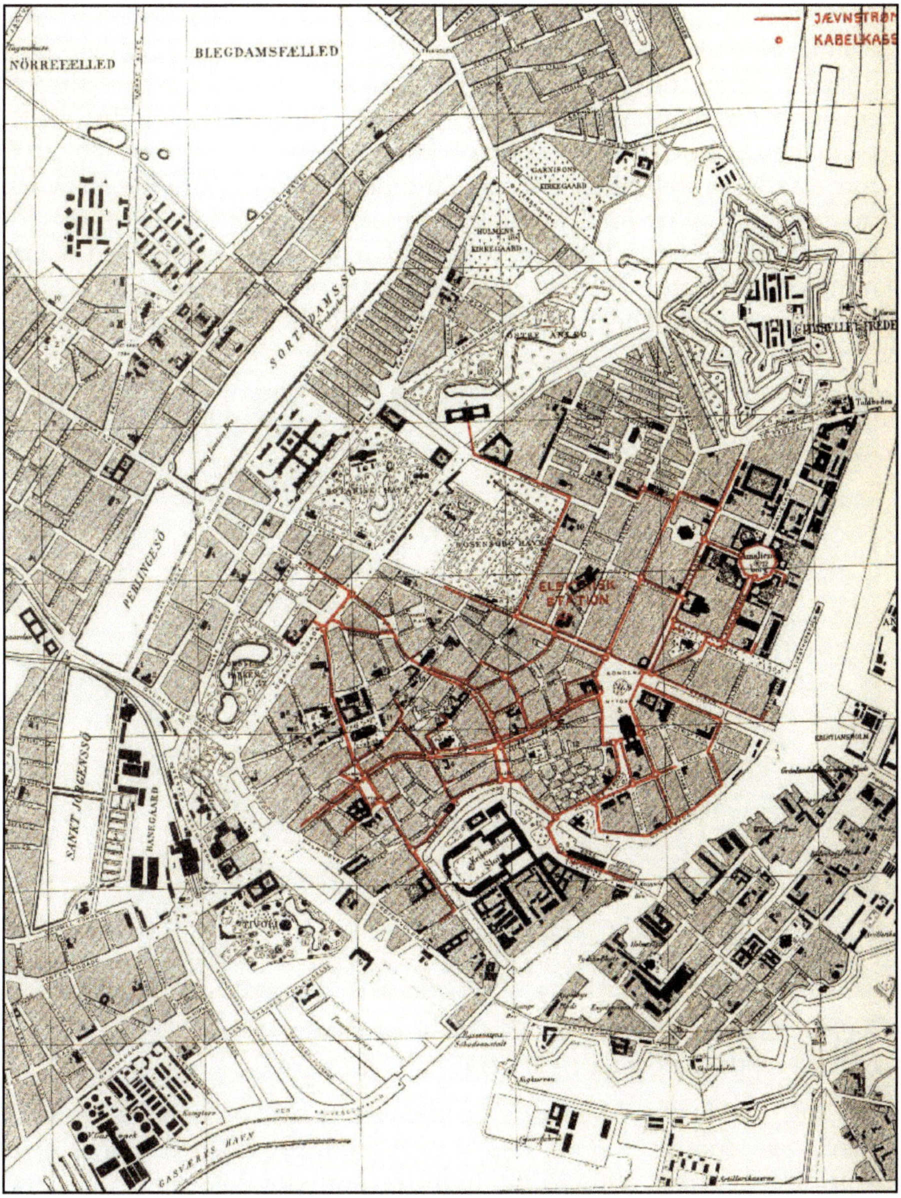


*Figure 9.26. Electricity grid in Copenhagen, 1892. The turbine station is noted in the middle of the map. Most of the network was damaged by a short-circuit incident in 1899, but repairs were accomplished in a matter of days and the network was soon expanded considerably (Andersen et al., 1942).*

From 1877 some private businesses in Copenhagen had installed their own dynamos and arc-lights, and in 1882 the first systems using incandescent lamps were installed in the Copenhagen Telegraph Company's reading rooms and at the Carlsberg breweries. Applications for establishing a general power supply for Copenhagen were made by an industrial group, including the city's gasworks, on the basis of the concession rules that had been used abroad, notably in Berlin. But the Copenhagen Municipality insisted on charging fees that the private investors saw as prohibitive, so it took another decade before the gaswork's manager J. Howitz finally convinced the city magistrate that if they did not soften their position, there would instead be established a large number of individual power plants on the premises of businesses and at the new Royal Theatre that saw electric light as the obvious solution for its performances.

*Figure 9.27. Turbine hall at Eastern Copenhagen Power Plant, 1908, with controls on the wall to the right (Andersen et al., 1942).*

By 1892, the first public power grid could finally be established in Copenhagen (Andersen *et al.*, 1942; Fig. 9.26). It provided two lines of 120 volt direct current. Due to the high transmission losses for direct current at this voltage, additional power plants had to be built in other parts of the city as demands grew (Fig. 9.27), and the suggestion of raising the voltage to 220 V and shifting to alternating current soon surfaced. Power stations were now being built in several other Danish cities. Safety rules for electric equipment and handling had been established. The further development of the electricity supply sector in the 20th century is the subject of Chapter 10, as is the fate of wind energy utilization. Here it should just be mentioned that the number of windmills at farms and in villages increased steadily (Fig. 9.28), whereas the new industries, with their tightly scheduled work, did not appreciably use wind energy, because it could not be regulated and adapted to the work pace as long as energy storage or other ways of handling intermittency had not been developed.

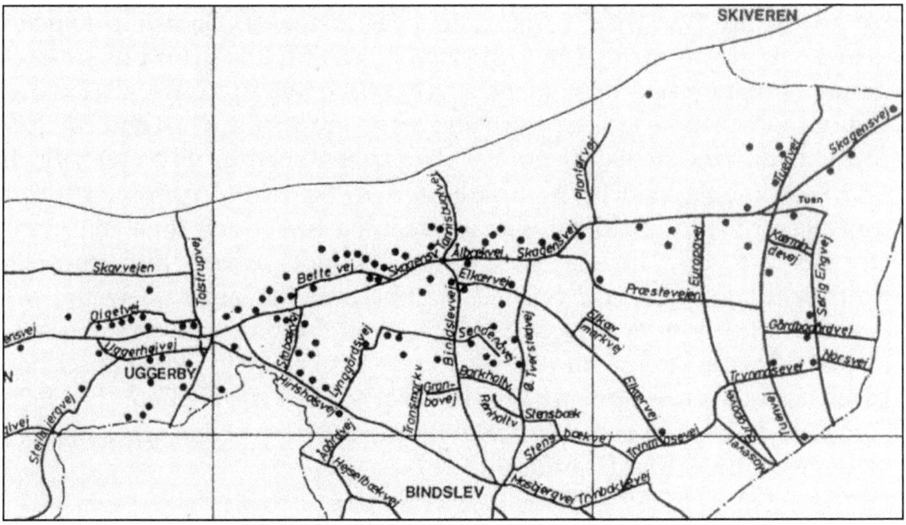


*Figure 9.28. Presence of windmills (denoted by dots) in the village of Tversted at the northern Jutland coast, any time between 1875 and 1948, but mostly during the early part of the period (Local History Archives, Tversted; Christensen, 2001).*

## 9.7 Social conditions towards the end of the 19th century

The 1849 constitution had spurred sentiments of freedom in the population, making way for many new initiatives, not only in production and business, despite the difficulties caused by the German wars. Science and arts flourished, and individuals started to explore the new boundaries to what could be said in meetings and rallies, and what could be printed in books, newspapers and magazines. The political debate was placed in a more rigorous framework of political parties, where the Liberal party (at the time mostly populated by farmers), in contrast to the Conservatives, in 1872 demanded parliamentarian cabinet formation (i.e. that governments should be backed by a majority in parliament). This development happened at the expense of the constitutional assurance that elected parliament members were bound only by their own conscience. The negative effects of replacing pure democracy by party democracy are felt as clearly today as in the 1870 debates. On the other hand, the procedure of the conservative prime minister J. Estrup (1875–1894), to govern through provisional laws (using a loophole in the constitution allowing a government in this way not to follow the majority of the parliament) was equally questionable. The conservative government finally stepped down in 1901, after their presence in the parliament had been reduced to 8 out of 114 members (Dybdahl, 1975).

A law from 1799 had formalized censorship of Danish publishing (ironically it was termed the "liberty of printing law"), and its practical implementation had been increasingly restrictive up to the 1849 constitution, which now guaranteed liberty of the press. This had not prevented the emergence of critical magazines such as *The Police Friend* (1798–1846), claimed by its first editor, K. Seidelin, to promote sound practice of justice. Seidelin was condemned to lifelong total censorship according to the "liberty law", which did not prevent him from also starting a newspaper *Dagen* in 1803. It became the largest newspaper in Denmark until its demise in 1843 and was the only one trying to deal seriously with foreign affairs. By 1835 it peaked with 3900 subscribers. Even a gutter press appeared in the form of *The Crowing Cock* (Fig. 9.28), a tabloid published in 1827–1829 with the purpose of smearing personal enemies of the editor (Davidsen, 1910).

*Figure 9.29. Frontispiece from the tabloid* The Crowing Cock, *1827–1829. The droppings from the horses were said to represent gold coins (as in a fairy-tale popular at the time; Davidsen, 1910).*

The most significant satirical magazine was *The Corsair* (*Corsaren*, 1840–1855), the editor of which, M. Goldschmidt, focused on literary criticism and in connection with publishing a cartoon came into a feud with Søren Kierkegaard (1813–1855), despite the fact that Goldschmidt actually held the philosopher in high esteem. The censors impounded several issues of *The Corsair*, which discussed the possibility of making Denmark a republic, and at one time jailed Goldschmidt. Saddened by the Kierkegaard affair, Goldschmidt left *The Corsair* in 1846 (Hansen, 1886–1902, vol. 3).

*Figure 9.30. Artists at a party in Skagen, by P. Krøyer, 1888 (Göteborg Kunstmuseum; public domain photo at Wikipedia, http://da.wikipedia.org).*

Several significant Danish writers appeared during the 19th century, notably Steen Blicher (1782–1848), who described life in the heath-covered parts of Jutland in a dense narrative language, often rendering the local dialects in written form. Another was Hans C. Andersen (1805– 1875), basically a poor writer who suddenly found a world-class niche in writing double-layered fairytales aimed at children, but with food for thought addressed to the adults reading the stories to their children. In the second half of the century, Jens Jacobsen (1847–1885) is noted for superb word handling in novels and short stories, and is acknowledged to have substantially influenced German, Austrian and British writers such as Thomas Mann, Stefan Zweig and David H. Lawrence. A full range of music composition (opera, orchestral and chamber music, as well as piano music) in a restrained or humorous romantic style was furnished by Johan Hartmann (1805–1900) and continued by his son-in-law, Niels Gade (1817–1890). Painting and sculpture saw equally productive years, notably with the "beach-painters" favouring light conditions in Danish beach areas, such as Peder Krøyer (Fig. 9.30).

Public health in some ways improved during the 19th century: epidemics of plague and malaria (called "cold fever" in Denmark) decreased, the latter despite increased livestock populations. A contributing reason was that the few malaria-carrying mosquito varieties (such as *Anopheles atroparvus*) living in northern countries with cold winters prefer blood from cattle, implying less probability for humans to be bitten, when cattle stables were separated from living quarters. Smallpox was eliminated, with help from vaccination introduced from about 1800 (Kjærgaard, 1991; Johansen, 2002).

On the other hand, the denser living and poor hygiene in cities such as Copenhagen paved the way for cholera, of which a major epidemic arrived in 1853, after having swept devastatingly across other Baltic countries over the preceding years (Christensen, 1912). The establishment of fresh water supply started by Ole Rømer (Fig. 7.35) and the primitive sewage channels carrying away untreated sewage only gradually improved. Throughout the 19th century, most quarters in Copenhagen had toilets in the yards between buildings, and although an emptying service was established, many found it too expensive and repeated the behaviour of previous centuries by rather digging deep wells below the toilets, and moving the toilet by a metre or two when the hole was full (Troels-Lund, 1914; Callisen, 1807-1809). By the mid-19th century, streets were swept twice a week and the dirt disposed of in dumps outside the city bastions, as food for pigs. However, the city did not yet have underground sewage systems, and the open gutters had a total length of 45 km, which together with the latrine basins below houses constituted a formidable breeding ground for infectious diseases. There were regulations for annual emptying of these sumps, followed by transport by carts to dumps outside the city, but they were not always followed, explaining that only 11 men were employed by the city to perform this job. Upon recommendation from Professor Hummel, attached to the city's Hygienic Commission, the public water pumps were in the summer of 1851 operated at full capacity every morning as an experiment, which, however, was given up when the stench from what was stirred up became unbearable, while water supply dwindled (Christensen, 1912). The Romans had been better organized nearly 2000 years before. It was not until the end of the 19th century that piped sewage systems were laid down in Copenhagen, often together with water supply pipes and sometimes with gas pipes or electricity cables, although the responsible authorities and supply companies did not always agree on coordinating the digging effort.

A new health issue was the growing incidence of tuberculosis (at the time called "tæring", "svindsot" or "brystsyge" in Danish), having reached 23% of all deaths by 1808, when Callisen (Fig. 9.31) made his survey, and peaking at 35% in 1835 (Faber, 1926). By 1870 the share in deaths had fallen to under 14%, which was before the systematic attempts to curb the disease by long stays at seaside sanatoria (exploiting the positive effect of fresh and slightly salty air) had really taken off. In 1920, the incidence was only 0.1% (Kjærgaard, 1991). A similar behaviour of the disease was seen in most of Europe (Schofield *et al.*, 1991). Tuberculosis is caused by a contagious infection, but its epidemiology is strongly influenced by social conditions and general health. Although overall mortality was declining in the 19th century, diets must have lacked important components of nutrition, as evidenced by the fact that the average height of humans in Denmark reached its lowest value during the first half of this century (see Fig. 4.4). That many were exposed to inferior food quality rather than insufficient food is suggested by the presence of relief programmes for the poor, and at the opposite end of the social hierarchy by the fat- and sugar-rich food eaten by the rich (as evidenced by available cookbooks and menu examples from restaurants and wealthy households).

*Figure 9.31. Henrich Callisen (1740–1824), born in Holstein, chief surgeon of the Danish navy and lecturer at Copenhagen University, keen to contribute to the bettering of health and hygiene in the city (Etching from Hansen, 1886–1902; available at http://commons.wikipedia.org).*

The demographic data of increasing quality available from the late 18th century tell us that fertility remained high (some 5% higher during the 19th relative to the 18th century) and that women had their first child at a rather high age (early twenties in rural areas and late twenties in cities), but that several changes occurred in death rates. Some of these are evident already from the overall death rates, shown in Fig. 9.32.

Looking more closely at the data on age dependence of death rates (Fig. 9.33a-d), a more complex picture emerges. The 1787 data have a very low mortality for the 10–19 years age groups, for both males and females. This could be the effect of high mortality (say from an epidemic disease) in the preceding period, particularly having affected the age group below (here 0–9 years). A similar behaviour is seen at 1850 (Fig. 9.33c), but only for females, possibly associated with the high disease death rate around 1834, seen in Fig. 9.32 and possibly caused by a smallpox epidemic (Johansen, 2002).

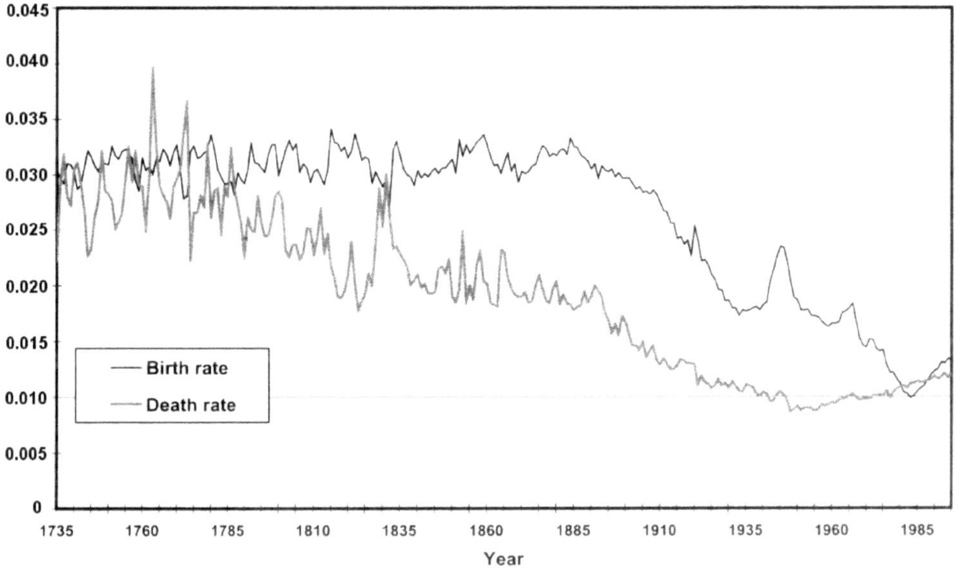


*Figure 9.32. The development from the late 18th century of a gap between Danish birth and death rates, to be closed towards the end of the 20th century (based on Matthiessen, 1976; Johansen, 2002; South Danish University Press).*

The overall behaviour is one of fairly small variations, where people over 40 acquire a consistently lower death rate, but people under 40 generally have higher probability of dying than people had in the 18th century. Causes could include higher population densities, and thus more risk of contagious diseases, but could also be the poor sanitary conditions of expanding cities and the new working conditions, with clear tendencies towards declining standards of life for the emerging working class. The death rate for children below 9 years falls during the first part of the 19th century, but then rises again by 1890. A very detailed polynomial smoothing of the death rate series between 1835 and 2000 has been made by Andreev (2003).

The life expectancy variations corresponding to the life-tables from which Fig. 9.33 is derived are shown in Fig. 9.34a–b, for females and males. The increased industrialization and the resulting changes in social conditions are seen to lower life expectancy for the young at 10–18 years of age, and particularly for the boys, while life expectations for adults and older people (ages 20 to 70 years) at first are relatively unchanged, and at 1890 substantially improving.

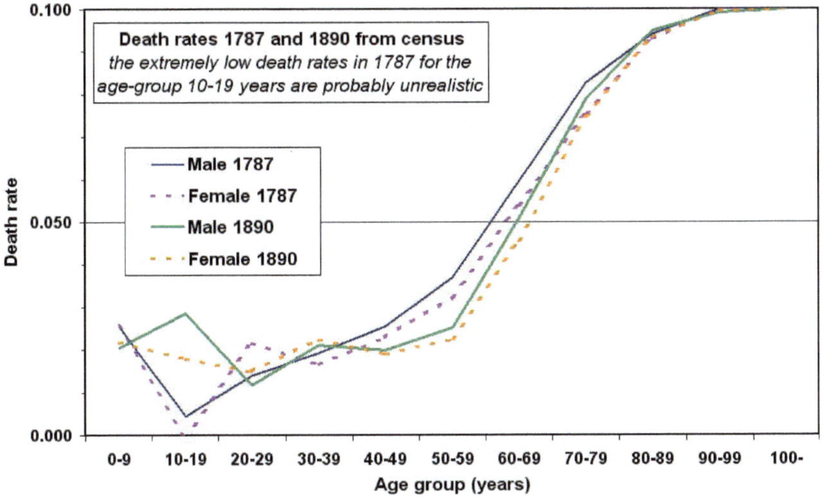


*Figure 9.33a. Age distribution of death rates in 1787, compared with those for 1890 (Danmarks Statistik, 2010; based on various earlier collections).*

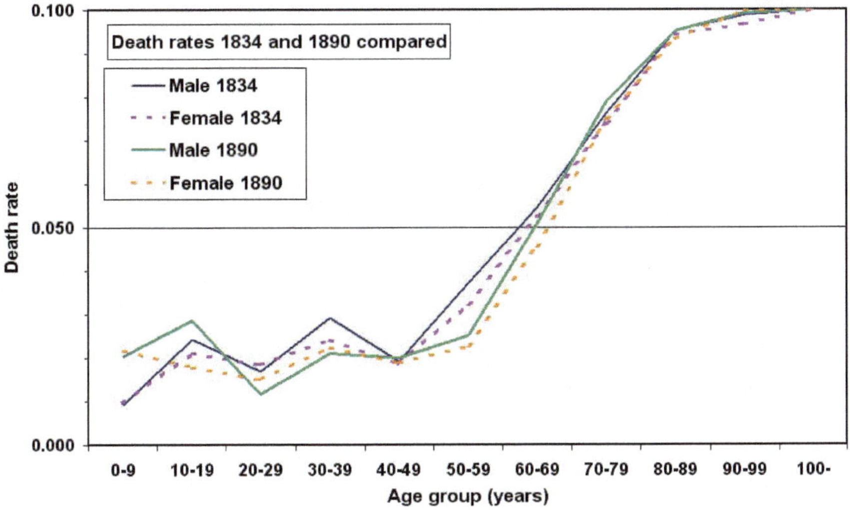


*Figure 9.33b. Age distribution of death rates in 1834, compared with those for 1890 (Danmarks Statistik, 2010).*

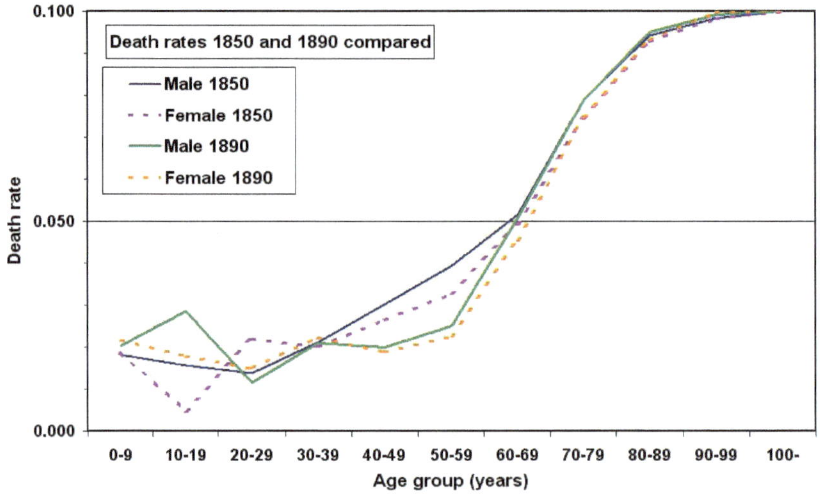


*Figure 9.33c. Age distribution of death rates in 1850, compared with those for 1890 (Danmarks Statistik, 2010).*

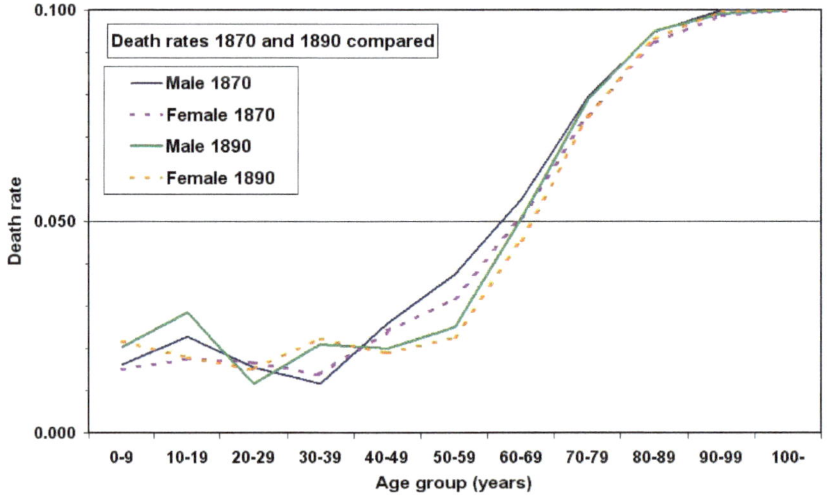


*Figure 9.33d. Age distribution of death rates in 1870, compared with those for 1890 (Danmarks Statistik, 2010).*

The 20th century demographic behaviour will be discussed in Chapter 11. It continues the decreasing trend of death rates, while at the same time exhibiting a dramatic decrease in birth frequency, following several centuries of more or less constant fertility profiles. Introducing the rather modest parameter modifications implied by Fig. 9.33 into the demographic model used in the previous chapters back to Neanderthal times, it is possible to check the overall appropriateness of the model, although not, of course, to prove its validity. Figure 9.35 shows the results of extending the model with death and birth rates representing the period 1780 to 1880, the input data that involve a smoothing of the behaviour shown in Fig. 9.33 being displayed in Fig. 9.36.

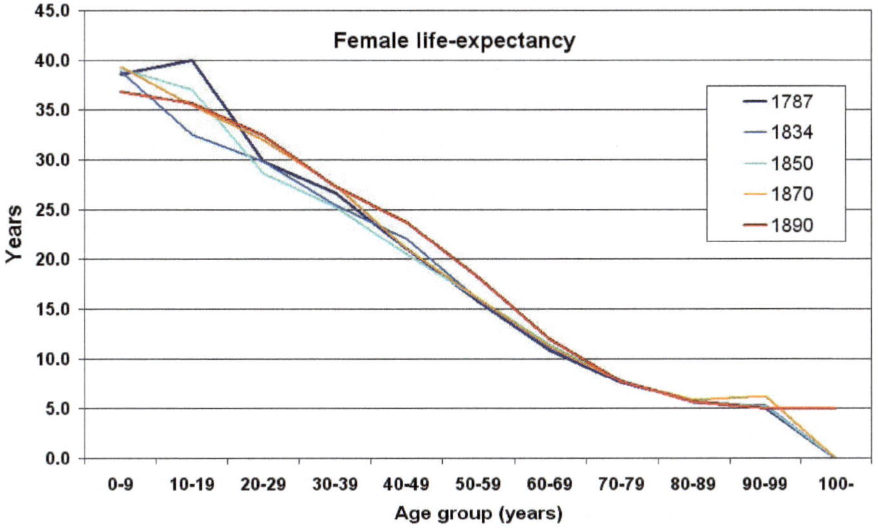


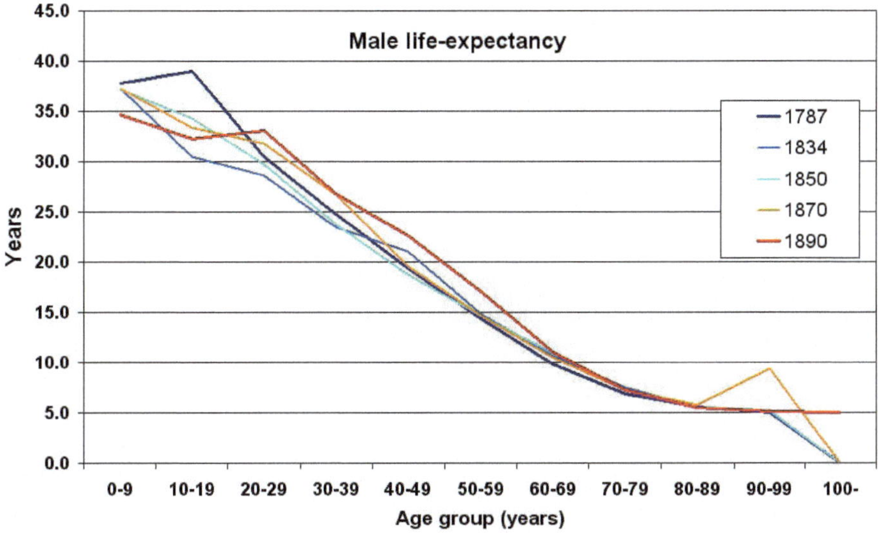


*Figure 9.34a (at top) and b (below). Life expectancy variations between 1787 and 1890, for female and male Danes (Danmarks Statistik, 2010).*

One additional factor that has to be taken into account in discussing population development to the end of the 19th century is the substantial emigration that took place from about 1860, due to the strong Danish population increase as well as to the poor working conditions, relative poverty and increasingly poor health for children. The emigration was towards North America, and notably the USA. Figure 9.37 shows people at a Copenhagen pier waiting to board ships to New York, and Fig. 9.38 gives the statistical data collected on the number of people emigrating–still much fewer than from Ireland or Sweden.

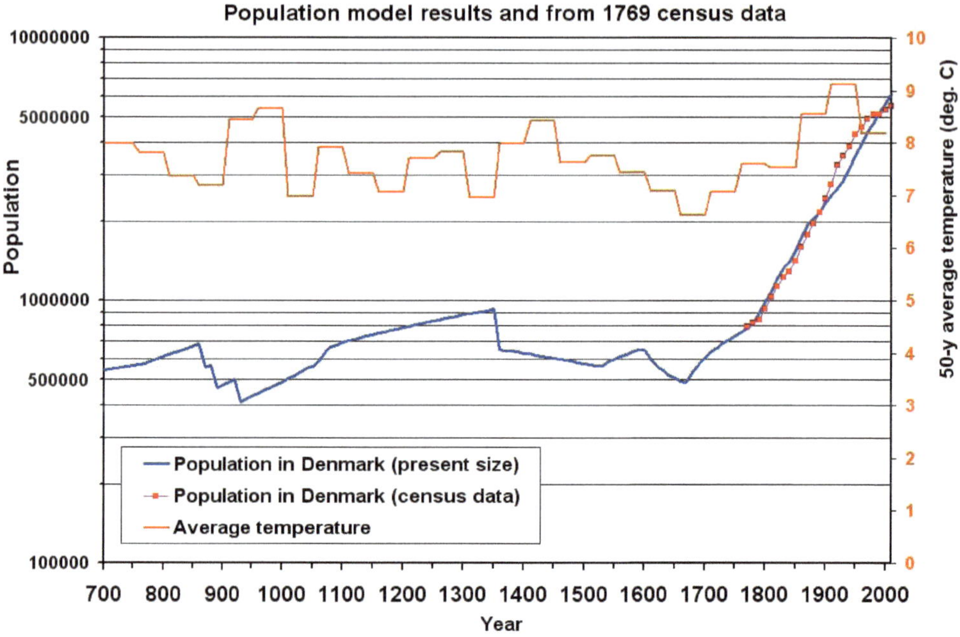


*Fig. 9.35. Danish population development 700–1880, continuing the demographic model of Fig. 6.22, and census data from 1769 onward. The census data do not include south Jutland (north Schleswig) up to 1920 (about 5% of the population). The extrapolated result of the demographic model up to year 2010 is shown, with disregard of changes in birth and death rates after 1880 (to be considered in Chapter 11), but with due subtraction of the fairly substantial number of people emigrating between 1860 and 1930.*

European countries were fortunate that large land areas with rich resource options but low population densities happened to be accessible to them in the Americas and in Australia, just at the time when their own economic expansion, however brilliant, could not keep up with the population explosion caused by advances in healthcare. Current overpopulated areas are less fortunate, as few unexploited settling land areas remain.

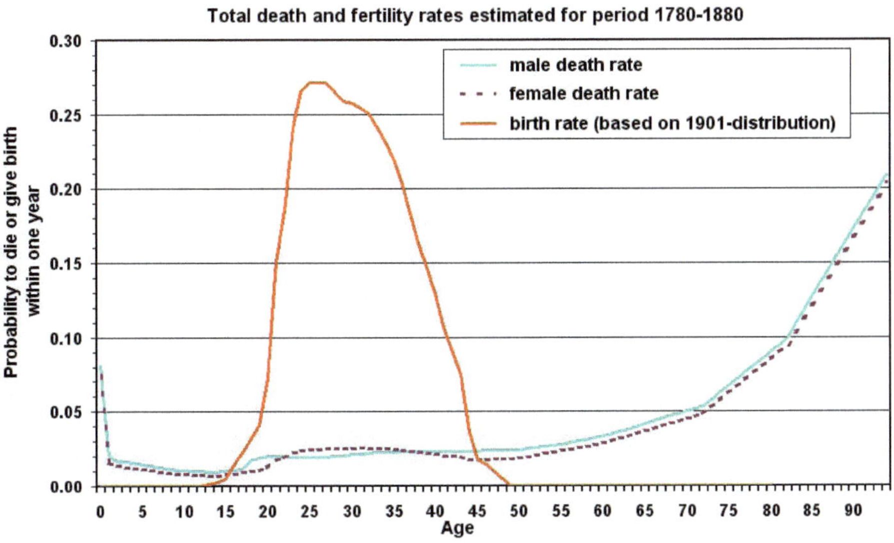


*Fig. 9.36. Death and birth rates used from 1780 (updating Fig. 6.15).*

*Figure 9.37. Emigrants waiting to embark on ships bound for North America at a pier in central Copenhagen (drawing by I. Tuxen in Illustreret Tidende, 1873; available at http://www.illustrerettidende.dk, a website maintained by the Danish Royal Library).*

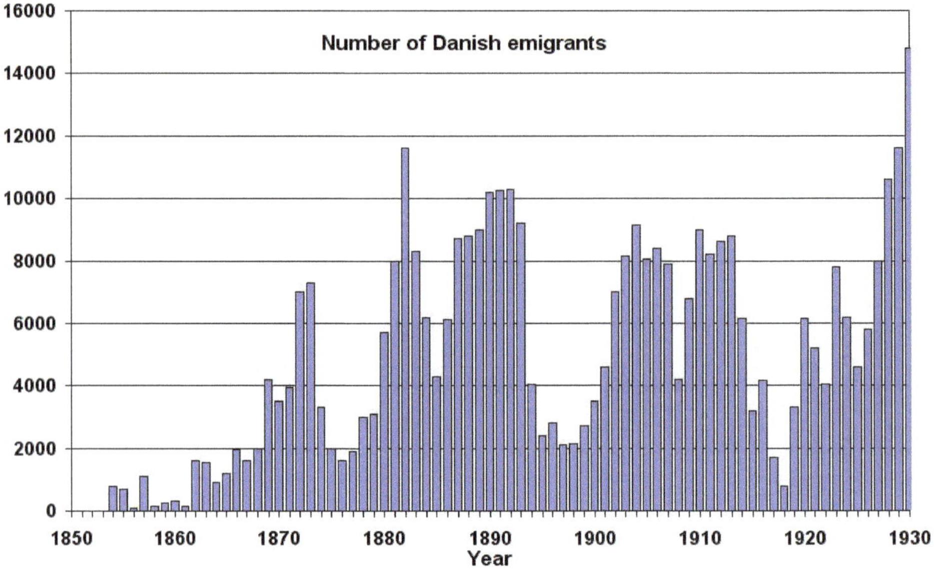


*Figure 9.38. Emigration from Denmark 1850–1930 (1850–1870 data based on US Immigration Office, the rest from Danmarks Statistik, Johansen, 2002). A total of 400 000 people left Denmark during this period–of the same order of magnitude as the exodus during the Viking times.*

## 9.8 Must what is lost externally be gained internally, at any price?

After the loss of south Jutland, the phrase then formulated as "What is lost outwardly must be gained inwardly" is repeatedly heard (around 1880 by the poet H. Holst and in contemporaty newspapers; Nielsen, 2011), as a call for increased industriousness and hard work. The saying was for a while attributed to Enrico Dalgas, who initiated an effort resulting in transforming heathland (Fig. 7.10), lakes and reclaimable shallow estuaries into productive agricultural fields, thereby increasing the area under plough by as much as the land lost to Germany. However, Dalgas's effort started during the 1830s, and about 60 (of a total of 94) areas had already been transformed before 1864 (Hansen, 2008). Statements similar to the pseudo-Dalgas one have appeared from Prussian King Friedrich Wilhelm III after the 1806 defeat at Jena (Rerup, 2011) and from the Swedish poet Esaias Tegnér in one version of his 1811 poem "Svea", deploring the loss of Finland (Tegnér, 1817). In any case, the motto has become associated worldwide with the Danish transformation of unproductive into productive land, and Dalgas has been quoted as a role model for Japanese recovering from the defeat in World War II (Iwanami, 1984). It is interesting that while some efforts in Japan were directed at recovering self-esteem by a return to the rather empty samurai virtues of honour, it was the Christian minority in the country that turned to Denmark for inspiration.

It is surprising that the agricultural expansion took place without debate on the possible side-effects, considering that the previous overexploitation of coastal land had cost nearly three centuries of hard restoration work and considerable economic losses (Chapter 8). Less surprising, perhaps, is that the environment still totally evaded the agenda. Side-effects of drying out lakes known today are loss of habitats and loss of biodiversity. The same is the case when moors and heather are made into cereal-growing fields. The area occupied by agriculture had now become a substantial fraction of the total land area, and the large monoculture plots may have had further negative implications, such as leading to increased intensity of plant diseases and insect attacks, eventually causing the perceived need for introducing chemical pesticides (in actuality used increasingly during the 20th century, along with chemical fertilizers). One apparent contradiction in the 19th century is that on the one hand, the romantic and later naturalistic waves in literature and art created an unprecedented interest in both nature and history/archaeology (in Denmark carried by poets such as Adam Oehlenschläger, 1779–1850, and Holger Drachmann, 1846–1908), but on the other hand did not lead to much effort to protect the natural environment from disappearing.

Among the attempts to transform estuaries and lakes into agricultural land are three large projects on Sealand: Lammefjord, Amager and Søborg Lake. The reclaimed Lammefjord area became a profitable potato-growing enterprise, the reclaimed areas on the island of Amager near Copenhagen now host an international airport (and the ground is finally becoming sufficiently firm for large buildings to be erected), while the well-documented Søborg project largely failed. It has some interesting features, which is why a few more words will be devoted to it here. During the Middle Ages, Søborg Lake was protecting the castle built on a small island in its middle, possibly by the archbishop Eskil responsible for church tax collection, and thus naturally fearing for his life (cf. Figs. 6.8 and 7.8). A first attempt to drain water from the lake was made in 1796, when the seasonal stream taking excess water from the lake to the sea at Gilleleye (formerly Gyldeleye, today Gilleleje) was dug out to become a permanent canal (Petersen and Meyhoff, 1986; see Fig. 9.42). Before, a ridge about a metre higher than the water level in the lake had barred the stream near the lake outlet, except at high water. The fact that the lake did not have any other outlets and had been regulated by evaporation, except for periods of strong rainfall, due to poor penetration downwards through its clay bottom, should become a major problem. Although most water left the lake through the new canal, up to 20 cm remained during most of the year and made it very difficult to sow any crop. Attempts to graze cattle in the lake floor also failed, as the cattle soon sank into the dung.

The idea had been to parcel out the reclaimed area to smallholders, who could then start by growing hay or reed, until the soil was ready for conventional farming. There was a substantial market for hay to feed horses, but initially less for reed, mainly used for roofing. However, the lake bottom soon became too wet for hay, probably because of settling of the dried lake floor, requiring the establishment of a new drainage system with ditches. As the price of hay declined during the 19th century, the cost of major additional work on the lake no longer appeared attractive, although some were indeed carried out during the 1830s and 1870s. On the other hand, the reed harvest became increasingly competitive, as demand for thatched roofs were on the rise. The remaining problem was for those who bought hay or reed on the annual auctions to get the purchase away, as the deal was for the buyer to be responsible for transportation out of the area. This was very often impossible, because even summer rains would set the area under water, and as this became known, bidding prices during following years went down. The lake was subsequently sold to a large proprietor, but yields and revenues continued to decline.

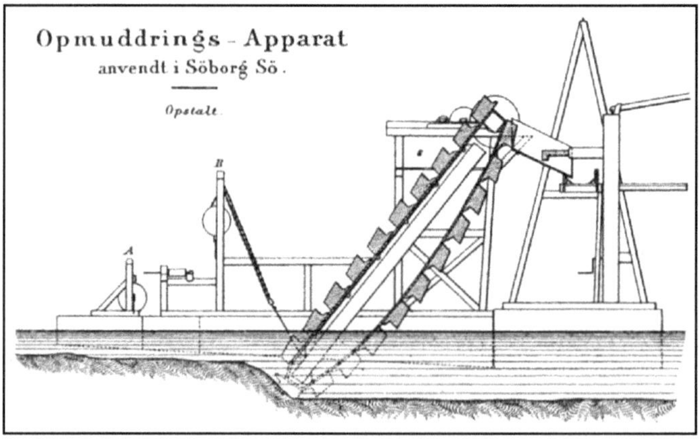


*Figure 9.39. Mud-removing machine built for the 1874 canal expansion work at Søborg Lake and Canal (from Tidsskrift for Landøkonomi* **4** *(9), 1875).*

In 1872, another new owner realized that a novel major drainage effort had to be made and called in some professional advisors. With at times nearly 300 workmen, the canal to Gilleleje was deepened, its sides stabilized and bridges established where needed. The work continued in 1874, including further digging of ditches and canals in a mesh across the muddy lake floor itself. The owner conceived a new plan for accomplishing this work quickly, having seen steam-powered mud-digging machines employed in Holland. However, he found them too expensive, and settled for a manually powered mud-digger of a type that had been used in building the Suez Canal, and more locally to free the harbour at Gilleleje from mud and sand accumulation (paid by a gift from the king to the fishermen, plus a compensation they had received for allowing the Søborg Canal to end in the harbour basin). The mud machine was built and operated by 8 men–4 for turning the crank handles, and 4 for water pumps and barge steering (Petersen and Meyhoff, 1986; see Fig. 9.39).

*Figure 9.40. Commemorative stone erected for Søren Kierkegaard at the coastal cliff west of Gilleleje, where Kierkegaard in 1835 used to sit each day before noon and let the sea and landscape inspire him for the afternoon's writing (photo by author).*

Despite these well-meant initiatives and a further effort to improve the soil quality of the reclaimed land during the 1890s by spreading soil and animal dung from neighbouring areas on the lake floor, and despite a new ditch system with circumferential canals excavated in 1945, when the lake bottom again became flooded, the situation today remains one of a very soft soil unsuited for manoeuvring agricultural machinery or for cattle to graze on without sinking in. An early eye for the destruction of a wild goose habitat is expressed by the philosopher Søren Kierkegaard, who spent the summer of 1835 in the area (Fig. 9.40). In his diary, and later in the book *Stages on the Road of Life* (Kierkegaard, 1845, part 3), he wrote about the "battle between lake and land" and deplored the loss of the natural water habitat by human intervention.

### 9.9 Fuel transportation experiences logistical transformations

As cities like Copenhagen grew, it became increasingly difficult to secure the supply of fuelwood, which could no longer be derived from forest areas at the outskirts of the city. The road systems were poorly suited for the task, and a basic consideration prevailing at least until the railways had been sufficiently developed was that transportation by sea was much cheaper than transport over land. Thus, the early 19th century problem of supplying fuelwood to Copenhagen was solved by intricate ways of minimizing land transport. The route selected is now described.

*Figure 9.41a, b. Fuelwood transport on barges at Esrum Lake (using horses, above; from an anonymous drawing, ca. 1830, in Danish Royal Library), and at Silkeborg Lake, Jutland (ca. 1890; photo from Silkeborg Archive of Local History).*

Main areas from which fuelwood could be obtained were in the north of Sealand, notably in the Crown-owned forests surrounding the city of Hillerød. A load of fuelwood arriving in Copenhagen had typically travelled through the following stages. Trees felled for the purpose were cut up on the spot in lengths of about one metre and stored in trailside stacks, not too far from the Esrum Lake, where they were subsequently loaded on barges (Fig. 9.41) to be transported northwards to the sea through a canal newly dug out during the period 1802–1805, based on a project proposed by chief surveyor Diderich von Recke (Fig. 9.42). There had been a ditch before, but it was deepened and made ready for the horse-pulled barges, with flat banks on both sides.

*Figure 9.42 (left). Sketch of the course of Esrum Canal, drawn in 1806 by Councellor of Justice Christian Pram in his report on the project. The Søborg Lake drainage canal is also shown to the left (in Danish State Archives).*

*Figure 9.43 (below). Reloading slope area between upper and lower Esrum canals. Only a narrowed lower canal is present today, but the upper canal floor depression and the height difference of some 5 metres is still noticeable in the landscape (photo by author).*

A complication was the height difference between the upper and the lower stretches of the canal, too large for the barges. Instead of installing sluices, a wooden ramp was made (at the first bend of the canal, cf. Fig. 9.42), down which the wood was rolled (Fig. 9.43), while the water went through a bypass. This meant unloading the wood and reloading it on another barge waiting in the lower canal. Once the barges reached the coast at Dronningmølle, the wood again had to be reloaded, even twice, to small boats (Fig. 9.44) taking the fuelwood out to the larger ships carrying the cargo to Copenhagen (Fig. 9.45).

*Figure 9.44. Seashore wood-fuel operation: first by horse-drawn carts, then by small sail or rowed boats to the large cargo ships (1873 photo, in Friis, 2006).*

*Figure 9.45 (right). Cargo ship with a deckload of fuelwood, similar to the ones sailing fuel from Dronningmølle to Copenhagen (copperplate engraving by Niels Truslew for his 1805 series "Ships at sea"; Friis, 2006).*

*Figure 9.46 (below). After arrival in Copenhagen the fuelwood was unloaded and stored (watercolour by Christian Hetsch, 1871; Danish State Archives).*

*Table 9.2. Energy use per capita in Denmark at year 1900, for an adult person (with use of Danmarks Statistik, 1959, 2010).*

| *Denmark, 1900* | *Human energ* | *Energy from fuels and external energy flows* | | | | | | |
|---|---|---|---|---|---|---|---|---|
| *Rate of energy use (average W for an adult person)* | Food energy | Low-temperature heat | Medium-temperature heat | High-temperature heat | Mobile power (excl. food) | Mechanical power (stationary) | Electric power (stationary) | TOTAL |
| **Liveable surroundings** | | | | | | | | **496** |
| Basic human metabolism | 70 | | | | | | | **70** |
| Indirect energy (e.g.through windows) | | 15 | | | | | | **15** |
| House and garden tools, equipment | 4 | | | | | 1 | | **5** |
| Active space heating | | 360 | | | | | | **360** |
| Artificial light | | | | 45 | | | 1 | **46** |
| Active air exchange, air cooling | | | | | | | | **0** |
| **Food and water provision** | | | | | | | | **190** |
| Hunting, gathering, gardening | 1 | 2 | | | | | | **3** |
| Fishing, aquaculture | 4 | | | | 4 | | | **8** |
| Agriculture | 20 | 2 | | | 40 | 10 | | **72** |
| Slaughtering, food processing | 12 | 1 | 1 | | | 3 | | **17** |
| Preparing, cooking, kitchen utensils | 10 | | 40 | | | | | **50** |
| Inhouse hot and cold water provision | | 40 | | | | | | **40** |
| **Security, assault, national image nursing** | | | | | | | | **21** |
| Weapons, delivery vehicles, etc. | 2 | | | | 2 | | | **4** |
| Government and legal institutions | 1 | 4 | | | | | | **5** |
| Administration and police | 1 | 8 | | | | | | **9** |
| Sports and public information facilities | 1 | 2 | | | | | | **3** |
| **Health-related efforts** | | | | | | | | **35** |
| Food preservation,clothes washing,etc. | 8 | 16 | | | | | | **24** |
| Water and waste management | 1 | 1 | | | | | | **2** |
| Doctors, medicine | 1 | 4 | | | | | | **5** |
| Hospitals | 1 | 3 | | | | | | **4** |
| **Relations and leisure** | | | | | | | | **19** |
| Music instruments, concert halls, etc. | 1 | 1 | | | | | | **1** |
| Theater, dance, TV broadcast, etc. | 1 | 1 | | | | | | **2** |
| Toys, furniture, arts and crafts | 2 | | | | | | | **2** |
| Written material, games, software etc. | | | | | | | | **0** |
| Computers,multimedia,communication | | | | | | | | **0** |
| Vacations, social transportation | 2 | 2 | | | 10 | | | **14** |
| **Occupational and learning activities** | | | | | | | | **336** |
| Schools, further education, research | 2 | 16 | | | | | | **18** |
| Construction and building | 8 | 2 | | | 10 | 6 | | **26** |
| Manufacture | 6 | 60 | 20 | 15 | 10 | 20 | | **131** |
| Trade, private services (excl. transp.) | 8 | 80 | | | | 5 | | **93** |
| Transportation of goods (incl. Internat,) | 8 | | | | 40 | | | **48** |
| Transport(delivery,services,commuting) | 4 | | | | 16 | | | **20** |
| **Raw materials procurement** | | | | | | | | **104** |
| Mining and minerals extraction | 4 | | | | 2 | 3 | | **9** |
| Forestry resources | 2 | | | | 10 | 1 | | **13** |
| Industrial raw or recycled materials | 8 | 1 | 10 | 20 | 6 | 10 | | **55** |
| Energy provision | 6 | | | | 10 | 10 | | **26** |
| Energy conversion/distribution losses | 0 | 1 | | | | | | **1** |
| **TOTAL** | **198** | **622** | **71** | **80** | **160** | **69** | **1** | **1201** |

There were no harbours on the north coast yet, and the sand banks in the shallow water were treacherous, so the ships were fairly far out on the water. The amount of work and the associated horse and human muscle energy put into this testifies to the difficulties of direct land transport traversing the roughly 50 km to Copenhagen. Reaching Copenhagen, the fuelwood was unloaded and stacked in warehouses or directly on the quay before passing though a chain of retail vendors. It is perhaps surprising that fuelwood was not imported by way of the sea, say from Sweden or Norway, as this would appear cheaper than the cumbersome land and canal detours, but Danish wood imports remained small (cf. Table 9.1 around 1900), perhaps due to the difficult political relations to Sweden-Norway. Once heating by coal became common, import by sea routes was the natural economic choice.

*Table 9.3. Estimated end-use energy per capita in Denmark at year 1900, for an adult person (with use of Danmarks Statistik, 1959, 2010), cf. Table 8.2.*

| ***Denmark, 1900*** | ***Human energy*** | ***Energy from fuels and external energy flows*** | | | | | | |
|---|---|---|---|---|---|---|---|---|
| *Rate of energy end-use (average W for an adult person)* | **Food energy** | **Low-temperature heat** | **Medium-temperature heat** | **High-temperature heat** | **power (excl. food)** | **Mechanical power (stationary)** | **Electric power (stationary)** | TOTAL |
| **Liveable surroundings** | | | | | | | | **194** |
| Basic human metabolism | 68 | | | | | | | **68** |
| Indirect energy (e.g.through windows) | | 15.0 | | | | | | **15** |
| House and garden tools, equipment | 1.3 | | | | | 0.7 | | **2** |
| Active space heating | | 100.0 | | | | | | **100** |
| Artificial light | | | | 8 | | | 1 | **9** |
| Active air exchange, air cooling | | | | | | | | **0** |
| **Food and water provision** | | | | | | | | **77** |
| Hunting, gathering, gardening | 0.3 | 0.7 | | | | | | **1** |
| Fishing, aquaculture | 1.3 | | | | 1.0 | | | **2** |
| Agriculture | 7 | 0.7 | | | 12 | 7 | | **26** |
| Slaughtering, food processing | 4 | 0.3 | 0.6 | | | 2 | | **7** |
| Preparing, cooking, kitchen utensils | 3 | | 20 | | | | | **23** |
| Inhouse hot and cold water provision | | 17 | | | | | | **17** |
| **Security, assault, national image nursing** | | | | | | | | **8** |
| Weapons, delivery vehicles, etc. | 0.7 | | | | 0.7 | | | **1** |
| Government and legal institutions | 0.3 | 4.0 | | | | | | **4** |
| Administration and police | 0.3 | 1.3 | | | | | | **2** |
| Sports and public information facilities | 0.3 | 0.7 | | | | | | **1** |
| **Health-related efforts** | | | | | | | | **12** |
| Food preservation,clothes washing,etc. | 2.7 | 6.0 | | | | | | **9** |
| Water and waste management | 0.3 | 0.3 | | | | | | **1** |
| Doctors, medicine | 0.3 | 1.3 | | | | | | **2** |
| Hospitals | 0.3 | 1.0 | | | | | | **1** |
| **Relations and leisure** | | | | | | | | **6** |
| Music instruments, concert halls, etc. | 0.3 | 0.3 | | | | | | **1** |
| Theater, dance, TV broadcast, etc. | 0.3 | 0.3 | | | | | | **1** |
| Toys, furniture, arts and crafts | 0.7 | | | | | | | **1** |
| Written material, games, software etc. | | | | | | | | **0** |
| Computers,multimedia,communication | | | | | | | | **0** |
| Vacations, social transportation | 0.7 | 0.7 | | | 3 | | | **4** |
| **Occupational and learning activities** | | | | | | | | **184** |
| Schools, further education, research | 0.7 | 5.1 | | | | | | **6** |
| Construction and building | 2.7 | 0.7 | | | 3 | 4 | | **10** |
| Manufacture | 2 | 30 | 14 | 13 | 3.3 | 15 | | **77** |
| Trade, private services (excl. transp.) | 2.7 | 27 | | | | 3 | | **33** |
| Transportation of goods (incl. Internat,) | 2.7 | 36 | | | 13 | | | **52** |
| Transport(delivery,services,commuting) | 1.3 | | | | 5 | | | **6** |
| **Raw materials procurement** | | | | | | | | **58** |
| Mining and minerals extraction | 1.3 | | | | 0.7 | 2 | | **4** |
| Forestry resources | 0.7 | | | | 3 | 0.8 | | **5** |
| Industrial raw or recycled materials | 2.7 | 0 | 7 | 17 | 2 | 7 | | **36** |
| Energy provision | 2 | | | | 3 | 8 | | **13** |
| Energy conversion/distribution losses | | 0 | | | | | | **0** |
| TOTAL | **111** | **249** | **42** | **38** | **50** | **50** | **1** | **540** |

## 9.10 Energy status at the end of the 19th century

The per capita energy use at the end of the 19th century, shown in Table

9.2, is 25% higher than the average a hundred years earlier (Table 8.1) and has a considerably different composition, now backed by a range of collected statistical data (however, with notable omissions, such as the energy provided by draft animals). The corresponding estimates of end-use energy are shown in Table 9.3.

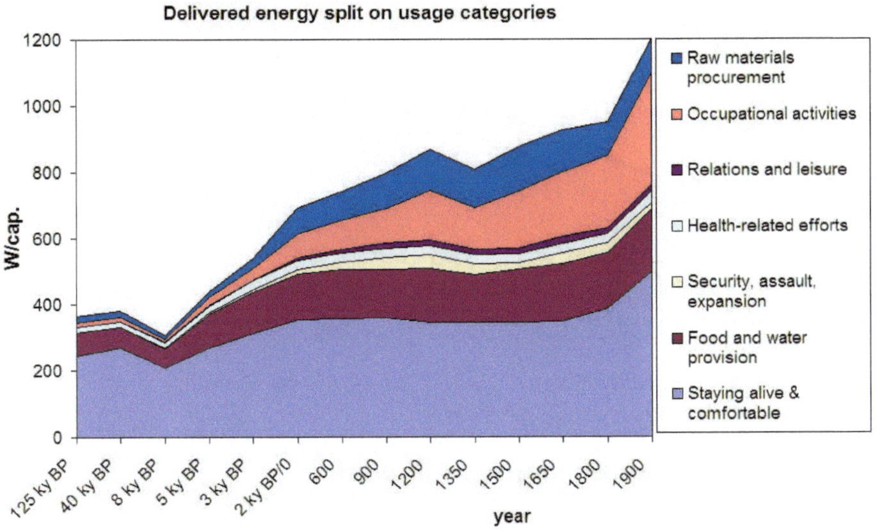


*Figure 9.47a. Estimated energy use per capita in Denmark from 125 ky BP to 1900, distributed on usage categories.*

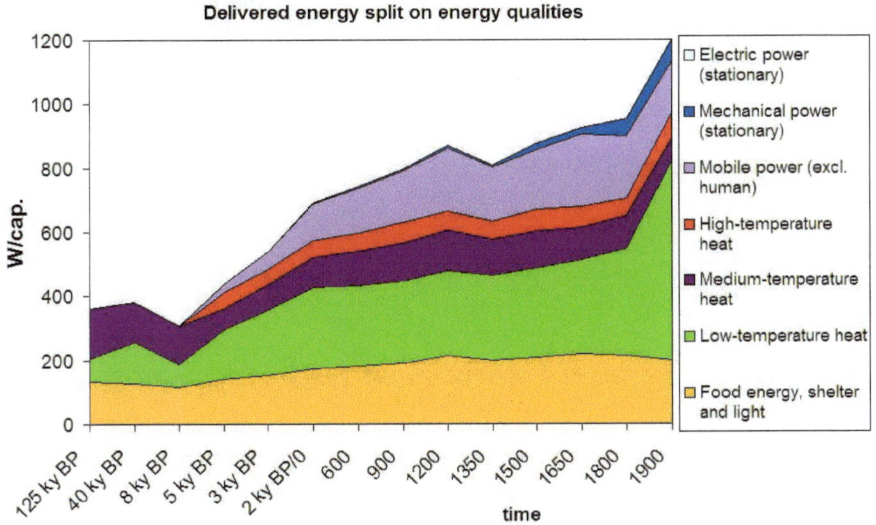


*Figure 9.47b. Estimated energy use per capita in Denmark from 125 ky BP to 1900, distributed on type of energy.*

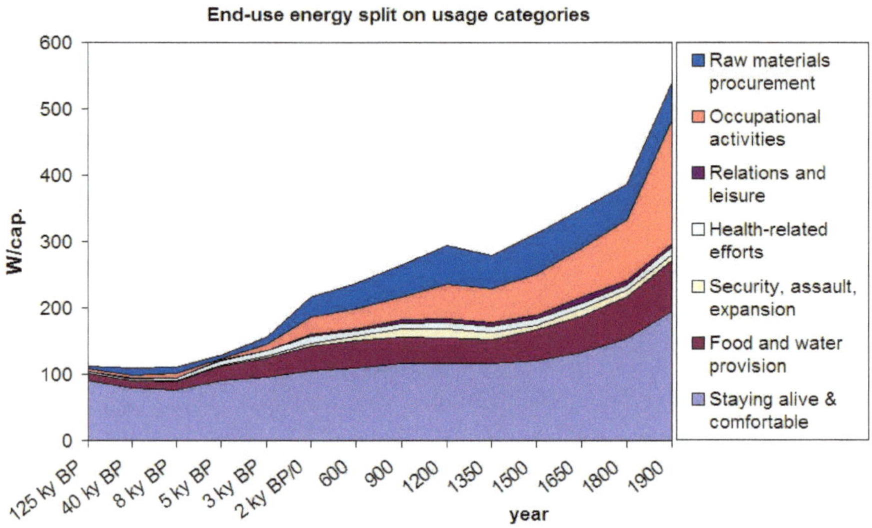


*Figure 9.48a. Estimated end-use energy per capita in Denmark from 125 ky BP to 1900, distributed on usage categories.*

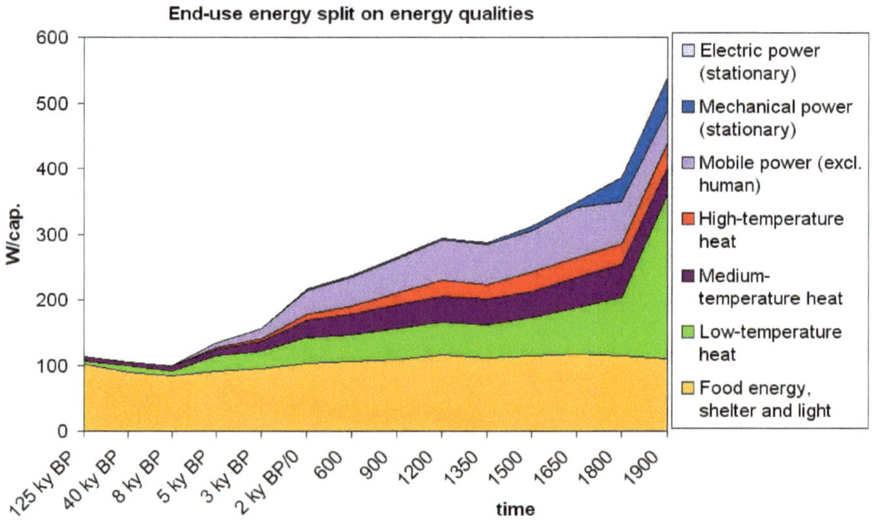


*Figure 9.48b. Estimated end-use energy per capita in Denmark from 125 ky BP to 1900, distributed on type of energy.*

The amount of human work, expressed in energy units, is 7% lower than in 1800, due to increased mechanization, but also because of the weakening of people's stature and health mentioned in connection with Fig. 4.4 and Fig. 9.34. Also the entry "mobile power" is lowered, because more efficient equipment is used for transportation, whether involving draft animals or steam-powered trains or ships. However, it should be mentioned that the method of evaluation of end-use energy changes between the periods before and after the 19th century: for the newer period, actual measured performance (e.g. for road vehicles) is used to estimate the efficiency relative to the best technology currently available, whereas for the earlier period, a flat factor of three was assumed between end-use (muscle work) and primary energy (food). This may overestimate the end-use energy because it does not take into account the ratio between muscle-work employed in the past and the best technology that may be considered today. In the fuelwood transportation case looked at above, the best-technology end-use will certainly be much lower than the muscle-work delivered for all the hauling and reloading processes described.

The total use of low-temperature heat energy for space heating, hot water, manufacture and retail trade has increased substantially since 1800, and there is an increasing amount of high-temperature heat used in industry, including gasworks and metal industries, but less medium-temperature heat, because efficiency considerations have caused the temperature levels used to increase (the division is taken at 500 °C).

The overall behaviour of delivered energy (equal to primary energy except for a minuscule amount of electricity entering around 1900) with time is shown in Figs. 9.47a and b, and the corresponding behaviour of end-use energy in Figs. 9.48a and b. The overall increase is much faster during the 19th century than before, and Figs. 47b and Fig. 48b show that this happens despite the reduced energy going into the improved-efficiency transport sector, and that it is chiefly caused by an explosive increase in the use of low-temperature heat in all sectors, including a larger building stock being heated and hot baths and clothes-washing playing an increased role in social behaviour. Figure 9.48a indicates that the end-use energy added to occupational usage is increasing more than the primary energy input.

Figure 9.49 continues the behaviour of the total energy use for the Danish population, showing how the resumed growth during the 18th century, following extended medieval and renaissance decline, is now accelerated to a 3.4-fold increase of primary energy and a 4.2-fold increase in the amount of energy made useful, during a single century.

Danish trade increased dramatically during the last decades of the 19th century. Exports of butter and pork to England increased from about £ 10 million worth each in 1866 to £ 150 million each in 1899 (Henriksen *et al.*, 2004). Of course, then as now, a country relying heavily on globalization and trade must have a sound economic home basis, capable of absorbing the effects of international price variations that are typical in the markets of both industrial and agricultural products.

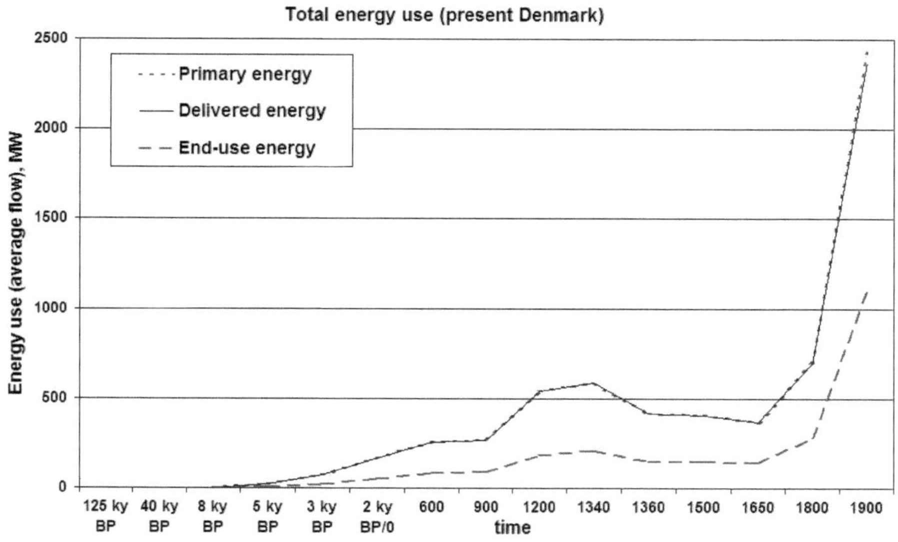


*Figure 9.49. Total Danish energy use (delivered/primary or end-use) for the total population of all ages at a given time (Figs. 8.15 and 9.35), in the present Danish territory (continuation of Fig. 6.22).*

## *References:*

Andersen, L., *et al.*, 1942. *Københavns Elektricitetsværker 1892–1942.* Københavns Belysningsvæsen, Copenhagen.

Alexander, T., 1918. *The Prussian elementary schools.* The Macmillan Co., New York. Available from Cornell University at the website http://www.archive.org/details/cu31924032699542

Andreev, K., 2003. *Evolution of the Danish Population from 1835 to 2000.* "Monographs on Population Aging", No. 9, Odense University Press. Available from the Max-Planck-Gesellschaft at http://www.demogr.mpg.de/books/odense/9/

Anthony, P., 1977. *The Ideology of Work.* Tavistock Publ., London.

Becker-Christensen, H., 1988. *Protektionisme og reformer 1660–1814.* Danish Society of Customs Duty History, Copenhagen.

Boye, P., 1976. *Det industrielle miljø 1840-1940.* Akademisk Forlag, Copenhagen.

Bredsdorff, N., 1969. *Læsestykker til den danske folkeskoles historie.* Gyldendal's Pedagogical Library, Copenhagen.

Buk-Swienty, T., 2009, 2010. *Slagtebænk Dybbøl. Dommedag Als.* Gyldendal Publ.,

Copenhagen.

Callisen, H., 1807–1809. *Physisk Medizinske Betragtninger over Kiøbenhavn.* 2 vols., Copenhagen University. At web http://www. kobenhavnshistorie.dk

Christensen, B., 2001. *Mindre danske vindmøller 1860–1980.* Danmarks Vindkraft-historiske Samling, Lem. http://www.vindhistorie.dk

Christensen, J., 2002. *Fabriksarbejdere og funktionærer 1870–1972.* Vol. 6 in *Dansk industri efter 1870.* Odense Universitetsforlag, Odense.

Christensen, V., 1912. *København i Kristian 8. og Frederik 7. tid 1840–1857.* G. Gads Forlag, Copenhagen. Available at http://eremit.dk

Danish Songbook, 1895–1942. *Danmarks Melodibog.* Vol. 1–5 published 1895-1928, a shortened version in 1942 by Wilhelm Hansen Music Publ., Copenhagen. Online at http://www.kb.dk

Danmarks Statistik, 1959, 1967. Danmarks Energiforsyning 1900–1958, 1948–1965. *Statistiske Undersøgelser,* No. 2 and 20, Copenhagen.

Danmarks Statistik, 1973–1989. *Statistisk tiårs-oversigt* 1973, 1982 and 1989, Copenhagen.

Danmarks Statistik, 2010. Statistikbanken – data og tal. Web resource at http://www.statistikbanken.dk (accessed Nov. 2010).

Davidsen, J., 1910. *Fra det gamle København.* Gyldendal, Copenhagen. Available on http://www.eremit.dk/ebog/fdgkk/ (accessed January 2010).

Dybdahl, V., 1975. *Industrialisering og folkestyre 1849–1939.* In "Hovedlinier i Nordens historie" (R. Thomsen, ed.), vol. 3, Gyldendal Publ., Copenhagen.

Edelberg, J., *et al.*, 1932. *Københavns Gasværker 1857–1932.* Københavns Belysningsvæsen, Copenhagen.

Essen, A. von, 1800. *Portions of the diary of a foreigner.* As quoted by Simon, 1989.

Faber, K., 1926. Tuberkulosen i Denmark. Gyldendal Publ., Copenhagen. See also Faber's "Tuberculosis and nutrition", *Acta Tuberculosis Scandinavia* **12** (1938) 289.

Falbe-Hansen, V., Scharling, W., 1885. *Danmarks Statistik.* Forlagsbureauet, Copenhagen.

Fode, H., 1977. *Fra reformation til folkestyre 1523–1849.* In "Hovedlinier i Nordens historie" (R. Thomsen, ed.), vol. 2, Gyldendal, Copenhagen.

Friis, B., 2006. *Esrum Kanal.* Esrum Sø Forlag (N. Richter-Friis), Nødebo.

Hansen, K., 2008. Det sagde han ikke. *Samvirke,* October issue. Danish Coop Publ., Albertslund.

Hansen, P., 1886-1902. *Illustreret dansk litteraturhistorie,* 3 vols. Det Nordiske

Forlag, Copenhagen. Available at http://runeberg.org/ildalihi/

Hansen, S., 1976–1977. *Økonomisk vækst i Danmark.* Vols. I and II. Akademisk Forlag, Copenhagen.

Hastrup, B., 1979. *Håndværkets økonomiske historie 1879-1979.* Håndværksrådets Forlag, Copenhagen.

Henriksen. I., Lampe, M., Sharp, P., 2004. *The strange birth of liberal Denmark: Danish trade protection and the growth of the dairy industry since the mid-nineteenth century.* Discussion paper No. 10-04 from Copenhagen University Department of Economics, available at http://www.econ.ku.dk (accessed January 2011).

Henriksen, O., Ølgaard, A., 1960. *Danmarks Udenrigshandel 1874–1958.* Copenhagen University Economic Institute/G. Gads Forlag, Copenhagen.

Hogben, L., 1938. *Science for the Citizen.* Unwin Brothers Ltd., London. The following editions 1942–1966 published by Allen & Unwin.

Holm, E., 1903. *Danmarks Riges Historie 1699–1814.* Det Nordiske Forlag, Copenhagen.

Hyldtoft, O., 1996. *Teknologiske forandringer i dansk industri 1870–1896.* Vol. 4 in *Dansk industri efter 1870.* Odense Universitetsforlag, Odense.

Iwanami, 1984. *Classics* No. 58, Iwanami Press, Tokyo (ISBN 978-4-03-265-4111, in Japanese).

Johansen, H., 1988. *Industriens vækst og vilkår 1870–1973.* Vol. 1 in *Dansk industri efter 1870.* Odense Universitetsforlag, Odense.

Johansen, H., 2002. *Danish population history 1600–1939.* University Press of Southern Denmark, Odense.

Jørgensen, A., 1904. *Danmarks Riges Historie 1814–1864.* Det Nordiske Forlag, Copenhagen.

Kierkegaard, S., 1845. *Stadier på livets vej.* Hilarius Bogbinder, Copenhagen (reissued by Martins Forlag, 1926).

Kjærgaard, T., 1991. *Den danske revolution 1500–1800. En økologisk tolkning.* Gyldendal, Copenhagen.

Kühle, E., 1993. *Danmark Historie Samfund.* Gyldendal, Copenhagen.

Lehrer, T., 1965. The folk song army. On *That was the year that was,* Warner Bros B000002KO7, Burbank, CA.

Matthiessen, P., 1976. *Befolkningens vækst.* Berlinske Leksikon, Copenhagen.

Møller, E., 1958. *Helstatens fald 1855–1864,* in 2 vols. Gad Publ., Copenhagen.

Møller-Holst, E. (ed.), 1877–1883. *Landbrugsordbog for den praktiske landmand.* Vol. 1–6. Gyldendal Bookshop, Copenhagen.

Montessori, M., 1912. *The Montessori Method.* English translation by A. George. Frederick Stokes Co., New York. Available at website http://digital.library.upenn.edu/women/montessori/method/ method.html

Müller, L., 1849. Letter in *Dansk Kirketidende.* Grundtvigsk Forum, Copenhagen.

Nielsen, N., 2011. Sagde Dalgas "Hvad udad tabes, skal indad vindes"? http://videnskab.dk/composite-4886.htm (accessed Jan. 2010).

Norrild, S., 1956. *Dansk Litteraturhistorie.* Gyldendal Publ., Copenhagen.

Paulinyi, A., Troitzsch, U., 1997. *Mechanisierung und Maschinisierung 1600 bis 1840.* Vol. 3 in *Propyläen Technikgeschichte,* Ullstein Buchverlage, Berlin.

Pearsall, A., 1985. *Steam enters the North Sea.* In *The North Sea* (A. Bang-Andersen, B. Greenhill, E. Grude, eds.), pp. 195–216. Norwegian University Press, Stavanger.

Petersen, F., Meyhoff, G., 1986. *Søborg Sø – gennem sump til agerland.* Græsted Bogudvalg, Græsted.

Petersen, H., 1983. *Planmæssig ankomst. Danske dampskibe indtil 1870.* Fiskeri- og Søfartsmuseet, Esbjerg.

Reiser, C., 1858. *Historiske Beskrivelsse over den mærkværdige og meget fyrgterlige store ildebrand 1728: fra 20de: og til 23de: october, inklusive.* Reprint by H. Nissen/H. Møller Forlag, Copenhagen. Available at http://eremit.dk

Rerup, L., 2011. The development of nationalism in Denmark, 1964–1914. Text from Humboldt-Universität zu Berlin, deposited at https://www.ni.hu-berlin.de/

Schieldrop, E., 1934. *Teknikkens vidundere.* 2 vols. Gyldendal Norsk Forlag, Oslo.

Schofield, R., Reher, D., Bideau, A. (eds.), 1991. *The decline of mortality in Europe.* Clarendon Press, Oxford.

Simon, E., 1963. Réveil national et culture populaire en Scandinavie. La genèse de la hojskole 1844–1878. *Annales Économies, Sociétés, Civilisations* **18** (5), 1028–1032.

Simon, E., 1989. *–og solen står med bonden op.* Translated from a manuscript in French by V. Nielsen, Askov Folk High School Publ. English translation by K. Parke: *And the sun rises with the farmer.* Folk School Association of America - Institute for People's Education and Action, Florence, MA, 1998. Accessed January 2011 at http://www.peopleseducation.org/resource/ericasimon.htm

Smith, A., 1776. *An inquiry into the nature and causes of the wealth of nations.* Methuen & Co., London 1904; University of Chicago Press, 1976; available online at http: //www.econlib.org/library/Smith /smWN.html

Sørensen, B., 2010. *Renewable Energy,* 4th edition. Academic Press-Elsevier,

Burlington.

Sørensen, B., 2011. *Life-Cycle Analysis of Energy Systems. From Methodology to Applications.* RSC Publishing, Cambridge, UK.

Tegnér, E., 1817. Svea. Poem awarded the Swedish Academy Prize for 1811. Printed in *Svenske Akademis Handlinger ifrån 1796*, part 6, pp. 155–169. Carl Deleen Publ., Stockholm. Available at http://books.google. com (some other published versions of the poem do not have the lines about regaining a substitute for Finland).

Thestrup, P., 1997. *Dampen binder Danmark sammen.* Vol. 1 in *På sporet 1847–1997.* Danish State Railway's Museum, Odense.

Troels-Lund, T., 1914. *Dagligt liv i Norden i det sekstende aarhundrede.* Gyldendal/Nordisk Forlag, available at http://runeberg.org/ dagligt/

# 10

# Electricity: wind or coal?

When electricity is generated in thermal power plants, the conversion efficiency is limited by the laws of thermodynamics. When a fuel is combusted, yielding a substance such as water steam at a certain temperature to drive a turbine, the rotational energy can then be fed into an electromagnetic generator that delivers the power output desired. The temperature of the heat delivered to the turbine (e.g. by steam), determines the efficiency, and the second law of thermodynamics states that there is a maximum, which is 50% at a steam temperature 300 °C above the ambient one (Sørensen, 1993).

In contrast, supplying an electric generator with rotational energy from a wind turbine, there are no unavoidable losses. There is, however, a fundamental limit (in practice some 59%) to how much of the energy in the wind itself that can be transformed into rotational energy by a turbine, because one cannot take out all the kinetic energy and bring the air movement to a complete standstill. The atmosphere must accept the slowing of the wind and eventually regenerate the air motion by temperature and pressure differences elsewhere, which is what for a free-stream wind turbine entails the conversion limit of $16/27 \approx 0.59$ (Betz, 1927; Sørensen, 2010a). In economic terms, this is less important than the conversion efficiency for fossil fuel steam plants, because one has to pay for the fuel regardless of conversion efficiency, while one does not pay for the wind, but only for the subsequent conversion equipment.

Wind power exhibits a more fundamental challenge, shared with several other types of renewable energy (such as direct solar energy). It is the characteristic intermittency and variability. It is worst for solar energy: there is no solar energy at night, less on overcast days, and in any case, a dependence on the angle of the Sun relative to zenith, which affects the losses of light by absorption and scattering in the atmosphere. Furthermore, solar radiation changes with seasons, particularly at high latitudes, and in a way opposite the variations in electricity demand, which tend to be largest during winter and darkness (at least in regions not requiring electricity for space cooling). For wind, the variability is connected to passage of weather systems with temperature and pressure variations ("fronts"), with typical timescales of a few days. To this come rapid wind variations at a scale of seconds, being of relevance to the operation of a single, isolated wind turbine, but not to a grid-connected system consisting of many turbines. The seasonal variations in wind energy are often in rough agreement with demand variations, as is the case in Denmark.

*Figure 10.1 (above). One of the very few vertical-axis windmills operated in Denmark, on the island of Christiansøe, northeast of the larger island of Bornholm in the Baltic Sea (left: from Pontoppidan, 1767; right: from the Danish architect Thurah, 1756; cf. discussion in Notebaart, 1971).*

*Figure 10.2 (left). Dutch-type windmill for grinding cereals (mill-wheels at bottom; Skenkelsøe in north Sealand), built 1891 and still in working condition (photo by author).*

This problem had not affected early penetration of windmills used for grinding cereals. Milling could be done when the wind was adequate. It did not have to be possible continuously, and waiting a few days for good wind conditions was rarely a problem. To some extent, the same was the case for agricultural applications such as for pumping water, threshing machinery, or operating other mechanical systems not in constant use. A different situation arose when application of wind for industrial equipment became considered by the late 19th century. Many farmers used the simple solution of replacing the older windmills (that had gone from the post mills of Figs. 7.32-7.33 to alternative designs, including the vertical-axis machines of Fig. 10.1, and then to Dutch windmills with only the upper part of the tower yawing, Fig. 10.2) by American multi-blade wind turbines (such as the one in Fig. 10.3). They start at a lower wind speed than the old mills, and therefore have more operating hours through the year. Although reducing the intermittency problem, the advantage was paid for by a substantial lowering of the conversion efficiency. For performing tasks that only required small amounts of energy, such as lifting water or doing a little threshing, that was acceptable, but not for use in the new production industries.

*Figure 10.3. Multi-bladed wind turbine manufactured and erected in Kenya near the "African Farm" of the Danish writer Karen Blixen (photo 1981 by author). Placing a windmill behind a tree is not a recommended practice. In Denmark, the manufacture of multi-blade windmills took place in several factories operating between 1890 and 1920 (Christensen, 2001).*

Poul la Cour, who was physics professor at the Askov Folk High School (third from right in Fig. 9.6), was the first to address this problem seriously. He experimented partly with what at the time seemed the most obvious solution–batteries–and partly with novel ideas such as hydrogen storage to overcome the intermittency of wind power production. Based on test windmills built for the high school, of which the 1891 turbine (shown on the right-hand side in Fig. 10.4) was the first in

Denmark to produce electricity rather than just shaft power (similar developments took place in England and the USA; Sørensen, 2010b, c). He installed an electrolyser based on a design by the Italian Pompeo Garuti (Fig. 10.5; Cour, 1900; Hansen, 1985), and contemplated a Danish production of electrolysers, but the project never got off the ground (Cour, 1903). The hydrogen produced was used for lighting the school by gas-candles. La Cour's government funding, allowing the erection of a larger windmill at Askov in 1896 (to the left in Fig. 10.4, the 6-bladed rotor was in 1900 replaced by a 4-bladed one), was earmarked for studying windmills and improving their efficiency, including wind-tunnel work on aerodynamics, and his interest was to provide farmers with electric and mechanical power for the new types of equipment coming on line. In 1902, he formed an electricity association (*Dansk Vind Elektricitets Selskab*) aiming to assist formation of decentralized electricity systems, and managed to help establish 418 plants by 1918 country-wide, of which 120 were based on 10–20 kW wind turbines, and the rest on gas–and later diesel–powered generators (Arnfred, 1964).

*Figure 10.4. Experimental electricity-producing windmills erected by Poul la Cour at Askov Folk High School (number 3 in Fig. 10.23), the larger one delivering power to the entire village through a battery storage (see Fig. 10.6) with back-up by a kerosene generator (1897 photo in la Cour Museum, Askov).*

Poul la Cour was an inventor, but not a remarkable scientist, despite his title in physics. It did not occur to him to apply conventional physics to calculate the limit to wind power extraction, as Betz did later (1927). Less surprising is his lack of aerodynamic understanding, which he shared with many colleagues before aviation made airfoil theory a scientific focus. This may still appear strange, because the knowledge that wind turbines are mostly turning because of the lift forces on the blades (rather than the drag caused by the resistance of the blade) is the same that explains the motion of sail ships, a technology that ought to have been understood by year 1900. Instead, la Cour insisted that the important parameter had to be the area of the airfoil, among other things leading to the false belief that 4-broad-bladed turbines had to be the optimal ones. He therefore interpreted his own wind tunnel experiments incorrectly, trying to fit them into his wrong perceptions.

*Figure 10.5 (left). Electrolysis unit in Poul la Cour's basement laboratory at Askov. Hydrogen and oxygen outlets are seen. The hydrogen was used for flame-burning lamps at the school (from Cour, 1900).*

*Figure 10.6 (below). Lead-acid battery back-up for wind power at a rural plant (Hansen, 1985).*

The world was full of inventors around year 1900, and many of them suffered from the same lack of knowledge of existing theoretical knowledge as la Cour. Lack of familiarity with alternating current made inventors such as Thomas Edison (and la Cour) delay progress by insisting on using direct current for electric power supply, despite the inability at the time to transmit it for more than a few hundred metres due to resistance losses (Engstfeld, 2004). Nicola Tesla had already established the viability of alternating current (AC) for long-range transmission, and direct current (DC) applications such as battery charging or electrolysis could be performed by locally transforming the AC power to DC. The first such transformers were used in Milano in 1886 (Wistoft *et al.*, 1991–1992). Stabilizing wind turbine rotations is easily done for alternating currents (by use of asynchronous generators), but difficult for direct current, and la Cour spent many years of his life trying to invent mechanical systems to smooth the rotation of wind turbines feeding into a direct current generator, with limited success (Hansen, 1985).

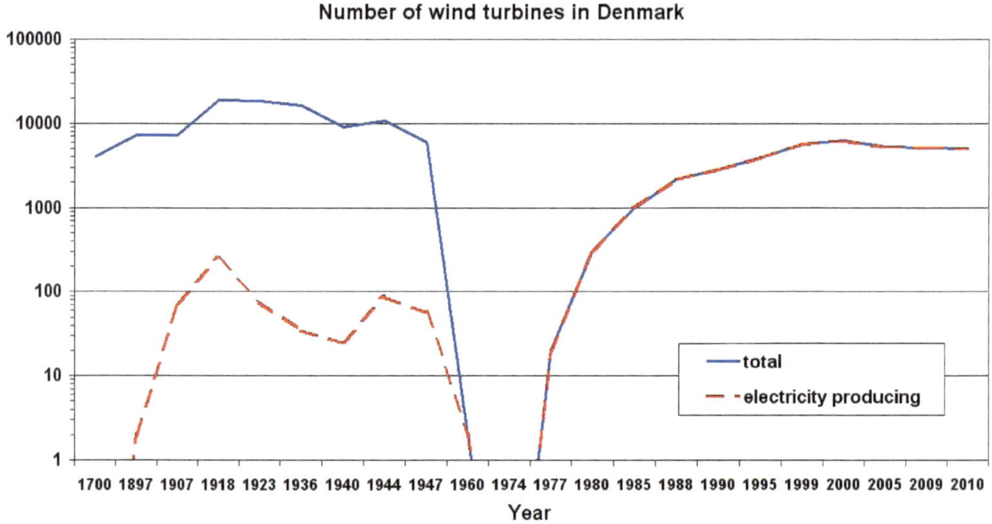


*Figure 10.7. Number of wind turbines installed in Denmark, given for selected years 1700 to 2010 (not equally spaced). Electricity-producing turbines are singled out. Sources: Chapter 7 estimates based on tax registration, Danmarks Statistik (1910, 1925, 1943, 1949), Warming (1913), Arnfred (1964), Pedersen (1948), Danish Energy Agency (1999, 2010a); see discussion in text.*

The general effort of establishing electricity supply in Denmark proceeded chiefly through municipal and cooperative companies established in towns and other agglomerations of potential customers (Wistoft *et al.*, 1991–1992). Due to the transmission problems with direct current, alternating current was introduced in 1906-7 by the electricity utilities in Copenhagen and northern suburbs (Andersen *et al.*, 1942), from the start using three phases of 220 volt 50 Hz AC, which was a good choice that several other European countries only made after initial investments in less suitable specifications, and which the USA still has not made, probably because Edison thought 220 V would be too dangerous. Early introduction of AC allowed a rapid Danish expansion of power supply, and the old direct current areas in central Copenhagen were soon supplied by direct current transformed from alternating current. In the country as a whole, the power supply was similarly organized except for the smaller village areas, where co-operative power plants were common (Danmarks Statistik, 1923) and recommended by advocates of decentralization such as la Cour. In terms of electricity supply, the demands of growing industry soon made wind marginal, whereas farms continued to see wind as a reasonably competitive supply option, because they would otherwise be expected to pay the cost of grid extensions. These observations are supported by the available statistical data, shown in Figure 10.7, to be further discussed below.

*Fig. 10.8. Agricco, the first Danish AC-producing turbine on the market (from manufacturer's brochure, 1921; number 6 in Fig. 10.23). The blades are made of steel fastened on aerodynamically profiled wooden shapes, and the efficiency of converting the wind energy could reach 50% at 8 m/s wind speed, more than twice as good as that of the la Cour design in Fig. 10.9 (Christensen, 2001; cf. the overall performance shown in Fig. 10.19b). Neither blade number nor lengthwise shape is optimized.*

Wind turbine manufacturers were slow in catching up on alternating current, first suggested for generators on wind turbines by Povl Vinding and R. Jensen as late as 1919 (Arnfred, 1919), and produced by Hans Larsen's wind turbine factory in Frederikssund on Sealand in 1921, with the trade name *Agricco* (Fig. 10.8). However, it was really too late, as the market for electricity had already been captured by the power plants based on fossil fuels. Also, the aerodynamic understanding of how a turbine works, and the associated theory for optimizing blade profiles, did not materialize in Denmark until the early 1920s, as endless debates in the magazine of the Danish Association of Engineers demonstrate (Vinding, 1919; Vogt, 1919; Jensen, 1920; Schou, 1921; Bjerknes, 1924). Neither the Polytechnical Highschool (currently renamed *Denmark's Technical University*) nor the professional engineering schools around the country placed much weight on theoretical knowledge. Engineering was supposed to consist of practical experience and rules of thumb. This went well for designing houses and bridges, thanks to the use of large safety factors (often implying an unnecessary increase in cost), but it made the competition between wind and fossil power fall out to the advantage of the latter. The engineering textbook by Schrøder (1924), commonly used at the time, simply repeated la Cour's incorrect description of the functioning of a wind turbine.

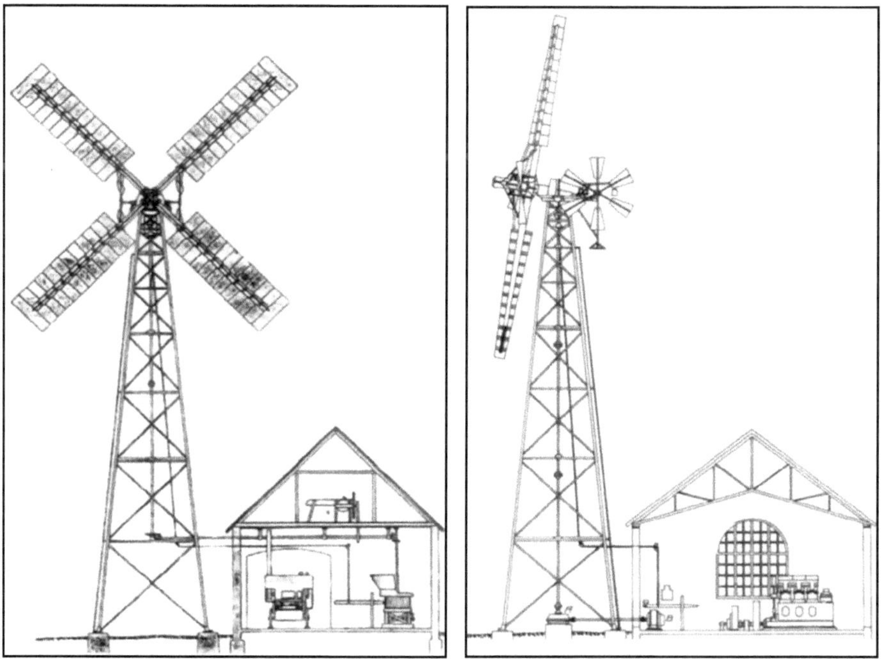

*Figure 10.9a, b. Between 1901 and 1953, the manufacturer Lykkegaard at Ferritslev (Funen) produced wind turbines adhering closely to la Cour's outdated design. From the company's sales brochures, the left arrangement caters to mechanical power application, while that on the right produces electricity (e.g. number 7 in Fig. 10.23; pictures courtesy of Knud Lykkegaard, 1974).*

A few words on the attempt in Fig. 10.7 to follow the development in numbers of wind turbines erected. From the late 19th century and into the 20th century, windmills were placed either near a farmhouse (e.g. with bevel pinion arrangements for right-angle transfer of mechanical power from the turbine tower shaft to machinery in the farmhouse; Fig. 10.9a), or away from the farm, on a hill with good wind conditions and a milling stone directly attached to the vertical shaft, or at a water pool or stream from which water could be hoisted into barrels or other containers. It was therefore a minor alteration to make an electric generator replace the mechanical equipment, possibly with a ground-level gearbox to obtain a suitable speed of rotation (Fig. 10.9b). This may explain the difficulty in finding precise statistical data on the number of electricity-producing wind turbines, beyond the ones used in "electric power stations" such as those commissioned by la Cour's society. Figure 10.7 includes only those that with certainty are producing electric power. Over ten times higher numbers are quoted by Ølgaard (1979), stated to be based on Pedersen (1948) and Danmarks Statistik (1949), but this is probably an error, perhaps due to assuming the word "power" to only mean electric power. Wind energy disappeared briefly from the Danish energy system, but after the oil supply and price crisis of 1973–74, a rapid expansion has taken place, with over 6000 wind turbines installed by year 2000 (compared to some 20 000 around 1920), but during the last decade decreasing to about 5000, as more turbines over 20 years old have been decommissioned than new ones added. The recent lack of government interest may of course not prevail.

Looking at wind capacity installed (Fig. 10.10) and actual electric power produced (Fig. 10.11), the substantial difference between the current wind power era and the previous ones becomes apparent. The installed capacity is presently some 50 times higher than in 1920, and the power production is more than 200 times higher than at any time before 1970.

The Danish history of energy experienced several rapid changes during the 20th century. A largely indigenous energy production (wood) became replaced by a nearly 100% dependence on imported fuels. As expanded on below, the aftermath of a particular one among several negative incidents related to import dependence during wars and during international and economic disruptions finally caused a change in policy, albeit for a period less than one might have expected, because of the finding of oil and gas resources in the Danish part of the North Sea, which would last a few decades and thus offer a perfect opportunity for a stable transition to a long-range solution. But first, wind lost to coal, the electric car lost to the oil-based car, and only when the shortcomings of fossil-fuel-based energy systems began to be felt did the future of energy supply enter the arena of public and political discussions.

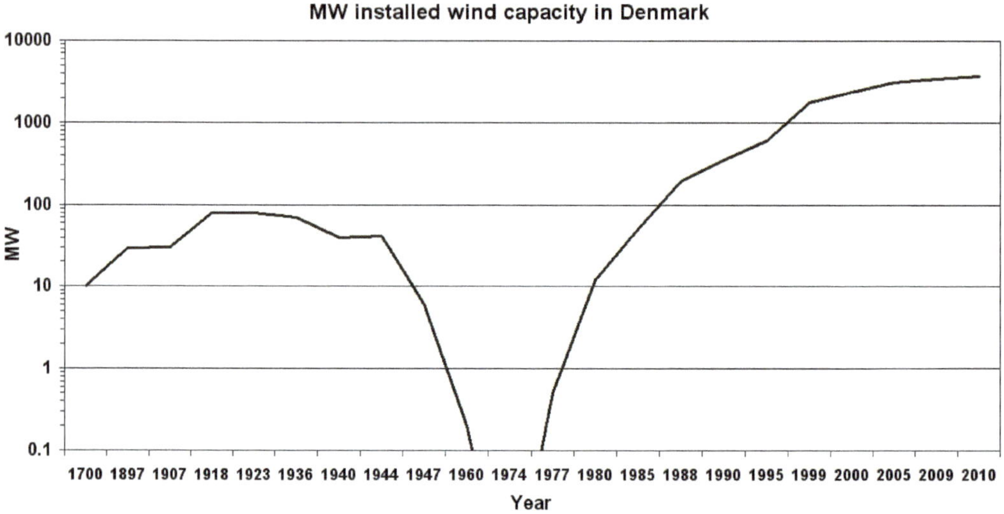


*Figure 10.10. Installed wind capacity in Denmark, 1700–2010 (sources are as in Figure 10.7, including estimates applied in periods of insufficient data).*

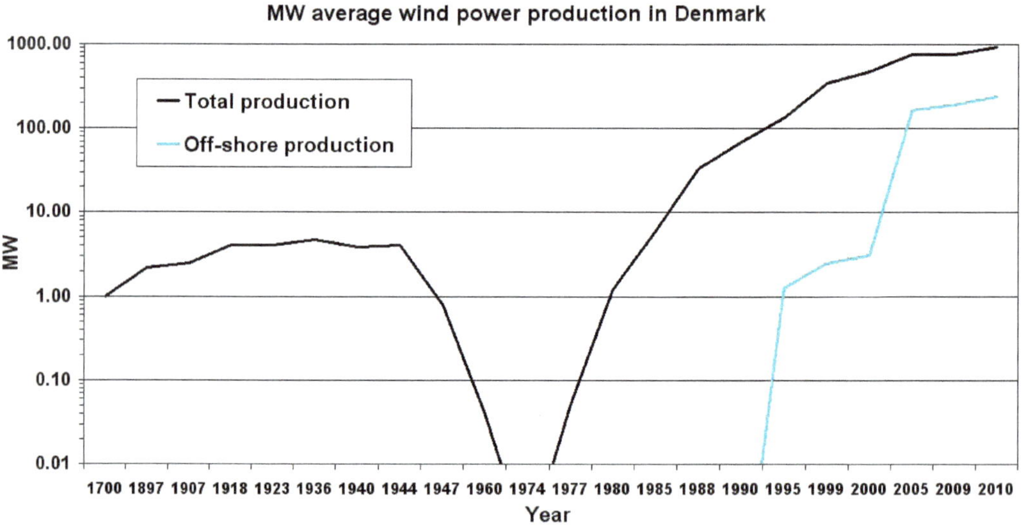


*Figure 10.11. Average power production by wind in Denmark, 1700–2010 (recent offshore activities are singled out). Sources are as in Figure 10.7, including estimates applied in early periods of insufficient data (cf. text).*

An important energy policy initiative tool place after the Suez crisis in 1956, a conflict that briefly led to the closing of the Suez Canal for international freight carriers. The Suez incident, although not having hurt the oil transport nearly as much as feared (because new cheap bulk carriers were able to transport oil from the Middle Eastern wells south of Africa to Europe without significantly increasing the oil price), made some people think about the vulnerability of Europe's recent reliance on imported energy. Johannes Juul (1887–1969), an engineer at the South Sealand Electricity Utility (SEAS), felt that a country like Denmark, with nearly 100% imports of energy, a commodity of evident strategic importance, ought to build up some emergency measures to counteract supply disruptions that might occur in the future. He suggested that domestic wind energy should be introduced at a level of at least 10% of demand, providing energy for essential tasks and also buying some time to change the energy system if a new crisis should happen (Juul, 1964a). He did more than that. Based on experiments on a smaller machine already carried out, he obtained government funding to build an experimental 200 kW wind turbine in 1957 at Gedser at the Baltic coast, larger than any previous machine and using the best technology available in Denmark at the time (Fig. 10.12 and 10.13; the constant-width blades are not quite of optimum shape, as known from early versions of the work of Hütter, 1964). The machine operated to specifications documented in the final reports (Juul, 1964b; available to UN New Energy Conference participants in 1961, and the Danish Wind Power Committee's report by Poulsen-Hansen *et al.*, 1962), but in 1967 operation was halted because no interest in wind energy was seen.

*Figure 10.12 (left). Gedser 200 kW wind turbine erected 1957, producing AC electricity through an asynchronous generator and having stall control of wing speed (number 11 in Fig. 10.23; photo by Peter Lise, 1979).*

*Figure 10.13 (below). Detail of over-speed control in Gedser mill, showing that low-technology solutions were not rejected: should the tower vibrate too much, the ball will fall down from its position on top of a tube and operate the lever shutting the turbine down (photo courtesy of L. Divone, 1975).*

Yet the Gedser windmill project turned out to be the experience that gave Denmark the world lead in wind technology after the next energy crisis, in 1973–74. A few months before that crisis, I had published a small paper reminding readers about Juul's view of wind as an "insurance" against supply interruptions and suggesting that Denmark had enough wind for not just 10% but 100% coverage if desired (Sørensen, 1973), and less than a year after the crisis, the first of the new generation of small Danish wind turbines had been manufactured and made ready for sale by Christian Riisager at Skærbæk (east Jutland). Riisager had simply taken the Gedser mill design and scaled it down with a few improvements, first to a 7 kW prototype tested in 1975 and then to the commercial 22 kW version sold from 1976, mainly to farmers who had not forgotten the old virtues of wind machines (Figs. 10.14-10.15).

*Figure 10.14. Farm-installed 22 kW Riisager wind turbine with rotor diameter of 10 metres (photo author, 1978; cf. number 12 in Fig. 10.23).*

Naturally, the first in this new generation of wind turbines were not quite competitive with coal power, but the Danish Academy for the Technical Sciences correctly thought that the government would be willing to chip in some subsidy money for the supply independence gained. A committee formed in 1974 produced a report showing that wind energy had the potential to become economically competitive, if the government would offer some seed money to have the new industry reach a sizeable production level (Fischer *et al.*, 1975, based on background material by Broe *et al.*, 1975, and Sørensen, 1975, repeating the prospect of 100% renewable energy coverage of Danish energy demand with inclusion of suitable energy storage options, such as hydrogen).

*Figure 10.15. 30 kW Wind-matic turbines at the Ikea furniture shop near Copenhagen, one of the first examples of new wind turbines not serving agricultural customers (photo by author, 1979).*

*Figure 10.16a, b. Left: 30 kW Nordtank turbine using a steel tube tower in contrast to the steel truss or concrete towers common at the time (photo by author, 1981). Below: 500 kW Vestas turbine as seen from oat field below. Allerød, north Sealand (photo by author, 1998).*

The Danish government bought the idea and formed a second committee (still under the Academy of Technical Sciences, as no Department of Energy had been formed yet, and similar to the first committee except that industry people had been omitted due to their possible vested interests), with the charter to advise on how best to shape the subsidy scheme. The report came in 1976 (Pedersen *et al.*, 1976) and the government subsequently followed its recommendations. The interesting thing is that the plan of action and its formulation of the rules for obtaining subsidy were quite different from the ones emerging in several other countries interested in wind energy (notably Sweden, Germany, the Netherlands and the USA). The Danish scheme would give the subsidy not to industry but to individuals who bought wind turbines, and only if the manufacturer could make a plausible argument that the payback time for the investment would be under 8 years, at least in the future where the advantages of mass production had been harvested thanks to the initial subsidized sales. At the time, the payback time was over 8 years, not just because of the early stage of technology, but also due to a poor investment environment after the oil crisis, with borrowing interest rates near 20% and inflation at 10–12%. Therefore the manufacturer could not just adjust prices to get a share of the subsidy, but had to work on improvements of his turbine concept. The wind turbine subsidy was at first 30% of the installation price, and over the next 15 years gradually reduced to zero, as the cost of wind-produced power became close to that of coal-based power. Another factor in making wind acceptable was the introduction of a buyback scheme, where utilities would buy excess wind power from private owners at a government-decided price, lower than the consumer electricity cost but higher than the fossil power production cost before transmission (arguing that dispersed power generation would reduce transmission costs).

Several more companies joined Riisager in the manufacture of wind turbines based on the Gedser concept, and they gradually scaled up in size (Fig. 10.16a shows an example with a steel tube tower). Wind power reached by the early 1980s the level it had around 1920, but now entirely with electricity-producing turbines. The philosophy of the Danish manufacturers was to increase turbine size very gradually, in order to minimize the risk of unforeseen problems that seemed to mar attempts in those countries trying to jump to megawatt-sized turbines in one step (e.g. German Growian project and the US two-bladed Mod-1 and Mod-2, all of which turned out dead ends not leading to any industrial breakthrough). However, one megawatt wind project was successful, namely the Tvind windmill erected at the Tvind People's College at Ulfsborg i west Jutland (Fig. 10.17). It had a 2 MW generator (obtained cheaply), but the rest of the turbine was more suited for 1 MW and it would not have made a noticeable difference if the generator had been rated at 1 MW. The Folk High School people had received spare-time assistance from engineers, but the construction was made in a craftsman style, with lots of the school's students as volunteers (and without too much concern over the health risks posed by epoxy used in the glass-fibre blade–winding process).

*Figure 10.17. 2 MW Tvind wind turbine built in 1977–78 by volunteers at the Tvind Peoples College at Ulfsborg (photo courtesy of Mark Diesendorf). The performance of the mill was measured and verified by a team from the Danish Shipbuilding Laboratory (Jensen and Bjerregaard, 1980).*

The nascent Danish wind industry found a substantial market in California (Figure 10.18), where a subsidy scheme introduced by the late 1970s allowed investors tax benefits for investing in wind energy, independent of whether the systems subsequently produced any energy. This led to a "gold rush" by decent and less decent manufacturers, and when the scheme was abandoned in 1986, some of the companies, including several Danish ones, went bankrupt. However, after some restructuring, the Danish wind industry again functioned, now having learned that dispersal of activities to customers in many different countries was a good thing, given that national wind energy support fluctuated, often with little predictability. During the 1990s, several Danish wind companies were bought by foreign investors, and currently only Vestas is in Danish hands and still possessing the largest market share of any producer in the world. The advanced fibre blades used for the turbine rotor wings are produced by subcontractors, with the Danish company LM Glasfiber dominating the market (and suspected of possibly causing health risks for its workers).

*Figure 10.18. Danish wind turbines at Tehachapi Pass, near Mojave in south California (Nordtank 55 kW machine in foreground; photo by author, 1986).*

Technological advances have made it possible to continue to increase the turbine size, currently to 6 MW per unit, with tower hub heights of 60–100 m (taking advantage of the general increase in wind energy with height; Sørensen, 2010a) and rotor diameters of 50–80 m, requiring quite advanced materials. The three-bladed concept prevails, despite the fact that two or one blade would do the job as far as energy production is concerned. The reason for avoiding the economic advantage of fewer blades is the stability offered by the three-bladed structure, not having the undesired resonance and stability problems of two-bladed structures or one-bladed ones with counterweights. Figures 10.20 and 10.21a show examples of late 20th-century vintage machines and current offshore arrays of turbines, while Fig. 10.19 gives an impression of the development in turbine size and in average efficiency over the last few centuries, measured in terms of the annual average power level produced per square metre swept by the rotor, a quantity that depends both on the wind conditions at the turbine location and on the design characteristics of the turbine. Locations are shown in Fig. 10.23.

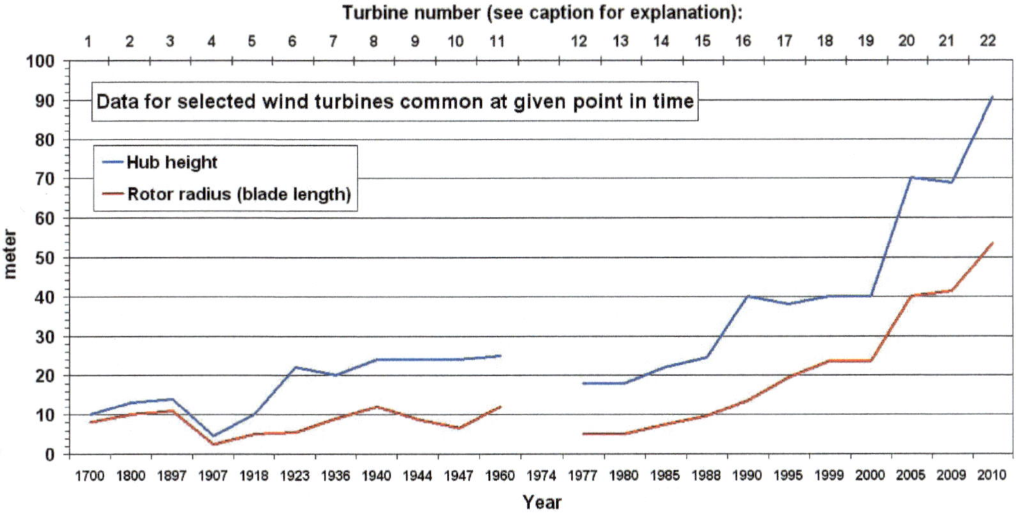


*10.19a, b. Above: Development in hub height and rotor radius for selected turbines 1700-2010. Below: Development in specific power production and efficiency in converting power in the wind, estimated from the location of the mills using the data shown in Fig. 10.22. Each turbine is identified by a number on top of the graphs: 1: Postmill (Fig. 7.33; 2: Dutch mill (Fig. 10.2); 3: Askov (Fig. 10.4 left); 4–5: Hurup and Dandy multiblade mills; 6: Åkirkeby (Fig. 10.8); 7: Tranebjerg (Fig. 10.9a); 8–9: F. L. Smidth mills at Frederikshavn and Bogø; 10: same rebuilt; 11: Gedser (Fig. 10.12); 12: Abiltrup (Fig. 10.14); 13: Sonebjerg mill at Samsø; 14, 17–20: Vestas mills at Veksø, Grurup, Nr. Herlev (Fig. 10.20); Fanø, Horns Rev (offshore, Fig. 10.21a); 15, 21–22: Bonus mills at Vestervig, Nysted (offshore) and Lemvig; 16: Nordex mill at Bovlund.*

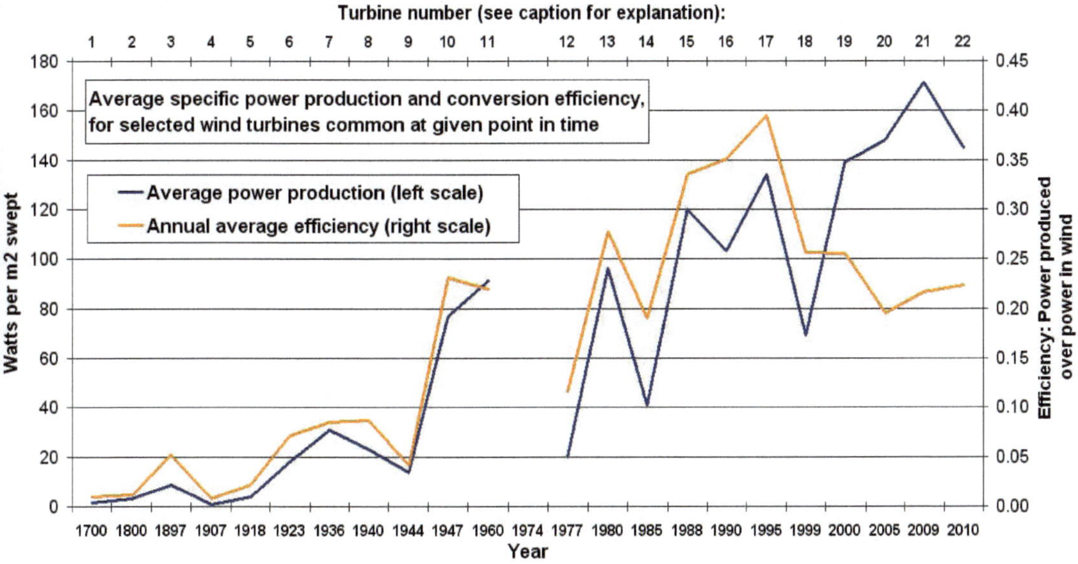

The hub height development shown in Fig. 10.19a is important, because the energy in the wind increases with increasing extraction height, and also diminishes the disturbance of any obstacles near the ground (called "roughness" and implying a slowing of the wind). The energy collected naturally also increases with the area swept by the rotor, and much development has centred on enabling construction of long blades of sufficient strength. There are two "dips" in the development: one around year 1900, when the American multi-blade windmills became popular (e.g. for water-pumping), despite their lower efficiency, compensated for by more operating hours during the year. The other is after the 1973–74 oil crisis, where Danish manufacturers had to "start again", slowly daring to increase size and efficiency. Fig. 10.19b shows that the efficiency of the early ones in this new generation of wind turbines was not impressive, but that it did not take long to improve it.

Compared with the early 20th century or even older mills, the efficiency has improved significantly since aerodynamic blades were introduced, but it peaked in 1995 and has since gone down, although the power produced per square metre swept by the rotor has continued to increase. The reason for the latter is that the larger turbines with larger hub heights experience better winds, particularly offshore, but the former observation is more surprising. There can be several explanations. The design objective during the 1980s and early 1990s was to maximize annual production, and this determined the blade profiles to use and the pitch angles, if they were not variable. By 1995, the offshore market evolved, and the manufacturers responded by offering the largest of their standard turbines for offshore wind farms. There were few wind measurements offshore, and it came as a surprise how much better the wind conditions were (as later verified by experience and by theory; cf. Sørensen, 2008). It would be possible to redesign the turbines to take advantage of the better wind conditions offshore, and also at larger hub heights inland, and thus to maintain the efficiency of 40% or even increase it, but it is not done, seemingly for the following reasons.

With larger penetration of wind energy in the electricity system, many operating hours may appear more advantageous than maximizing annual production (just as in the reasoning on multi-blade mills a century ago), so leaving a less efficient system may defer the need for trading surplus power or establishing expensive energy stores (Sørensen, 1981). A second reason is structural strength and stability. The rapid move towards very long blades has challenged the available materials technology, but also poses risks that are difficult to model beforehand, such as excess vibrations and resonance in the tower–rotor structure. When the upper and lower blade tips at any time are well over 100 m apart, they experience different wind regimes, and this creates bending stresses that have to be managed. Not allowing the top performance may be one way of playing safe, until more practical experience has been gathered with the multi-megawatt machines. A third reason for not optimizing the turbine is that manufacturers today see markets all over the world, and therefore may not wish to speak too loudly about the fact that wind conditions vary enormously and would therefore invite separate design optimization for each new location. This could be a process introduced in the future, using "smart" manufacture, where the advantages of mass production are not lost just because the wind turbines have to be adapted to individually sited users.

During the 1970s, Danish manufacturers established a market dominance, having been able to start rapidly with a concept already proven (the Gedser mill), and they were able to keep the position for a couple of decades, by a very conservative strategy with no radically new technology introduced, and hardly any new industrial approaches, but a slow scaling up from a proven design with some modest improvements along the way. Some factory managers even boasted of the fact that no scientific input had been required (forgetting that such had indeed been made by Juul, and with government financial support). It is becoming apparent that the future is more complex, requiring skills to harvest more than the mere 20% of the power in the wind characterizing the 2010 vintage turbines. Possibilities are plenty: gearless machines with variable output that is made to conform with demands by computerized after-treatment; adjustable geometry beyond the pitch control that many turbines do not even have today; entirely new composite materials for blades and rethinking of tower structure; and finally, manipulations of the wind field before it reaches the rotor area (shrouds and tip-vanes were briefly researched but then abandoned; Sørensen, 2010a). Use of advanced carbon fibres is likely to allow longer blade lengths.

The turbines shown in Figs. 10.20 and 10.21a–c support the claim that the visual impact of wind turbines does not increase as fast as the power output, not just because of better wind conditions at larger hub heights, but also because the blades of a typical large wind turbine appear to move more slowly than those of a small turbine (although in reality, the tip speed is much higher). The industry has learned to use appealing slender designs and unobtrusive colouring, causing most people to be more annoyed by the silhouette of the "black" overhead power lines and associated towers than by the wind turbines. Anyway, power lines are increasingly buried underground, and as long as there are more church towers (some 8000 in Denmark) than wind turbines, there may not be any real need to push the wind turbines further offshore (at higher foundation costs). Presently, offshore wind farms such as the one in Fig. 10.21a have become prime tourist attractions.

*Figure 10.20. Three 660 kW Vestas wind turbines operating since 1998 at the inland site Nr. Herlev near Hillerød in north Sealand (number 18 in Fig. 10.23; photo by author, 2011).*

*Figure 10.21a. Tourists inspecting offshore wind park at Hornsrev (2 MW Vestas turbines operated since 2002; number 20 in Fig. 10.23; the turbines are in a 4 × 4 km² array, with the closest about 15 km from the beach in the foreground; photo by author, 2003).*

*Figure 10.21b, c. Model photos (by author, 2007) showing the visual impact difference between wind turbines of size 50 kW (upper) or 500 kW (lower).*

The production data for the newer wind turbines in Fig. 10.19b are taken from the Danish Energy Agency online database with GIS links (2010b).

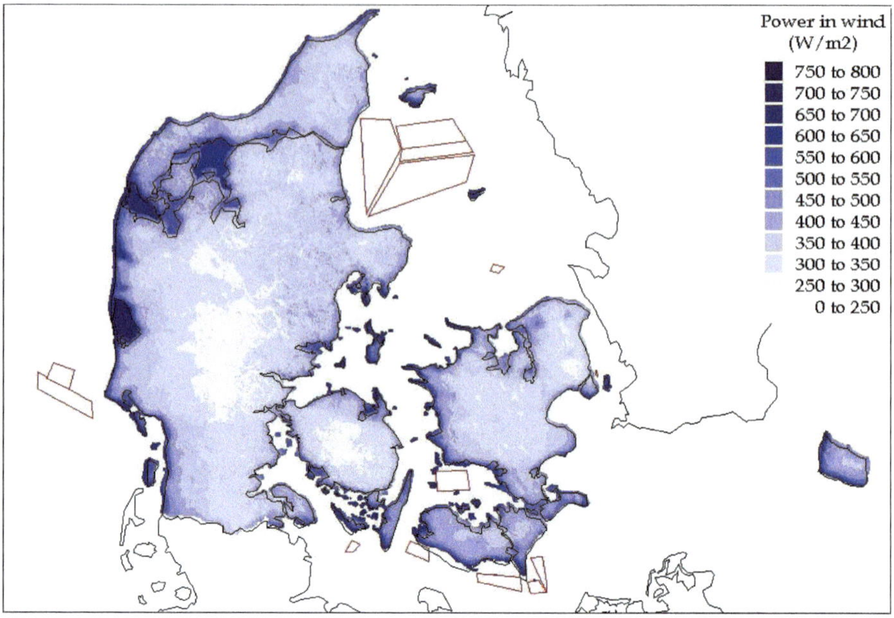


*Figure 10.22. Map of power in wind at about 70 m height over land. Sites for possible offshore wind park use are indicated (based on Sørensen et al., 2001; Sørensen, 2005; Energi & Miljødata, 1999; Danish Power Utilities, 1997).*

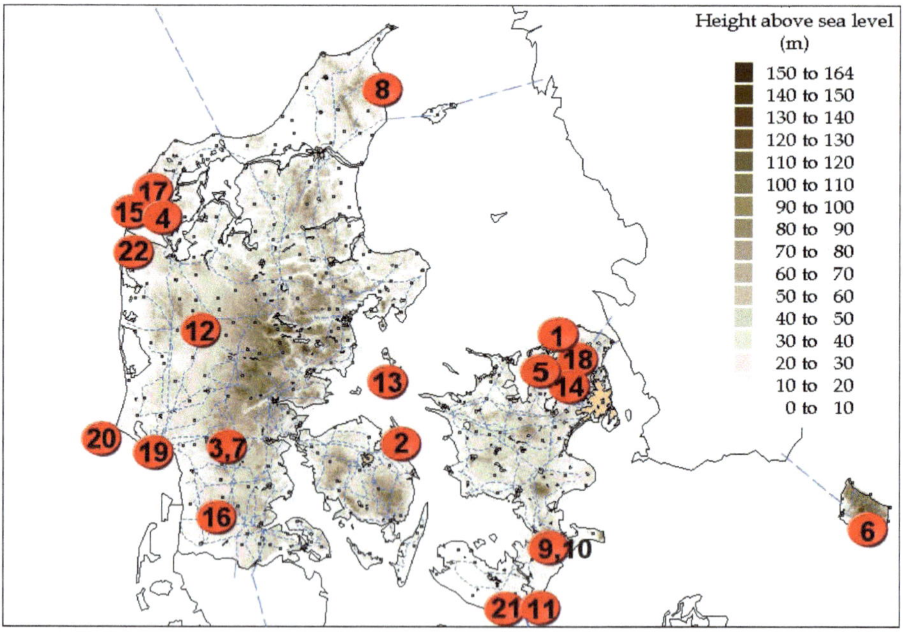


*Figure 10.23. Location of wind turbines used for Fig. 10.19. The base map indicates heights over mean sea level, cities and trajectories of electricity transmission lines (cf. Fig. 10.22; GIS design by author).*

The geographical information system (GIS) and database of power production with precise location for each turbine include some 80% of the installed wind turbines, missing most of those constructed before the Energy Agency was formed in the early 1980s. Some of these have contributed data to the magazines *Wind Power Monthly* and *Windstats Newsletter*, published in Denmark by Torgny Møller from 1985 but later transferred to a US publisher. The data for the period before the Gedser turbine are somewhat uncertain, comprising an interview study for the period 1940–1947 (Pedersen, 1948) and some information from Danmarks Statistik (1923, 1925, 1943, 1949), except for the first considering only wind turbines used at farms, and using different definitions for different years, such as house mills and "other", or mills used for water pumping, agricultural work or lighting. The reports after 1923 just state that the data are for "farms using wind machines", not precisely whether the mill is on the farm or if a communal mill is used (e.g. for milling). Similar data exist for 1906, comprising only agricultural windmills, merged with watermills (Warming, 1913).

In order to estimate the efficiency of wind energy conversion, statements for individual mills for particular years have been used (Christensen, 2001; DVS, 2001–2002). For all turbines selected for Fig. 10.19, the published production is then compared with the power in the wind at the location for a typical year, obtained by wind profile scaling to the actual hub height (Sørensen, 2010a), based on 56 m data available on a fine grid (500 × 500 $m^2$) from an earlier project (Sørensen *et al.*, 2001; Energi & Miljødata, 1999). The actual hub heights were estimated from pictures or from similar turbines (such as Jespersen and Loeber, 1971, for the Dutch mill) in cases where they were not immediately available. Similar studies have previously been made for shorter periods (Sørensen, 1995; Lauritsen *et al.*, 1996; Sørensen, 2010a).

In the years of competition between fuel-based and wind electricity, a decisive factor was the cost framework adopted. Several of the electricity utilities were established with a co-operative structure, but also on a quite decentralized, local basis (Wistoft *et al.*, 2001–2002). This meant that electricity prices were different in different parts of the country, reflecting the actual production and distribution costs. On isolated islands, wind power was most competitive, because of the cost of transporting fuels to the islands, while in cities, where the electricity customers were densely located, the distribution costs were smaller and favoured fossil power, as did the usually poor wind conditions near cities. In the countryside, wind fared better, but because distances between customers in Denmark are still fairly small, several of the power companies, whether co-ops, municipally of privately owned, did not differentiate prices within their supply areas. The bottom line is that the declining prices for imported fossil fuels in most cases made wind the less favourable economic choice. It has been argued in other geographical settings, such as France (Gouvello and Matarasso, 1998), that the adoption of a solidarity-based pricing policy with similar prices for all customers of a given category (households or industry), whereby city customers actually subsidize rural customers, may cause unfair rejection of renewable energy solutions in less densely populated areas. This does not seem to have played a role in the phasing out of wind energy in Denmark during the 1920s, although some examples of price equalization can be found.

Danish Labour governments in the 1980s did not want wind power to become an investment object of the Californian type, or even that ordinary people could make a little money on wind energy, and therefore they curbed the popular co-operative ownership of turbines (the arrangement selected when wind turbines became too big for a single family) by requiring that the co-op shareholders lived in proximity to the wind turbine and that the shares owned could not exceed what corresponded to the family's own electricity consumption. By the 1990s, things became worse, as the government now ordered all local governments in Denmark to return a statement saying which areas of land they were willing to set aside for wind turbines. Naturally the replies were usually "none", because local governments do not want to tie up their future freedom of making decisions on land use, and in any case the requirement was irrational, since individual wind turbines take up only a tiny footprint of land area, and do not preclude a number of other uses of the land around them (Lauritsen and Svendsen, 1995).

The result was that individual establishment of wind turbines stopped, leaving expansion to the wind projects developed by the electric power utilities. When these were privatized about a decade later, further expansion of wind energy was made part of a "public service obligation", meaning that the government could dictate how much minimum share wind (or renewable energy generally) should have in the electricity mix. This led to the establishment of the first offshore plants, but the subsequent Liberal governments neglected any expansion of renewable energy production until the recent economic crisis, which spurred an employment-stimulating resurrection of limited offshore wind power plant construction. The reason that government intervention still appears to be required is a marginal price difference (particularly for offshore installations) relative to coal-based power, which would not be there if the operators had to pay the full externality cost of fossil fuel pollution and greenhouse gas emissions (Sørensen, 2010a; 2011).

## 10.1 Structural changes in the power industry

Selection of power technology has been affected by the recent liberalization of the electric power industry (Olsen *et al.*, 2000) and establishment of a Nordic auction arrangement for electricity trade (a "power pool"; Meibom *et al.*, 1999). The Danish deregulation of the electric utility sector has involved establishing a state-owned transmission company, arguing that competition is not meaningful for transmission, plus freely competing private power supply companies under no control beyond the mentioned "public service obligation". To make this competition meaningful, the net company has to allow any producer to offer power to customers at any location, implying a management of the transmission grid to avoid unfair preferences for, say, nearby suppliers, a task that may not be possible with a finite transmission capacity. Even if suitable expansion of the transmission grid is an obligation of the net company, it cannot be carried out instantaneously when the need arises. The largest problem of the selected liberalization model is that the electricity distribution has not been transferred to a publicly controlled entity similar to the net company. Evidently, the final distribution of power to the end users is a business where no competition can be established (short of having cables from ten companies extended to each building). As it is, the Danish government has left the distribution on private hands, which in practice means the electricity producers. They have responded by large increases in the price charged for distribution (where no competitors can get in) combined with lower prices for the power production, where full competition has been enforced.

The main power utilities have consequently been sold, chiefly to foreign companies or to the currently state-owned Danish natural gas company DONG. This company was established when the exploitation of North Sea resources became operational by the early 1980s, in order to pipe the gas to the customers from the companies (with Maersk as the largest Danish partner) given the license to extract oil and gas offshore. The government establishing DONG asked the Danish taxpayers to bear the additional expense from establishing the gas network, in return for the promise to deliver cheaper energy when DONG would break even around year 2000. Instead, current governments have declared that they want to sell DONG to foreign investors as soon as possible. The taxpayers will lose by this arrangement, because the sales price can never compensate for the benefits derived during the time remaining until the gas is exhausted. Bids to buy DONG will necessarily be lower in order leave the new owner a return on his investment.

North Sea oil and gas, available for less than 50 years, is a fantastic gift.

It could have allowed Denmark to establish a sustainable energy system without suffering or hardship, rather than delaying serious efforts until introducing new energy supply options become a necessity.

### 10.2 Changes in use of energy

Coal did not just win the early 20th century battle of electricity production over renewable energy, but also became the primary fuel for low-temperature heating and hot water in buildings for living, commerce and production. Only the transportation sector got in the hands of oil producers, with gasoline for passenger cars and diesel oil for lorries. The old wood-burning stoves placed in each important room of a building became replaced by a heat distribution system, called "central heating", in which pipes carried hot water from a central boiler (e.g. in the basement) to radiators placed in each room. The fuel was coke, due to its high energy density and less dusty nature compared with hard coal. It was delivered to a local storage area in each building basement by lorry, from where an amount was fed into the boiler daily by hand (Fig. 10.24). Not until the decade from 1950 to 1960 did the use of coke for heating become replaced by fuel oil, the declining price of which had finally made it competitive relative to coal. The transition went very fast, because the person responsible for shovelling the coal into the furnace was very happy to see a fuel that would flow into the burner by a simple device set into the boiler that did not even have to be replaced.

*Figure 10.24. Basement home boiler room with coke storage area and central heating boiler to be kept at adequate fuel feeding level (1955 photo by A. Andersen, permission purchased from Scanpix). The person in the picture, Hans Hedtoft, in 1947 became prime minister in Denmark.*

Energy for cooking continued to be supplied by town gas where available. Rural households used wood stoves (Fig. 10.40b) and hot plates integrated into heating stoves for hot water provision. More options for leisure were offered, including an increasing appreciation of the use of nature for recreational purposes. The beaches that surround large parts of Denmark had been appreciated by artists for their colour-enhancing light (actually a feature of medium- and low-angle solar radiation reflected from the sand), but they now became excursion targets for city people (thanks to the new railroad lines extended to the resort areas) who had learned about sun- and ocean-bathing from the similar movements in other parts of Europe, such as France (cf. Figs. 9.30 and 10.25).

*Figure 10.25. Bathing party at Gilleleje beach around 1905 (postcard at Gilleleje Museum; cf. Terslin, 1959).*

*Figure 10.26. The first ocean-going ship equipped with a diesel engine* (Selandia, *built in 1911 by Burmeister & Wain in Copenhagen; public domain photo from Wikipedia).*

Communication and transport facilities continued to be expanded. At sea, diesel ships took over from the simple steamboats, and Denmark developed an export industry based on shipyards and diesel engine manufacturers, notably Burmeister & Wain (Fig. 10.26). Train tracks continued to be expanded, but road vehicles gradually increased in importance. A few early ones were produced in Denmark (Fig. 10.27), but after 1900, practically all automobiles were imported. Their numbers rose, at first slowly but later with more vigour. One reason for the slow start was the rules passed by state and local governments, at first requiring a person to walk ahead of an automobile, waving a flag, and imposing in most places a speed limit of 15 km/h (in the city of Vejle even 7.5 km/h). The motorists formed their own organization, lobbying for relief from such rules (Ipsen, 1926). After World War II, due to the prevailing trade imbalance, imported car purchase was allowed by law only in special cases (such as for doctors or businesses with demonstrated need). In this situation, an attempt to produce automobiles in Denmark was again made, but coming into the 1950s, the balance of foreign trade improved and car importation wiped out the Danish producer Erik Sommer (FDM, 1984). The required expansion in asphalted road networks led to the formation of sizeable contractors specializing in road construction (an activity requiring substantial energy inputs).

*Figure 10.27. Danish automobile produced 1887 by Hammel (photo from Ipsen, 1926).*

Bridges also became a speciality of Danish consulting engineers and the construction industry, using the home market of the many Danish islands as a basis for establishing further business abroad. An early example is shown in Fig. 9.16; several more followed, and this type of activity is still going strong, with several 25 km bridge/tunnel projects (in 2000 the Great Belt road and railway bridge and tunnel, Fig. 10.30; a few years later a similar bridge and tunnel from Copenhagen to Sweden; and currently plans for a connection from the islands south of Sealand to Germany).

Denmark joined the pioneering efforts in aviation in 1906, when Jacob Ellehammer, as the first in Europe, flew 42 m at a height of 47 cm over the ground (Fig. 10.28), in an airplane with a 13.4 kW star engine, three years after the Wright brothers made their first flight in the USA.

Art and science flourished. Composer Carl Nielsen and songwriter H.

S. Pedersen wrote a Promethean cantata for the Polytechnical School for its 100-year celebration in 1929, not quite the masterpiece of Nielsen's 5th Symphony (1922). Niels Bohr collected atomic scientists around him at the Blegdamsvej Institute (Sørensen, 1989) and writers such as Jeppe Aakjær and Johannes Jensen (see Chapter 7) reached international fame.

*Figure 10.28. Jacob Ellehammer in 1906 makes the first European flight with an aircraft heavier than air on Lindholm island (Schieldrop, 1936).*

*Figure 10.29. Sterling charter plane at Copenhagen airport (photo by author).*

Regular aviation was commenced in Denmark by the company DDL in 1918, and in 1946 it was merged with the other Scandinavian airlines in the company *Scandinavian Airlines* (SAS). The relatively undamaged situation of Scandinavia after World War II made it possible for this company to achieve a leading position during the following years, with worldwide routes including the polar route via Greenland to Los Angeles, much shorter, and therefore faster, than the existing routes across the Atlantic Ocean via New York. *Scandinavian Airlines* focused on first class and business travel, and when the post-war surge in popular interest for vacationing in subtropical climates befell the Danes, this market was captured not by SAS but by charter airlines, notably *Sterling*, formed in 1962 by the travel arranger Eilif Kroager, who since 1950 had taken Danes on bus trips as far as to Spain (Fig. 10.29). Together with his competitor, *Spies Travel*, Kroager's *Tjæreborg Travel* flew millions of Danes to all-inclusive stays in southern European resort locations. These two large service-sector enterprises continued until the death of the founders, after which both companies found foreign owners. Another form of vacation becoming popular after the car had become more commonly owned was camping in tents or trailers ("campers") ranging in size from a few bunk beds to entire mobile homes.

*Figure 10.30. Tunnel-boring machines leaving tunnel entrance after having completed excavation at the 25 km Great Belt bridge/tunnel connection, 1990 (contractor: Monberg & Thorsen; photo by author).*

*Figure 10.31. In 1910, the electricity-driven vacuum cleaner produced by Fisker & Nielsen (Nilfisk) found many applications, such as cleaning taxi cars (picture at Energimuseet, Tange-Bjerringbro).*

The Danish production industry grew, interrupted by World War II, and reached its largest size around 1970, including both a large cement, iron and steel industry, forming the basis for construction work, and many smaller industries directed at consumer products, such as batteries, vacuum cleaners (Fig. 10.31) and other specialized, mostly electrical devices, many of which reached a large market for a limited period. They were based on good ideas (e.g. electric hair curlers for women) but did not last, as the small Danish manufacturers did not have the research tradition necessary for continuing to be a market leader once international competition set in. Danish earnings became increasingly based on small enterprises lasting only a short while, but being continually replaced by others, owing to the general education system (see Chapter 9) that rewarded creativity rather than volume of curriculum.

*Figure 10.32. Bicycles remained a favourite tool for commuting in cities, and in the Copenhagen Town Hall Square advanced light signals were installed to control the traffic of mainly cycles. (photo 1938 by Holger Sørensen).*

*Figure 10.33. The first stop signal had appeared in Copenhagen in 1920, but it did not operate on electric power (Ipsen, 1926).*

Despite electric trams and later subways constructed in Copenhagen and a few other cities, the bicycle continued to be the main vehicle of transportation for many people commuting to and from work (Fig. 10.32), as well as for weekend trips with the children to the forests, beaches and amusement facilities in the surroundings. There were also buses, but relatively few cars in the city, so traffic flowed fairly smoothly, helped by the electric light signals replacing manned signalling (Fig. 10.32 and 10.33).

## 10.3 From decentralization via centralization to decentralization

By 1930, the Copenhagen gasworks delivered 130 m$^3$ of town gas per inhabitant, or 2.2 GJ/cap., up from 0.4 in 1850 and 1.7 in 1900 (Edelberg *et al.*, 1932). The large number of power plants supplying electricity to the Danish customers in the early 20th century (Fig. 10.34) was gradually replaced by a smaller number of larger power plants, located in groups at very few locations (Figs. 10.35, 10.37). The expectation by the power utilities as late as the period 1970–1980 (e.g. Danish Power Utilities, 1974) was that this state of affairs would continue, probably with 5–10 nuclear power stations serving the entire country. They were wrong. A decade later, the power system consisted of several thousand decentralized combined heat and power (CHP) plants in the 1–50 MW size bracket, plus thousands of wind turbines, reducing the market for the large power plants quite dramatically (Fig. 10.36).

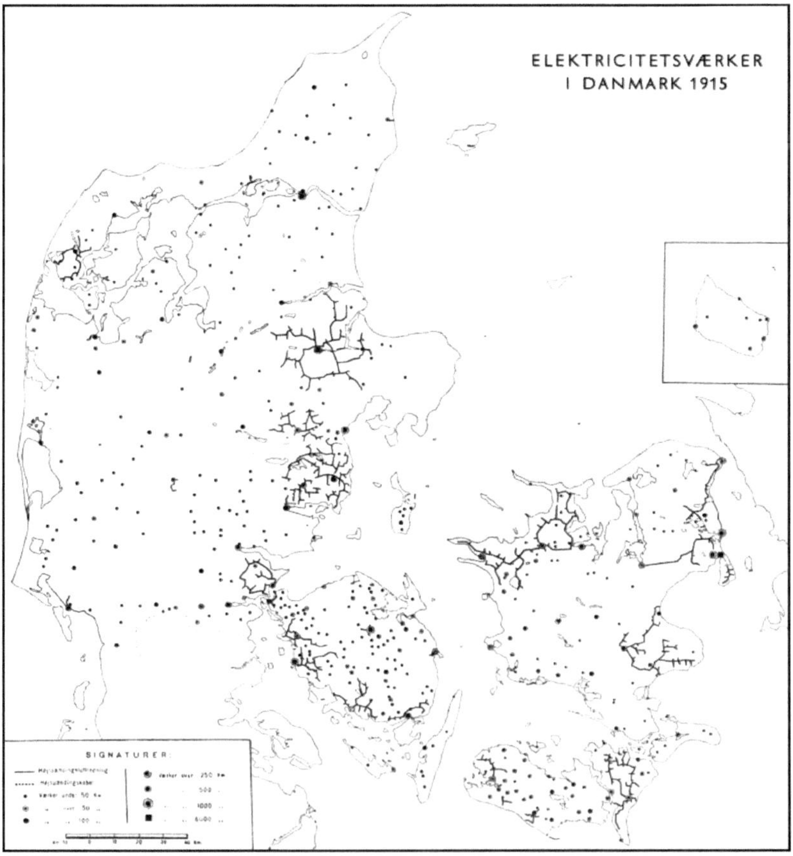


*Figure 10.34. Power stations, 50 kW to 5 MW, and transmission lines in Denmark, 1915 (Wistoft et al., 1991–1992; Energimuseet, Tange-Bjerringbro).*

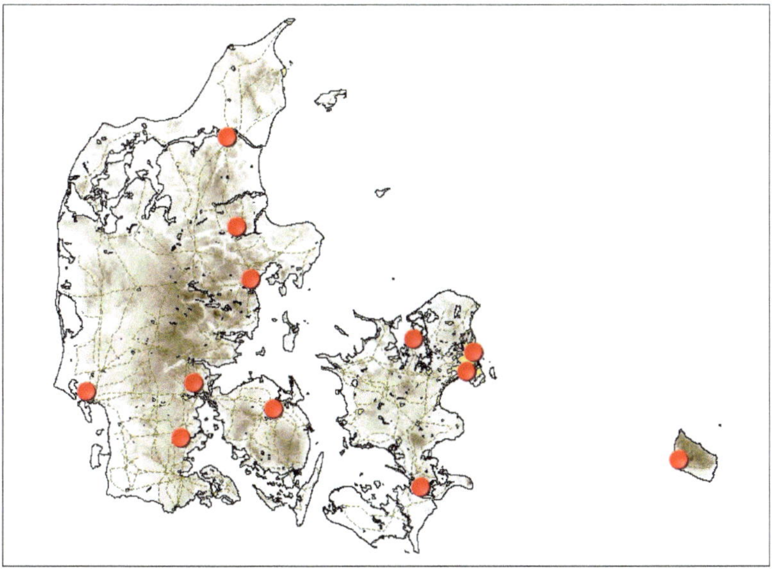

*Figure 10.35. Power plants, 1970 (based on Danish Power Utilities, 1974).*

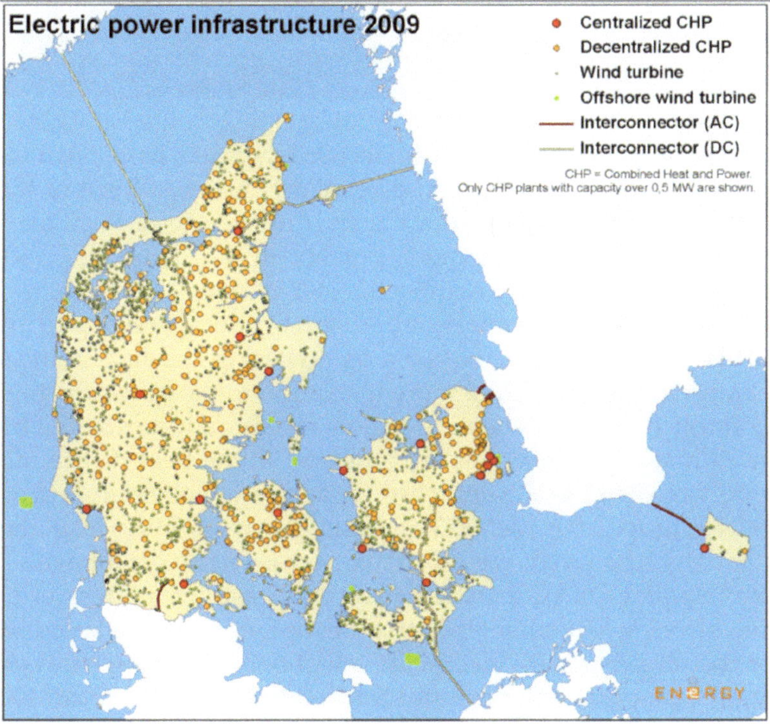


*Figure 10.36. Power stations and international network connections in Denmark, 2009 (courtesy of Danish Energy Agency).*

*Figure 10.37. H. C. Ørsted coal-fired power plant located in Copenhagen city (8mm film still by author, 1961).*

*Fig. 10.38. Hillerød municipal CHP plant in N Sealand, with natural gas store shown to the left (photo by author, 2011).*

The characteristics of the Danish electricity system have been strongly influenced by the widespread use of district heating. Most of the large central power plants were from the start delivering not just electricity, but also heat to local customers, notably in the dense demand areas such as Copenhagen city. During the last 40 years, it has been considered economical to transmit heat in district heating lines up to about 30 km. This has been made possible by lowering the heat transmission temperatures and by adding more insulation to the heat-carrying pipes. Many municipal district heating systems were built during the first half of the 20th century, implying that the cost of the heat distribution network 50 years later is a "sunk cost", apart from maintenance work. The lower heat usage after the 1973 oil crisis, where building insulation was substantially improved, would in many cases not have allowed district heating pipelines to be installed afresh (at least not in areas of existing houses – in new settlements, district heat pipes could be laid down at a lower cost together with electricity, telephone, water and sewage lines, during the preparation of the area for buildings). When natural gas was introduced in the 1980s, it became advantageous for the district heating plants to also produce power, so most pure heat plants were converted into combined power and heat plants (both abbreviations CHP and

CPH are in use). Because direct air pollution from natural gas is modest, this was considered acceptable even for plants located in city centres. An example of a decentralized CHP plant is shown in Fig. 10.38, and one of a recent large centralized plant in Fig. 10.39.

*Figure 10.39a, b. Avedøre power plant in south Copenhagen (top), with turbine hall (bottom). In addition to the highly efficient coal-fired unit, a unit for natural gas and fuelwood is available (photos by author, 1998).*

*Figure 10.40. Jutland farmer Niels Laustsen poses on his new tractor (a: upper panel). The new mechanized farming equipment often had a higher priority than installing an electric range in the kitchen (b: lower panel, from a north Sealand farmhouse). Eventually, rural kitchens did get either electric ranges or ranges using bottled gas. Both photos were taken by the author in 1958.*

## 10.4 Role of the traditional agricultural sector in Danish economy

Despite industrial growth, agriculture and food industries for a long time continued to be the largest profit-earners and the largest exporting businesses. The introduction of chemical fertilizers and pesticides during the 20th century, along with machinery to administer it and mechanize weeding, sowing and harvesting processes (Fig. 10.40a, 10.41), made the yield per unit of agricultural area go up. Much of the hay and grain harvest was fed to livestock animals, along with imported high-protein fodder, and despite the higher losses associated with producing animal-based food relative to vegetarian food, the resulting quantities allowed an increasing export of primarily meat and dairy products.

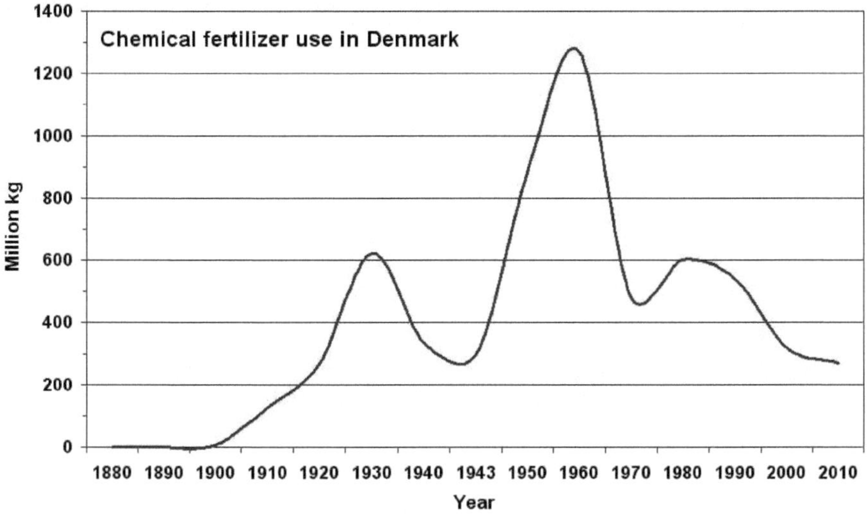


*Figure 10.41. The use of chemical fertilizers in Danish agriculture (based on Henriksen and Ølgaard, 1960; Danmarks Statistik, 1993; Olsen and Vinther, 2008). Use of natural fertilizers (manure and residues) was estimated as 350 million kg for 1970, which is probably its highest value. Recent concern over waterway pollution caused by overflow of fertilizers has sharpened the control of dosage, and the emerging ecological farms do not use chemical fertilizers.*

The agricultural production in Denmark is large compared with the food need of the indigenous population. Figure 10.42 gives the development of the output of some key products derived from agricultural activities over time. By 1880, cattle were the most important livestock animals and beef the largest production outcome, but from about 1925, pork took over. Today, there are 13 million pigs in Denmark, 1.5 million pieces of cattle and 15 million poultry animals (down from 22 million in year 2000), all in a country with 5.3 million people (Danmarks Statistik, 2010b). The production of beef, butter and eggs has declined in recent decades, while that of cheese and poultry is levelling off after a substantial growth during the 1980s. Increasingly loud protests over pollution and smell from an increasing number of pig farms have not yet curbed the growth, but there are signs that this may happen over the next decade.

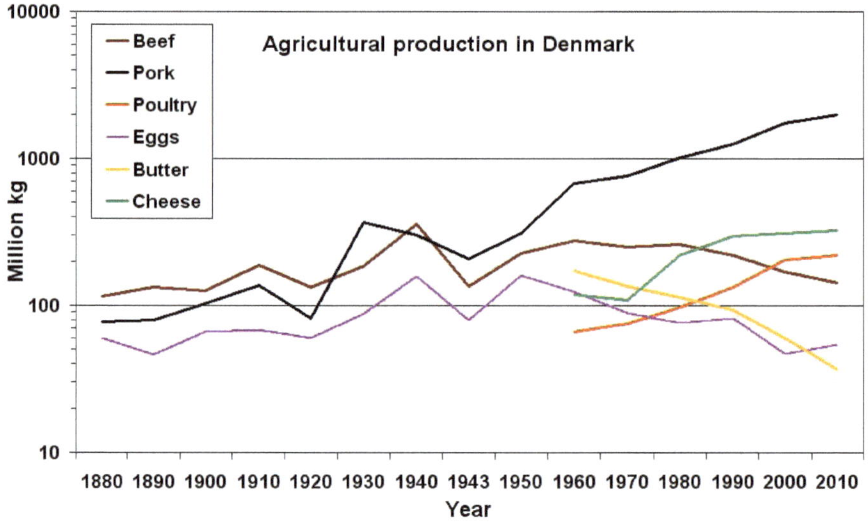


*Figure 10.42. Danish production of important commodities derived from livestock agriculture (based on Henriksen and Ølgaard, 1960, using prices from Hansen, 1983, to translate monetary values into physical amounts before 1960; from 1960, Danmarks Statistik, 1993, 2010b, was used). The beef, pork and poultry entries include production of meat as well as whole animals for sale.*

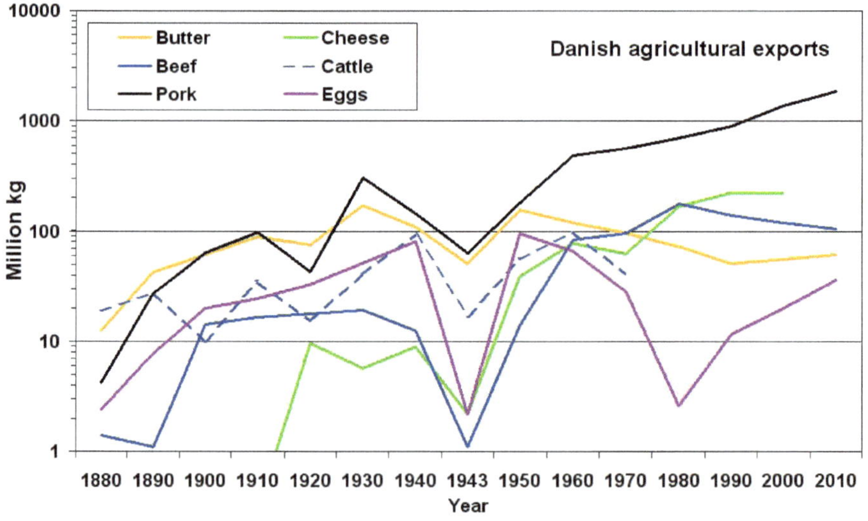


*Figure 10.43. Export of some important animal-based agricultural products (based on Hansen, 1983; Danmarks Statistik, 1993, 2010b).*

The primary vegetable-type production comprises cereals (grains, stalks and root residues), tubers, pulses, grasses and horticultural crops. Some of these, like grasses, are not necessarily harvested but grazed directly by livestock animals. Figure 10.44 gives imports and exports of grain. It is seen that except for the most recent 50 years, there is hardly any export. Instead, imports follow variations in Danish harvest yields as affected by climatic variations and epidemiology of plant diseases and insect attacks. It is also seen that additional import of seeds for next year's sowing is large in those periods where import is needed. The situation has changed dramatically in recent years, with a large surplus of grain produced for export (the smaller import persisting is primarily of plant types not grown in Denmark, such as rice).

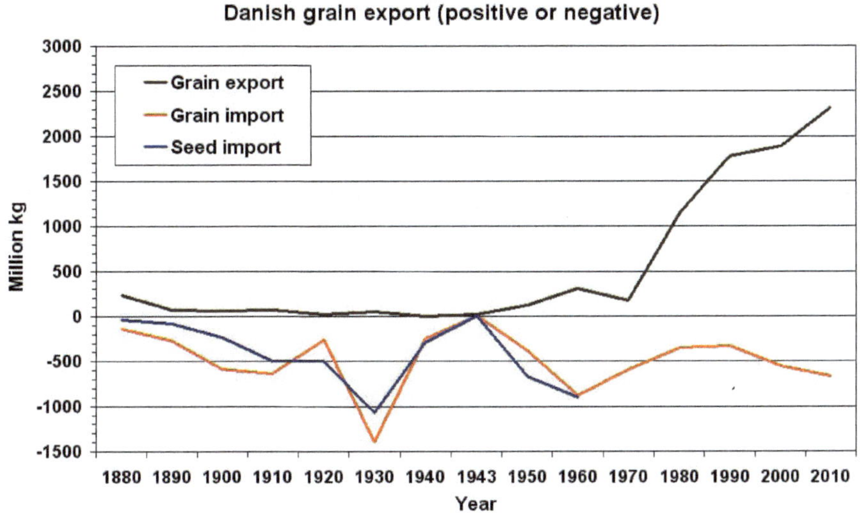


*Figure 10.44. Danish export and import (negative) of grains and seeds (wheat, rye, barley and oats; based on Hansen, 1983; Danmarks Statistik, 1993, 2010b).*

During the last decades of the 20th century, imported high-protein fodder has been replaced with winter or spring protein crops like lucerne, and some effort has been made to reduce the use of pesticides and replace chemical fertilizers with recycled straw and manure from stables, or from biogas plants, particularly in ecological farms. Unfortunately, the widespread use of straw in power plants, starting around 1990, has reduced the recycling of nutrients in plant residue material, but not as much as one might think, because before the practice was banned in about 1980, farmers often burned surplus straw (after setting some aside for stable usage) and residues in open fires on the fields.

Figures 10.43 and 10.44 show that agriculture is a very important ingredient in the foreign trade balance of the Danish economy, despite the small number of people in recent years who work directly on farms (see Fig. 9.8, including forestry and fisheries). For the country itself, provision of food for the indigenous population is a further reason for the key importance of agriculture. One might expect that these observations would be reflected in the national accounts of the economy.

Figure 10.45 shows the development in economic activity, measured by the gross national product (GNP), calculated as the sum of all monetary compensations for activities in the society. This means the *added value* that characterizes each step in handling a commodity or delivering a service, as long as the transaction involves exchange of money. Much domestic work is excluded from the GNP, as are voluntary activities and a good deal of the activities done for pleasure or recreation. Danish national accounts give various versions of GNP-related measures, depending on whether subsidies, taxes and payments to foreign wage-earners are included. Because subsidies from Denmark and from the European Union to farmers have been substantial during recent decades, inclusion of these can change the activity measure to the tune of some 30%.

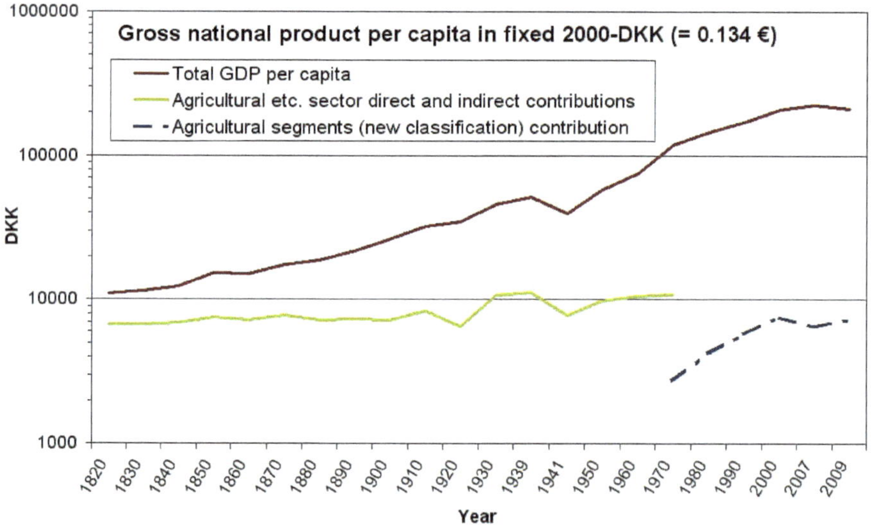


*Figure 10.45. Development in Danish per capita gross national product in year 2000 prices and the agricultural sector contributions to it (with subsidies added and taxes subtracted where relevant; based on Hansen, 1983; Danmarks Statistik, 2010b). See text for discussion of the discontinuous agricultural sector definition.*

The GNP indicator of overall economic growth exhibited in Fig. 10.45

shows a tenfold increase over the first 150 years, except for an extra-high growth rate in 1950–1970, making up for the World War II decline, but after year 2000 a flattening out (or even a decline following the 2008 worldwide financial crisis). The income earned by the agricultural sector has not increased until around 1970, except for the wartime opportunities. The number of people employed in the Danish agricultural sector was also constant until 1970 (Fig. 9.8), but then fell dramatically. The precise behaviour is difficult to estimate, due to several changes occurring in the way in which Statistics Denmark has defined the agricultural sector, prompted partly by European Union harmonization efforts and partly by transition to use of economic input–output models with increasing numbers of compartments describing the different segments of what initially was lumped together as the agricultural sector, not only including forestry, horticulture and fisheries, but also draft animals, farm machinery, and the food processing industry and trading enterprises. Many of these tasks were initially performed on the farms, but with time became separated as independent businesses, with requirements for transporting agricultural products between them and trading companies for taking care of selling the goods on suitable markets. The current economic model segmentation has several hundred segments of activity linking to agriculture, but difficult to separate, because they also do business not related to agriculture (e.g. transporting and trading other types of goods). In Fig. 10.45, only direct activities in agriculture, horticulture, forestry, dairies and slaughterhouses have been included after 1970, whereas the numbers before 1970 supposedly include all direct and indirect contributions to the agricultural sector as defined above.

The use of fixed year 2000 prices is partly necessary to dispose of general inflationary effects over the long time span, but also serves to flatten out short-term fluctuations in the prices of agricultural goods. Because such goods are characterized by a flex-price mechanism, variations from year to year can be quite substantial, such as for the latest five years, where increasing prices have been replaced by substantial price declines for the most important Danish agricultural commodity, which is pork. The decline in running-price added values is thus not seen in the graph using fixed year 2000 prices.

An implication of the above is that it is not possible to assess the development in the importance of agriculture in the Danish economy by looking at its Fig. 10.45 contribution to GNP, currently appearing to be only about 2.5%. One might suggest using an input–output model to calculate total activity levels, expressed as GNP, with and without the agricultural activity included, in order to expose the contributions hidden in other segments of the economy. However, this brings in new caveats. Most economic models would try to substitute the missing food commodity by imports, and at current prices, imported food would appear no more expensive than food produced in Denmark (which is the reason, in Denmark and in the EU, that agriculture is subsidized). But this type of argument could be applied to most nations currently producing agricultural goods, and if several of them totally abandoned agriculture, advised by the economic models telling them that it would make no difference, then surely food prices would rise dramatically. Therefore, the model exercise should include a price model capable of predicting the food price changes implied by traditional food-producing countries giving up that production. Unfortunately, prices are treated exogenously by most economic models, and substitution of local food production is anyway not a straightforward process. This digression may be taken as an example of the limitations of current economic modelling. Fortunately, economists are rather intelligent people, who have refrained from giving decision-makers the advice to close down agricultural activities by arguing how little they contribute to overall GNP, while occupying large areas of land valuable for other purposes (due to the low efficiency of photosynthesis). Advice to remove subsidies has appeared, but the effect would be somewhat similar, by first moving food production to countries with low prices but later to be forced to accept higher prices, especially if local production has stopped and the land been diverted to other uses (this happened recently to Mexico, after having been offered US surplus corn for some years at dumping prices, until the US farmers found other options such as biofuels giving them better prices).

To these arguments comes the strategic point of view that self-sufficiency in food is valuable in a world that has seen international trade-stopping wars in every century. Similar arguments may be brought into play regarding the energy balance of agricultural activities, where machinery for land treatment (tilling, weeding, sowing, harvesting, etc.), production of chemical fertilizers and pesticides, transport and processing require much energy in comparison with the energy we derive from eating the food (Ansbæk *et al.*, 1973; Sørensen, *et al.*, 1994). While economic theory might tell us that higher profits can be derived by investment in activities other than agriculture, most people disregard the advice and continue to eat food. However, this should not rule out a discussion of better ways to organize agricultural production, with smaller energy inputs from non-solar energy and less negative impact on the natural environment.

One effort in this direction is the introduction of ecological agriculture (sometimes called organic agriculture), where the farmer cannot use pesticides or chemical fertilizers and the food processor cannot use additives to replace taste or colour lost when substituting inferior raw materials. Currently, about 5% of Danish farming land is cultivated and certified according to ecological norms, and ecological products constitute some 5% of Danish food purchases (in 2009 1% of cheese, 4–5% of meat, 9% of milk and 15% of eggs, according to Danmarks Statistik, 2009). Most Danish supermarkets carry a number of imported ecological products, such as fruit and vegetables.

## 10.5 Social developments and a changing society, interrupted by war

The early 20th century was also the period when women began to be restored to the status in society that they had lost when the Christian religion was introduced over 1000 years earlier. There had been individual voices expressing the dissatisfaction over the discriminating treatment of women, with the writer Mathilde Fibiger one of the first to emphasize women's right to develop their talents (Fibiger, 1851). Other feminist precursors were not exactly promoting liberalization. The strongly conservative and religious Natalie Zahle had in 1852 established a school for women, emphasizing education as a prerequisite for any human development. The noblewoman Astrid Stampe urged her friend, female magazine *Woman & Society* editor Elisabeth Grundtvig, to speak at an 1887 meeting for women only, about sexual morality based on monogamy, chastity before marriage and no prostitution. However, not all members of the movement wanted women's liberation to be based on religion.

Around 1900, questions regarding the status of women were incorporated into the drive to update the Danish constitution, partly to obtain a true parliamentarian system and partly to secure basic human rights. An obvious part of this was to extend the right to vote to both sexes. Equal access to inheritance was given to women already in the 19th century, but elections in 1901 and 1910 had shown the inadequacy of election laws, giving the growing city populations a gross underrepresentation (Dybdahl, 1975). The new constitution was passed in 1915. It included parliamentary voting for women (having already local voting rights since 1906) and a combination of proportional representation and constituency election. The Women's Movement, formed in 1871 with partial roots in the Folk High School circles, was keen to celebrate the victory (Fig. 10.46). The following years saw several further steps (equal salaries in and equal access to public sector jobs, 1919–21, and improved legal status in marriage, 1922). Salary discrimination continued in the private sector.

*Figure 10.46. The Society of Danish Women celebrates the 1915 constitution at the king's residence in Copenhagen (Dybdahl, 1975).*

Several male writers and politicians had supported the women's rights movement, including Fredrik Bajer, who in 1887 first proposed to let unmarried women vote in local elections (the proposal was passed in the Lower House, but was stopped in the Upper House, for fear that this might lead to all women obtaining general suffrage). The leading early 20th century Danish historian, Johannes Steenstrup (1913, 1917, 1930) provided an accurate history of the discrimination against women in earlier periods. Gyrithe Lembke (1918) wrote in *Woman & Society* that the goal of the movement should not be simple equality but winning the right to develop female qualities just as the men had developed theirs. It is worth mentioning that along with women's rights, the movement worked hard on questions of peace, in consonance with other Scandinavian women's organizations at the time of World War I.

Acceptance of cabinet responsibility (that governments should always be backed by a parliamentary majority) paved the way for replacing Conservative and Liberal Danish governments first by Centrists (1909) and with the new election rules also by governments led by the urban-dominated Labour party (based on a labour union formed 1871 by Louis Pio, modelled after French and English socialist movements; Pio was later bribed, allegedly by industry or by the Danish police, to emigrate to the USA; Kühle, 1993). Thorvald Stauning (Fig. 10.47) of the Social Democratic Party formed the first Labour government in 1924, which over the following decades would become the vehicle for development of the Danish welfare concept (free education, free healthcare, social compensations for unemployed, retired or disabled persons, all paid for by redistribution of collected taxes, plus insurance-type associations based on fees that in many cases later were made compulsory).

*Figure 10.47, Thorvald Stauning, Danish prime minister 1924–26 and 1929–40 (photo from 1930s; Nørlund, 1955).*

The international economic situation after the 1929 US stock exchange crash was favourable to such social reforms, or at least the welfare model was one way of responding to the problem. Other countries like Germany used military build-up as its response to unemployment. A strong critic of the Nazi activities in Germany, and instrumental in making the first steps to carry out the social reforms in Denmark, was Karl Steincke (1938), a thoroughly liked (but for reasons of perceived rough edges also often disliked) minister of social affairs 1929–35 and minister of justice during several periods from 1924 to 1950, but kicked out in 1939 to please the German neighbours and then occupiers.

Danish neutrality during World War I (1914–1918) had allowed increasing exports to countries diverting their manpower to the battlefields. Included were some thrifty operators making money on selling meat waste and animal fodder in cans as human food (Lindeberg, 1966). After the war, these canned food companies went bankrupt and in 1922 brought a large Danish bank down with them.

During World War II (1939–1945), Denmark was from 1940 occupied by German forces, primarily to prevent allied forces from landing on Danish coasts and to control the Danish straits. The Danish government (Stauning) for a while embarked on a policy of measured collaboration, in order to avoid bloodshed. Danish industry and agriculture went along with (and probably had encouraged) this policy, but the growing Danish resistance movement recruited many of its members from industry (engineers are a good choice for solving the technical problems of sabotage) and much of the rest from members of the small Communist Party, which also seemed to have some knowledge of guerrilla warfare. Danish farmers (and fishermen) played a double role, on one side providing the Germans with the food imports they badly needed, and on the other hand often helping the resistance movement and harbouring individuals on the run from Gestapo agents (e.g. the many Jewish people fleeing in 1943). It is estimated that the absence of German men from farms in Germany reduced the harvest yield by at least a third, and that this deficit was made up for by food imports primarily from occupied countries such as France (over 40%) and Denmark (near 10%; Trommer, 2009; Videnskab.dk, 2011). As Fig. 10.42 and 10.43 show, the Danish agricultural production and export (during the war restricted to Germany) was still lower than before and after the war, where many more markets were open. The decline in production can be explained by loss of possibilities for importing chemical fertilizers and lack of fuel for agricultural machinery during the war. The occupying German forces treated farmers and fishermen very cordially, aware of their interest in keeping the production up, and until the breakdown of the collaboration policy in 1943, a blind eye was turned to the assistance to the Danish resistance movement (e.g. in transporting resistance fighters wanted by the Germans by boat to neutral Sweden).

However, the number of conflicts with the German occupiers increased, and on 29th August 1943, the German "regent" Werner Best, who had administered the relatively friendly treatment of Denmark (as compared with other countries occupied by Germany), was over-ruled by Hitler, and had to dismiss the Danish government and declare martial law in Denmark. His anti-racist adviser Georg Duckwitz managed to warn the resistance movement (through two Danish politicians, one of whom was Hans Hedtoft; cf. Fig. 10.24) that all Danish Jews would be captured in a campaign starting on the night between 1st and 2nd October and sent to German concentration camps. However, the resistance movement managed during the two preceding days (with the help of Copenhagen taxi drivers) to move nearly all to coastal locations, where they were hidden in attics, summer houses and farm barns, until boat transportation to Sweden could be arranged (Tortzen, 1970).

Some 7200 Jews were safely brought to Sweden, and despite the short time available, very few operations went wrong. One was the capture of 110 Jews hiding in the attic of Gilleleje church (Fig. 10.48). Their attempt to embark on fishing boats in Gilleleje harbour had been spotted by a Gestapo team sent to assist the too-complaisant German military officer placed there, and the Jews were supposed to stay temporarily in the church, until an alternative route could been established, using rowing boats to sail the refugees out to the fishing boats at sea from a sheltered location called the Beach Hills east of the city. A young engineering student, Jean Fischer (the same that later headed the Wind Energy Committee, 1975, discussed earlier in this chapter), was sent by the resistance movement to the church to warn the Jews, but in the hurry he had not been told the password, and the Jews refused to let him in, so he could only watch when the Germans forced their way into the church and took the Jews away. Half of the captured people were not 100% Jewish and they were subsequently released. Several hundred more Jews hiding that night in other attics near the church escaped safely to Sweden.

*Figure 10.48. Gilleleje church, site of a 1943 Jewish refugee capture (photo by author).*

However, unlike their co-religionists from other occupied countries, even most of these captured Jews survived their detention in German concentration camps, presumably again due to the German interest in keeping tolerable relations with Danish food suppliers (3 from the Gilleleje church incident and in total 54 out of 472 Danish Jews, most of them captured before managing to get out of Copenhagen, died in German concentration camps; Tortzen, 1970). The story of the events in Gilleleje church has recently been retold by Martin (2000), because a lock of Beethoven's hair, passed on through generations from the person cutting it off at Beethoven's deathbed, was given as thanks by one of the Jewish refugees to the family that sheltered him in Gilleleje and then later sold to two Americans, who were able by X-ray analysis of the hair to suggest that Beethoven had died from lead poisoning.

Many commodities that had earlier been imported became scarce during the war. Energy supply was hit hard, and both coal and oil supply suffered badly. Coal was partially replaced by indigenous sources; Denmark does have some deposits of brown coal and peat, but this was at the expense of high pollution from these inferior sources, including corrosion that destroyed many boilers over the following years. Oil products for the transportation sector were more difficult to find a substitution for. Lorries for freight transportation, taxis and doctor's cars were equipped at considerable expense with Lurgi-type gasifiers, converting mostly fuelwood to generator gas fed into an attachment in front of the combustion engine (Fig. 10.49).

*Figure 10.49. Lorry equipped with wartime gasifier, 1943 (Ibsen, 1951).*

*Figure 10.50. Public school class picture, 1951. Nearly everyone is dressed in domestic knitting or creative remakes of the grandparent's clothes (4th grade at Katrinedals School, Copenhagen; the author is fourth from right in top row).*

Living conditions reached their lowest level not during the war but in the years after 1945. As mentioned, food exports to Germany had partly compensated for the lost markets in England and elsewhere during the war. Afterwards, most European countries were so devastated that imports were ruled out, at least until indigenous industries had been re-established. For this reason, Danes continued to exercise the wartime virtues of repeated reuse and remake of old clothes (Fig. 10.50). For a while, reuse and recycling became large industries.

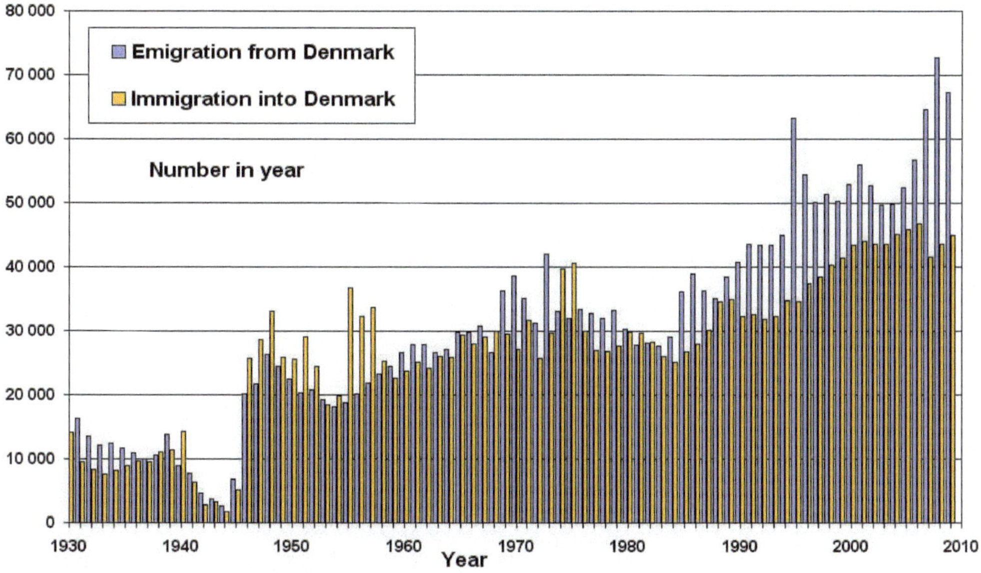


*Figure 10.51. Emigration from and immigration into Denmark, 1930–2009 (continuation of Fig. 9.38; Danmarks Statistik, 2010b).*

However, it did not take many years before industrial production and export had recovered and entered a period of unprecedented growth, recovering not only what was lost but attaining exponential growth at a rate higher than at any previous time (see Fig. 10.45). The quiet first years after World War II lead to an increase in both emigration and immigration. Some Danes went to the USA for better opportunities, but slightly more Europeans came to Denmark to improve their situation. Fig. 10.51 continues the presentation of Fig. 9.38. It is interesting that the increased rate of immigration started before the high growth period 1950–1970, where Denmark made up for insufficient labour by inviting "guest workers", notably from Turkey and the Middle East. This is similar to what happened in other European countries with growth, in particular Germany. The UK chose a different model based on a favourable citizenship offer to immigrants from former colonies, open for a limited period. It is similarly remarkable that Danish immigration continued to rise through the recession of the late 1970s and until about 2006. However, emigration also increased, and after 1985, the number of emigrants has considerably surpassed the number of immigrants, though both have been increasing. The current Danish government is trying to raise the retirement age (as are other European governments), claiming that otherwise jobs in the future will stand empty. Figure 10.51 suggests that creating conditions where fewer choose to emigrate away from the country may be a better idea.

*Figure 10.52. Lamp designed by Poul Henningsen, who in 1924 started to design lamps with optimized distribution of light from an incandescent source (photo by author).*

*Figure 10.53. Chairs designed by the furniture architect Hans Wegner in the 1950s (photo by author).*

The multi-talented Poul Henningsen–known among many other creations for the wartime musical song text "They tie up our mouth and hand", miraculously passing the German censorship, and for his lamps (one example in Fig. 10.52), movies and architecture–had thoroughly criticised the cultural policy accompanying the Labour (Social Democrat) government's social reforms (Henningsen, 1933). After World War II, the very conventional labour outlook on culture became increasingly challenged by artists of great talent. Some, like architect Jørn Utzon, had to work in other parts of the world in order to realize their most ambitious projects (construction of Sydney Opera House, 1957–73), as had the filmmaker Carl Drejer earlier (Jeanne d'Arc, Paris, 1928), but, increasingly, Denmark could offer stages for exceptional talents. This was the case for the Royal Danish Theatre ballet corps in Copenhagen, with world-class dancers and choreographers such as Harald Lander, Flemming and Vivi Flint, and Henning Kronstam. Unfortunately, their finest performances (e.g. in Béla Bartók's *Miraculous Mandarin*) have not been preserved on film for the benefit of posterity, because the stage workers demanded unacceptable extra pay to allow recording: a quite typical effect of the union-based work environment at the theatre.

Labour unions had been instrumental in the early battle for decent salaries and work conditions, and later played an important role in promoting Labour views in government. However, as the number of large factory-type industries have diminished in Denmark, the role of the unions has become less clear. The inhomogeneous variety of smaller businesses has requirements not easily resolved by centralized negotiations. For a long time, unions focused on salaries, and paid little attention to health and environment. In the late 20th century, the largest employer has been the state (see Fig. 9.8), often a very harsh employer, and union negotiations and strikes took a rather theatrical form, as outcomes were often over-ruled by parliamentary prescription.

Danish furniture production benefited from a number of very gifted designers, such as Hans Wegner (Fig. 10.53), and the Danish pharmaceutical industry was growing very fast after World War II, with Novo (Fig. 10.54) as the flagship. This science-based company, becoming a world leader in producing insulin for diabetes patients, benefited from the fact that more natural science students were graduating from Danish universities during the 1960s than ever before (or since).

*Figure 10.54. Safety exit from Novo production plant, in the 1960s located in the middle of Copenhagen (8mm film still by author, 1961).*

In 1953, the constitution was changed to allow general referenda as a way to settle particular questions. Basic human rights concerning political and religious freedom became emphasized, and it was decided that only a qualified majority would allow a government to cede parts of its power to international organizations. A prime reason for wanting the constitution changed at this particular time was the opportunity to allow female regents, spurred by the inter-marriage degeneracy of the Danish royal family, where only the Princess Margrethe (soon to become Queen Margrete II and marry a non-royal, insisting that her children did the same) was seen as having a normal intelligence level.

One flaw that did not receive any remedy in the 1953 constitution was that Denmark still had the teaching of the Protestant Christian Church as a state religion, a compulsory religion for the regent and accorded its own ministry in governments. Despite constitutional "freedom of religion", this particular religious institution was still part of the school curriculum and had dedicated time set aside for it in radio and television (when that became common during the 1950s, see Fig. 10.55). It was possible to opt out from the religious school education (after application by the parents), but it was still some years before "Christianity" on the school class schedule was replaced by a general survey of religions.

*Figure 10.55. At a 1950 exhibition, Bang and Olufsen presented one of the first Danish television receivers aimed at commercialisation (photo Bang & Olufsen Mediacenter). Danish amateurs had constructed television receivers since 1930 and watched German television.*

Two examples may serve to illustrate these issues. The vicar at Blovstrød church, Harald Søbye, considered his pulpit an appropriate forum for political debate (similar to the historic Jesus and the Friday practice of Islamic imams). He supported the move to allow female vicars, he was fined for speaking against apartheid in South Africa, and he expressed negative views on nuclear arms. In 1962 he was accused by the Danish military of advocating pacifism to his confirmation candidates, and despite overwhelming support from his congregation, the Minister for the Church discharged him in 1964 (Wikipedia, 2010).

The other example is Inger Nordentoft, head of the Katrinedals Public School in Copenhagen (cf. Fig. 10.50) from 1945–1960. She had in 1946 given birth to a child that (in the jargon of the time) was "illegitimate", and contrary to many others she did not hide it, but was happy and fully convinced that she could bring it up a single parent. She told this to the public school director, who responded that her attitude was "damaging to society". Newspapers covered the story intensely, and when a majority in the local school board expressed full satisfaction with her work, the outcome was that a second school had to be built in the district, so that parents had a choice (Nørgård, 2010). Inger Nordentoft represented the Communist Party in parliament and had the school conduct experiments in teaching, introducing project and group work (as in the Montessori schools mentioned in Section 9.2).

After recovering from the war, increased affluence also led to a more relaxed relationship between parents and children, which paved the way for a fairly broad acceptance of the formation of specific youth subcultures. Then as now, music played an important role in defining the singularity of these cultures, making sure that overlap with the preferences of the parents was as small as possible. For a while, jazz music played this role (Fig. 10.56), but it was soon replaced by rock music, termed "barbwire music" by parents noticing the disharmonies emitted from the electronics of early guitar amplifiers. So Fred Astaire had long been succeeded by John Coltrane, who had to cede the way for the Beatles and Jethro Tull, and later for Kraftwerk and Depeche Mode. But because there is a certain degree of cyclic behaviour not only in economics but also in musical preferences, the current business-oriented liberal yuppies have largely returned to the soft entertainment style of historical popular music, readying the scene for the next generation of protest music (if not protest songs).

*Figure 10.56. Inside the Copenhagen jazz joint* Laurits Betjent *around 1960. The extreme concentration of tobacco smoke found on these premises, notably from pipes and cigarettes, may induce surprise in an observer from the current epoch (photo by author).*

*Table 10.1. Energy use per capita in Denmark in 1948, for an adult person (with use of Danmarks Statistik, 1959, 1967, 2010b; cf. Table 9.2).*

| *Denmark, 1948* | *Human energy* | *Energy from fuels and external energy flows* | | | | | | |
|---|---|---|---|---|---|---|---|---|
| *Rate of energy use (average W for an adult person)* | **Food energy** | **Low-temperature heat** | **Medium-temperature heat** | **High-temperature heat** | **Mobile power (excl. food)** | **Mechanical power (stationary)** | **Electric power (stationary)** | **TOTAL** |
| **Liveable surroundings** | | | | | | | | **432** |
| Basic human metabolism | 70 | | | | | | | **70** |
| Indirect energy (e.g.through windows) | | 20 | | | | | | **20** |
| House and garden tools, equipment | 10 | 1 | | | | 1 | | **12** |
| Active space heating | | 315 | | | | | 0 | **315** |
| Artificial light | | | | 1.5 | | | 13 | **15** |
| Active air exchange, air cooling | | | | | | | 0 | **0** |
| **Food and water provision** | | | | | | | | **263** |
| Hunting, gathering, gardening | 1 | 2 | | | | | 0 | **3** |
| Fishing, aquaculture | 4 | | | | 4 | | | **8** |
| Agriculture | 12 | 2 | | | 20 | | 8 | **42** |
| Slaughtering, food processing | 10 | 5 | 2 | | | 2 | 3 | **22** |
| Preparing, cooking, kitchen utensils | 2 | | 35 | | | | 1 | **38** |
| Inhouse hot and cold water provision | | 150 | | | | | 0 | **150** |
| **Security, assault, national image nursing** | | | | | | | | **32** |
| Weapons, delivery vehicles, etc. | 2 | | | | 5 | | 0 | **7** |
| Government and legal institutions | 1 | 8 | | | | | 1 | **10** |
| Administration and police | 1 | 8 | | | | | 1 | **10** |
| Sports and public information facilities | 1 | 4 | | | | | 0 | **5** |
| **Health-related efforts** | | | | | | | | **32** |
| Food preservation,clothes washing,etc. | 5 | 10 | | | | | 1 | **16** |
| Water and waste management | 1 | 1 | | | | | 0 | **2** |
| Doctors, medicine | 1 | 5 | | | | | 0 | **6** |
| Hospitals | 1 | 5 | | | | | 2 | **8** |
| **Relations and leisure** | | | | | | | | **47** |
| Music instruments, concert halls, etc. | 0.3 | 2 | | | | | 0 | **2** |
| Theater, dance, TV broadcast, etc. | 0.5 | 2 | | | | | 1 | **4** |
| Toys, furniture, arts and crafts | 2 | | | | | | 0 | **2** |
| Written material, games, software etc. | | | | | | | | **0** |
| Computers,multimedia,communication | 0 | | | | | | 1 | **1** |
| Vacations, social transportation | 8 | 20 | | | 10 | | 0 | **38** |
| **Occupational and learning activities** | | | | | | | | **696** |
| Schools, further education, research | 2 | 40 | | | | | 1 | **43** |
| Construction and building | 8 | 2 | | | 2 | 20 | 0 | **32** |
| Manufacture | 10 | 160 | 35 | 25 | 1 | 40 | 8 | **279** |
| Trade, private services (excl. transp.) | 10 | 80 | | | | 12 | 4 | **106** |
| Transportation of goods (incl. Internat,) | 8 | | | | 104 | | | **112** |
| Transport(delivery,services,commuting) | 4 | | | | 120 | | | **124** |
| **Raw materials procurement** | | | | | | | | **377** |
| Mining and minerals extraction | 4 | | | | 8 | 18 | | **30** |
| Forestry resources | 2 | | | | 20 | | 0 | **22** |
| Industrial raw or recycled materials | 10 | 5 | 25 | 30 | 5 | 10 | | **85** |
| Energy provision | 1 | 10 | | | 10 | 5 | 0 | **26** |
| Energy conversion/distribution losses | 2 | 212 | | | | | | **214** |
| TOTAL | **194** | **1069** | **97** | **57** | **309** | **108** | **45** | **1878** |

Table 10.1 shows the per capita energy use in 1948. Compared with year 1900 (Table 9.2), the new role of electricity stands out, as does energy use in transportation and low-temperature heating, both about doubled. The total energy use is some 50% higher than half a century earlier, which is not unexpected. Most of the increase is in the production sector, plus of course the new type of energy loss characterizing electric power plants. This obviously does not exclude that the conver-

sion efficiency may be much higher than in 1900, and that the energy demanding activities therefore may have increased more than the totals suggest. This becomes more clear when looking at the end-use energy expenditures given in Table 10.2. The industrial activity has indeed risen by some 80%, and in the transportation sector the efficiency has risen quite substantially (e.g. in progressing from horse- and oxen-driven vehicles to motorized road vehicles).

*Table 10.2. Estimated end-use energy per capita in Denmark in 1948, for an adult person (with use of Danmarks Statistik, 1959, 1967, 2010).*

| *Denmark, 1948* | *Human energy* | *Energy from fuels and external energy flows* | | | | | | |
|---|---|---|---|---|---|---|---|---|
| *Rate of energy end-use (average W for an adult person)* | Food energy | Low-temperature heat | Medium-temperature heat | High-temperature heat | Mobile (excl. food) | Mech. power (stat.) | Electric power (stat.) | TOTAL |
| **Livable surroundings** | | | | | | | | **200** |
| Basic human metabolism | 68.0 | | | | | | | **68** |
| Indirect energy (e.g.through windows) | | 20.0 | | | | | | **20** |
| House and garden tools, equipment | 3.3 | 0.3 | | | | 0.5 | | **4** |
| Active space heating | | 105 | | | | | | **105** |
| Artificial light | | | | 0.3 | | | 2.7 | **3** |
| Active air exchange, air cooling | | | | | | | | **0** |
| **Food and water provision** | | | | | | | | **105** |
| Hunting, gathering, gardening | 0.3 | 0.7 | | | | | | **1** |
| Fishing, aquaculture | 1.3 | | | | 1.3 | | | **3** |
| Agriculture | 4.0 | 0.7 | | | 6.7 | | 7.0 | **18** |
| Slaughtering, food processing | 3.3 | 1.7 | 0.9 | | | 1.5 | 3.0 | **10** |
| Preparing, cooking, kitchen utensils | 0.7 | | 21.0 | | | | 1.0 | **23** |
| Inhouse hot and cold water provision | | 50.0 | | | | | | **50** |
| **Security, assault, national image nursing** | | | | | | | | **12** |
| Weapons, delivery vehicles, etc. | 0.7 | | | | 1.4 | | | **2** |
| Government and legal institutions | 0.3 | 2.7 | | | | | 1.0 | **4** |
| Administration and police | 0.3 | 2.7 | | | | | 1.0 | **4** |
| Sports and public information facilities | 0.3 | 1.3 | | | | | | **2** |
| **Health-related efforts** | | | | | | | | **13** |
| Food preservation,clothes washing,etc. | 1.7 | 4.5 | | | | | 1.0 | **7** |
| Water and waste management | 0.3 | 0.3 | | | | | | **1** |
| Doctors, medicine | 0.3 | 1.3 | | | | | | **2** |
| Hospitals | 0.3 | 1.3 | | | | | 2.0 | **4** |
| **Relations and leisure** | | | | | | | | **17** |
| Music instruments, concert halls, etc. | 0.1 | 0.7 | | | | | | **1** |
| Theater, dance, TV broadcast, etc. | 0.2 | 0.7 | | | | | 1.0 | **2** |
| Toys, furniture, arts and crafts | 0.7 | | | | | | | **1** |
| Written material, games, software etc. | | | | | | | | **0** |
| Computers,multimedia,communication | 0.0 | | | | | | 1.0 | **1** |
| Vacations, social transportation | 2.7 | 7.0 | | | 3.3 | | | **13** |
| **Occupational and learning activities** | | | | | | | | **282** |
| Schools, further education, research | 0.7 | 13.0 | | | | | 0.5 | **14** |
| Construction and building | 2.7 | 0.7 | | | 0.7 | 15.0 | | **19** |
| Manufacture | 3.3 | 53.0 | 21.0 | 15.0 | 0.3 | 30.0 | 7.0 | **130** |
| Trade, private services (excl. transp.) | 3.3 | 27.0 | | | | 8.0 | 3.0 | **41** |
| Transportation of goods (incl. Internat,) | 2.7 | | | | 34.0 | | | **37** |
| Transport(delivery,services,commuting) | 1.3 | | | | 40.0 | | | **41** |
| **Raw materials procurement** | | | | | | | | **80** |
| Mining and minerals extraction | 1.3 | | | | 2.7 | 10.0 | | **14** |
| Forestry resources | 0.7 | | | | 6.7 | | | **7** |
| Industrial raw or recycled materials | 3.3 | 2.0 | 17.0 | 17.0 | 1.7 | 7.0 | | **48** |
| Energy provision | 0.3 | 3.3 | | | 3.3 | 3.0 | | **10** |
| Energy conversion/distribution losses | 0.7 | 0.0 | | | | | | **1** |
| **TOTAL** | **109.1** | **299.9** | **59.9** | **32.3** | **102.1** | **75.0** | **31.2** | **710** |

*References:*

Andersen, L., *et al.*, 1942. *Københavns Elektricitetsværker 1892–1942.* Københavns Belysningsvæsen, Copenhagen.

Ansbæk, J., Bennekou, G., Jensen, E., Jensen, G., Johansen, P., Rand, P., Schroll, H. 1973. *Dansk landbrug – økologisk belyst.* Copenhagen University Student Council, Copenhagen. (first group-authored Master's Degree thesis allowed at a Danish University).

Arnfred, J., 1919. Om vindkraften. *Ingeniøren,* (40), 251-258 (including discussion part with *inter alia* P. Vinding).

Arnfred, J., 1964. Developments and improvements in wind power utilization. In *Proc. UN Conference of New Sources of Energy, 1961,* vol. 7, W/1, pp. 376–381. United Nations E/Conf. 35/8, New York.

Betz, A., 1927. Die Windmühlen im Lichte neuerer Forschung. *Die Naturwissenschaften* **15** (46), 905–914.

Bjerknes, V., 1924. Om aeroplaners bæreevne og motstanden mot deres bevægelse. *Ingeniøren,* (48), 549–556.

Broe, G., Kofoed, S., Johansson, M., Sørensen, B., Hvelplund, F., 1975. *Vindkraft Bilagssamling.* Danish Academy of Technical Sciences, Lyngby.

Christensen, B., 2001. *Mindre danske vindmøller 1860–1980.* Danmarks Vindkrafthistoriske Samling, Lem. http://www.vindhistorie.dk

Cour, P. la, 1900. Om vandets elektrolyse og anvendelsen af brint og ilt. *Den Tekniske Forenings Tidsskrift,* 1899–1900 (3) 1–8.

Cour, P. la, 1900–1903. *Forsøgsmøllen* (I to III). Reports on the Danish Government test mill (in Danish). Det Nordiske Forlag, Copenhagen.

Danish Energy Agency, 1999. *Energistatistik.* Copenhagen.

Danish Energy Agency, 2010a. Energy statistics in tabular form 1972–2009. Spreadsheet available at http:/www.ens.dk

Danish Energy Agency, 2010b. Interactive energy info maps: Wind turbines. Map and database available at http:/www.ens.dk

Danish Power Utilities, 1974. *Koordinaret kraftværksudbygning I 1980'erne.* Attachment to government policy statement, Ministry of Commerce, Copenhagen; cf. Chapter 11 below. More detailed information may be found in annual reports from the now defunct power companies ELSAM and ELKRAFT, and in electricity production power plant databases held at the Danish Energy Agency, including current updates.

Danish Power Utilities, 1997. Havmølle-handlingsplan for de danske farvande. Offshore Wind Working Group, S Sealand Electricity Company, Haslev.

Danmarks Statistik, 1910. Anvendelsen of landbrugsmaskiner m.v. i Danmark. *Statistiske Meddelelser*, Series 4, Vol. 34, No. 2, Danish Bureau of Statistics. Bianco Luno Printer, Copenhagen.

Danmarks Statistik, 1923. Elektricitetsværker i Danmark 1921–22. *Statistiske Meddeleleser*, 4. Series, Vol. 68, No. 2, Danish Bureau of Statistics. Bianco Luno Printer, Copenhagen.

Danmarks Statistik, 1925, 1943, 1949. Tables of machinery in agriculture. Statistiske Årbog, Danish Bureau of Statistics. Thiele, Bianco Luno and Gyldendal Nordic Publ., Copenhagen.

Danmarks Statistik, 1959, 1967. Danmarks Energiforsyning 1900–1958, 1948–1965. *Statistiske Undersøgelser*, No. 2 and 20, Copenhagen.

Danmarks Statistik, 1993. *Statistisk Tiårsoversigt 1993.* Danish Department of Statistics, Copenhagen. Other year intervals have also been consulted.

Danmarks Statistik, 2009. *Landbrug 2009.* Danish Department of Statistics, Copenhagen.

Danmarks Statistik, 2010a. *Statistisk Årbog 2010.* Danish Department of Statistics, Copenhagen.

Danmarks Statistik, 2010b. *Statistikbank.* Sections Befolkning, Landbrug, Nationalregnskab. http://www.statistikbanken.dk, Danish Department of Statistics, Copenhagen.

Divone, L., 1975. Future federal plans. *Proc. 2nd Workshop on Wind Energy Conversion Systems, Washington, DC* (F. Eldridge, ed.), pp. 509–522. The Mitre Corp., VA.

DVS, 2001-2002. Vindmøller i landbruget 1907, 1923 og 1936. Faktablad 1–7. Dansk Vindkrafthistorisk Samling, http://www.vindhistorie.dk

Dybdahl, V., 1975. *Industrialisering og folkestyre 1849–1939.* Vol. 3 in "Hovedlinier i Nordens historie" (Thomsen, R., ed.). Gyldendal, Copenhagen.

Edelberg, J., *et al.*, 1932. *Københavns gasværker 1857–1932.* Københavns Belysningsvæsen, Copenhagen.

Energi & Miljødata, 1999. Windresource Mapper on CD-ROM, Aalborg.

Engstfeld, A., 2004. *The electricity war.* Documentary, Zweites Deutsche Fernsehen, ZDF, Mainz.

FDM, 1984. *75 år med FDM og bilen* (J. Andersen, ed.). Association of Danish Motorists, Forlaget Motor, Copenhagen.

Fibiger, M., 1851. *Clara Raphael.* Letter novel available at http://www.kb.dk

Fischer, J. (chair), Estrup, C., Hvelplund, F., Jensen, M., Johansson, M., Kofoed, S., Lykkegaard, K., Meyer, N., Myhre, T., Pedersen, B., Sørensen, B., 1975. *Vindkraft.* Wind Energy Committee, Danish Academy of Technical Sciences, Lyngby.

Gouvello, C., Matarasso, B., 1998. Bottlenecks and obstacles – elements of a methodology for national diagnosis. Ch. 5 in *Long-term Integration of Renewable Energy Sources into the European Energy System* (LTI.Research Group, ed.), pp. 215–238. Physica-Verlag, Springer, Heidelberg.

Hansen, H., 1985. *Poul la Cour – grundtvigianer, opfinder og folkeoplyser.* Askov Højskoles Forlag, Askov.

Hansen, S., 1983. *Økonomisk vækst i Danmark.* Vol. II Tables. Akademisk Forlag, Copenhagen.

Henningsen, P., 1933. *Hvad med kulturen.* Mondes Forlag, Copenhagen.

Henriksen, O., Ølgaard, A., 1960. *Danmarks udenrigshandel.* Copenhagen University Economic Study 2, Gads Forlag, Copenhagen.

Hütter, U., 1964. The aerodynamic layout of wing blades of wind-turbines with high tip-speed ratio. *Proc. UN Conference of New Sources of Energy, 1961,* vol. 7, W/31, pp. 217–227. United Nations E/Conf. 35/8, New York.

Hyldtoft, O., Johansen, H., 2005. *Teknologiske forandringer i dansk industri 1896-1972.* Vol. 7 in *Dansk industri efter 1870.* Syddansk Universitetsforlag, Odense.

Ibsen, K., 1951. *KDAK og Motorsagen 1901–1951.* Special edition of *Auto,* No. 9. Royal Danish Automobile Club, Copenhagen.

Ipsen, E., 1926. *Kongelig Dansk Automobil Klub Historie gennem 25 år, 1901–1926.* Royal Danish Automobile Club, Copenhagen.

Jensen, J., 1920. Hvorvidt kan nyttegraden forbedres ved vingehjul, der tjener til udnyttelse af dynamisk strømenergi i luft eller vand? *Ingeniøren* (39), 315–326; (41), 327–331.

Jensen, S., Bjerregaard, E., 1980. Tests performed on the 2 MW Tvind WECS. *Proc. 3rd Int. Symposium on Wind Energy Systems* (H. Stephens, C. Stapleton, eds.), pp. 391–400. BHRA Fluid Engineering, Cranfield.

Jespersen, A., Loeber, E., 1971. Terminologia molinologiæ. *Transactions of the 2nd Int. Symposium on Molinology, 1969* (A. Jespersen, ed.), pp. 195–210. Danske Møllers Venner, Brede.

Juul, J., 1964a. Economy and operation of wind power plants. *Proc. UN Conference of New Sources of Energy, 1961,* vol. 7, W/21, pp. 399–408. United Nations E/Conf. 35/8, New York.

Juul, J., 1964b. Design of wind power plants in Denmark. Recent developments and potential improvements in wind power utilization for use in connection with electrical networks in Denmark. *Proc. UN Conference of New Sources of Energy, 1961,* vol. 7, W/17 and 20, pp. 229–239, 396–398. United Nations E/Conf. 35/8, New York.

Lauritsen, A., Svendsen, T., 1995. *Vindmølleudbygningen i Danmark*. 4th Semester Report, Roskilde University. This and the following paper are summarized by Lauritsen, Svendsen and Sørensen in *Proc. 1996 European Union Wind Energy Conf.*, pp. 67–69, H. Stephens & Associates, Cranfield.

Lauritsen, A., Svendsen, T., Sørensen, B., 1996. *Wind power in Denmark*. Final report of the European Commission Project RENA-CT94-0012, "A Study of the Integration of Wind Energy into the National Energy Systems of Denmark, Wales and Germany as Illustrations of Success Stories for Renewable Energy". Roskilde University.

Lembke, G., 1918. Article in *Kvinden og Samfundet* (20), p. 190. The Danish Society of Women, Copenhagen. See also text of 1912 open air speech available at the Århus University Institute of History website http://danmarkshistorien.dk

Lindeberg, L., 1966. *Danmark under 1. verdenskrig*. Skrifola Forlag, Copenhagen.

Martin, R., 2000. *Beethoven's Hair*. Bloomsbury Publ., London.

Meibom, P., Svendsen, T., Sørensen, B., 1999. Trading wind in a hydro-dominated power pool system. *Int. J. Sustainable Development* **2**, 458–483.

Nørgård, E., 2010. Nordentoft, Inger Merete. In *Leksikon for det 21. århundrede*, http://www.leksikon.org

Nørlund, A., 1955. *Danmarks historie i billeder*. Gyldendal, Copenhagen.

Notebaart, J., 1971. The windmills. Systematics as a foundation for a typology. *Transactions of the 2nd Int. Symposium on Molinology, 1969* (A. Jespersen, ed.), pp. 165–179. Danske Møllers Venner, Brede.

Ølgaard, P., 1979. On the number of wind machines in Denmark from 1900 to 1950. *Solar Energy* **22**, 477–478.

Olsen, J., Fristrup, P., Munksgaard, J., Skytte, K., 2000. *Konkurrence i elsektoren?* Jurist- og Økonomforbundets Forlag, Copenhagen.

Olsen, P., Vinther, F., 2008. *Næringsstofbalancer og næringsstofoverskud i landbruget 1987-2007*. Report DLF Markbrug 20, Agricultural Faculty, Århus University.

Pedersen, B. (chair), Hvelplund, F., Jels, H., Kofoed, S., Meyer, N., Petersen, H., Sørensen, B., 1976. *Vindkraft 2*. Wind Energy Committee, Danish Academy of Technical Sciences, Lyngby.

Pedersen, M., 1948. Oversigt over vindelektricitetsproduktionen fra 1940 til 1948. *Elektroteknikeren* **44**, 151–152.

Pontoppidan, E., 1767. *Den danske Atlas*. Vol. III (completed and published by A. Godiche in Copenhagen). Unaltered reprinting 1968, by Rosenkilde og Bagger, Copenhagen.

Poulsen-Hansen, P. (chair), Billeschou, H., Hanning, W., Juul, J., Kauffmann,

E.v., Lykkegaard, H., Meldahl, A., Rambøll, B., Raun, H., Vester, B., 1962. *Vindkraftudvalgets betænkning*. Danske Elværkers Forening, Copenhagen.

Schieldrop, E., 1936. *Teknikkens vidundere.* Vol. 2. Gyldendal Norsk Forlag, Oslo.

Schou, E., 1921. Det moderne grundlag for konstruktionen ad vindmøller. *Ingeniøren* (31), 209–220.

Schrøder, P., 1924. *Lærebog i Maskinlære.* Vol. III, 2nd ed., N. Olaf Møller, Copenhagen.

Sørensen, B., 1973. Vindkraft i Danmark. *Gamma* **17**, 29–41. Niels Bohr Institute, Copenhagen.

Sørensen, B., 1975. Energy and resources. Feature article, *Science* **189**, 255–260.

Sørensen, B., 1981. A combined wind and hydro power system. *Energy Policy* **9** (1), 51–55.

Sørensen, B., 1989. *Blegdamsvej 17. Et essay om fysik og virkelighed.* Munksgaard Nysyn, Copenhagen.

Sørensen, B., 1993. What is life-cycle analysis? *Life-cycle analysis of energy systems*. Expert Workshop 1992, pp. 21–53. Organisation for Economic Cooperation and Development, International Energy Agency, Paris.

Sørensen, B., 1995. History of, and recent progress in, wind-energy utilization. *Annual Reviews of Energy & Environment* **20**, 387–424.

Sørensen, B., 2005. *Hydrogen and Fuel Cells.* 1st edition. Academic Press/Elsevier, Burlington.

Sørensen, B., 2008. A new method for estimating off-shore wind potentials. *International Journal of Green Energy* **5**, 139–147.

Sørensen, B., 2010a. *Renewable Energy. Physics, engineering, environmental impacts, economics & planning.* 4th edition. Academic Press/Elsevier, Burlington.

Sørensen, B., 2010b. Editorial Introduction. *Renewable Energy Origins and Flows.* Vol. I of *Renewable Energy. Earthscan Reference Collection* (B. Sørensen, ed.), pp. 1–20. Earthscan, London.

Sørensen, B., 2010c. Editorial Introduction. *Renewable Energy Technologies II.* Vol. III of *Renewable Energy. Earthscan Reference Collection* (B. Sørensen, ed.), pp. 1–11. Earthscan, London.

Sørensen, B., 2011. *Life-Cycle Analysis of Energy Systems: from Methodology to Applications*. RSC (Royal Society of Chemistry) Publishing, Cambridge.

Sørensen, B., Nielsen, L., Pedersen, S., Illum, K., Morthorst, P., 1994. *Fremtidens vedvarende energisystem – et lysegrønt og et mørkegrønt scenarie.* Danish Technology Council, Report 1994/3, Copenhagen. Available at http://

www.tekno.dk/pdf/projekter/943.pdf

Sørensen, B., Petersen, A., Juhl, C., Pedersen, T., Ravn, H., Søndergren, C., Simonsen, P., Jørgensen, K., Nielsen, L., Larsen, H., Morthorst, P., Schleisner, L., Sørensen, F., 2001. Scenarier for samlet udnyttelse af brint som energibærer i Danmarks fremtidige energisystem. Final report to Danish Energy Agency. *Tekster fra IMFUFA* No. 390, Roskilde University. Available in Danish from http://rudar.ruc.dk/handle/1800/3500 as a file entitled IMFUFA_390.pdf. English summary in *Int. J. Hydrogen Energy* **29**, 23–32, 2004.

Steenstrup. J., 1913, 1917, 1930. *Den danske kvindes historie 1701–1917* (2 vols.), Reprinted together with *Kvindens stilling i Danmark i Renæssancens og Reformationens tidsaldre* and *Nogle strejflys over Christian IV's tidsalder særligt med hensyn til kvindens stilling*. Selskabet for Udgivelse af Kilder til Dansk Historie, Copenhagen, 1982.

Steincke, K., 1938. *Tegnestifter*. Essays. Gyldendal, Copenhagen.

Terslin, H., 1959. Gilleleje opdages. In *Fra det gamle Gilleleje.* Yearbook, pp. 3–33. Gilleleje Museum Publishers.

Thurah, L. de, 1756. *Omstændelig og tilforladelig beskrivelse over ... Bornholm, og ... Christiansøe.* C. Glasings Eftf., Copenhagen. Unaltered reprinting 1968, Rosenkilde og Bagger, Copenhagen.

Tortzen, C., 1970. *Gilleleje oktober 1943.* Society for Publications in the Newest Danish History. Fremad Forlag, Copenhagen.

Trommer, A., 2009. Danmarks historie (1940–45). Besættelsen. In *Den store danske encyclopædi,* Gyldendal, Copenhagen, and http://www.denstore-danske.dk (accessed 2011).

Videnskab.dk, 2011. Danske fødevarer holdt liv i tyskerne. Journalistic web portal financed by Danish Department of Research at http://videnskab.dk

Vinding, P., 1919. Vindmøllen. *Ingeniøren* (63), 401–408; (64), 409–411.

Vogt, H., 1919. Vindmøllen. *Ingeniøren* (93), 601–603.

Warming, J., 1913. *Haandbog i Danmarks statistik.* Gads Forlag, Copenhagen.

Wikipedia, 2010. Harald Søbye. http://da.wikipedia.org (accessed 2011).

Wistoft, B., Petersen, F., Hansen, H., Thorndahl, J., 1991–1992. *Elektricitetens aarhundrede. Dansk elforsynings historie* (B. Wistoft, ed.), 2 vols., Danske Elværkers Forening.

# 11

## Oil, nuclear (no thanks!) and natural gas

The reconstitution of Danish economic activity from the end of World War II to about 1960, was followed by a period of exceptional growth from 1960 to 1973, with several implications for energy usage. The 1950s had seen a very rapid substitution of coal with oil products (Fig. 11.1), in power plants, in building heating and hot water systems, and in industry. The reason was greater convenience, but also the stable or modestly declining price of oil products (Fig. 11.2). The low energy prices led to energy being used less efficiently. Little attention was paid to the conversion efficiency of appliances or road vehicles. This may be illustrated by the energy intensity, defined as the amount of energy spent in earning one unit of GNP (despite the many omissions of contributions to the GNP mentioned in Chapter 10). Figure 11.3 shows the gross behaviour of the energy intensity over the 20th and most of the 19th century. First, the ratio declines as the new technologies replace, for example, inefficient use of draft animals, then it increases again. The behaviour will be discussed further below.

The individual types of energy supply for the Danish energy use in Fig. 11.1 show a number of interesting features. As noted, oil imports were barred during World War II, and the period around 1960 saw a substitution of coal with oil. Subsequently, the oil embargos in 1973–74 and 1979–80 led to resuming coal use, notably in power plants. However, when Danish oil and gas began to flow from the North Sea extraction fields, coal use was again diminished, due to the large power plants now getting competition from small CPH plants, mostly using natural gas, and from wind turbines (Fig. 10.36). This period also saw more export of surplus electricity during years where low precipitation caused Nordic hydro reservoirs to have low filling levels, or where nuclear reactors were out of use, as happened frequently in Sweden.

The most striking feature is the behaviour of oil in the energy mix. After the oil-price jumps in 1974 and 1979, Danish oil use declined, and by

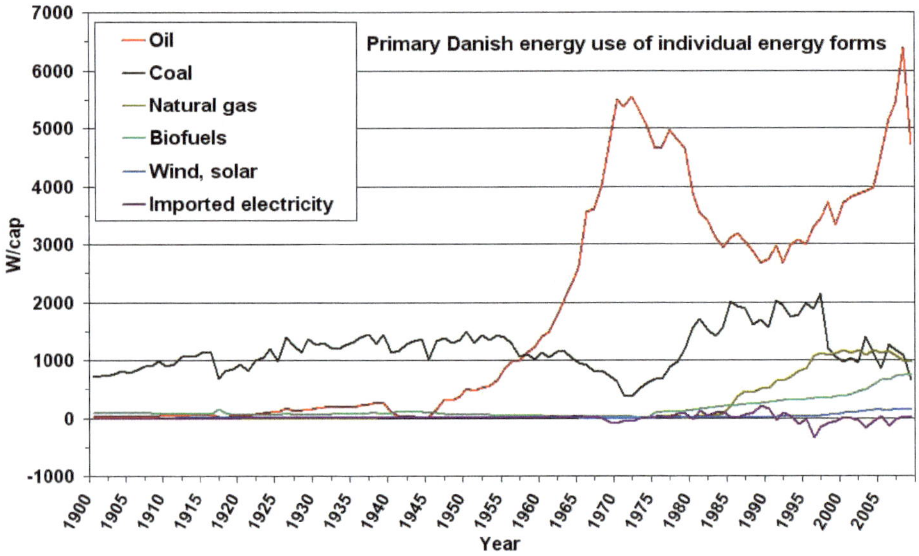


*Figure 11.1. Development in the primary Danish use of energy by type since 1900, determined as primary energy production within Denmark plus imports minus exports. Energy purchases used in Danish international shipping and transportation are included, independent of the locality of purchase (based on Danmarks Statistik, 1959, 1967, 2010).*

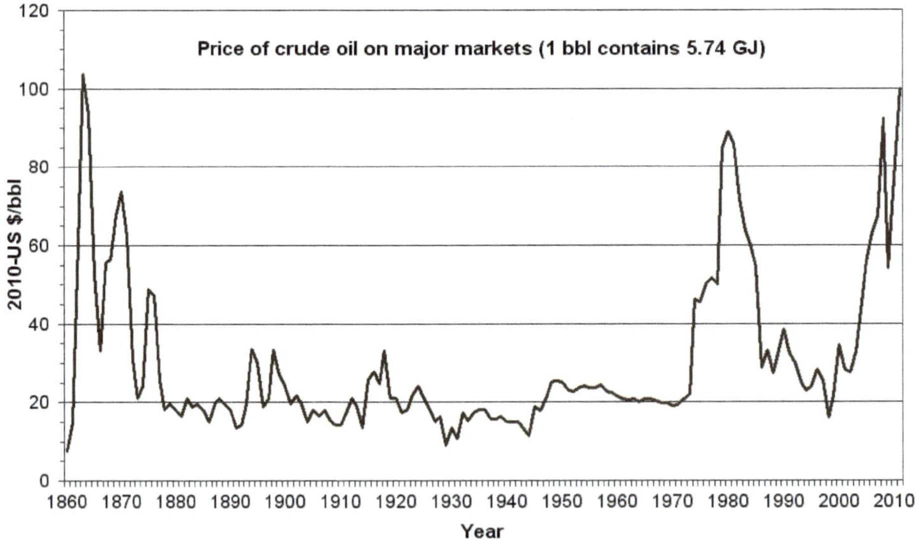


*Figure 11.2. The price development for crude oil in fixed (inflation-corrected) 2010 US $ per barrel. The first half is based on US price data and the later part on trade at international markets (Lippold et al., 1995; US EIA, 2011).*

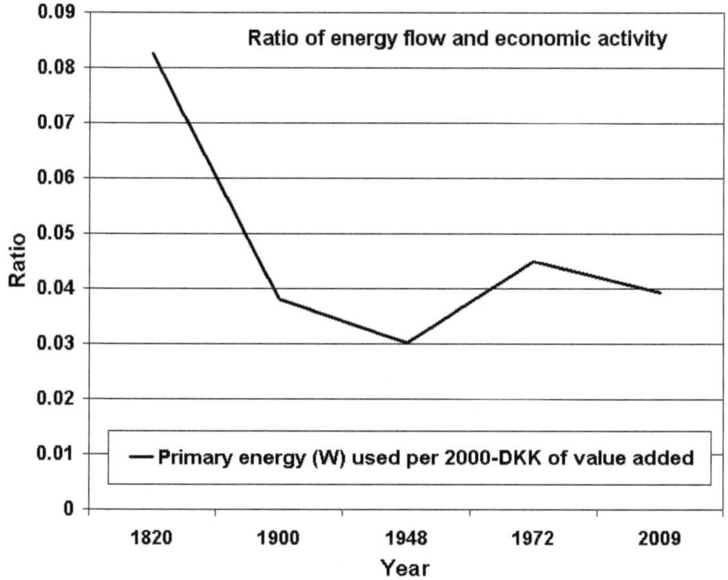


*Figure 11.3. Gross behaviour of the ratio between the rate of Danish primary energy use and the additional income created if the associated activity is in the monetized sector of the economy (based on Hansen, 1983, Figs. 9.49, 11.27).*

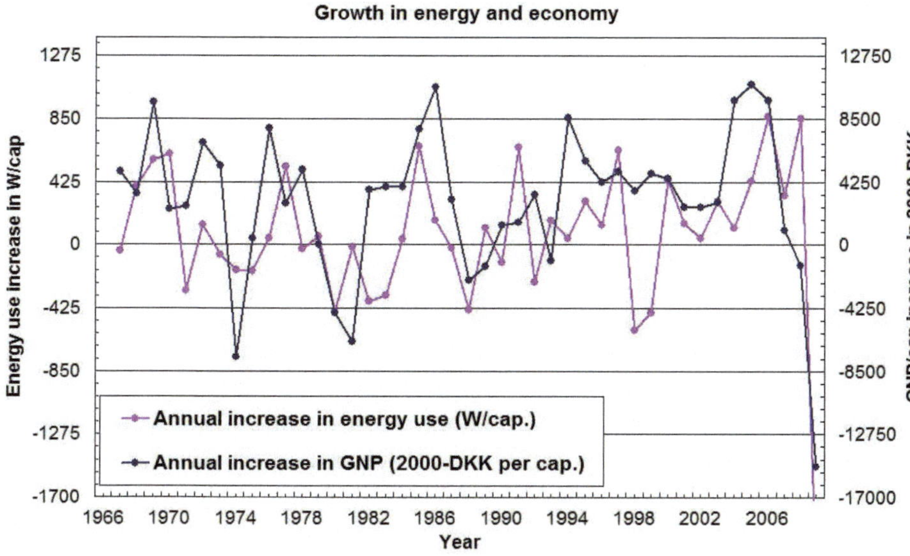


*Figure 11.4. Annual growth in energy use and gross national product through four recent (fairly exceptional) decades (based on Danmarks Statistik, 2010). Although the two curves are highly correlated, averaged over the entire period the economic growth is significantly larger than the growth in energy use.*

1990 had reached about half of its magnitude in 1970. However, during the most recent period there is a dramatic increase in the use of oil, almost entirely due to Denmark's capture of new markets in international shipping. This will be discussed in more detail in Chapter 12. The data from Danmarks Statistik used in Fig. 11.1 represent primary energy production plus fuel imports minus exports (raw or transformed into various energy products). No losses in extraction, refining or converting energy have been deducted from the production data, but for imports only the energy reaching Denmark or used in shipping has been included. In contrast to the detailed energy usage tables and figures presented throughout this book (e.g. Table 10.1 and Fig. 9.49), the data from Danmarks Statistik do not include food energy or animal draft power.

Because the Danmarks Statistik database is ordered according to segments of the economy, it contains double counting of energy by giving the energy-product production of each sector of the economy, even if the production is based on fuels extracted by another sector, and thus already counted. This is why only primary energy is included here (e.g. crude oil, unrefined gas and coal, but not coke). Energy used by Danish companies or citizens but purchased outside Denmark are included to the extent that they are known. The philosophy is that, since they are counted in the economic activity measure (GNP), they must also be included in the energy used to create that income (Danmarks Statistik, 2007). Figure 11.1 shows that coal has played a fairly constant role in Danish per capita energy use from 1900 to 2010, apart from yielding to oil during the 1970s and doubling its share during the 1980s and 1990s.

Since the late 1990s, natural gas has reached the same per capita consumption levels as coal, and renewable sources contribute somewhat less, and mostly through combustion of biomass in power plants and inefficient woodstoves. Wind use is small overall but has a 20% share in electricity production. The fossil fuel that would generate the same amount of electric power would be more conspicuous. From 2008, oil use dropped abruptly, presumably connected to the financial crisis, but also due to the decline in the Danish maritime freight hauling service operations required by the US military. Maersk, the company providing these services, believes that the loss of military commissions will be made up for by transport needs of the Asian growth nations, and following the company's largest annual profit in 2010 (after a loss in 2009) they have in 2011 ordered an expansion of their fleet of container ships by several large and highly energy-efficient carriers (DR, 2011; cf. Chapter 12).

Oil prices are part of the explanation for the large variability in oil usage. As Fig. 11.2 indicates, the initial 19th century price reduction based on extraction technology improvements was followed by 90 years of largely constant oil prices (in inflation-corrected currencies). The exceptional growth period in the 1960s was thus not (as sometimes stated) caused by declining oil prices, but was the result of structural changes in society and realizing the lesser convenience of handling coal (the price of which was as stable as that of oil). The oil prices had some surges during the world wars and in the late 1920s, but it was in late 1973 that a dramatic change in prices first occurred. It is interesting that the US oil companies for some years had urged the Middle Eastern oil producers to raise their prices (Friberg *et al.*, 1974), in order that the onshore US oilfields could continue profitable operation while adopting higher-cost enhanced recovery techniques (only the first 30–40% of oil *in situ* will flow easily to a well; in order to gain more from an oil well, injection of gas or other enhancing methods must be used). The OPEC (oil–producing) countries had refused to raise prices until the Yom Kippur War, when they decided for the first time to use supply withdrawal (embargo) and cost as a political weapon. The second price increase in 1979 was purely economic, achieved by reducing production in the Middle Eastern oilfields. This was possible as long as Europe and the USA had not sufficiently curbed their energy use by efficiency measures, and as long as the non-OPEC oilfields (on both sides of the Atlantic Ocean) had not been fully developed to the productive stage.

When the offshore extraction technique finally came in place and efficiency was increased, at least in Europe, the strength of OPEC was diminished, because now further reduction in Middle East oil production would not hurt the Western world as much as it would hurt the new, extravagant lifestyles in the leading OPEC countries. This explains the 1995 to 2005 dip in oil prices (Fig. 11.2), but as there is relatively little oil in the Atlantic Ocean, and as other oil producers such as north Africa and Iraq have reduced production due to on-going insurgence, revolution and war, the oil price is again rising to an extent that even the current financial crisis cannot counter. Oil prices by far exceed typical extraction costs and are thus subject to the fluctuations of nervous market dealers, influenced by every newspaper notice of unrest in a producing country. The large excursions up and down seen in the last five years of Fig. 11.2 are therefore what may also be expected to characterize the next decades. However, resource depletion will necessarily make the average rise to levels where substitution is the only option. Already, at an oil price of US $100/bll, there are many so far little employed alternatives available. The problem is that societies will have to adjust their energy-related activities to higher prices, which has far-reaching consequences for structural arrangements (influencing optimum transportation distances and discouraging globalization) and for the optimum efficiency of manufactured goods with both short and long replacement times (such as motor vehicles or houses).

The oil used in Denmark has for a short period been domestically produced at the North Sea offshore wells, but as these are now producing less due to gradual resource depletion, imports again account for most of the oil usage, as seen in Fig. 9.17. For natural gas, there is still a surplus to export, but diminishing in this case also after 2008.

Turning now to the relationship between energy use and economic growth, Fig. 11.3 shows that there is no fixed ratio between the two, at least not in the long run. After the initial substitution of muscle power by fuel-based energy, a slower decline in the energy-to-GNP ratio prevails from 1900 to 1948 (the early economic data are simple value-added figures, whereas the latter include indirect taxes and government subsidies as mentioned in connection with Fig. 10.45). Then follows the exceptional growth period 1948–1972, where energy use grows faster than the national income. A possible reason is that the cost of energy was a very small overall cost-fraction for most of the new or increasing activities contributing to the economic growth. After the oil price hike in 1973–74, the energy-to-GNP curve again starts to decline, at roughly the same slope as before the exceptional growth period. This reflects measures taken to use energy efficiently and to produce equipment without extravagant energy use.

Figure 11.4 looks more closely at the recent periods of varying energy prices. Rather than giving the ratio of energy and GNP, Fig. 11.4 gives the growth in these quantities (i.e. their values for a particular year minus the value for the previous year). Economic growth has been negative during several time intervals, notably at the two energy crises and the recession starting in 2008. The growth in energy use was negative during exactly the same periods, and has large excursions up and down from 1995 to 1998, but is generally highly correlated with the economic growth. This is true on short timescales, indicating that efficiency improvements cannot be achieved overnight in response to economic and political events in the Middle East or elsewhere. However, efficiency improvements are happening with delay in response to the price changes, as seen in Fig. 11.3 and also in Fig. 11.4: if one averages over longer periods, the energy growth rate is clearly below that of the GNP. The phenomenon is called "decoupling" and it suggests that, at least in the long run, it should be possible to maintain economic growth without having growth in energy use. For instance, purely non-physical activities may be counted in the gross national product and yet not contribute any energy expenditure at all.

### 11.1 Roots of an information society, or just more computing

After World War II, a rapid technological development based on microelectronics took place and deeply influenced conditions of human societies over the subsequent decades. Computers at first appeared to be a very small wheel in this development, initially judged to be of interest mainly to the military (for calculating cannonball trajectories) and for manipulating databases, such as the Hollerith punch cards invented in 1725 by Basile Buchon in France for controlling textile looms, and which since the 1890s had been used for data storage (e.g. in connection with elections or census registration) by a US company formed by Herman Hollerith that was later renamed IBM (Hollerith, 1889; cf. Fig. 11.5).

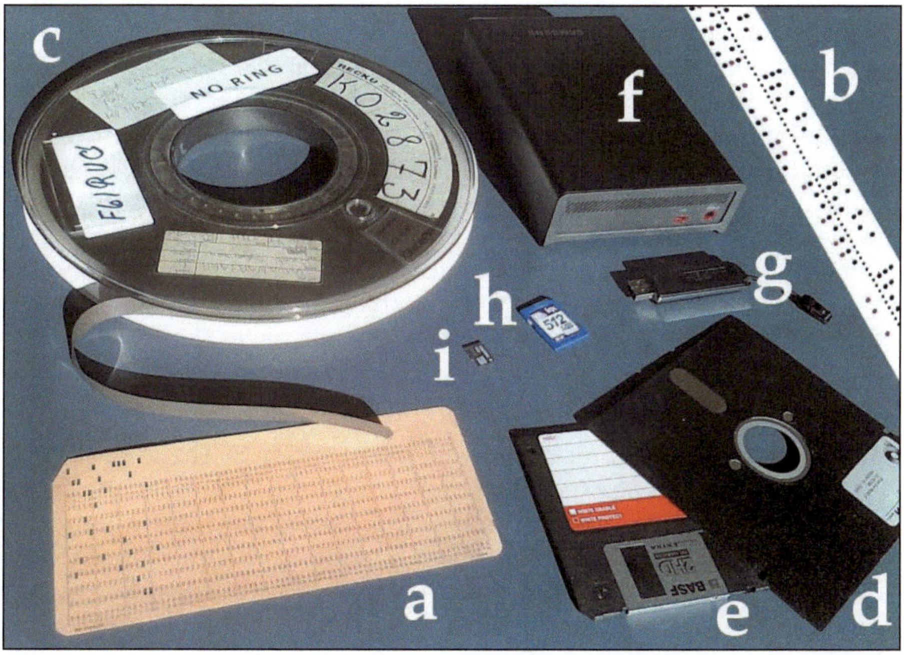


*Figure 11.5. Some technologies used for data storage and handling at different times. a: Punched cards; b: punched paper rolls; c: magnetic tapes; d, e: magnetic floppy disks; f: magnetic hard disks; g: solid-state memory sticks; h, i: solid-state memory cards (photo by author).*

Computers based on mechanical rather than electronic devices have a long history, the earliest finding being a Greek navigation instrument found in a sunken ship and dated to slightly over 2100 years ago (Freeth *et al.*, 2006, 2008). An earlier mention of "animated figures" is found in a poem by Pindar (about 2500 years ago), and in a text by Lie Zi (2300 years ago) describing an automaton constructed for a Chinese king 700 years earlier (Wikipedia, 2011a). Pneumatic devices, sometimes driven by wind energy, were described by Ctesibius (2250 years ago) and Heron (2050 years ago), both in Alexandria. Several intricate automata for entertaining, for turning statues and for dispensing drinks were constructed in Turkey and Middle Eastern countries in the 8th to 12th century, mostly based on readings from the Greek precursors. This rich tradition included Abu ibn Hayyan (8th century, mostly known for his work on chemistry), the three Banu Musa brothers (850) and Badi Al-Jazari (1206; cf. Hill, 1984). Continuing into medieval times, spring- or hand-operated crank-driven automata (e.g. with spikes on a revolving drum operating musical instruments or animating puppets) were constructed in Europe, for entertainment such as playing music or driving visual spectacles, but also for more practical purposes such as operating water closets or dispensing soap for hand-washing. Of course the scientific community had good use of mechanical automata, such as watches and astronomical visualizing models, but there was also commercial use for navigation of ships, in line with the Greek origin of automata. The idea of constructing all-purpose automata, in contrast to those performing a single task, was pursued in the 19th century, as witnessed by the never completed mechanical computing device of Charles Babbage (in London Science Museum). Present practice is to reserve the term *computer* for devices that can be programmed, rather than performing a fixed set of operations. This makes the looms capable of producing cloth in different weaving and colour patterns according to the punched cards inserted qualify as computers.

The first electronic computers were constructed in the 20th century. Again, entertainment was a driver of technology. The magnetic tape or wire recorders of Valdemar Poulsen (Fig. 9.19c; used, for example, by Bela Bartók to record Hungarian folk singing from villagers before the melodies were forgotten) and the groove-based turntables were initially driven by springs wound by hand, but later became powered by electricity. The same is the case for the player pianos popular from late 19th century to about 1930, but in their latest decade, sophistication increased and in both Germany and the USA organ automata, and even orchestra simulators, were built and sold. These could well be called computers, as the programmed piano or organ rolls were operating a complex set of relay-controlled air supply valves driving, for example, the organ pipe sounds (Fig. 11.6). They were still single-purpose computers, like the loom-pattern-controlling or election-vote-counting devices, but programmable within the intended application area.

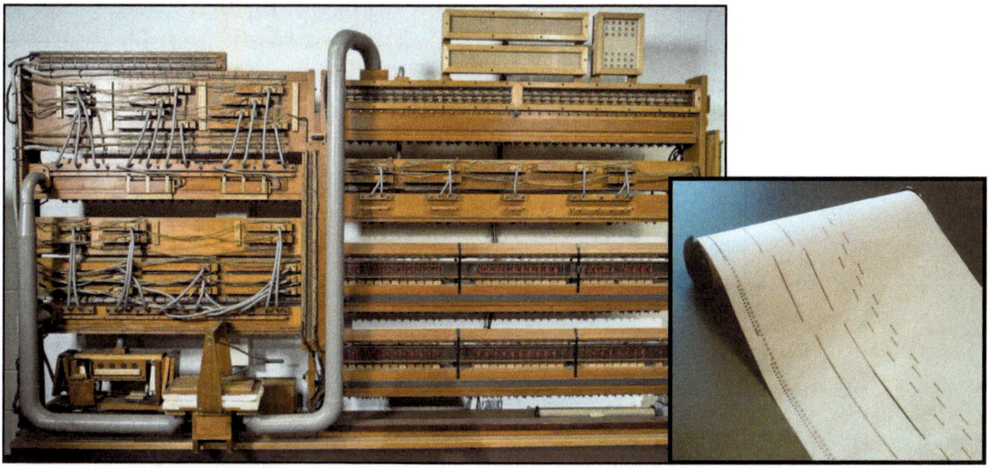

*Figure 11.6. Skinner 1928–29 player organ (opus 764, installed at Boston University) relay switching mechanism and a music program roll (left photo courtesy of Nelson Barden Associates Inc., right photo by author).*

General-purpose electronic computers, which could be programmed and reprogrammed for different tasks, were built at the time of World War II, notably with the purpose of cracking the coded messages that the enemy sent by radio or telegraph. In Germany, Konrad Zuse constructed computers in 1938–1941, moving from simple relay operated machines (similar to the organ in Fig. 11.6) towards general programmable devices. Zuse was also the first to suggest in 1948 the use of a high-level programming language translated into machine-understandable instructions by the computer itself (Wikipedia, 2011b). In England, several computers were built for code-cracking purposes, from the mechanical Bombe in 1941 to the radio valve Colossus in 1944. In the USA, the Atanasoff-Berry Computer was built in 1942 at Iowa State University, followed in 1946 by ENIAC (Pennsylvania University), a fully programmable computer using radio valves (in US English "tubes"). However, due to a high failure rate of the radio tubes, it could not carry through longer calculations. Despite this, several radio tube computers were built over the following decade, becoming slight more reliable, but it was when the transistor technology was introduced that really useful computers first emerged.

In Denmark, this development was intensely followed by a working group established by the Academy of Technical Sciences in 1946. One purpose of the academy was to form government-owned nascent industries that later might become private commercial companies. One such enterprise was *Regnecentralen*, which in the late 1950s tested its skills by building a copy of a Swedish radio tube computer and in 1961 had its first working prototype of a fully transistorized modular board computer named GIER ("Geodesic Institute Electronic Calculator", as the Geodesic Institute was one of the initiators; Fig. 11.7). Among the group of scientists at *Regnecentralen* was the astronomer Peter Naur, who was an active member of the team specifying the programming language ALGOL-60 (Naur, 1963; with precursors 1951–59), considerably more powerful than other then-existing languages, and the core in later developments such as Niklaus Wirth's PASCAL and C. The first commercial GIER versions (1963) from *Regnecentralen* came with an ALGOL compiler, which gave them a leading position among scientists.

*Figure 11.7. The transistor board arrays making up the core of the 1961 Danish GIER computer, delivering 660 floating point operations per second (which is 10 orders of magnitude below fast computers in 2010) and costing some 6 million 2010 DKK (Danish kroner; Krarup and Svejgård, 1961; photo from Isaksson and Petersen, 1961; Gram et al., 1963; Dansk Datamuseum, 2010).*

Another of the scientists at *Regnecentralen* was Peter Lindblad Andersen, around 1970 one of the first to experiment with a personal computer, consisting of a single transistor board, a television screen, a keyboard and a cassette recorder as program and data store. He unfortunately died in 1973, and *Regnecentralen*, having achieved exports chiefly to Eastern European countries, went out of business by 1979, so it was other companies that reaped the market for personal computers (Clive Sinclair in England, and a number of US companies such as Apple, Commodore, Atari and IBM). Denmark has preserved a strong presence in global software development and has come up with interesting technologies such as Skype telephony over the internet. The internet is a resilient network solution based on splitting a message into packets travelling their separate ways through a large network (of cable or microwave connections), thereby to an extent avoiding congestion, and finally being reassembled at the recipient's end (as first proposed by Donald Davies at the British National Physical Laboratory in 1968). The internet based on this idea was soon brought to life by the US military (DARPANET) and by scientists at the European Organization for Nuclear Research (CERN in Genève).

Considering the sector of communication in general, it has developed in parallel to that of computing, from 18th century telegraphs to wireless radio communication, being made portable during World War II (where the sender and receiver equipment could be carried in an attaché case), and advancing from allowing to be dragged to becoming lightweight and pocket size with the advance of mobile telephones and later smart phones, taking over many of the tasks performed by portable computers (audio-visual entertainment, calculation, GPS (general positioning system) navigation, portable internet, etc.). E-mails between computers proceed as for internet communication, while mobile phones initially used a simplified protocol for small messages (SMS). The phone capability is currently based on a dense network of ground-based aerials, and the phones are not operational in wilderness areas, where their use for emergencies would seem most required. This is remedied by communicating via satellite, as with the GPS technologies that are likely in the near future to be combined (e.g. to work in collision-avoidance systems for individual road vehicles).

The use of computers, peripherals and generally microprocessor-based devices has become a major factor in energy use. Each device uses relatively little energy, not least because manufacturers have to choose low-energy designs to avoid heat damage from the still more miniaturized components of an integrated circuit board or processor (present microprocessors contain features down to a size of 20 nm, that is $2\times10^{-8}$ m). Thus, going through the phases of relay switching, radio valves, transistors, printed boards and microchips, the specific energy expense has generally decreased. However, this has been more than compensated for by the growth in number of devices sold and used. The present level of electricity use for computers, screen displays, sound and related technology may in Denmark be estimated as around an average 1 GW, or about a third of the entire electricity use, the rest being used for motors, illumination, heat (for cooking, washing, heat pumps, etc.) and battery charging not just associated with (portable) computers or mobile phones (Center for Energibesparelser, 2011). A significant reduction in specific (i.e. per device) energy use was associated with the replacement of cathode ray television and display screens by solid-state devices. By 2010, the per capita computer-related electricity use (including audio-visual broadcast, game-playing and other entertainment, all of which are increasingly integrated into computers) amounts to about 200 W, averaged over time.

The current type of society, with extensive use of computers and what generally has been termed knowledge-based activities (presumably to underline a contrast to the skill-based activities of earlier societies), has been proposed to constitute an "information society" or "information economy" (Porat, 1977). It was easy to see that, for example, US employment in agriculture had dropped dramatically, and industry seemed to go the same way. For the service sector, the situation was less clear, but Porat managed to create a new category of knowledge-based work that he defined as the rapidly growing information sector. It comprised all kinds of educational activities, all manufacture of computers and related products (although it is difficult to see a difference between these and other industrial products), and finally all enterprises that handle information in a major way (including activities such as advertising for products that otherwise would not be bought). Today, nearly 100% of the GNP-earning activities employ computers or handle information. For these reasons the proposed concept of an information society seems to have lost its usefulness, in addition to the questions posed regarding the consistency of assigning activities to this sector.

In Denmark, it follows from Fig. 9.8 that the public sector has risen from employing 10% to employing 35% of the workforce between 1950 and 2009. Jobs in the public sector are in part performing tasks of control and administration, and in part services for health, education and care for elderly or deprived citizens. Many of the latter activities, such as caring for the elderly, were also carried out earlier, but to a large extent by family, as is the case in many so-called less-developed countries today. In other words, not all the public activities are new, they are just now being compensated for by salary payments where before they were in the informal economy. As for increasing administration and control, these are characteristics of an organization of society with less solidarity and mutual trust. Such features may be necessary, as social relations largely based on ethical norms are being replaced by encouragement of competition and selfishness. Still, there are probably also empty overhead costs in some of the administration and control jobs.

Fears have been expressed that societies will not be able to maintain a large public sector, referring to the need for someone to remain in productive enterprises in order to furnish the food and material goods required, but such fears seem exaggerated, because basic human needs can, with current technology, be covered by just a few per cent of the workforce, and the remaining activities are (or should be) based on our agreed preferences for secondary social goals, which may well not be materialistic in nature. The real question will be how to distribute the proceeds of societies with wealth far above what is required for basic subsistence. In rich countries, there are today great differences in the wealth of individual citizens, and in emerging high-growth countries such inequality is rising rapidly. The theory of the liberalistic market economy states, that despite the minority becoming very rich, there will be a gradual lift in the conditions for the poorer strata of society, from massive deprivation of means to a middle-class position, where members should be grateful to the operators occupying the most lucrative positions in society for having made this development possible. In practice, the recurrent financial setbacks (usually caused by incompetence of the ruling strata) will hit the middle class most severely, making it questionable how satisfied they should be with the pursuit of this particular economical dogma. The least-developed countries are often ruled by dictators or corrupt regimes that will not even allow establishment and sharing of their wealth with a middle class.

As regards the service sector, which in Denmark is also growing quite rapidly (Fig. 9.8), the Swedish economist Staffan Linder already in 1970 wondered why he and most other people felt that the level of services was declining at the same time as the service sector increased its GNP earnings ("nobody opens the door for you anymore"). The reason would appear quite simple, in that salaries for human work have been increasing substantially (e.g. in Scandinavia), but at the same time such work cannot much benefit from the technology and mass production advantages allowing the price of industrial products to diminish. Relatively speaking, one gets less and less for a deflated euro or dollar spent on service, and purchasing a new appliance or other product is in most cases cheaper than having the old one repaired.

In an early discussion of the possible developments of computer usage in societies, Valaskakis and Fitzpatrick-Martin (1980) noted that general-purpose computers hold considerable promise for a democratic and non-exploitative organization of society. This is on the condition that each citizen is free to write or use any kind of software he or she prefers, both for control of the computer (the operating system) and for performing specific tasks (application software). Such an option appeared possible when personal computers were first introduced, with the open software architecture developed during the 1970s at Xerox in Palo Alto (Raymond, 2010), a graphic software system with mouse control later (from 1985) used in the Commodore Amiga computers. The operating system contained a collection of routines written in the UNIX style (based on a freely available operating system developed at Bell Laboratories in the late 1960s), but where the end-user was free to change and replace any piece of code by use of a programming language such as C, or basic machine language if speed requirements made it necessary. There was a smooth transition from operating system routines to applications, and the user had straight access to everything, both physically and also mentally, with a minimum of education, because all code was organized in a simple and easily learned style. This opening for creativity was in plain contrast to the operating systems and programming environments of the IBM/Microsoft or Apple personal computers, where the system routines were mostly inaccessible to the common user, either because of their volume and complexity or because access was not permitted. As a result, around 1990, when the Amiga held around 1% of the market, it produced over 90% of the public domain software (i.e. software written by users and put at the disposal of others without compensation). IBM and later its clones using the Microsoft operating system by then had 90% of the market but under 1% of the public domain software written for their PC's.

Valaskakis and Fitzpatrick-Martin (1980) also note a number of other possible scenarios, in which the users have less and less freedom, and in the end face computing technology used entirely for control and creation or maintenance of totalitarian societies. Important parameters for the scenarios are whether an unrestricted internet exists (called a "central electronic highway" in the 1980 paper) and whether there is centralized control of the contents and options of computer usage that makes the technology a vehicle for centralized one-way information and trivial entertainment, or if individuals and small groups are free to follow different tracks. Today, the internet exists and has played an important role as a vehicle for freedom, such as during the 1991 Russian revolution making the Soviet totalitarian state fall, or during the Arab revolutions in 2011, as yet with an unknown outcome. However, there are increasing efforts by political or fundamentalist forces to censor the content made available (much beyond the evident need to ban criminal content), and the freedom to control the computers has, as mentioned, also been curbed by the software providers, leaving only non-system programming to be easily performed by users. If they try to approach the heart of the computers, common users are termed "hackers", while the software providers unashamedly install hidden surveillance programs and sell the information to vendor companies. Further threats are in the pipeline, such as "cloud computing", meaning that users no longer possess the applications they want to use, but have to submit their tasks to software located at a company site such as Google's, allowing the users to compute over the internet but without having control over privacy and the use of what they are doing.

How could the IBM PC concept gain such a high market share? Probably because IBM had better sales strategies than their competitors. They aimed marketing entirely at the business sector, and their tactic was to approach the top managers of the companies, avoiding as much as possible the people in computing departments who knew about the capability of different hardware on the market. They told the managers that the competitors ("bamboo machine" manufacturers) would probably be out of business in a year or two, and the investment then lost, as no one could furnish parts to repair the computers. IBM, on the other hand, had provided typing machines to businesses for a century and would exist forever: "no one has ever been fired for buying big blue" (the nickname for IBM). Most managers at that time were completely computer-ignorant and quickly decided not to risk their jobs.

A highly ironic event accompanied IBM's entrance into the personal computer business. It was easy for them to pull hardware components out of their mainframe assembly lines and rearrange them, and the easy replacement of such modular components (on print plates, small component boards) turned out to offer major upgrading advantages over the complete systems of some competitors, but IBM had little experience with simple operating systems. Their mainframe machines had operating systems more complex than those of the competitors, so these were not suitable for the broader PC user populations. Therefore IBM bought a PC operating system from a small company, Microsoft, consisting of little more than one programmer (with experience in writing interpreters for BASIC, a simplified software language) and perhaps a part time lawyer. But Microsoft used the same tactics as IBM, dealing only with top management, and sold a simple but already outdated operating system, QDOS, that they themselves bought for US $ 50 000 (and which was again based on another outdated but at the time common operating system, CP/M, that IBM had attempted to buy but failed; Bellis, 1996). The scoop was that Microsoft made the IBM managers believe that they had bought the operating system including all rights to market the system under IBM's own name (PCDOS), but there seems to have been a clause that any updates had to be renegotiated. So IBM did all the marketing for version 1.0 but then realized that the DOS (disk operating system) needed updating, after which its name suddenly changed to MSDOS, with high revenues now going only to Microsoft. Many observers enjoyed this David and Goliath game, but the fact was that MSDOS delayed personal computing by ten years, until the PCs (PC standing for the IBM-Microsoft hardware-software concept, not for personal computers) finally in 1995 got the fully multitasking, graphical interface that Macintosh and Amiga already had, and computer users could finally skip the months of attending courses in rote learning of keyboard combinations that would do tasks obvious by a graphical interface. IBM later tried to write and market its own operating system, but too late, as they had already secured the position for Microsoft. Today, even managers are slightly computer-literate, and private users are probably at least as numerous as business users, so it will be interesting to see how the software wars develop in the future.

### 11.2 The nuclear debate: the power of grassroots

The period of the 1960s characterized by exceptionally high growth of economic and energy-using activity (Fig. 11.3) changed the views on fossil energy resources, and particularly oil, because reserves were known to be fairly modest and incompatible with exponential consumption growth. As shown in Fig. 11.3, the average energy intensity (energy per unit of activity) increased by 50% between 1948 and 1960 (Sørensen, 1975). Exponential growth means that each year's use is a fixed percentage (in Denmark some 5% for all energy during the 1960s, but over 10% for electricity) of the use in the previous year, rather than of some fixed level. Government economists published "perspective plans" where the only perspective was the infinite continuation of straight lines on logarithmic paper (Erik Smidt, 1971, 1974; Smidt was head of the Department of Finance, the perfect bureaucrat but with a past as enthusiastic planned-economy Communist).

With this outlook, the energy future looked obvious to the engineers and economists in government and energy supply enterprises: only nuclear light-water reactors (based on once-through nuclear fuel) could sustain the energy growth during the rest of the 1970s, only nuclear breeder reactors could sustain the growth expected during the 1980s, and only nuclear fusion reactors could sustain the expected growth during the 1990s.

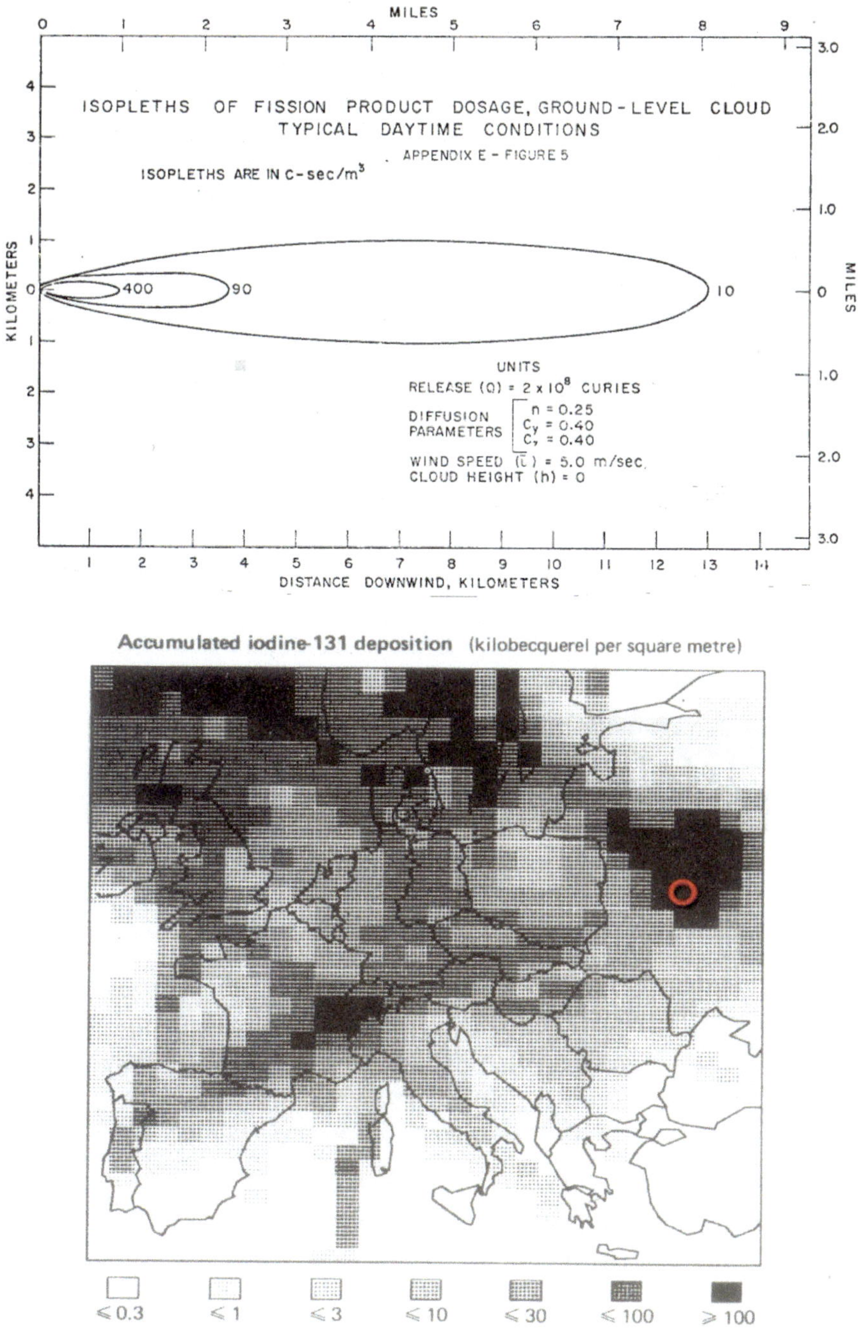


*Figure 11.8a, b. Dispersal of radioactive emissions from nuclear accident, according to simplistic stack model (a: top; WASH-740, 1957) and actual dispersal after Chernobyl accident near Kiev in 1986 (b: bottom; IAEA, 1986).*

In 1955, upon the recommendation of Niels Bohr, the Danish government formed the Danish Atomic Energy Commission and the associated Risø National Laboratory (inaugurated 1958), charged with "promoting peaceful use of nuclear energy to serve the interests of society", based on fissile material furnished by the USA through the "Atoms for Peace" campaign of President D. Eisenhower, who had decided no longer to keep nuclear secrets and fissile materials from the allies of the US, now that the Soviet Union had its own bombs. It is important to understand the motivation of Niels Bohr for furthering this activity. His view of nuclear technology was related to his basic ideas of complementarity, and he would stress that nuclear science could be used both for making bombs and for medical or energy applications. His recommendation of nuclear power was based on his wish to provide a positive application to balance his work on nuclear weapons at Los Alamos during World War II, an involvement that had affected his conscience (as indicated by his proposal not to reserve nuclear secrets for one nation; Bohr, 1950). Now he wanted to promote the introduction of a positive application of atomic science to provide energy, not because he had studied energy science and knew anything about the qualities of nuclear power compared with alternatives, but just because he saw nuclear energy as the positive counterpart to nuclear bombs.

In the Danish context, there was a problem. The Danish Electric Utility companies boasted in the 1960s being the world's best coal traders, ensuring coal power at a very low production cost. Coal was no longer imported from Newcastle but came from wherever it was cheap, such as apartheid South Africa, USSR or mines in South America with underpaid workers. Moral aspects had not yet entered the vocabulary of Danish trade, and the electric utilities assumed that everything was allowed that was not explicitly forbidden by Danish law. Because of their perceived mastering of the fossil fuel markets, the electric utility industry had demonstratively played down any relations to Risø or nuclear energy, reasoning that experience from the early nuclear plants being operated in the USA and England clearly indicated higher production prices than that of Danish coal-based power.

From 1958 to 1971, Risø was therefore unable to undertake work related to practical application of nuclear reactors. Risø became the last choice for employment considered by scientists or engineers when the possibilities of interesting jobs had failed, and in consequence, practically no work qualifying to be published in the international scientific or engineering journals and no development of new ideas in nuclear reactor science was produced at Risø during these years. A few Risø scientists did decent work in areas not directly related to nuclear power, such as monitoring nuclear weapons fallout or undertaking meteorological measurements (facilitated by the availability of excellent facilities, such as a meteorological measurement tower over 100 m high, paid for by the generous funding received by Risø). However, using the meteorological insights to simulate detailed radioactivity dispersal after a nuclear accident did not occur to the Risø scientists until after the public debate of nuclear power started in 1972. Reasons could be that, even as late as 1975, Risø reports stated that the worst accident that could happen at a nuclear power station would at most cause release of radioactivity through a pinhole-sized containment failure and would have no significant effect outside the facility fence (Thykier-Nielsen, 1974, 1975). The disregard of larger amounts of radioactivity escaping was at the time explained in a report to the government (Korsbech and Ølgaard, 1974) stating that if large amounts did escape (thought possible only in cases of earthquakes or direct hits by a falling airplane), then the heat created by the radioactive release would make it diffuse upwards in the air and cause no harm at the ground. In reality, as the 1986 Chernobyl accident showed, heated radioactive matter can travel large distances, but eventually it will be brought down to people and crops at the ground, for instance by rain. The simple dispersal models used by Risø during the following years when they started to look at long-range transport following a nuclear accident situation assumed (by analogy with the much earlier WASH-740, 1957; Fig. 11.8a) that the concentration dropped quadratically with the distance from the point of origin. As seen in Fig. 11.8b, this is simply wrong, as atmospheric transport follows wind-based trajectories, sometimes with little dilution, and where different trajectories later meet again or have multiple passages over a location, it can cause concentrations at large distances that are higher than closer to the point of release. One did not have to wait for the Chernobyl accident to know this, because it had already been demonstrated by the radioactivity fallout over Europe following the 1957 nuclear accident at Windscale, UK (UK Atomic Energy Office, 1957; Lindell and Löfveberg, 1972).

The Danish electric utility market was by 1970 dominated by two concessioned enterprises, one in the west of Denmark and one in the east, each network connected to neighbouring countries but not to the other Danish consortium. The utilities had started as municipal or co-operative enterprises, and "concessioned" meant that their pricing had to be approved by a government overseeing body. As in other European countries (e.g. CEGB in England, EdF in France), the utilities had become semi-private companies, only reluctantly having issues discussed outside the company walls. By 1971, the west Danish company

ELSAM decided that with the on-going growth in electricity use, time had finally come to introduce nuclear energy (ELSAM, 1972). Having little confidence in Risø, they announced that they would buy turnkey nuclear plants in the USA, and formally asked the government to approve the first installation. In fact, only approval of the Minister for Education was required, according to the legislation passed earlier (Danish Law No. 170, 1962), but the government soon realized that a more serious scrutiny of the application was needed, and formed the necessary framework for this (Atomenergikommissionen, 1973).

Risø may have felt squeezed out by this decision, considering that with three research reactors behind them, they could well have been asked to make the engineering design of the power reactors required, but Risø chose to resign to the role of helping in the dissemination of information to the public, explaining why nuclear power was needed and was nothing to be afraid of. Neither the power utilities nor Risø had realized that "the lights had changed": the student revolt in 1968 had taught Danes that important questions leading to decisions on a national level could be seriously debated in public, in contrast to the earlier style of decisions being made by bureaucrats inside ministerial departments with high walls around them. This new spirit was actually not very foreign to Danes, as it constituted a natural extension of the Folk High School debates already familiar to most Danes (see Chapter 9), as well as of the 20th century women's liberation campaigns.

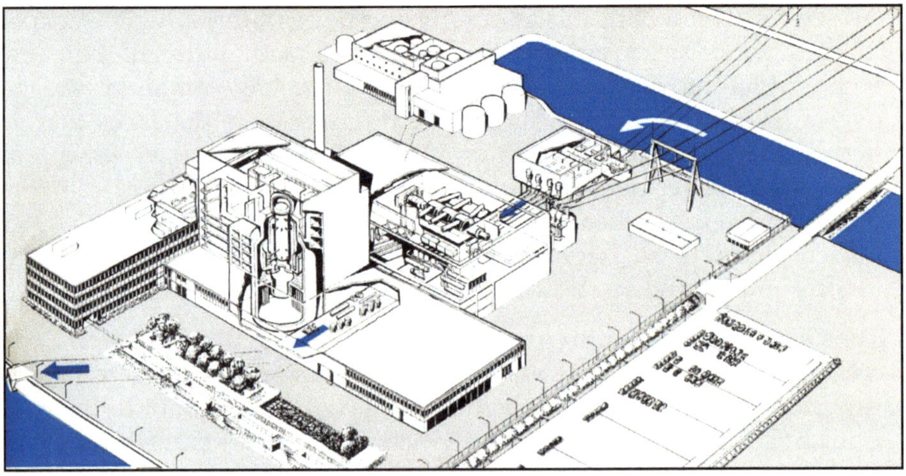

*Figure 11.9. Sketch of the first nuclear reactor unit under construction at Barsebäck, Scania, 20 km from Copenhagen (pamphlet from Sydkraft, 1971).*

The nuclear debate was launched in 1972, focusing not on Danish energy policy, but on a Swedish nuclear reactor already under construction at Barsebäck, 20 km from the Danish capital of Copenhagen (Fig. 11.9). A feature article in a major newspaper (Sørensen, 1972) asked if it was reasonable to built installations with a perhaps small risk for accidents but with large consequences so near the administrative capital of another country not having any say in the decision (at least not formally; the Swedish government had actually informed the Danish one of its plans, and the Danish government had expressed no objections, presumably based on the Risø advice; Bernhardt, 1977). Inspiration for trying to raise this discussion in the public came from Rachel Carson's *Silent Spring* (1962), presenting the view that bureaucratic deliberations regarding damage from persistent pesticides had not reduced the use of DDT, and that Carson felt that taking the issue to the public and possibly thereby creating a pressure to forbid DDT was an approach more likely to bring results. The question of safety of nuclear reactors had at the time surfaced in several other countries, notably the USA, where the later winner of the Nobel Prize in physics, Henry Kendall, and his team at the Union of Concerned Scientists in Boston (Forbes *et al.*, 1971) had found a flaw in the emergency cooling system of the US light-water nuclear reactors, showing that it simply might not work if called upon. Scientists eventually found ways to mend the problem (Hannerz, 1983), but they were never implemented, because the US reactor manufacturers feared legal suits from those utilities that had already bought reactors, if a design change admitted that the early types were flawed. Attempts to ridicule Henry Kendall were made by the nuclear proponents (Saxe and Murray, 1972). Following the March 2011 accident at the Fukushima light-water reactor compound in Japan (where the plant designers had falsely assured the government that the reactors were earthquake- and tsunami-proof), not a single one among a large number of emergency core-cooling systems worked (NHK, 2011), leading to substantial core meltdown and release of radioactivity to the surroundings (ZAMG, 2011).

While it is difficult for a single scientist to be heard in a political environment such as the Danish one, where, during the early 1970s virtually all the members of parliament supported the introduction of nuclear energy, help soon came from the establishment of a grassroots nuclear information organization OOA (*Organisationen til Oplysning om Atomkraft*), formed in 1974 by a group of students on a suggestion from Jens Brøndum, student clergyman at the Danish Technical University. Brøndum had studied grassroots movements during a visit to the USA and wanted to engage young Danish people in similar activities. The subject was to him unimportant, but it had to be concrete and not politically biased. OOA considered several issues, such as "big brother" behaviour using computer registration of the activities of citizens, but ended up deciding that the nuclear power issues raised by Sørensen (1972) offered the best possibilities. The three leading members of the OOA formation group were theology student Siegfried Christiansen (later engaged in work for Actionaid–in Danish *Mellemfolkeligt Samvirke*), geology student Tarjei Haaland (currently at *Greenpeace*, Denmark) and biology student Jørgen Steen Nielsen (now an extremely active journalist at the–in its own words "least inferior" Danish newspaper *Information*). These three people and their many helpers created the most successful grassroots organization ever, achieving both its initial goal of postponing the Danish nuclear decision by three years to get time to have a meaningful public debate of all relevant issues, and in 1985 its ultimate goal of getting nuclear energy off the agenda in Denmark.

*Figure 11.10. The OOA smiling sun with the text "Nuclear power? No thanks" was designed by Anne Lund in 1975 and has become the preferred logo for anti-nuclear movements around the world (OOA Foundation, used with permission).*

Three issues concerning nuclear power were raised for debate: the risk of large accidents, the long-term disposal of radioactive waste, and the connections between nuclear power and nuclear weapons. OOA was not the first grassroots organization in Denmark, but it operated with a set of rules that gave it the success that others had not reached. The environmental group NOAH, formed in 1969, was an association honouring the grassroots principle of democratic decision-making by the members appearing at scheduled general assemblies, and although most decisions were drafted by the secretariat, this structure did not allow new issues to be taken up by the secretariat between national meetings. Furthermore, many of the leading people in NOAH had known or alleged connections to the political left, so opponents could use the rhetoric of political extremism to dismiss the NOAH points of view without serious argumentation. OOA, in contrast, stressed political independence as a fundamental rule and had a much more flexible structure, allowing new items to be taken up and allowing the secretariat to enter into day-to-day public debates on any item. OOA was not a membership organization, but anyone could use it in public energy debates (e.g. sending letters to newspaper editors) if they just signed themselves "active in OOA". There was no censorship, but only the secretariat could sign interventions plainly by "OOA". If novel actions were not approved at the next national meeting, the secretariat could be criticized or replaced. In reality there was a remarkable consensus on means and methods, such as discussing in an urban fashion (cf. Fig. 11.10) based on scientific facts and interpretation, no screaming and not mixing the nuclear issue with political colours. This was probably a major factor in the success of OOA, because its opposition from the pro-nuclear constellations was considerably more emotional in debates.

By 1974, the oil crisis and price hike had deeply influenced Danish society. The electric utility company ELSAM believed that the nuclear decision would now be forthcoming without debate, so that they could start digging for the foundation already the same year. Thanks to OOA that did not happen. The political parties realized that OOA had supporters all over the country and in all political segments, and they decided to have a real debate, accorded a reasonable amount of time. The government financed elements of such a debate by asking the Folk High School teacher Uffe Geertsen to help elevate the level of information by producing six books exposing the aspects of the main issues with inclusion of both sides of debaters. Six eminent books emerged and were diligently used in the public debates unfolding over the following years in any village hall or other meeting venue across the country. Of course, the pro-nuclear side complained that it did not think it was getting its 50% of the word count, but the fact is that few arguments in favour of one or the other side cannot be found in these books (Geertsen, 1975).

The level of public engagement in the Danish nuclear debate was overwhelming. Several thousand formal and informal meetings were held all over the country, often using the style and facilities of the Folk High Schools as their frameworks. The power utilities also staged meetings to promote their view. Over the following years, more than half the Danish population acquired knowledge at a considerable depth on the issues. This was precisely the claim of the OOA: that the nuclear issue was not too technical or requiring special qualifications unavailable to the common people, and that in fact it was possible to understand the problems sufficiently well to be able to use the tools of democracy to reach a decision (such as having a general referendum on the issue).

The uncertainty in estimating the consequences of a large nuclear accident was the content of an approach by two scientists to the parliament on the occasion of its first debate on the nuclear question (Sørensen and Ulfbeck, 1974). This should remain a key question in the Danish discussions taking place over the next ten years. A new safety study commissioned by the US Atomic Energy Commission (WASH-1400, 1975; Fig. 11.11) was heavily drawn upon in the Danish debate. The draft report, circulated for about a year, showed that fallout from major nuclear reactor accidents were considerably larger than anticipated, and the US Atomic Energy Commission subsequently took the work out of the hands of the team head, engineer Norman Rasmussen (who for a while was unavailable due to a nervous breakdown), redid all the calculations and mysteriously destroyed the documentation for the draft work. The final report appeared in 1975 and was altered in different ways, but still showed results that disturbed many readers. The physics community in the USA made an independent investigation (Lewis *et al.*, 1975), and an official assessment (Lewis *et al.*, 1978) later concluded that the methodology of the WASH-1400 study was flawed and the data used in many cases had no empirical basis, but were often free guesswork (the paraphrase used by Rasmussen was "sound engineering estimates"). In Denmark, Hoffmann-Jørgensen and Væth (1979) noted that WASH-1400 was using statistical methods incorrectly.

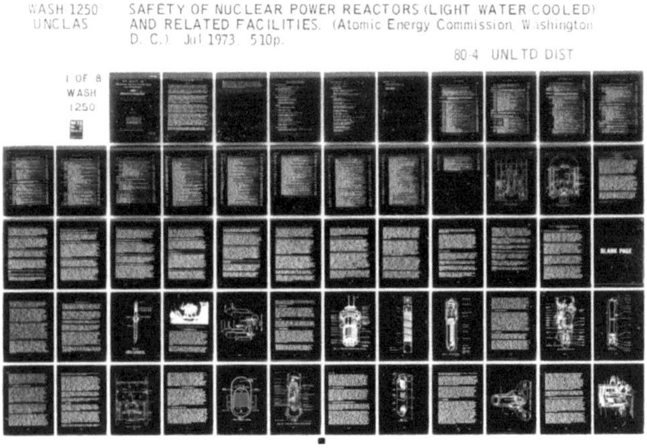


*Figure 11.11. The US Atomic Energy Commission's reactor safety reports WASH-740, WASH-1250 and WASH-1400 were mailed on microfiche cards, readable at many libraries, to anyone around the world who asked for it.*

Nuclear energy was debated in several countries beyond Denmark, and eventually a moratorium on new plants was decided or tacitly effectuated (e.g. in Germany, Sweden and the UK). In France, where imperialistic ambitions fuelled a major nuclear weapons programme, a rapid expansion of civilian nuclear power was taking place in a likely planned effort to blur the distinction between costs of the military and civilian programmes. The public had not been asked its opinion, but scientific safety questions had been raised by analogy with the discussions elsewhere (Barrère, 1975). However, when a schoolteacher in 1977 was killed by police brutality at a peaceful demonstration at the Creys-Malville experimental breeder reactor site, the anti-nuclear organizations made the decision that the nuclear issue, although important, was not a cause for which it was reasonable to risk the lives of dedicated people, and in consequence they closed their operations (Falk, 1982).

*Figure 11.12a, b. Demonstration 18th April 1979 in front of the Danish parliament, organized by OOA. Participants from all layers of Danish society took part (photos: above: Jesper Stormly Hansen, permission bought from POLFOTO; below: courtesy of photographer Per Folkver). The speakers addressing the crowd were Tine Bryld and the author.*

An interesting side-issue in the nuclear debate was the role of experts, or perhaps this was really one of the central themes. The pro-nuclear side claimed that only experts should be listened to, and one debater went so far as to suggest that there were no experts outside the Risø National Laboratory (Hansen, 1975). He arrived at this surprising conclusion by stating that only experts in an institution with access to many similar experts could be sure to draw the right conclusions. Now, in normal language, an expert is a person with a narrow field of deep knowledge, here knowing the details of, say, the assembly of nuclear fuel elements, but unlikely to have the broad insights required for determining if the nuclear solution is superior to other energy solutions. Niels Bohr (who did not like philosophers) generalized the definitions to the extreme, saying that an expert knows everything about nothing, a philosopher nothing about everything, but a scientist at least knows something about something (Nielsen, 1962). As mentioned, the argument of the OOA was that anyone willing to learn could master the problems posed by even a technology as complex as nuclear energy to an extent allowing her or him to cast their democratic vote on the issue.

Figure 11.12 from a 1979 anti-nuclear demonstration in front of the Danish parliament building shows that this complicated technical issue could master as many attendants as the most serious social questions. A part of the Danish debate has been documented by Danielsen (2005), namely the part that appeared in print. However, this was only one dimension of the activity. As mentioned, a very large part of the debate took place in public meetings around the country, and further in radio and television. Somewhat late in the controversy, Danish television featured a few debates at prime time, in the form of hearings or tête-à-tête discussions between proponents and sceptics (Fig. 11.13).

*Figure 11.13. Nuclear power television duel between Peter Skovbye and Bent Sørensen (to the right) in the television channel of the Danish Broadcasting Corporation (6th October 1982, entitled "The truth about nuclear power"; host Werner Svendsen). Permission purchased from DR Arkivsalg.*

The 1979 accident at the Three Mile Island nuclear reactor in Pennsylvania showed that even with the chance rescue of the reactor containment after a substantial core meltdown, preventing large external radioactivity releases, the economic costs of an accident could still by far exceed the cost of building the reactor in the first place (Booth, 1987, found that clean-up already by then had cost US $ 1 billion, even before the dismantling of the wreck for final disposition had begun). For the Chernobyl accident in 1986, temporary remedies at the site have been still more expensive, and final dismantling will be more difficult than at Three Mile Island. In this case, the extensive fallout all over Europe (Fig. 11.8b) has caused damage to both soil and crops, and also directly and indirectly to humans. Many difficult issues arise in assessing this damage and particularly in trying to express it in monetary terms (Sørensen, 2011).

However, the Danish government halted the plans of nuclear power generation in Denmark before the Chernobyl accident. In February 1985, the parliament passed a resolution requiring the government not to employ nuclear energy in its plans or its permissions to power plant operators. The Conservative Prime Minister, Poul Schlüter, welcomed the decision by saying that industry could not continue to live with uncertainty about the future supply situation. The energy industry in particular needed to know what conditions it was to work under, say, before investing in efforts to market wind power or decentralized natural gas-powered CPH plants. At the time, a referendum on the nuclear issue would have given the same result, but the government was not in the mood for another referendum after the one accepting Danish membership of the European Union in 1972, made necessary because the 5 out of 6 majority needed in parliament for yielding sovereignty according to the 1953 constitution was not present.

In consequence of abandoning nuclear energy, the mission of the Risø National Laboratory was in 1986 changed from being charged with promoting nuclear energy to work on a broader range of energy issues. A wind power test centre created after the oil crisis as part of a special funding programme for unemployed engineers had been incorporated into Risø, and the new Ministry for Energy formed in 1979 put all its fact-finding and planning scenario work out as mandated projects for Risø (despite the fact that more qualified teams at the time existed at Danish universities). The reactor people at Risø were, of course, not very happy with this development, and earlier, when the closing of the Risø nuclear division looked like the outcome of the nuclear debate, they had tried to secure their survival by applying to the US authorities to supply them with radioactive high-level waste material for various studies, but instead of the previous practice that such material had to be returned to the USA when the project ended, they asked to be allowed to keep the radioactive substances at Risø (Olesen, 1983).

Needless to say, the US–inspired Danish intelligence services surveyed the nuclear protest movements closely (PET, 2009). However, at the time the police respected the right of people to engage in peaceful demonstrations, in contrast to the situation at the recent 2009 UN climate meeting in Copenhagen, COP 15, where the police used "preventive arrests" in violation of constitutional rights, as subsequently ruled by the Copenhagen City Court (Dahlin, 2010).

## 11.3 North Sea oil and gas for a while, plus a little solar energy

The production of oil and natural gas from the Danish part of the North Sea commenced during the 1970s but did not really take off until the 1980s (Fig. 11.14). After 2000, the production started to decline (Danmarks Statistik, 2010), so one may view this interim access to reasonably cheap energy sources as a one-time opportunity allowing a smooth transition to a sustainable energy supply (Fig. 11.15). Unfortunately, this has happened only to a very small extent. As is clear from Fig. 11.1, the use of wind power has not appreciably displaced coal use, which nearly doubled between 1980 and 2000 and then fell back to the previous level primarily due to use of natural gas for power production, and to a smaller degree the use of wood scrap and straw. This behaviour was due to natural gas at first being reserved for heating private homes, leading to the establishment of an expensive natural gas distribution network. The government listened too late to the energy scientists who all along had said that using the gas in decentralized combined power and heat plants would be the best solution (Hvelplund *et al.*, 1983).

In the years following the 1973–74 oil supply problems, the Department of Energy established in 1979 was able to put considerable amounts of research and development money into energy research, focusing mainly on the short-term fossil production opportunities but also on technologies for using renewable energy. Production of biogas from manure was developed first at farm installations but subsequently in larger units connected to waste treatment, and finally for handling municipal waste (Danish Energy Agency, 1995a; 1996). However, a major project for transforming source-sorted household waste into biogas in north Sealand failed, which led to halting most new biogas activities. This was unfair, as the problem was largely poor engineering, exposed by the fact that a similar facility had worked well in Jutland by just adding a bit of agricultural waste to the urban waste. Currently, most biomass residues are simply combusted in power plant boilers, and although there is research in using biomass residues to produce liquid biofuels, the prospects are bleak because biofuels have acquired a bad reputation after the US use of the food grains, earlier exported to Mexico, for example, to produce ethanol for cars.

*Figure 11.14. Oil and gas platform in the North Sea (courtesy of Maersk).*

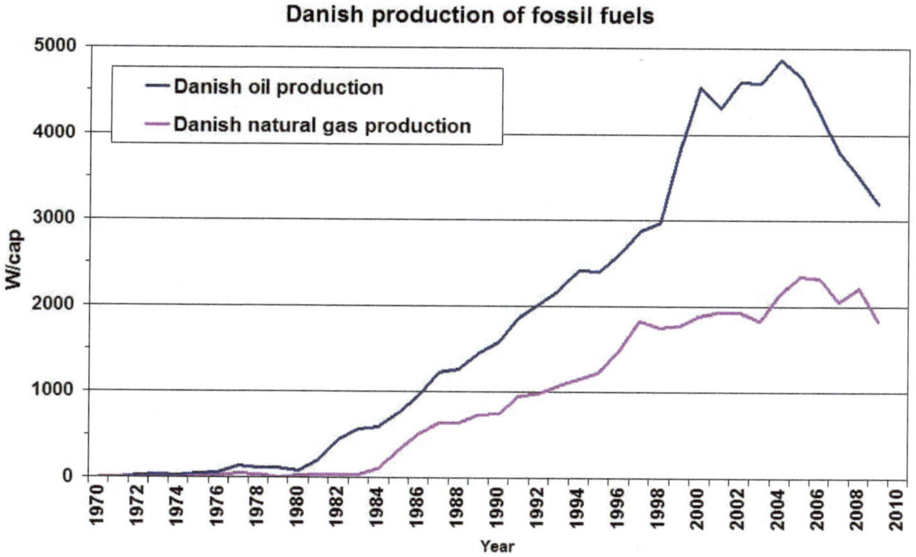


*Figure 11.15. Danish production of oil and natural gas from the North Sea, currently in rapid decline (Danmarks Statistik, 2010).*

Solar heating and later solar electricity were given a 30% subsidy through the Energy Agency, and a number of Danish solar panel manufacturers became established, but the single photovoltaic cell manufacturer soon went out of business in a recessive market. The problem is that solar energy in Denmark is available only in summer, which was fine for the early few square metre solar hot water systems placed on detached houses to enable the boiler (around 1980 mostly oil-fired) to be shut off. The oil crisis had made many house-owners add insulation, implying that their boilers were over-dimensioned for the new conditions and very inefficient during summer, where load was only for hot water. However, during the following years, when natural gas replaced oil in many private heat installations, the inefficient equipment was replaced and the advantage of a few months of solar energy marginal. Figure 11.16 shows a family dwelling with a substantial solar heat system, and Fig. 11.17 a municipal district heating system.

*Figure 11.16. Solar home built 1982 by Jean Fischer (head of the Wind Energy Committee mentioned in Chapter 10). Standing to the left is the H. C. Andersen biographer, Elias Bredsdorff, and at the far right Piet Hein, designer of the super-egg and poet writing "grooks" in Danish and English (photo by author).*

For solar electricity, the opposite seasonal behaviour of demand and supply was equally problematic, particularly because wind energy offered a much cheaper form of renewable energy, with a seasonal behaviour very similar to the variations in demand (plus of course in both the cases of sun and wind the short-term variations due to weather variations, and in the solar case unavailability at nights). For this reason, photovoltaic installations remained a showcase for demonstrating "greenness" (Fig. 11.18). One interesting idea was to combine solar heat and electricity, by collecting the "waste" heat from the back-side of a solar cell system, but the system was never made commercial in Denmark, due to the climate-related limitations of solar energy use, but for countries at lower latitudes the idea could be taken up (Sørensen, 2010).

*Figure 11.17. Solar district heating array built 2010 in Hillerød, north Sealand (photo by author).*

*Figure 11.18. Detached dwelling with photovoltaic (left) and thermal solar collectors (Gilleleje, photo by author, 2011).*

Research and demonstration programmes for electric vehicles started in the late 1970ies (Jensen, 1978) and similar programs for hydrogen energy and fuel cell vehicles 20 years later (Sørensen, 1998). Both of these programs faded away in Denmark after some years. Current international efforts suggest that hybrid electric–diesel and later electric–hydrogen vehicles may play a role in the post-oil period.

### 11.4 Energy policy: plan or leave to the market?

Before the 1973–74 oil crisis, energy planning had been left by the Danish government to the energy companies, such as coal and oil providers and electric utilities, whether with private or public ownership. The only use of "plans" was the economic planning based on straight extrapolations mentioned at the beginning of Section 11.2. The oil crisis awakened the Danish government (and many others) to the strategic importance of energy supply, and hence the need for a certain amount of planning. The first outcome of this was the establishment of oil storage facilities that would at least give the country some months to reflect in case of another supply cut-off. Both Sweden and Denmark tried to negotiate with Norway for delivery of oil from their vast offshore reserves, but without success, as in exchange for oil Norwegians wanted control over key industries in its neighbouring countries. However, as regards electric power, a Nordic collaboration was already in place following the formal collaboration agreement NORDEL, formed in 1963 with the purpose of transmitting power to the other participants, in cases of power plant failures and on a planned basis (e.g. during dry years where the 100% Norwegian hydro power was low and could be predicted to be so long before the winter snow started to melt). New undersea cables between Denmark and both Norway and Sweden had been established by 1965, in addition to the existing short connection across the Sound to Sweden (see Fig. 10.23).

In 1974, the Ministry of Commerce issued the first Danish energy statement, based on an expected doubling of energy demand reached in 1988, 14 years ahead, exactly as projected by the "perspective plan" mentioned in Section 11.2 (Smidt, 1974). As an appendix to the statement, the electric utilities had been asked to provide their expectations for the electricity demand, which was a tripling of power demand over the same period, or 8.5% per year growth (Danish Ministry of Commerce, 1974). A second statement from the Ministry of Commerce had the growth in demand from 1975 to 1995 at 50%, provided that the government's proposed planned regulation was applied (Danish Ministry of Commerce, 1976; see Fig. 11.19).

The interesting thing here is the departure from simple forecasting based on extrapolation, the tool used at the time by nearly all economists. Because economic theory is based on the assumption that economic actors behave rationally (and have the necessary information to do so), then there is, if not a deterministic development, then at least a "most likely future" that the trick of forecasting is to predict. Against this paradigm, the new concept of "scenarios" was under development, acknowledging that there are a multitude of different possible futures, each of which requires certain actions to be taken in order to become realized. Such analysis of normative visions of future society had been introduced in the energy policy field (Eriksson, 1974; Sørensen, 1975) and soon became the preferred way of debating energy futures, in Denmark as well as internationally. Every interest group or political spokesperson could suggest such a normative scenario, and the task of the scholarly community (here energy scientists) would be to assess the consistency of the scenario assumptions, and to check whether there existed an implementation strategy that would take the system from the current state to the scenario layout. The implementation strategy would typically consist of a succession of decisions to be taken, primarily in governments, but often also involving attitude changes in the general population, in order for the scenario to materialize. If these decisions are not taken, one of course cannot blame the scenario people if the scenario does not come true (this needs to be said because the previous forecasters were often ridiculed when their forecasts turned out not to materialize). A broad range of internally consistent energy scenarios depicting different attitudes common in present societies have recently been discussed (Sørensen, 2008).

Figure 11.19 shows both the actual development and a number of government forecasts and plans, as well as scenarios from scientists and organizations outside the government. It is seen that in 1974, the government thought that the energy crisis had just caused a few years of delay in the exponential growth, which would soon resume its flight towards the skies. In some quarters outside the government, more realistic views on the prospects prevailed, as seen by the 1975 scenario (Sørensen, 1975) assuming that energy efficiency improvements would provide a good way of dealing with the problems of energy supply, because at the energy prices after 1974, a lot of energy efficiency measures would be cheaper than the energy they displaced. This particular scenario was also daring in suggesting that the entire Danish energy supply could be delivered by renewable sources, and the intermittency dealt with by using hydrogen (produced by wind power through electrolysis), an energy carrier capable of being stored, just like fossil fuels.

In 1976, the Danish Ministry of Commerce, at the time responsible for energy, reduced its growth expectation slightly, provoking a group of scientists to produce an alternative, published by the anti-nuclear organization OOA (Blegaa *et al.*, 1976; Fig. 11.20). The purpose was to show that the same standard of living and economic activity as assumed in the official plan could be achieved without nuclear energy, and the alternative plan therefore called for more use of natural gas and slightly more coal, in order to avoid the need for nuclear energy. After the publication of the 1981 government plan, a group of scientists again felt the need to provide an alternative, despite the fact that the new Ministry of Energy in fact had taken onboard many of the points of view expressed by the university scientists in 1975 and 1976 (e.g. regarding the need to provide a range of different scenarios).

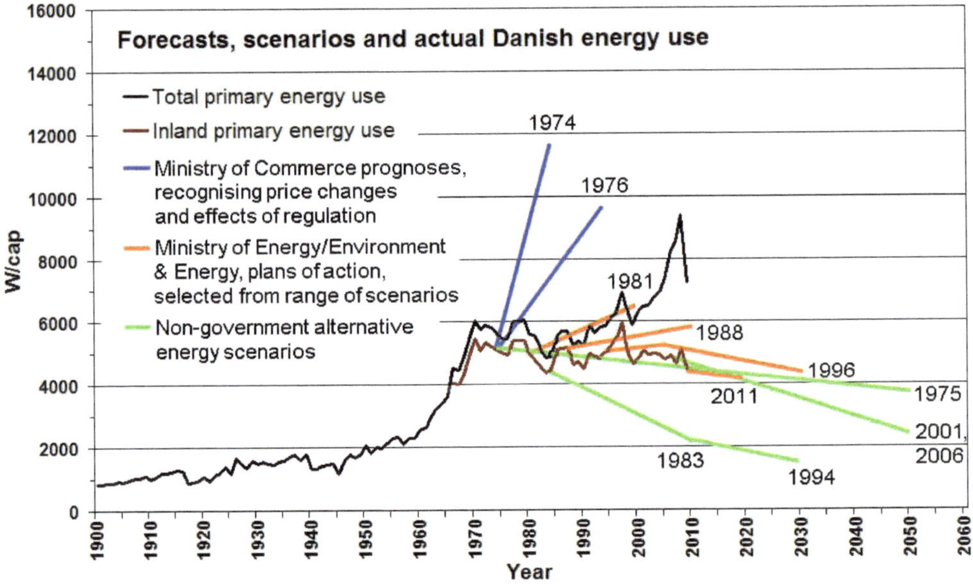


*Figure 11.19. The actual primary energy use in Denmark, with and without oil for international shipping (based on Danmarks Statistik, 2010), compared with a number of forecasts, plans and scenarios: Danish Ministry of Commerce (1974), 1976); Danish Ministry of Energy (1981, 1988); Danish Ministry of Environment and Energy (1996); Danish Government (2011); Sørensen (1975); Hvelplund et al. (1983); Sørensen et al. (1994, 2001); Lund and Mathiesen (2006). The data do not include food and muscle work by draft animals, and none of the planning work depicted takes into account international bunkers.*

The main objection brought forward by the group of independent scientists was that the government had underestimated the potential of energy efficiency measures. During the late 1970s, several technical studies had shown that the potential was not just a few per cent, but in many cases as high as a factor of 3–5, a result that was later confirmed internationally (Sørensen, 1982; Weizsäcker, 1998). This low-energy line of scenario work was continued by the Danish Technology Council

(Sørensen, *et al.*, 1994). The Danish Energy Agency, responsible for preparing the Government energy statements and plans of action, adopted the scenario method of the alternative energy plans, first a cautiouns presentation of 4 slightly different developments (Danish Ministry of Energy, 1981), but later by scenarios spanning from "business as usual" scenarios with no government intervention to strongly regulated low-energy scenarios. In the work leading up to the 1996 plan (Danish Energy Agency, 1995b), the highest variant had a primary energy increase of 19% from 1995 to 2030, while the lowest had a reduction by 31%.

*Figure 11.20. Covers of government (unknown designer) and OOA alternative energy (designer Annikki Tiirikkala) plans 1976, playing important roles in the debates for and against nuclear energy taking place all over Denmark.*

The latest government paper on strategic energy policy (Danish Government, 2011) suggests more wind power over the next ten years and 2–3% decline in primary energy. The statement neglects the shipping contribution mentioned (and detailed in Fig. 12.4) but stresses the need for substantial reduction in greenhouse gas emissions. However, nothing is suggested for the transport sector, considered outside the scope of Danish policy, as there are no automobile manufacturers in Denmark. The fact is that a lot could be done to improve the efficiency of vehicles used for transportation by regulation and taxation that would make people buy the most energy-efficient vehicles, especially as there is no pressure from a domestic car-industry. After having neglected the energy issue for 9 years, it appears that this government has elected to pay only lip service to the problem, facing prospects of losing the end-of-2011 election, among other things due to accusations of corruption (Hansen and Borg, 2009; Nielsen, 2010, Metz, 2010) and violation of international laws (on immigration; see Geist and Dahlin, 2011), as well as possibly the Danish constitution (by joining the 2nd Iraq war without UN mandate; Klingsey et al., 2007; Aagaard, 2010; Balslev, 2010, noting that the Danish Supreme Court finds the 1953 Danish Constitution too vague to support a conviction).

The value-based liberalization of the Danish energy sector (by different governments around 2000) has made the available options less transparent, by transferring energy system ownership to international interests different from those of the Danish population (cf. Section 10.1), and by artefacts (such as tradable emission permits) that allow European countries to avoid $CO_2$ reductions by instead paying money to less–developed countries, often without any clear emission effect.

## 11.5 Final update of demography and energy use

The demographic model that has been used in this book from the Neanderthal period up to the creation of Fig. 9.35 will again have to be adjusted for use in the 20th century, where sanitary conditions and medical advances caused new changes in mortality, which, together with altered social structure, caused birth rates to decrease as a result of planned parenthood, helped by easier access to contraceptive means, changed position of women in society, as well as the basic fact that a large number of children became a less–needed requirement for economic subsistence and for care of the elderly. However, at the same time the modelling is less required, because now detailed statistical data exist year by year for all the input variables, making the calculation of population distribution and its time changes trivially agree with data, except for the correction for emigration and immigration (Fig. 10.51), which in any case must be added manually.

The development in mortality data is shown in Fig. 11.21 for the period from 1835 to 2000. One notes the male death tolls during the wars with Germany in the mid-19th century and during the World War II resistance fights, and for both sexes the tolls of the Spanish flu epidemic in 1919. For females, the surplus death rate component attributable to childbirth disappears after year 1900, and infant death diminishes dramatically from about 1870 to 2000. After World War II, the general health of both genders up to around age 40 improves strongly. At ages around 60, the female survival increases until about 1980 and then slightly declines (due to smoking?). The corresponding male survival is lower (presumably due to work-related deaths) until 1980 but then catches up with the female level. Above age 70, female death rates decrease from 1940 to 2000, while the male death rates stay roughly constant. The same is true above age 80, and by year 2000 the female life expectancy (at birth) is 79, five years higher than that of males.

Looking at the raw statistical data (e.g. for 2009), it is seen in Fig. 11.22 that spurious fluctuations appear, particularly in death rates, due to the variations in population between adjacent years caused by immigration and emigration. As the basic behaviour is understood, no detailed modelling has been carried out here, and only some average trends are sought by using two sets of fixed birth and death rates in the pre-existing demographic model, one averaged over the period 1880–1940, the other for 1940–2010 (Fig. 11.24). The resulting calculated population variations are compared with data in Fig. 11.23.

The data on energy use and estimates of end-use distributions were last made for the year 1948 (Table 10.1 and 10.2). In Tables 11.1 to 11.4, similar data are presented for the years 1972 and 2009, where 1972 represents the last year before the energy crisis that halted the period of exponential growth in energy use, and 2009 the most recent year with all data available, at the time of writing.

The 1972 data in Table 11.1 show a tripling of low-temperature heat usage for space heating, as well as substantial growth for other uses, such as in vacation houses, in building space used for industry and commerce, and of course the waste heat associated with increased electricity production, despite some use of such heat for district heating. The increase in medium-temperature heat is primarily in the energy sector, notably from oil refineries built during the interval between the years selected for Tables 10.1 and 11.1. A six fold increase in energy use took place in the transportation sector, both for transportation of people and goods, and also linked to increased repair and delivery services. A particular component is the energy associated with transport to and from vacation houses (second homes built in large numbers during the period from 1948 to 1972). Finally, there is in all sectors a nine fold increase in use of electric power.

Table 11.2 shows the primary energy use in Denmark by 2009. Higher efficiency obtained by building insulation and tightness retrofitting has caused the low-temperature heat usage to decline relative to 1972. The medium-temperature heat is up, e.g. due to the noted energy cost of the treatment of fossil fuels in refineries and gas treatment plants.

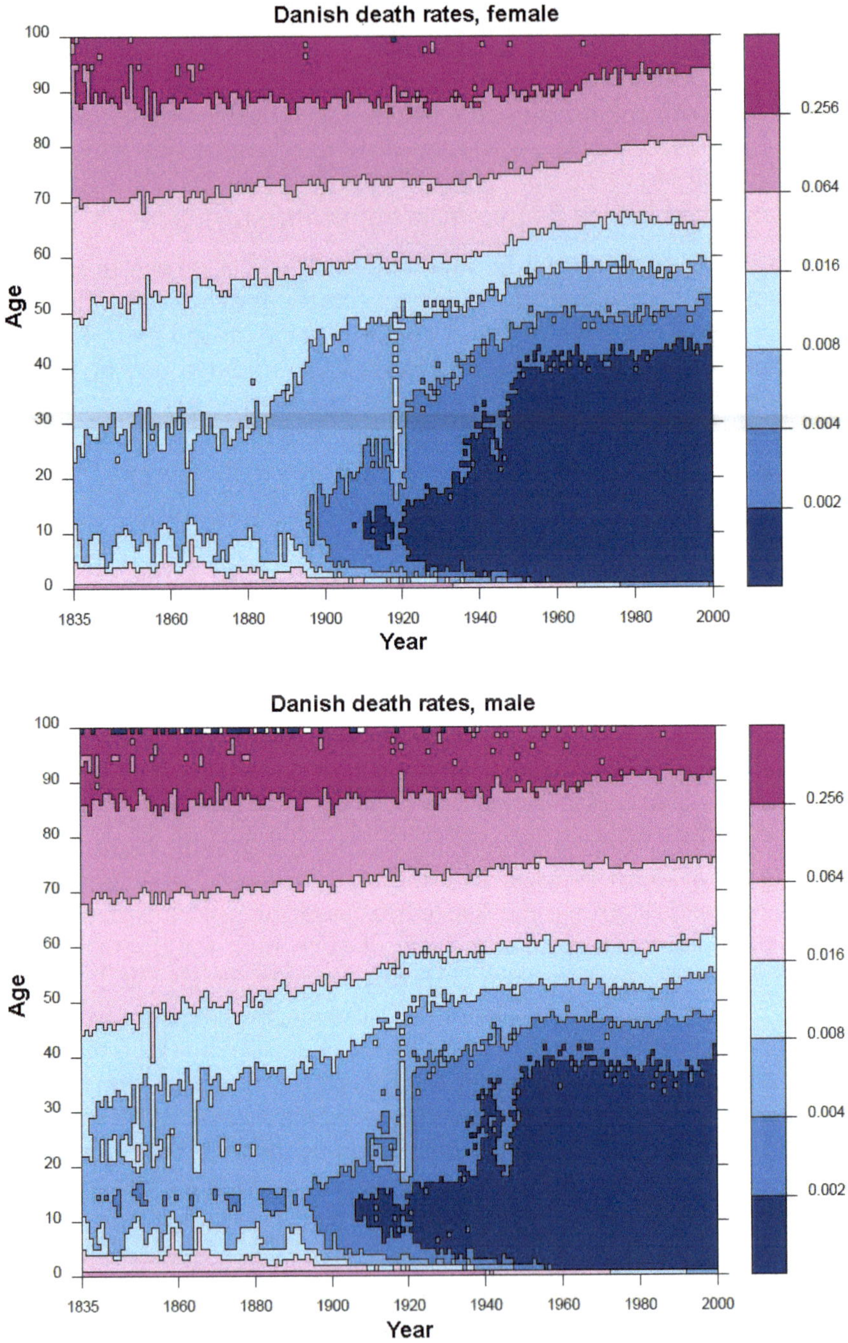


*Figure 11.21a, b. Female (top) and male (below) death rates as a function of age and time, from 1835 to 2000 (Andreev, 2003).*

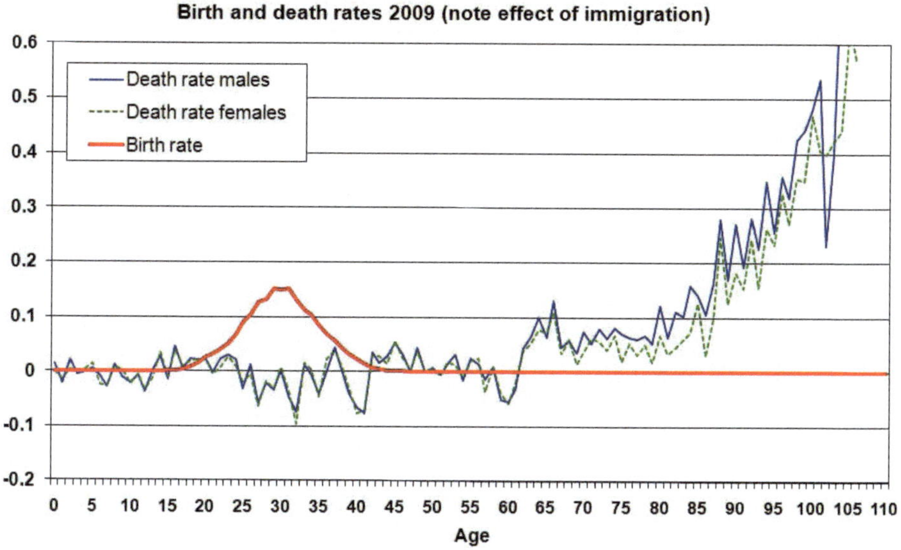


*Figure 11.22. Raw 2009 data for Danish birth and death rates as a function of age, relative to population one year earlier. Death rate data include fluctuations caused not by deaths but by emigration and immigration (Danmarks Statistik, 2010).*

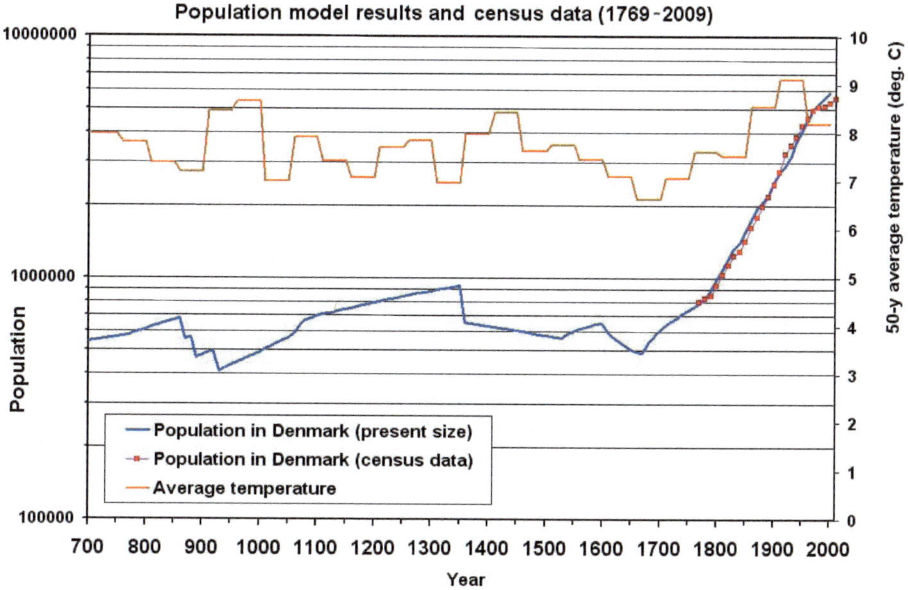


*Figure 11.23. Results of population model (cf. Fig. 9.35) using the two datasets shown in Fig, 11.24. Because the last set averages over the period 1940–2010, the calculated population is first slightly low and then slightly high.*

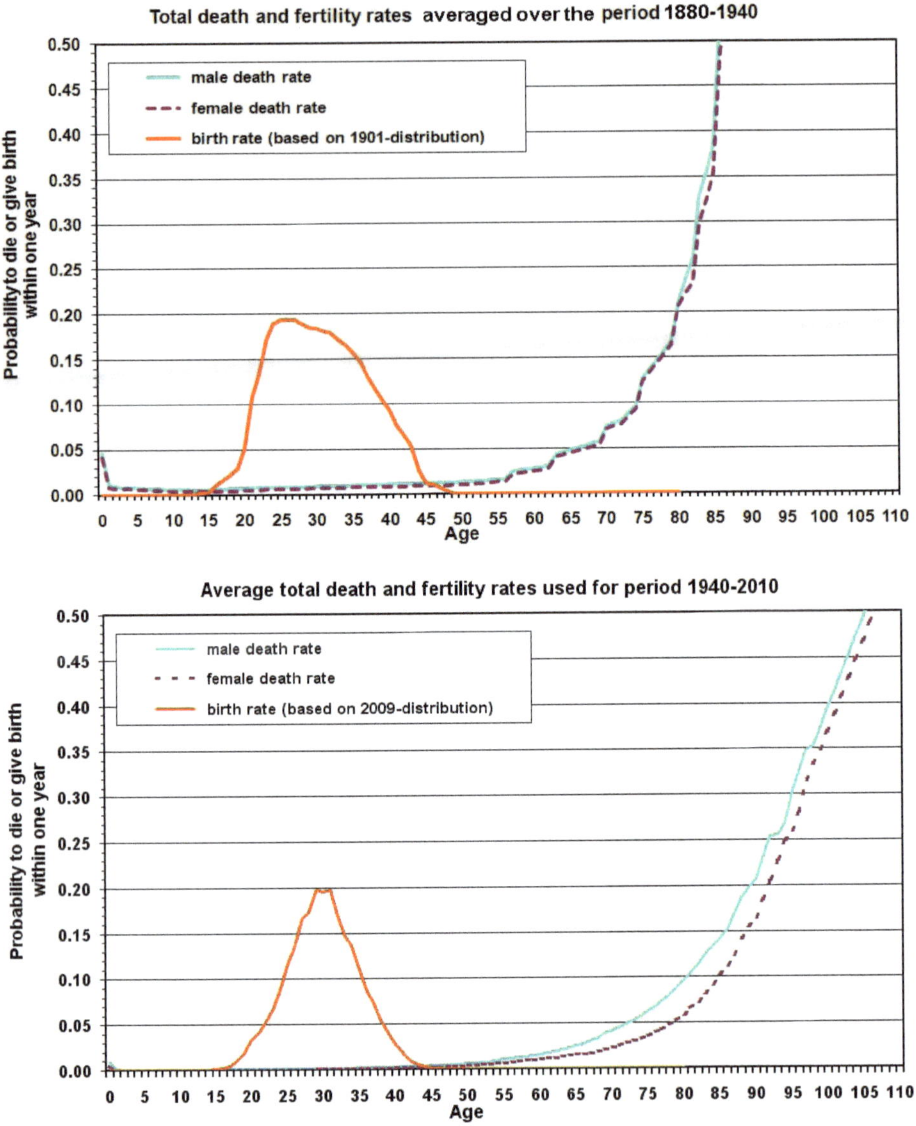


*Figure 11.24a, b. Average birth and death rates for the periods 1880–1940 (top) and 1940–2010 (below) used in the population model together with earlier rates (e.g. Fig. 9.36) to obtain the results shown in Fig. 11.23.*

Energy for transportation has increased by a factor of 2.6 between 1972 and 2009, partly due to more domestic transportation services, but primarily due to the dramatically increased activity on international shipping markets. Use of electricity is up by 90%, with increases in all sectors, including particularly large contributions from domestic services such as retail commerce.

In Tables 11.3 and 11.4, the end-use energy use is estimated by analogy with previous tables (the last being Table 10.2). In 1972, the total end-use energy derived is 2.75 times higher than in 1948, while the increase in primary energy is a factor of 3.38. By 2009, the end-use energy obtained is 1.65 times that in 1972, while the primary energy is 1.42

*Table 11.1. Energy use per capita in Denmark, 1972, for an adult person (based on Danmarks Statistik, 2010; cf. Table 10.1).*

| *Denmark, 1972* | *Human energy* | *Energy from fuels and external energy flows* | | | | | | |
|---|---|---|---|---|---|---|---|---|
| *Rate of energy use (average W for an adult person)* | Food energy | Low-temperature heat | Medium-temperature heat | High-temperature heat | Mobile power (excl. food) | Mechanical power (stationary) | Electric power (stationary) | TOTAL |
| **Liveable surroundings** | | | | | | | | **1205** |
| Basic human metabolism | 100 | | | | | | | **100** |
| Indirect energy (e.g.through windows) | | 30 | | | | | | **30** |
| House and garden tools, equipment | 5 | 2 | | | | 1 | | **8** |
| Active space heating | | 987 | | | | | 10 | **997** |
| Artificial light | | | 1 | | | | 68 | **69** |
| Active air exchange, air cooling | | | | | | | 1 | **1** |
| **Food and water provision** | | | | | | | | **663** |
| Hunting, gathering, gardening | 1 | 76 | | | | | 3 | **80** |
| Fishing, aquaculture | 2 | | | | 68 | | | **70** |
| Agriculture | 5 | 20 | | | 57 | | 31 | **113** |
| Slaughtering, food processing | 4 | 50 | 23 | | | 23 | 11 | **111** |
| Preparing, cooking, kitchen utensils | 2 | | 9 | | | | 28 | **39** |
| Inhouse hot and cold water provision | | 245 | | | | | 5 | **250** |
| **Security, assault, national image nursing** | | | | | | | | **96** |
| Weapons, delivery vehicles, etc. | 1 | | | | 59 | | 1 | **61** |
| Government and legal institutions | 1 | 10 | | | | | 1 | **12** |
| Administration and police | 1 | 10 | | | | | 1 | **12** |
| Sports and public information facilities | 2 | 8 | | | | | 1 | **11** |
| **Health-related efforts** | | | | | | | | **80** |
| Food preservation,clothes washing,etc. | 2 | | | | | | 45 | **47** |
| Water and waste management | 1 | 1 | | | | | 1 | **3** |
| Doctors, medicine | 1 | 11 | | | | | 3 | **15** |
| Hospitals | 1 | 11 | | | | | 3 | **15** |
| **Relations and leisure** | | | | | | | | **433** |
| Music instruments, concert halls, etc. | 0.3 | 4 | | | | | 1 | **5** |
| Theater, dance, TV broadcast, etc. | 0.5 | 7 | | | | | 1 | **9** |
| Toys, furniture, arts and crafts | 2 | | | | | | 1 | **3** |
| Written material, games, software etc. | 0 | | | | | | | **0** |
| Computers,multimedia,communication | 0 | | | | | | 18 | **18** |
| Vacations,2nd home,social transportation | 6 | 148 | | | 243 | | 1 | **398** |
| **Occupational and learning activities** | | | | | | | | **2444** |
| Schools, further education, research | 1 | 69 | | | | | 8 | **78** |
| Construction and building | 8 | 2 | | | 2 | 34 | 3 | **49** |
| Manufacture | 5 | 300 | 34 | 17 | 1 | 113 | 99 | **569** |
| Trade, private services (excl. transp.) | 8 | 319 | | | | 15 | 57 | **399** |
| Transportation of goods (incl. Internat,) | 8 | | | | 974 | | | **982** |
| Transport(delivery,services,commuting) | 10 | | | | 357 | | | **367** |
| **Raw materials procurement** | | | | | | | | **1436** |
| Mining and minerals extraction | 4 | | | | 9 | 11 | | **24** |
| Forestry resources | 1 | | | | 23 | | 2 | **26** |
| Industrial raw or recycled materials | 4 | 5 | 20 | 30 | 5 | 10 | | **74** |
| Energy provision | 1 | | 127 | | | | 6 | **134** |
| Energy conversion/distribution losses | 1 | 1177 | | | | | | **1178** |
| TOTAL | **189** | **3492** | **214** | **47** | **1798** | **207** | **410** | **6357** |

*Table 11.2. Energy use per capita in Denmark, 2009, for an adult person (based on Danmarks Statistik, 2010).*

| *Denmark, 2009* | *Human energy* | *Energy from fuels and external energy flows* | | | | | | |
|---|---|---|---|---|---|---|---|---|
| *Rate of energy use (average W for an adult person)* | Food energy | Low-temperature heat | Medium-temperature heat | High-temperature heat | Mobile power (excl. food) | Mechanical power (stationary) | Electric power (stationary) | TOTAL |
| **Liveable surroundings** | | | | | | | | **1047** |
| Basic human metabolism | 110 | | | | | | | **110** |
| Indirect energy (e.g.through windows) | | 36 | | | | | | **36** |
| House and garden tools, equipment | 5 | 4 | | | | 6 | 6 | **21** |
| Active space heating | | 800 | | | | | 21 | **821** |
| Artificial light | | | 1 | | | | 50 | **51** |
| Active air exchange, air cooling | | | | | | | 8 | **8** |
| **Food and water provision** | | | | | | | | **624** |
| Hunting, gathering, gardening | 1 | 40 | | | | | 2 | **43** |
| Fishing, aquaculture | 1 | | | | 42 | | | **43** |
| Agriculture | 4 | 25 | | | 70 | | 37 | **136** |
| Slaughtering, food processing | 4 | 30 | 25 | | | 25 | 24 | **108** |
| Preparing, cooking, kitchen utensils | 2 | | 7 | | | | 40 | **49** |
| Inhouse hot and cold water provision | | 230 | | | | | 15 | **245** |
| **Security, assault, national image nursing** | | | | | | | | **64** |
| Weapons, delivery vehicles, etc. | 1 | | | | 15 | | 1 | **17** |
| Government and legal institutions | 1 | 8 | | | | | 4 | **13** |
| Administration and police | 1 | 8 | | | | | 10 | **19** |
| Sports and public information facilities | 2 | 10 | | | | | 3 | **15** |
| **Health-related efforts** | | | | | | | | **99** |
| Food preservation,clothes washing,etc. | 2 | | | | | | 54 | **56** |
| Water and waste management | 1 | 1 | | | | | 2 | **4** |
| Doctors, medicine | 1 | 8 | | | | | 10 | **19** |
| Hospitals | 1 | 10 | | | | | 9 | **20** |
| **Relations and leisure** | | | | | | | | **623** |
| Music instruments, concert halls, etc. | 0.3 | 4 | | | | | 2 | **6** |
| Theater, dance, TV broadcast, etc. | 0.5 | 6 | | | | | 6 | **13** |
| Toys, furniture, arts and crafts | 2 | | | | | | 3 | **5** |
| Written material, games, software etc. | 0 | | | | | | 6 | **6** |
| Computers,multimedia,communication | 0 | | | | | | 49 | **49** |
| Vacations, social transportation | 6 | 200 | | | 333 | | 5 | **544** |
| **Occupational and learning activities** | | | | | | | | **5180** |
| Schools, further education, research | 1 | 56 | | | | | 22 | **79** |
| Construction and building | 8 | 2 | | | 2 | 37 | 9 | **58** |
| Manufacture | 5 | 140 | 30 | 10 | 4 | 79 | 180 | **448** |
| Trade, private services (excl. transp.) | 8 | 198 | | | | 7 | 203 | **416** |
| Transportation of goods (incl. Internat.) | 6 | | | | 3696 | | | **3702** |
| Transport(delivery,services,commuting) | 7 | | | | 470 | | | **477** |
| **Raw materials procurement** | | | | | | | | **1416** |
| Mining and minerals extraction | 1 | | | | 6 | 25 | | **32** |
| Forestry resources | 1 | | | | 31 | | 3 | **35** |
| Industrial raw or recycled materials | 1 | 10 | 25 | 30 | 10 | 10 | 1 | **87** |
| Energy provision | 1 | | 298 | | | | 0 | **299** |
| Energy conversion/distribution losses | 1 | 962 | | | | | | **963** |
| TOTAL | **186** | **2788** | **386** | **40** | **4679** | **189** | **785** | **9053** |

times higher, indicating the substantial improvement in energy conversion efficiency carried through by the measures following the 1973 oil crisis. The achievements in raising energy efficiency were possible by available technology and are independent of whether the activity level requiring energy is increased, which is the case for container ships, for example.

*Table 11.3. Estimated end-use energy per capita in Denmark, 1972, for an adult person (with use of Danmarks Statistik, 2010).*

| *Denmark, 1972* | *Human energ* | *Energy from fuels and external energy flows* | | | | | | |
|---|---|---|---|---|---|---|---|---|
| *Rate of energy end-use (average W for an adult person)* | Food energy | Low-temperature heat | Medium-temperature heat | High-temperature heat | Mobile (excl. food) | Mech. power (stat.) | Electric power (stat.) | TOTAL |
| **Liveable surroundings** | | | | | | | | **447** |
| Basic human metabolism | 68.0 | | | | | | | **68** |
| Indirect energy (e.g.through windows) | | 30 | | | | | | **30** |
| House and garden tools, equipment | 1.7 | 0.7 | | | | 0.5 | | **3** |
| Active space heating | | 329 | | | | | 2.5 | **332** |
| Artificial light | | | 0.1 | | | | 14 | **14** |
| Active air exchange, air cooling | | | | | | | 0.2 | **0** |
| **Food and water provision** | | | | | | | | **253** |
| Hunting, gathering, gardening | 0.3 | 25 | | | | | 1.5 | **27** |
| Fishing, aquaculture | 9,7 | | | | 18 | | | **18** |
| Agriculture | 1.7 | 7 | | | 15 | | 25 | **49** |
| Slaughtering, food processing | 1.3 | 17 | 12 | | | 12 | 10 | **52** |
| Preparing, cooking, kitchen utensils | 0.7 | | 4 | | | | 25 | **30** |
| Inhouse hot and cold water provision | | 77 | | | | | 0.4 | **77** |
| **Security, assault, national image nursing** | | | | | | | | **25** |
| Weapons, delivery vehicles, etc. | 0.3 | | | | 13 | | 0.3 | **14** |
| Government and legal institutions | 0.3 | 3.3 | | | | | 0.3 | **4** |
| Administration and police | 0.3 | 3.3 | | | | | 0.3 | **4** |
| Sports and public information facilities | 0.7 | 2.7 | | | | | 0.5 | **4** |
| **Health-related efforts** | | | | | | | | **54** |
| Food preservation,clothes washing,etc. | 0.7 | | | | | | 40 | **41** |
| Water and waste management | 0.3 | 0.3 | | | | | 1.0 | **2** |
| Doctors, medicine | 0.3 | 3.7 | | | | | 2.0 | **6** |
| Hospitals | 0.3 | 3.7 | | | | | 2.0 | **6** |
| **Relations and leisure** | | | | | | | | **149** |
| Music instruments, concert halls, etc. | 0.1 | 1.3 | | | | | 1.0 | **2** |
| Theater, dance, TV broadcast, etc. | 0.2 | 2.3 | | | | | 0.5 | **3** |
| Toys, furniture, arts and crafts | 0.7 | | | | | | 10.5 | **11** |
| Written material, games, software etc. | | | | | | | | **0** |
| Computers,multimedia,communication | | | | | | | 6 | **6** |
| Vacations, social transportation | 2.0 | 49 | | | 75 | | 0.5 | **127** |
| **Occupational and learning activities** | | | | | | | | **888** |
| Schools, further education, research | 0.3 | 23 | | | | | 4 | **27** |
| Construction and building | 2.7 | 0.7 | | | 0.6 | 17 | 2 | **23** |
| Manufacture | 1.7 | 100 | 17 | 12 | 0.3 | 60 | 80 | **271** |
| Trade, private services (excl. transp.) | 2.7 | 106 | | | | 7 | 20 | **136** |
| Transportation of goods (incl. Internat,) | 2.7 | | | | 325 | | | **328** |
| Transport(delivery,services,commuting) | 3.3 | | | | 100 | | | **103** |
| **Raw materials procurement** | | | | | | | | **132** |
| Mining and minerals extraction | 1.3 | | | | 3 | 6 | | **10** |
| Forestry resources | 0.3 | | | | 7 | | 2 | **9** |
| Industrial raw or recycled materials | 1.3 | 2 | 10 | 22 | 2 | 5 | | **42** |
| Energy provision | 0.3 | | 65 | | | | 6 | **71** |
| Energy conversion/distribution losses | 0.3 | 0 | | | | | | **0** |
| TOTAL | **97** | **787** | **108** | **34** | **558** | **108** | **258** | **1949** |

The long-term development in both primary, delivered and end-use energy distribution on categories of usage or on energy qualities can now be finalized to the present time, in continuation of the results shown in Figs. 9.47 and 9.48. Figure 11.25a shows the behaviour of primary and delivered energy, giving the central conversion losses on top of the distribution of delivered energy on usage sectors. The dramatic increase in energy used in connection with occupational activities stands out, and one notes the wasteful use of comfort heat in 1948–1972.

*Table 11.4. Estimated end-use energy per capita in Denmark, 2009, for an adult person (with use of Danmarks Statistik, 2010).*

| *Denmark, 2010* | *Human energy* | *Energy from fuels and external energy flows* | | | | | | |
|---|---|---|---|---|---|---|---|---|
| *Rate of energy end-use (average W for an adult person)* | Food energy | Low-temperature heat | Medium-temperature heat | High-temperature heat | Mobile (excl. food) | Mech. power (stat.) | Electric power (stat.) | TOTAL |
| **Liveable surroundings** | | | | | | | | **490.1** |
| Basic human metabolism | 68 | | | | | | | **68** |
| Indirect energy (e.g.through windows) | | 36 | | | | | | **36** |
| House and garden tools, equipment | 1.7 | 1.3 | | | | 2 | 5 | **10** |
| Active space heating | | 330 | | | | | 15 | **345** |
| Artificial light | | | 0.1 | | | | 28 | **28** |
| Active air exchange, air cooling | | | | | | | 3 | **3** |
| **Food and water provision** | | | | | | | | **286** |
| Hunting, gathering, gardening | 0.3 | 13.0 | | | | | 2 | **15** |
| Fishing, aquaculture | 0.3 | | | | 12 | | | **12** |
| Agriculture | 1.3 | 8.5 | | | 20 | | 34 | **64** |
| Slaughtering, food processing | 1.3 | 15.0 | 15 | | | 12 | 24 | **67** |
| Preparing, cooking, kitchen utensils | 0.7 | | 4 | | | | 20 | **25** |
| Inhouse hot and cold water provision | | 100.0 | | | | | 3 | **103** |
| **Security, assault, national image nursing** | | | | | | | | **24** |
| Weapons, delivery vehicles, etc. | 0.3 | | | | 4 | | 1 | **5** |
| Government and legal institutions | 0.3 | 2.7 | | | | | 2 | **5** |
| Administration and police | 0.3 | 2.7 | | | | | 5 | **8** |
| Sports and public information facilities | 0.7 | 3.0 | | | | | 2 | **6** |
| **Health-related efforts** | | | | | | | | **69** |
| Food preservation,clothes washing,etc. | 0.7 | | | | | | 50 | **51** |
| Water and waste management | 0.3 | 0.5 | | | | | 2 | **3** |
| Doctors, medicine | 0.3 | 2.7 | | | | | 5 | **8** |
| Hospitals | 0.3 | 3.3 | | | | | 4 | **8** |
| **Relations and leisure** | | | | | | | | **241** |
| Music instruments, concert halls, etc. | 0.1 | 1.3 | | | | | 1 | **2** |
| Theater, dance, TV broadcast, etc. | 0.2 | 2.0 | | | | | 3 | **5** |
| Toys, furniture, arts and crafts | 0.7 | | | | | | 2 | **3** |
| Written material, games, software etc. | | | | | | | 6 | **6** |
| Computers,multimedia,communication | | | | | | | 45 | **45** |
| Vacations, social transportation | 2.0 | 65 | | | 110 | | 3 | **180** |
| **Occupational and learning activities** | | | | | | | | **1906** |
| Schools, further education, research | 0.3 | 22 | | | | | 12 | **34** |
| Construction and building | 2.7 | 0.7 | | | 1 | 18 | 6 | **28** |
| Manufacture | 1.7 | 50 | 20 | 9 | 1 | 40 | 150 | **272** |
| Trade, private services (excl. transp.) | 2.7 | 70 | | | | 3 | 120 | **196** |
| Transportation of goods (incl. Internat.) | 2.0 | | | | 1232 | | | **1234** |
| Transport(delivery,services,commuting) | 2.3 | | | | 140 | | | **142** |
| **Raw materials procurement** | | | | | | | | **282** |
| Mining and minerals extraction | 0.3 | | | | 2 | 13 | | **15** |
| Forestry resources | 0.3 | | | | 10 | | 2 | **12** |
| Industrial raw or recycled materials | 0.3 | 4 | 15 | 25 | 3.3 | 5.0 | 0.7 | **53** |
| Energy provision | 0.3 | | 200 | | | | 0 | **200** |
| Energy conversion/distribution losses | 0.3 | 0 | | | | | | **0** |
| TOTAL | **93** | **734** | **254** | **34** | **1535** | **93** | **556** | **3299** |

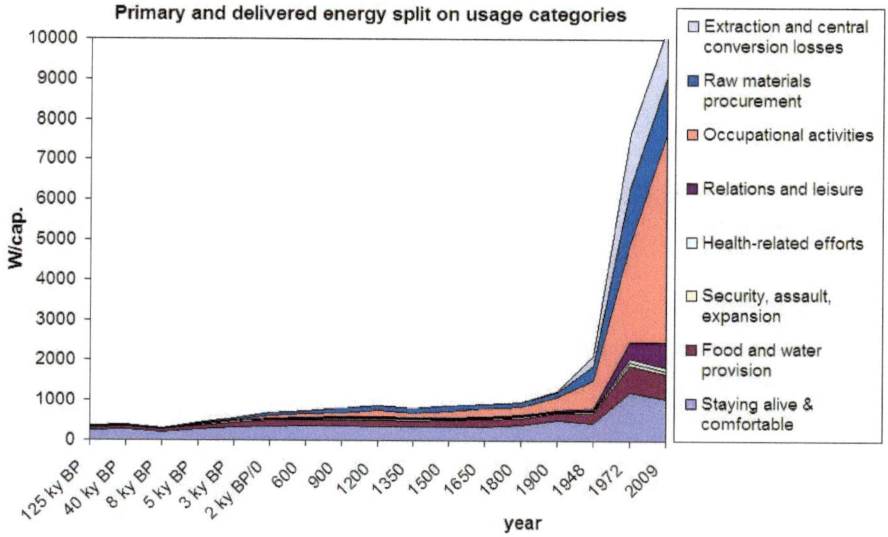


*Figure 11.25a. Estimated energy use per capita in Denmark from 125 ky BP to 2009, distributed on categories of usage.*

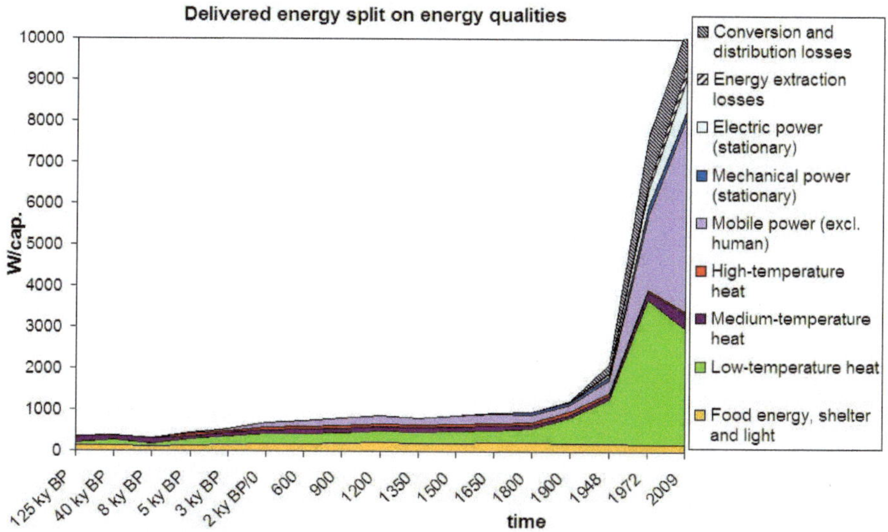


*Figure 11.25b. Estimated energy use per capita in Denmark from 125 ky BP to 2009, distributed on types of energy employed.*

Figure 11.25b shows the increase in energy used for transportation, not only in international shipping, and the end-use estimates in Fig. 11.26a and 11.26b emphasize this, as the general improvement in efficiency after 1972 is ultimately lower for vehicles used for transportation than for the appliances and other equipment used in other sectors. Fig. 11.27

summarizes the total energy use over the more than 100 000 years covered by this study. In Figures 11.25 to 11.27, the enormous energy use during the latest century hides most of what happened before that. Therefore, the same figures will be presented in Chapter 12 on a logarithmic scale.

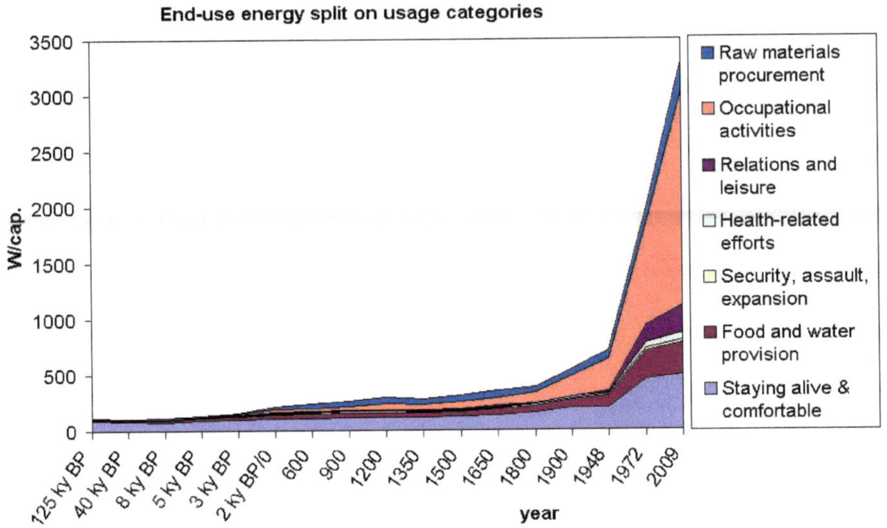


*Figure 11.26a. Estimated end-use energy per capita in Denmark from 125 ky BP to 2009, distributed on categories of usage.*

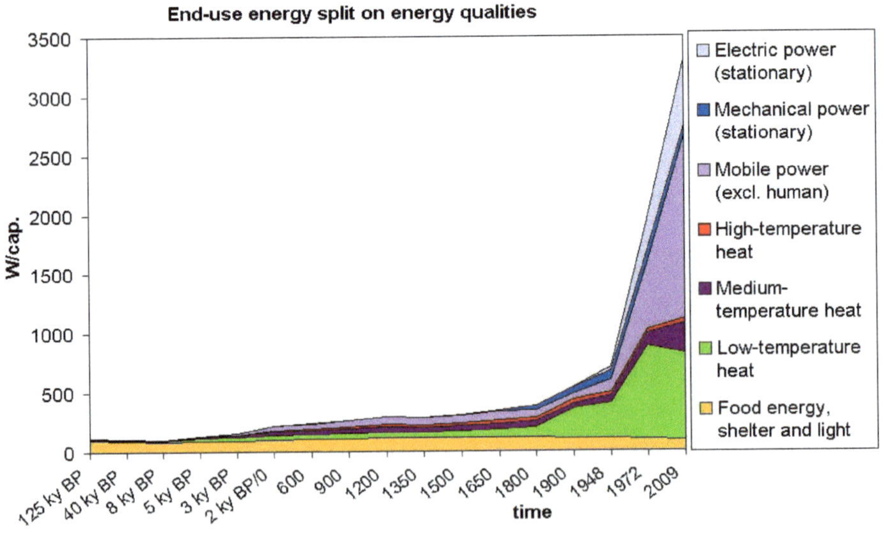


*Figure 11.26b. Estimated end-use energy per capita in Denmark from 125 ky BP to 2009, distributed on types of energy.*

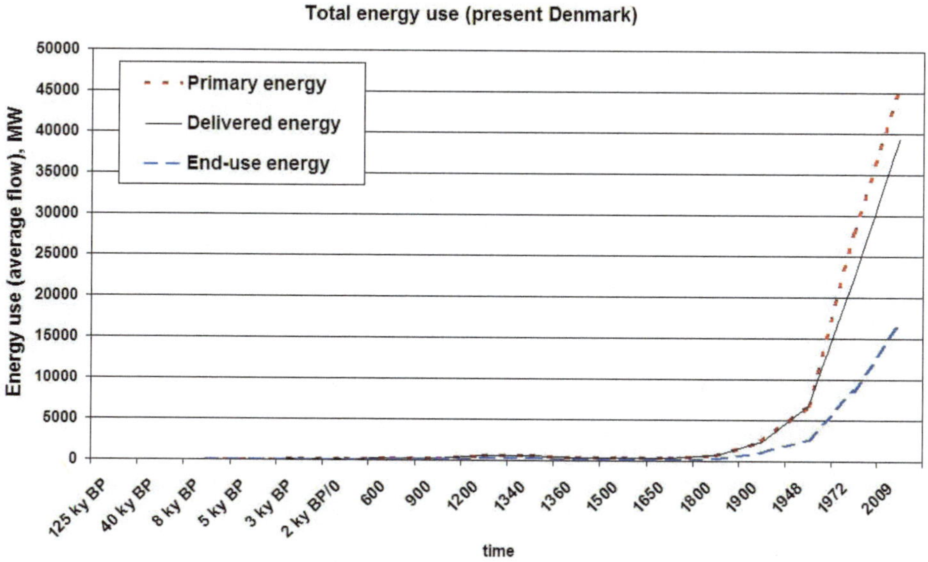


*Figure 11.27. The long-term development in total primary, delivered and end-use energy in Denmark, adjusted to usage within the present territory by the population occupying that area at a given time (cf. Fig. 9.49). The figure will appear on a logarithmic scale in Chapter 12.*

### *References:*

Aagaard, C., 2010. Danske jurister: Iraq-krigen var ulovlig. *Information* 29th January, p. 8.

Al-Jazari, B., 1206. *Book of Ingenious Mechanical Devices.* English translation by D. Hill, Riedel Publ. (1974), Dordrecht.

Andreev, K., 2003. *Evolution of the Danish Population from 1835 to 2000.* "Monographs on Population Aging", No. 9, Odense University Press. Available from the Max-Planck-Gesellschaft at http://www.demogr.mpg.de/books/odense/9/

Arley, N., Skov, H., 1956. *Atomkraft. Atomalderens tekniske, militære, medicinske og biologiske problemer.* Gyldendal, Copenhagen.

Atomenergikommissionen, 1973. Memorandum J. No. 18-7f from Danish Nuclear Commission, Copenhagen.

Balslev, J., 2010. Iraq-krigen sluttede I Højesteret. *Fyens Stiftstidende,* 18th March, p. 12.

Banu Musa, A., Banu Musa, M., Banu Musa, H., 850. *Book of ingenious devices.* English translation by D. Hill, Kluwer Academic Press, Dordrecht, 1978.

Barrère, M., 1975. La sûreté des centrales nucléaires. *La Recherche* **6** (55), 305–317.

Bellis, M., 1996. Inventors of the modern computer. Seen at http://inventors.about.com/library/weekly/aa33099.htm (accessed February 2011; similar stories can be found in the magazine *BYTE*, 1975–1998).

Bernhardt, S., 1977. VKR gav grønt lys for Barsebäck. Interview with Ivar Nørgaard, Minister of Commerce. *Ekstra Bladet*, 22nd September, p. 4.

Blegaa, S., Hvelplund, F., Jensen, J., Josephsen, L., Linderoth, H., Meyer, N., Balling, N., Sørensen, B., 1976. *Skitse til alternativ energiplan for Danmark.* OOA, Nuclear Power Information Organisation, Copenhagen (in collaboration with OVE, Renewable Energy Organisation, Copenhagen).

Bohr, N., 1950. *Open letter to the United Nations.* Schultz Publ., Copenhagen.

Booth, W., 1987. Postmortem on Three Mile Island. *Science* **238**, 1342–1345.

Carson, R., 1962. *Silent Spring.* Houghton Mifflin Company, New York.

Center for Energibesparelser, 2011. Consumer advisory institution at http://www.goenergi.dk

Dahlin, U., 2010. COP 15-dom er begmand til politi og lømmelpakke. *Information*, 16th December.

Danielsen, O., 2005. *Atomkraften under pres. Dansk debat om atomkraft 1974–85.* Roskilde University Press, Copenhagen.

Danish Government, 2011. *Energistrategi 2050 – fra kul, olie og gas til grøn energi.* Copenhagen.

Danish Energy Agency, 1995a. *Progress report on the economy of centralized biogas plants.* Copenhagen.

Danish Energy Agency, 1995b. *Danmarks energifremtider.* Copenhagen

Danish Energy Agency, 1996. Biomass for energy – Danish solutions. Copenhagen.

Danish Ministry of Commerce, 1974. *Danmarks energiforsyning.* Energy Policy Statement, with appendices including *Koordineret Kraftværksudbygning i 1980'erne,* written by a utility committee. Handelsministeriet. Copenhagen.

Danish Ministry of Commerce, 1976. *Dansk energipolitik 1976,* with 5 appendices. Handelsministeriet, Copenhagen.

Danish Ministry of Energy, 1981. *Energiplan 81.* Copenhagen.

Danish Ministry of Energy, 1988. *Statusnotat Energiplanlægning.* Copenhagen.

Danish Ministry of Environment and Energy, 1996. *Energy 21. The Danish Government's Action Plan for Energy 1996,* Copenhagen.

Danmarks Statistik, 1959. Danmarks Energiforsyning 1900–1958. *Statistiske Undersøgelser* No. 2, Danish Department of Statistics, Copenhagen.

Danmarks Statistik, 1967. Danmarks Energiforsyning 1948–1965. *Statistiske Undersøgelser* No. 20, Danish Department of Statistics, Copenhagen.

Danmarks Statistik, 2007. Energibalancer 2006, Miljø og Energi. *Statistiske Efterretninger* 2007:8, Danish Department of Statistics, Copenhagen.

Danmarks Statistik, 2010. *Statistikbank*. Energi og miljø. Energi. Energiforbrug efter branche. Data og tal. http://www.statistikbanken.dk, Danish Department of Statistics, Copenhagen.

Dansk Datamuseum, 2010. Historical collection under establishment by Dansk Datahistorisk Forening. http://www.datamuseum.dk

DR, 2011. Danish Television program DR1 Evening Newscast on 24th February, Danish Broadcasting Corporation, Copenhagen.

ELSAM, 1972. *Atomet i fremtidens el-forsyning*. Brochure on nuclear power, Collaborative Union of Power Plant Operators in Jutland and Funen.

Eriksson, B., 1974. Konsekvenser af nogra olike energiannvändingsnivåer i samhället. In *Energi – inta endast en fråge om teknik* (B. Jungen, ed.), pp. 116–122. Centrum för Tvärvetenskap, Göteborgs Universitet, Gothenborg.

Falk, J., 1982. *Global Fission. The Battle over Nuclear Power.* Oxford University Press, Melbourne.

Forbes, I., Ford, D., Kendall, H., MacKenzie, J., 1971. Nuclear reactor safety: an evaluation of new evidence. *Nuclear News,* September, 32–40.

Freeth, T., *et al.*, 2006. Decoding the ancient Greek astronomical calculator known as the Antikythera Mechanism. *Nature* **444**, 587–591.

Freeth, T., Jones, A., Steele, J., Bitsakis, Y., 2008. Calendars with Olympiad display and eclipse prediction on the Antikythera Mechanism. *Nature* **454**, 614–617.

Friberg, M., Hettne, B., Leksell, I., Tompuri, G., 1974. Utveckling, magt och energi i ett internationellt konfliktperspektiv. In *Energi – inta endast en fråge om teknik* (B. Jungen, ed.), pp. 51–69. Centrum för tvärvetenskap, Göteborgs Universitet, Gothenborg.

Geertsen, U. (ed.), *et al.*, 1975. *Energi 1–6*. Six primers: Energy policy; Nuclear Power?; Energy economics and planning; Energy use and environmental problems; Energy resources and alternative energy policies; Alternative energy sources–alternative energy policy. Danish Government Energy Information Committee, Copenhagen.

Geist, A., Dahlin, U., 2011. Statsløse: det bør Birthe Rønn svare på. *Information,* 29th January, p. 4; Eksperter: Rønns forklaring holder ikke. *Information,* 2nd March, p. 6.

Gram, C., Hestvik, O., Isaksson, H., Jacobsen, P., Jensen, J., Naur, P., Petersen, B., Svejgaard, B., 1963. GIER - a Danish computer of medium size. *IEEE Transactions on Electronic Computers* **EC-12** (5), 629–650.

Hannerz, K., 1983. Emergency cooling. *Nuclear Eng. Int.*, December, p. 41.

Hansen, H., 1975. Til forsvar for den akademiske tradition. *Ingeniørens Ugeblad*, 2nd May, expanded in *Information*, 16th Januar 1976.

Hansen, J., Borg, O., 2009. Små sko, ringe dømmekraft eller systemfejl? *Jyllands-Posten*, 29th March, p. 8.

Hansen, S., 1983. *Økonomisk vækst i Danmark*. Vol. II. Tables. Akademisk Forlag, Copenhagen.

Hill, D., 1984. *A History of Engineering in Classical and Medieval Times*. Routledge, London.

Hoffmann-Jørgensen, J., Væth, M., 1979. Om baggrunden for kritikken af de sandsynlighedsteoretiske og statistiske beregninger i WASH-1400, Reactor Safety Study ("Rasmussen rapporten"). Various Publication Series No. 31, Mathematical Institute, Aarhus University.

Hollerith, H., 1889. An electric tabulating system. *The Quarterly* **10** (16) 238–255. Columbia University School of Mines, New York (The company of Hollorith was later merged with two others to form IBM).

Hvelplund, F., Illum, K., Jensen, J., Meyer, N., Nørgård, J., Sørensen, B., 1983. *Energi for fremtiden. Alternativ energiplan 1983*. Borgen Publ., Copenhagen.

IAEA, 1986. Chernobyl article by staff writer. *IAEA Bulletin* **28** (3), 27.

Isaksson, H., Petersen, B., 1961. GIER. *Ingeniøren* (International edition) **5** (4), 9–15.

Jensen, J., 1978. *Hybrid vehicles for urban transport*. Odense University Press.

Klingsey, M., Sperling, A. von, Keiding, P., Aagaard, C., 2007. Regnskabets time er inde. *Information*, 13th August, p. 6.

Korsbech, U., Ølgaard, P., 1974. *Atomkraften og Danmark*. Energy Policy Report of the Ministry of Commerce, Copenhagen.

Krarup, T., Svejgård, B., 1961. GIER. *Ingeniøren* (International edition) **5** (4), 3–8.

Lewis, H., *et al.*, 1975. Report to the American Physical Society by the Study Group on Light Water Reactor Safety. April, University of California, Santa Barbara. Also in *Reviews of Modern Physics* **47** (S1), 1–123.

Lewis, H., Budnitz, R., Kouts, H., Loewenstein, W., Rowe, W., von Hippel, F., Zachariasen, F., 1978. *Risk Assessment Review Group report to the US Nuclear Regulatory Commission*. NUREG/CR-0400, Washington DC.

Lindell. B., Löfveberg, S., 1972. *Kärnkraften, människan och säkerheten*. Almänna

Förlaget, Stockholm.

Linder, S., 1970. *The Harried Leisure Class.* Columbia University Press, New York.

Lippold, K., *et al.*, 1995. *Mehr Zukunft für die Erde. Nachhaltige Energipolitik für dauerhaften Klimaschutz.* Report of German "Enquete-Kommission" to the German parliament. Economica Verlag, Bonn.

Lund, H., Mathiesen, B., 2006. *IDAs energiplan 2030.* Baggrundsrapport. Danish Association of Engineers, Copenhagen.

Metz, G., 2010. Intermetzo: niårsdagen. *Information,* 20th November, p. 18.

Naur, P. (ed.), 1963 (name misspelled as Nauer in article). Report on the algorithmic language ALGOL 60. *Communications ACM* **6** (1), 1–17.

NHK, 2011. Continuous television newscasts 12th to 28th March. World Channel of Japanese National Television, Tokyo.

Nielsen, J., 1962. Memoires of Niels Bohr. *Physics Today* **16**, 28.

Nielsen, K., 2010. Det beskidte dusin. *Frederiksborg Amts Avis,* 19th January, p. 4.

Olesen, M., 1983. Risø får brugt atomaffald. *Politiken*, 1st January.

PET, 2009. PET's overvågning af protestbevægelser 1945–1989. Danish Police Intelligence Service's Report, vol. 10. Schultz Publ., Copenhagen. Available at http://jm.schultzboghandel.dk/upload/microsites/ebooks/pet/pet_bind10/index.html

Porat, M., 1977. *The Information Economy* (5 volumes). US Department of Commerce, Washington DC.

Raymond, E., 2010. The first GUIs. http://www.catb.org/~esr/writings/taouu/html/ch02s05.html

Saxe, R., Murray, R. (writing under the pseudonyms P. Unch and J. Udy), 1972. Personal transportation safety: an evaluation of the evidence. *Nuclear News*, January, 69–70.

Smidt, E., 1971. *Perspectiv Plan I.* Danish Department of Finance, Copenhagen.

Smidt, E., 1974. *Perspectiv Plan II.* Danish Department of Finance, Copenhagen.

Sørensen, B., 1972. Danmark tvunget ind i risikoen ved atomkraft. *Politiken* Kronik 3rd September, Copenhagen. (As precursors to the Danish nuclear debate, concerns over routine emissions and waste from nuclear power plants had been expressed by Arley and Skov, 1956; Zimling, 1971; Vikstrøm, 1971).

Sørensen, B., 1975. Energy and resources. *Science* **189**, 255–260. An early version "Araberne skal vel ikke være skyld i, at vi får en energipolitik?", appeared as feature article in *Information,* 7th January.

Sørensen, B., 1982. Energy choices: Optimal path between efficiency and cost. In *Energy Resources and Environment* (Conference Papers, Beijing, China; S. Yuan, ed.), pp. 279–286. Pergamon Press, New York.

Sørensen, B., 1991. *Energy Conservation and Efficiency Measures in other Countries.* Australian Department of Arts, Sport, the Environment, Tourism and Territories, Greenhouse Studies No. 8. Canberra.

Sørensen, B., 1998. *Brint*. Strategy statement of the Danish Energy Agency Hydrogen Committee. Copenhagen.

Sørensen, B., 2008. A sustainable future: Construction of demand and renewable energy supply scenarios. *Int. Journal of Energy Research* **32**, 436–470.

Sørensen, B., 2010. *Renewable Energy.* 4th Edition. Academic Press/Elsevier, Burlington.

Sørensen, B., 2011. *Life-Cycle Analysis of Energy Systems: From Methodology to Applications.* RSC (Royal Society of Chemistry) Publishing, Cambridge.

Sørensen, B., Ulfbeck, O., 1974. *Uafklarede problemer omkring atomkraften*. Submission to Danish Parliament. Reprinted in *Atomkraft?* No. 2 (14 pp, 1974). Nuclear Power Information Organisation (OOA), Copenhagen.

Sørensen, B., Nielsen, L., Pedersen, S., Illum, K., Morthorst, P., 1994. *Fremtidens vedvarende energisystem.* Danish Technology Council, Report 1994/3, Copenhagen.

Sørensen, B., *et al.*, 2004. Hydrogen as an energy carrier: Scenarios for future use of hydrogen in the Danish energy system. *Int. Journal of Hydrogen Energy* **29**, 23–32.

Sydkraft, 1971. *Välkomna til Barsebäck*. Visitor's brochure, Malmö.

Thykier-Nielsen, S., 1974. *Modeller til beregning af eksterne gammadoser og inhalationsdoser fra frigørelse til atmosfæren af radioaktive stoffer*. Risø Report M-1726, 48 pp. plus 6 pp errata, Roskilde.

Thykier-Nielsen, S., 1975. *Vurdering af konsekvenser på dansk område af modelhavarier på Barsebäckværket*, Risø report stamped "may not be published or made available to outsiders". Who would qualify as an "outsider" was not defined, and the results from the report were quoted both in public debates and by government officials. With a subsequent change of director at Risø, the stamps disappeared and earlier reports were also made generally available. At the end of 1975 (19th December), a short internal memo (#1611309) from Risø compares the results with the WASH-1400 draft, noting that even the second-worst accident type in the US report leads to whole-body doses 20 km from the accident that are 1000 times above those of the Risø report, stated to be based on release data furnished by another internal memo from Henning Jensen at the Danish Nuclear Inspectorate.

UK Atomic Energy Office, 1957. *Accident at Windscale No. 1 pile on 10th October,*

*1957.* Her Majesty's Stationary Office, London.

US EIA, 2011. *World Crude Oil Prices.* Spreadsheet covering the period from 1946 to 11. February 2011. US Energy Information Agency, Washington DC. Available at http://www.eia.gov/nev/pet/pet_pri_wco_k_w.xls

Valaskakis, K., Fitzpatrick-Martin, I., 1980. Socio political impacts of the information economy. In *The Human Context for Science and Technology* (T. Schrecker, C. Hooker, eds.), pp. 299–335. Research Council of Ottawa, Canada.

Vikstrøm, M., 1971. Hvilken fare frembyder atomkraftværker? *Ingeniørens Ugeblad* (36), 24–25.

WASH-740, 1957. *Theoretical Possibilities of Accidents in Nuclear Power Plants.* US Atomic Energy Commission, Washington DC.

WASH-1250, 1973. *Safety of Nuclear Power Reactors (Light Water-Cooled) and Related Facilities.* US Atomic Energy Commission, Washington DC.

WASH-1400, 1975. *Reactor Safety Study. An Assessment of Accident Risks in US Commercial Nuclear Power Plants.* US Atomic Energy Commission, Washington DC (substantially different draft version circulated 1974).

Weizsäcker, E. von, 1998. *Factor Four: Doubling Wealth, Halving Resource Use.* Earthscan, London (followed up by *Factor Five* in 2009, also Earthscan).

Wikipedia, 2011a. *Automaton.* Description on http://en.wikipedia.org

Wikipedia, 2011b. *Konrad Zuse.* Description on http://en.wikipedia.org

ZAMG, 2011. Atomunfall: Ausbreitung von Radioaktivität in Japan (G. Wotawa). Zentralanstalt für Meteorologie und Geodynamik, Vienna. http://www.zamg.ac.at/aktuell (accessed 24th March 2011).

Zimling, B., 1971. Er atomkraftværkerne en uhyrlig forureningskilde? *Ingeniørens Ugeblad* (22), 8–9.

# 12

# Conflicting directions: Renewable energy and decentralization in a world of globalization and growing consumption

The interest in keeping track of energy use came surprisingly late in Denmark. The Danish Statistical Office was throughout the 20th century compiling hundreds of datasets on the number of newspapers and magazines sold, on consumption of cucumber and pumpkin, on turnover in flower shops, and so on, but until 1959 not a word on energy supply or its use. Other countries, such as the USA, had detailed energy supply and consumption data from the start of the 20th century (US Department of Commerce, 1975). The earliest serious Danish effort seems to have been at the start of World War II, when the Danish government wanted to ration the coal obtained by the limited wartime import options. A few large industries were asked how much coal they used to burn, and the available coal was then distributed among them according to these fixed ratios. After the war, the Danish Academy of Technical Sciences (1947) saw it as interesting to map the variations in energy supplies that had been taking place between 1938 and 1944. The coal import was found have declined threefold during the war, and the oil import to nearly zero (the few operating automobiles were converted to use coke or wood gasifiers; see Fig. 10.49). Indigenous sources such as peat, wind and hydropower were unable to replace the lost imports over the limited time horizon. On the demand side, half of the energy use supplied was unaccounted for, as available data only covered industry and central energy conversion plants, and the efficiency of these applications often had to be guessed.

One employee (Henning Normann) at the Statistical Office, perhaps inspired by this work, in 1959 was allowed to publish a similar analysis of Danish energy supplies back to year 1900. It was originally made for the Danish Atomic Energy Commission, who wanted to make a case for the insufficiency of non-nuclear energy to cover future Danish energy needs. Only commercial fuel purchases were included in the study and no usage distributions appeared except for that of the electric power and gas works (Danish Statistics, 1959). A later attempt to estimate the distribution on users was only able to go back to 1948 (Danish Statistics, 1967). This study was presumably spurred by the 1956 Suez Channel passage blockade, feared to negatively influence oil imports. No serious follow up on these studies were made until after the 1973–74 oil price hike, which eventually prompted the establishment of a Department of Energy and an Energy Agency, and from 1980 a more systematic collection and publication of data on energy supply, central conversion and distribution on end-users in Denmark ("delivered energy"; see Danish Energy Agency, 2010a). Data for the years from 1972 to the operational readiness of the new department, in the form of energy balances, were collected and published by the Danish Boilers Association and the Association of Danish Power Plant Operators (DK/DEFU, 1974). The Energy Agency has since collected and published its data from 1972 to 2009 in spreadsheet form (Danish Energy Agency, 2010b).

*Figure 12.1. In contrast to official Danish energy statistics, this study has included muscle work by humans and draft animals (photo by author, 1978).*

All of these official energy statistical sources are still covering only a subset of the real energy usage: until 2006, the estimates of non-commercial energy (such as fuelwood exchanged locally) were flawed (Evald, 2006, 2009). This type of energy usage has recently been increasing and now contributes the major part of primary air pollution from small particles in Denmark (Sørensen, 2011). Also, the energy provided by muscle work by humans or draught animals are not found in statistical compilations of Danmarks Statistik or those of the Energy Agency (Fig. 12.1). For this reason it falsely appears as if the transportation sector did not use any energy before coal-based steam locomotives and oil-consuming motors were introduced. Further, by concentrating on energy used within Denmark, the Energy Agency statistics, as mentioned (Fig. 11.1 and Section 11.4), omits the very substantial amounts of energy used in international shipping and air traffic by Danish carriers. This is inconsistent, as one either has to include direct and embedded energy in all imports and exclude those in exports, or alternatively include all energy used anywhere by Danish individuals or enterprises. This discussion had been raised in connection with life-cycle analysis and assessments (LCA) of the energy sector (see Sørensen, 2011). The problem in recent years has also been discussed internationally (IMO, 2000), and the Danish international shipping consumption (sometimes called *international bunkers,* a terminology left over from coal handling) is found to be among the largest in the world, due to the high Danish share in the global bulk cargo shipping market (Fig. 12.2), as shown by the alternative approach to national energy statistics suggested by Thomas Olsen in a publication by Danmarks Statistik (2007) and now used in the energy statistics of Danmarks Statistik, which is entirely ordered according to the codes for national economic sector activities. The point is, of course, that if the income from the large shipping company Maersk is included in the Danish gross national product, then the energy expended in earning the money should also be included.

*Figure 12.2. Energy efficient container ship Elly Maersk (courtesy of Maersk).*

Although having a common basis, the numbers furnished by Danmarks Statistik on its website databank (now extended back to 1966; Danmarks Statistik, 2010) are not easy to compare with those of the Energy Agency. The data from Danmarks Statistik do not allow identification of energy for transportation, because it is distributed on the large number of economic sectors transporting persons or goods, and Danmarks Statistik distributes conversion losses on all users in a way that is not very transparent (e.g. lumping electricity and district heating together in one category, despite the different quality of usage). The Energy Agency statistics also have some transportation energy unspecified, such as local transportation within industry, but generally, the data from the Energy Agency are in a form more suited for energy analysis.

As mentioned, there are other ways of cutting the energy cake: one might include the energy embedded in imported products (and correspondingly omit that in exports). This would give a more useful picture of the changes occurring during a period of increasing globalization, where many – and specifically the most energy-intensive products – are no longer produced in Denmark but imported. The industry statistics over the last 40 years show a dramatic decrease in medium and heavy industry economic activities in Denmark, almost certainly implying that the indirect energy use embedded in imported raw materials and products must have increased. This makes it difficult to maintain pride over the statement made by Danish political leaders since 1974 that Denmark has been extremely successful in using energy more efficiently and hence maintainrf a stable level of energy use after 1974, totally avoiding the previous trend of exponential increase. The validity of such statements have already been shattered by the Danmarks Statistik consideration of energy expended by international shipping, now furnishing the largest industrial contribution to the Danish national income, and this notably despite the fact that the Maersk fleet of freight ships are among the currently most energy-efficient ones operated anywhere (energy use by cargo ships is in any case low per ton of cargo moved, compared with other means of freight transportation–see Fig. 12.3–so the large energy expenditures are prompted by the huge amounts of cargo moved (as illustrated by the Danmarks Statistik data for the last decade in Fig. 12.4), including jobs performed by Maersk for the US military, moving equipment in connection with the Iraq and Afghanistan wars (Kaarsholm, 2007). If the further increase in embedded imports of energy were included, the increase in Danish energy use would be even more pronounced, even if the energy imbedded in exports were subtracted. So far, this type of study has not been undertaken, due to the difficulty of determining the origin of each component in a product and getting data on the energy used to manufacture the components at different locations around the world, including determination of the input energy mix in the country of origin used at the time of manufacture.

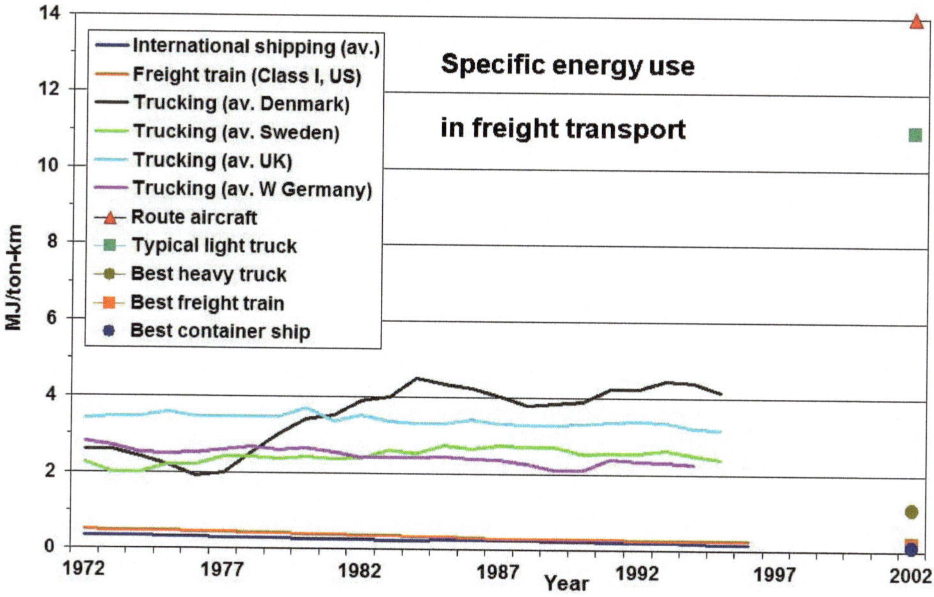


*Figure 12.3. Average efficiency of various modes of freight transportation as function of time, plus some recent "best-technology" values (with use of IMO, 2000; Schipper and Marie-Lilliu, 1999; Sørensen, 2007; US DoE, 2005).*

Figure 12.3 shows that freight transportation by rail or by ship is far more efficient than transport by road, and that although Denmark in the wake of the 1974 oil crisis had one of the most efficient lorry uses, this position has been dramatically replaced by one of the worst performances between 1982 and 1995. These numbers are national averages of all road freight hauling. To better understand the numbers, the right-hand side of Fig. 12.3 gives some 2002 figures for specific types of vehicles, showing that the most heavy trucks reach efficiencies considerably better than the average, but still a factor of ten higher than the efficient bulk carriers at sea. At the other end, light trucks have extremely poor energy efficiency, but this is again an average number, strongly affected by the capacity factors, and thus reflecting that light trucks are more likely to carry loads considerably below their rated freight capacity. The figure suggests that land transport of goods from an energy point of view should be made on trains, but the problem is of course that there are no tracks extending to the recipients of many goods, while there may well be roads. In consequence, a train transport operation may involve two reloadings–one after driving the goods by lorry to a train station and another (also by lorry) from a final train station to the destination. Use of standard-size containers has made the human work required for these reloadings fairly small, but due to the favouring of private road transport over (in Denmark) public train transport by governments focusing on privatization, transport by lorry appears to be less expensive than by rail. Decisive factors are of course road taxes (which at the moment do not fully pay for the wear on road surfaces, to which large trucks contribute much more than passenger cars), and safety concerns (where road traffic would be much safer if light passenger cars were alone on the roads). Fig. 12.3 clearly indicates that transport by sea should be preferred whenever feasible (exactly as in the 19th century, cf. the discussion accompanying Fig. 9.44), and that transporting goods by air is the least efficient option (but often selected at present in order to get, say, fresh fruit from South Africa to Denmark, taking advantage of the hemisphere's opposite growing season.

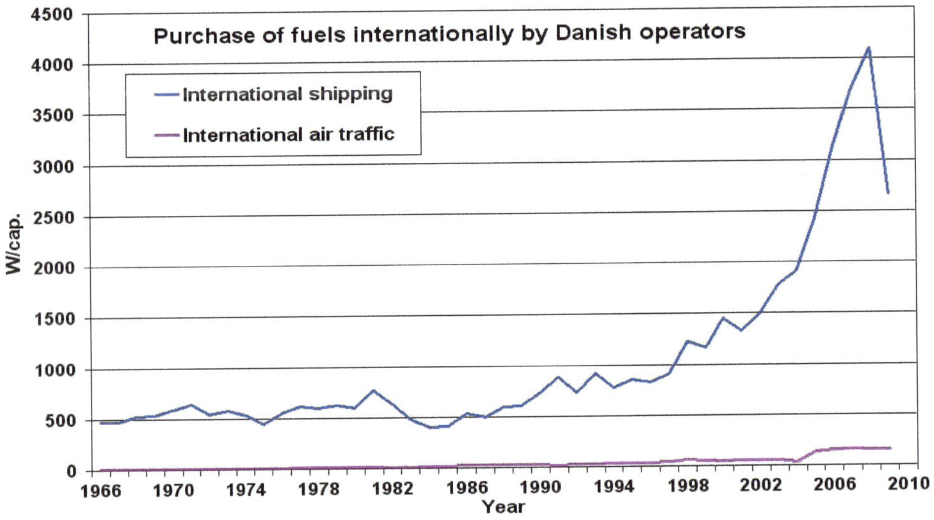


*Figure 12.4. International purchases of oil products by Danish air carriers or sea vessels, in watts (W) per Danish citizen (Danmarks Statistik, 2010).*

In Fig. 12.4, the part of the Danish energy usage that can be attributed to international shipping, led by the company Maersk (formerly A. P. Møller), is singled out. Maersk had obtained a 50-year concession to exploit Danish oil and gas in 1962 (Danish Energy Agency, 2006), and started North Sea oil extraction in 1972, natural gas extraction around 1980. A number of further North Sea concessions were awarded during the following decades, on terms generally favourable to the bidding consortia. The similar concessions given by the Norwegian state to the operators in the Norwegian part of the North Sea ensured that a high proportion of the revenues went to the Norwegian state, where it has subsequently been used to expand welfare and secure future social conditions. The Danish concessions were generally lenient regarding taxation, leaving deduction of exploration costs and country allocation of revenues to be based on estimates by the concession holders.

Seen from the perspective of Maersk, the revenues from the North Sea operation have been employed very sensibly. Recognizing that the oil and gas resources in the Danish part of the North Sea will soon be exhausted (production has already started to decline; cf. Fig. 11.15), the large "one-time" income has been wisely invested in ships for the core business of the company: international shipping (Fig. 12.2). By making these bulk carriers and container ships the most efficient on the market, Maersk has increased its global market share to a dominant position. This may well be seen as a valid insurance for future prosperity for the company. Even if the present low cargo prices (subsidized by being relieved from fuel and other taxation of its environmental impacts) should not prevail, Maersk will continue to earn money as long as there is a market for international transportation and the Maersk ships stay efficient. This transport market is fairly certain to increase, for obvious reasons: higher energy prices will probably affect the volume of international shipping negatively, but considering the growing economy in some parts of the world, the volume moved will almost certainly increase anyway, to the advantage of a well-established shipping country with efficient ships. What may be reduced are some current absurdities, where Europeans import food from distant locations that could have been grown locally at a slightly higher direct cost, or where manufactured goods have components that have been shipped back and forth around the world several times, in order to get particular components added or to get pieces assembled or painting done where the workers' salaries are lowest. Without the current subsidy to intercontinental transportation (emerging from the ideology of "globalization" and administered through the rules of the US-dominated World Trade Organization), such excessive transportation will hopefully disappear.

The preceding chapters have exhibited two kinds of energy tables, one with primary or delivered energy (the difference being important only after the introduction of centralized conversion plants some hundred years ago), and one with end-use energy, defined as the energy made useful at the end of the final energy conversion taking place at the end-user. This quantity (taking into account efficiency of refrigerators, light bulbs, automobiles, etc.) is of obvious importance and stresses the fact that there is often a choice between alternative methods providing the same energy service, but probably with quite different negative energy and environmental aspects.

For instance, this could be either to produce an additional unit of energy or to avoid it by improving the efficiency of conversion. Unfortunately, official statistical data do not generally provide data on the end-use energy (but to confuse matters, some countries use the term "end-use" for delivered energy), which may perhaps explain why energy efficiency has received much less political attention compared with new energy supply. The implication is that the estimation of end-use energy made here has not been able to benefit from any available statistical data even for the most recent period. It is instead based on the end-use methodology described in Sørensen (2010), where end-use is not taken as some absolute minimum of thermodynamic energy supply (according to "second-law efficiency", the implied energy called "exergy"), but rather as the efficiency characterizing the best technology available today or in the near future. The concept of end-use energy is necessarily vague, because there could be technological inventions coming online that totally change the practical minimum of future "available technology", or even totally eliminate use of energy for a particular task. This is because what human beings and their societies demand is never energy as such, but rather the products and services that can be produced with use of energy, but which sometimes turn out to be accessible by entirely different approaches requiring much less energy and in some cases no energy at all. An example of the first is the switch from cathode ray displays (for television or computers) to light-emitting diode flat screens. An example of the second type might be the switch from hot-water clothes washing to washing with cold-water detergents. The energy for heating water is avoided, but in this case at the expense of probably less environment-friendly chemicals in the washing powder.

The long-term graphs of energy use, split on energy qualities or on types of usage, which have been provided in the preceding chapters, have in their last update (shown in Figs. 11.25 to 11.27) exhibited the exceptional growth in energy use since 1950. At the same time, earlier variations have become more difficult to follow on the linear plots, and therefore the same graphs are in this chapter repeated on a logarithmic scale in Figs. 12.5 to 12.7. Figure 12.5a clearly shows the 20th century increase in energy used for raw materials extraction and conversion, for occupational as well as for leisure activities. Figure 12.5b similarly shows the losses in conversion associated with the centralized production of electricity, the largest energy growth component, followed by transportation fuels.

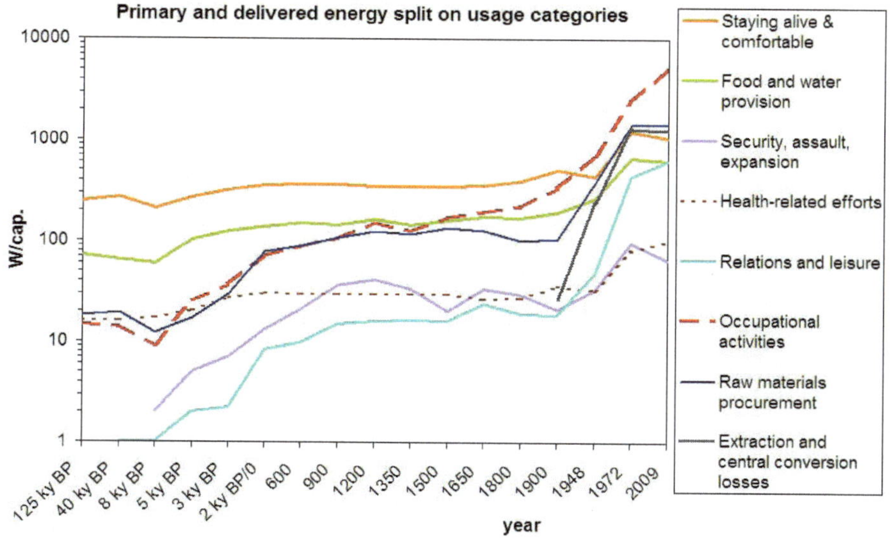


*Figure 12.5a. Estimated energy use per capita in Denmark from 125 ky BP to 2009, distributed on categories of usage (logarithmic scale, cf. Fig. 11.25a).*

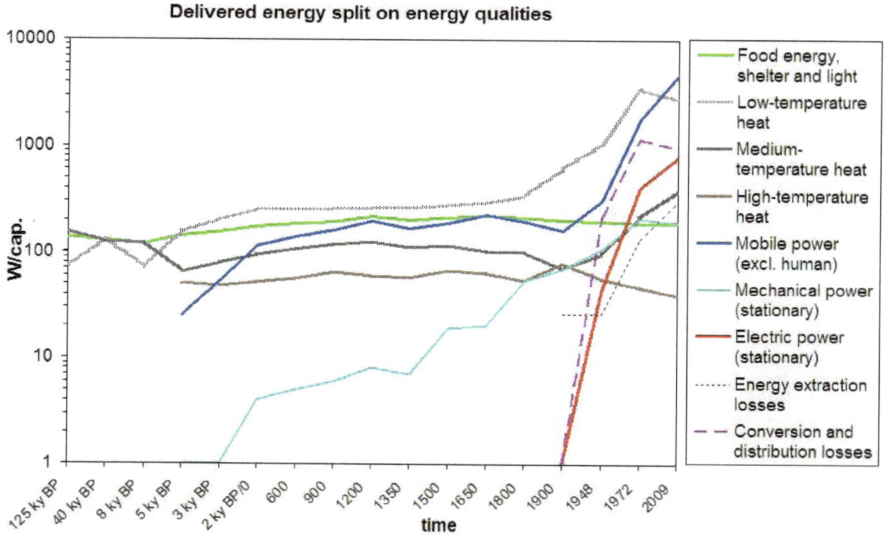


*Figure 12.5b. Estimated energy use per capita in Denmark from 125 ky BP to 2009, distributed on types of energy employed (logarithmic scale).*

The end-use numbers shown in Fig. 12.6 (a and b) bear the same message, but further emphasize growth in electricity usage and mobile power, for leisure activities, commerce and household appliances, due to higher exergy values relative to heat usage.

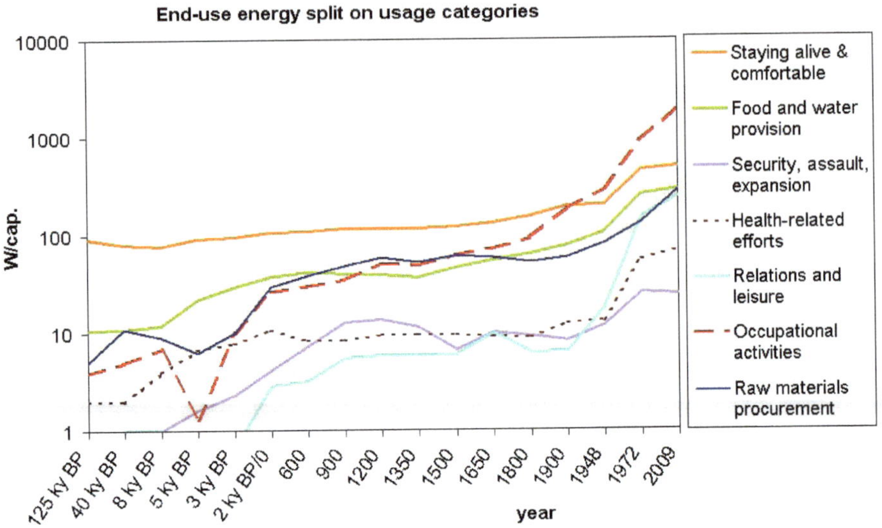


*Figure 12.6a. Estimated end-use energy per capita in Denmark from 125 ky BP to 2009, distributed on categories of usage (logarithmic scale, cf. Fig. 11.26a).*

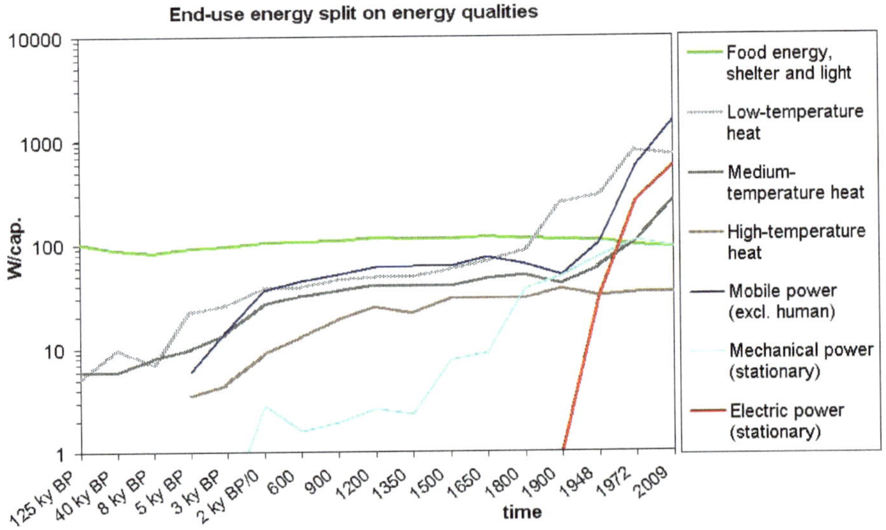


*Figure 12.6b. Estimated end-use energy per capita in Denmark from 125 ky BP to 2009, distributed on types of energy employed (logarithmic scale).*

The total Danish energy use shown in Fig. 12.7 is of course determined by a combination of variations in population size and in per capita energy usage, with the latter dominating only during the last century. Note that the scale of years is expanding as it gets closer to the present.

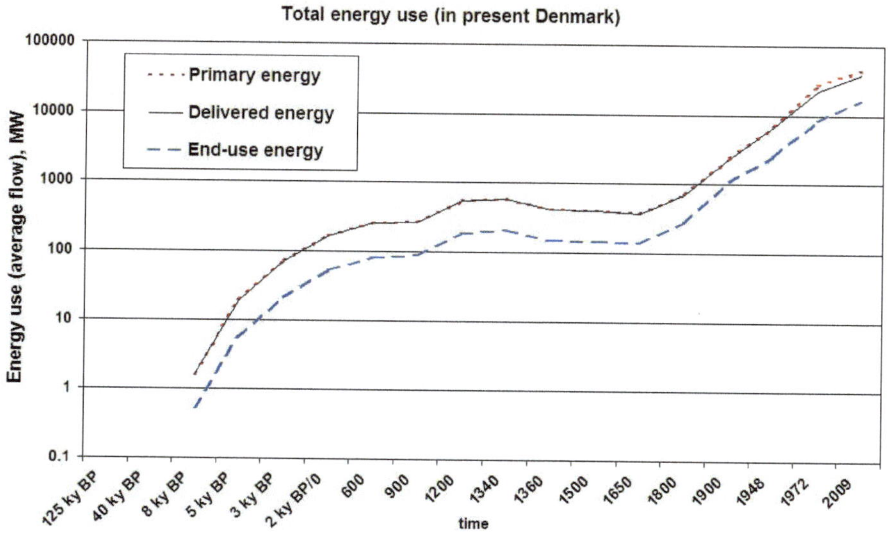


*Figure 12.7. The long-term development in total primary, delivered and end-use energy in Denmark, adjusted to usage within the territory presently constituting the nation (logarithmic scale, cf. Fig. 11.27).*

The author would summarize the situation as follows: as the fossil interlude is beginning to draw towards the end (Fig. 11.1), one has to ask why the governments elected by the Danes have not made sure that this period of unique access to inexpensive energy was used to build a lasting energy system, but lost it in offering extra everyday consumption to further their chances of re-election. The years of surplus on the balance of payments account were not exploited, and thus the upcoming transition to a dependence on more expensive energy has to be established along with financial tightness, a prospect not likely to advance the prosperity of Danish society. It does not offer much consolation that many other countries have made the same mistake.

## 12.1 Changes in the social fabric

Most Danish income earning has for some time been generated in small firms, including services providers as much as manufacturers. These small companies are often based on a good idea and some entrepreneurship, with the person initiating the enterprise risking his or her own money as well as what he/she can get in loans from banks. Naturally, only some of the initiatives are successful, and when they are, they tend to stay in Danish hands only for a few years, after which they are sold off to foreign investors (examples are abundant in the energy industry, and also for providers of computer software and communication technologies). Still, the multitude of such small-scale efforts of limited duration makes them net earners of money to the country, in amounts that make Denmark one of the richest countries in the world.

In order to promote an economy based on such creativeness, the education system should, as the Danish one traditionally has done, emphasize critical analysis, original reasoning and self-selected project work rather than learning as many facts and great theories of past efforts as possible by heart. It is therefore disturbing that recent Danish governments, for ideological reasons, have returned to the easily monitored skills, allowing simplified international comparisons by methods that are very poorly founded, from a scientific point of view, such as OECD's PISA ratings (Wikipedia, 2010) or the claimed "bibliometric" correlation between on one hand scientific journal readership and mutual quotation, and on the other hand scientific quality and novelty (for an early critique see Phillimore, 1989). A reason this dubious type of valuation has gained popularity is that it fits with the currently prevailing liberal ideologies, according to which universities are seen solely as service providers for industry. Governments and the EU Commission decide which subject should be researched, and scientists get funding if they can write detailed applications telling which results will be delivered each quarter throughout the project period. Lately the further condition that research projects are made jointly with industry partners ensures that no long-term ideas are pursued. It is already decades since young researchers were told that if the chance of success in an investigation were more than 50% then it could not be classified as research.

It would appear that this approach to satisfying scientific curiosity is the recipe for broad creativity no longer continuing to be part of the Danish make-up. Current Danish university funding is totally based on production (note the industry terminology) of candidates, which naturally leads to reduced requirements for passing exams. The positive way of describing this is to say that the aim is to give some education to as large a fraction of the youth as possible, rather than to offer a high education to those capable of consuming it. The author would think that both objectives could be satisfied in an appropriate education system. Such an appraisal may, however, be too pessimistic, because in practice it is very difficult to stop people from being creative if they have the urge.

A curiosity appearing in a different type of ranking study is the position that Denmark has gained in studies of satisfaction with life conditions in the population. An early study by Estes (1984) concluded that living conditions were better in Denmark than in any other country, based on a range of indicators including conventional economic prosperity, but not confined to materialistic signs of wealth. In a more recent study of similar nature (Legatum Institute, 2011; Zakaria, 2011), Denmark came in number two after Norway. The study augments economic prosperity with indicators of entrepreneurship and business opportunities, governance, education, health, safety and security, personal freedom and social capital. The problem with studies such as these is the use of interview studies as an important component in reaching the scoring. What does it mean Danes themselves are very satisfied with their situation? It could be caused by lack of knowledge of the conditions elsewhere, and it could be the result of political propaganda from politicians wanting to create the impression that they are doing well and that Denmark is far ahead in wealth and social services. For example, the health sector could be evaluated by means other than by the amount of money spent and by interviewing hospitalized citizens. Danish life expectancies are not impressive. Objective data on how the Danish government has contributed to the 2008 financial crisis by freezing property taxes at 2001 levels and thereby creating a surge in house prices that, because lending institutions were issuing loans according to the inflated market value of property put up as security, led to the bubble economy of overspending based on loans with no collateral once property prices started to fall. A very large amount of money had to be written off by banks, and the government chose to save the banks from collapse by use of public funds, a policy that will limit the options for future governance.

## 12.2. Sustainability and globalisation?

Ecological products (sometimes called "organic") have in recent decades moved from occupying a small consumer niche to become a significant player in retail shopping for food and skin products. Figure 12.8 gives the development in Danish production of animal and vegetable ecological food products. However, both exports and particularly imports of ecologically produced goods have increased during the period monitored, including many categories that cannot be made in a temperate climate (such as wine).

For the consumers, dairy products were the first food group for which ecological food became a preferred choice, particularly in families with children. It is no longer necessary to go to a specialist shop (Fig. 12.9a) to buy most ecologically grown products, as every supermarket offers a substantial selection (Fig. 12.9b). However, in a quantitative assessment, the market penetration of ecological products is still at the 10% level, but with large variations between types of products.

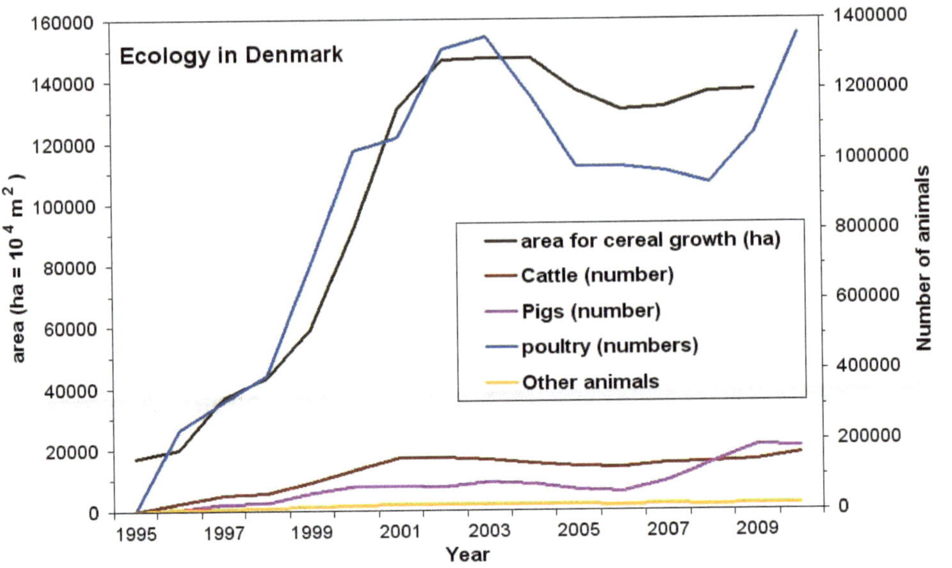


*Figure 12.8. Development in ecologically farmed area and number of animals raised on ecological feed (Danish Ministry of Food, Agriculture and Fisheries, 2010; Danmarks Statistik, 2010). Import/export statistics are not provided.*

The ecological products with high market penetration (dairy products) are only slightly more expensive than products of chemical agriculture (using pesticides and other chemical industry components), while those of low penetration (such as meat) may be 20–30% above the price of non-ecological products.

A discussion is on-going on the global implications of using ecological products, because farming yields, especially during the early period of ecological practice, were lower than those of chemical farming. This gap has since narrowed to under 10% lower yields, due to development of non-chemical pest control (e.g. using biological predator species) and advanced mechanical weeding techniques (also positively affecting the human labour required). The current talk of a food crisis, spurred by rising prices for some food products, could suggest that ecological farming is not consistent with global food security, but the historical development is more complex. The food crisis after World War II was countered by the "green revolution" advances in the efficiency of third world agriculture, and despite large population growth, the situation a few years ago was that world agriculture produced a surplus over demand, leading the European Union to subsidize abandoning farm land in member countries, converting it into fallow, and ultimately to recreational land and protected zones for saving diversity of species.

Also culinary preferences have changed, from the extravagance of the classical French dishes in Babette's kitchen (Axel, 1987) to more focus on smaller but more exquisite meals, compatible with health advice.

*Figure 12.9a, b. Specialist ecological shop (top) and typical supermarket (here a Nordic co-op outlet) ecology section (below) (photos by author, city of Gilleleje, 2011). Negative environmental impact from packaging is still an issue.*

Ecological production has a long history. In several German regions, medieval instances of the use of additives to beer (for foaming and preservation) by chemicals with negative health impacts led to legislative measures such as the 1516 Bavarian "Reinheitsgebot", a law forbidding use of additives in beer production, which was in effect until European Union complaints led to a European Court decision nullifying the beer regulation in 1988 (Wikipedia, 2011). Sale of hygienic lotions and other wonder remedies through special stores were common up to the 21st century, often based on natural herbal medicine discovered by inmates of convents in the medieval period and occasionally with a positive effect. Because of this background, the current ecological movement is dependent on a sober certification procedure, ensuring that the miracle doctor products can be identified by at least the consumers who are in the habit of reading labels and declarations. The ecology movement is today a powerful political factor, and a result of collaborating producers being as globalized in their outlook and trade as the traditional agents on the food and hygiene/cosmetics scene. At present, the ecology business is spreading to other areas, such as furniture production, and the aim is to one day encompass everything.

It is interesting that those promoting a sustainable development with emphasis on social solidarity and minimizing environmental impacts until recently also believed that this was possible only in a highly decentralized structure. Bad experiences with centralized power and decision structures in government and private enterprises had made the local focus a part of the paradigm, and the corporate motion towards increased globalization is seen as incompatible with achieving sustainability. A few examples of this have already been mentioned, such as the exemption of international transportation from environmental taxation. With this background, it is debatable how the present embracement of globalization by the ecological movement should be seen. We may prefer to eat organic food for our health and for its good taste, but is ecological foodstuff shipped by use of fossil energy from South America to Europe a sustainable avenue? Using the term "sustainability" implies consideration of resource use and negative impacts on climate, as well as influence on direct health and environment. Furthermore, globalization also implies removing decision-making power from local society to distant entities that may be difficult to identify or influence.

The dilemma is not only between globalization and local participation in decision-making, but also between nation states and super-national players on the international scene. National entities used to have what were termed "strategic sectors" (i.e. enterprises that were agreed to be too important for the national survival in times of crisis, such as the recurrent wars) to be left to foreign enterprises. Such strategic areas would comprise food supply, communication, energy supply, financial institutions, and the "law and order" institutions, in the legislative and juridical system, police and defence. Even such areas are currently being privatized and left to market-determined ownership and decision-making structure, often implying a passing to foreign ownership by investors having interests that might be very different from those of the national state selling out of its cultural treasures and strategic assets.

## 12.3 Changes in political structure

The political landscape in Denmark is presently dominated by short-term considerations. Some would say that issues discussed in parliament are selected on a day-by-day basis according to what the media are focussing on, but this is probably unfair, as there is occasionally an underlying ideological agenda, which is currently of liberal persuasion, such as the "minimal state" ideology announced by the first prime minister of the 2001-2011 Danish government coalition (Rasmussen, 1993). The agenda has changed dramatically since the creation of the Danish "welfare economy", notably by the Labour prime minister (1929–1942) Thorvald Stauning in the 1930s (see Chapter 10). Parties currently represented in the Danish parliament comprise an ultra-right-wing nationalist party ("Danish People's Party"), a right-wing liberal party (with the misleading name "Left", dating from over a century ago, where it was the party for the farmers, opposing the leading conservative party – today it primarily fights for the remaining large industry interests, using populist propaganda to attract middle-position voters), a conservative party (currently focused on the interest of small-scale industry and businesses), a centrist party (for historical reasons called the "Radical Party"), a social democratic party (currently having largely joined the liberal ideology, as have its sister parties, for example in the UK), a soft socialist party (called the "Socialistic People's Party"), and a hard-liner socialist party (called the "Unity List", less centralistic now that the concept of "Communism" is not much in favour), plus some individuals elected to parliament but subsequently shifting party or starting a new party (an current example is called "Liberal Alliance").

Attempts to resume the social debate leading to the welfare ideology and its political implementations from the 1930s to the 1970s, with interruptions, were made in 1978 when an attempt was made to revive the *guaranteed basic income* (or citizen's wage) proposals first aired by Thomas More (1516) in his *Utopia*, and later taken up and made more operational in the UK by Bertrand Russell (1918) and by Dennis and Mabel Milner (1918). The idea is to avoid poverty and exploitation of salaried people by paying all citizens a basic salary without any work or other requirements attached and large enough to allow a basic life. Financing these payments would be achieved by taxation of those choosing to increase their income by work, and evaluations suggest that there will probably always be enough people with a desire for more material goods to ensure the economic support of everyone at the required level, and not estimated to cost much more than the present cost of social welfare in countries such as Denmark. These ideas were introduced at the political level by the Australian Parliament in 1974 (Australian Government Commission, 1975) and subsequently discussed in Europe, particularly in the Netherlands (Kuiper, 1976) and in Denmark (Meyer *et al.*, 1978). The Danish author team consisted of Kristen Petersen (a politician from the Centrist Party), Villy Sørensen (a philosopher with a keen interest in utopian writing), and Niels Meyer (a technocrat with political inclinations). Although the basic income discussion has now faded in Denmark, it is still moving on in several other countries, including to some extent the UK (Citizen's Income Trust, 2011), Australia (Tomlinson, 2001), Belgium and South America (Basic Income Earth Network, 2008). A pleasant set of video lectures explaining the motivation for and objections to the basic income concept has been produced in the Ukraine by Jörg Drescher *et al.* (2010).

The Danish constitution of 1953, mentioned at the end of Chapter 10, has been criticized for lack of clarity or consistency, fulfilling Napoleon Bonaparte's requirement that constitutions should be "short and unclear", so that changing governments could bend them for their own ends (Ytzen, 2011). The parliamentary democracy in Denmark is partial, as the king or queen retains some political power, but with the tacit understanding that it should not be used. The Church and State were not separated as in many other European countries (e.g. Sweden), and Denmark was not prevented from entering attack wars (such as the 2002 Iraq war) without consent by the United Nations. Not even a public referendum is required for the Danish government to assault another country (Smidt, 2007). As mentioned in Chapter 11, the Danish Supreme Court formally decided that the Danish Constitution was too vague to permit the Iraq war to be deemed unconstitutional (Balslev, 2010). Clearly, the vagueness suits some political parties, who have repeatedly refused to consider an update of the constitution except if problems of royal succession should suggest it.

The Danish discussion of keeping the level of immigration to a prescribed level has brought up the idea of an examination to determine if applicants have sufficient knowledge of the Danish society to successfully integrate. A test of suitability for citizenship was proposed by the ultra-right-wing Danish People's Party and passed in parliament in 2005 (Munkers, 2010). The combination of keeping native traditions while adapting to Danish ways was played down in this debate, but clearly an attractive gain from immigrants can be their contribution of new dimensions to society. An interesting thought would be to extend the testing idea to all citizens, demanding that an exam on human rights and obligations (not just Danish culture) must be passed in order to gain the right to vote for parliament or to run for office. In this case the current age criteria for voting could be avoided, and those unsuited to take part in decisions might be sorted out. The problem is of course as with the present limited test (for immigrants only): that those formulating the questions may show bias (Vangkilde and Seidelin, 2010).

Pursuing the idea of basing a new constitution on human rights, such as the formulation contained in the United Nations Declaration of Human Rights (1948), which to an appreciable extent is founded upon the post-revolutionary constitution of France or that of the United States, one might consider the following main paragraphs (Sørensen, 2005a).

1. BACKGROUND

1.1. The rights and obligations described below pertain to all human beings, at all ages, independent of gender, outward appearance, lineage and social grouping.

2. HUMAN RIGHTS

2.1. To cherish, express and live by any personal views, opinions and convictions not harming fellow human beings or diminishing their rights, and not in conflict with human obligations as described below.

2.2. No world citizen can be forced or enticed by any individual or by any institution, be it political, religious or of any other nature, to act in specific ways or to adopt or express views that are not part of the basic fabric of human rights, obligations and social arrangements as described in this constitution.

2.3. To have access to food, shelter and, when fulfilling human obligations as described below, to human relations.

2.4. To have fair opportunities for learning and for enjoying life.

3. HUMAN OBLIGATIONS

3.1. Not to harm any other human being, except in cases of clear self-defence or as part of legal enforcement of the terms of this constitution.

3.2. To be considerate of the world's natural makeup, whether the immediate surroundings, the distant or global ones, and whether constituted by living creatures, environment or any other kind of resources.

3.3. To acquire knowledge and skills according to ability and use them to the benefit of society.

3.4. To contribute to the achievement of common goals in society.

4. GOVERNANCE

Additional paragraphs that may contain specific prescriptions for organisation of governance, such as the commonly employed definition of three independent branches for legislative, legal and executive tasks, use of representative versus direct (referendum) democracy, distributing government on spatial levels (local, regional, global), setting specific rules for term periods, elections and formation of governments, including appointment of overseers (in Danish "ombudsperson") to avoid corruption and misuse of power.

This proposal has incorporated the idea of connecting "rights" with "obligations", as surfaced in the Danish debate described above. It is an important feature that human rights are considered individual rights, just as in the UN human rights declaration. Thus it does not give religious or political institutions the right to tell people what to do, which is a common misunderstanding of religious or political freedom. Each citizen should have the unchallenged right to select her or his beliefs, even if they are different from those of the persons next door.

## 12.4 Bids for histories of the future

Energy will continue to play an important part in the satisfaction of human needs. The unprecedented growth in the recent century (Fig. 11.27 or 12.7) is almost over, and will probably be replaced by the type of development seen before–one of staying on a level plateau for a long time, and jumping to the next plateau only when exceptional events (e.g. technological breakthroughs) take place. Presently, we have witnessed a large growth in worldwide population and therefore have to question the nearly 100% reliance on energy sources that will soon become depleted: oil and natural gas in decades, unless the very low-grade shales and tar sands are utilized, with even more negative environmental impacts than those of the current greenhouse gas and air pollution emissions; and coal in a century or two, but requiring sequestration of the carbon dioxide, emitted at about twice the rate associated with similar amounts of energy from oil or gas. Nuclear fuels will last only some 50 years if used in current reactor types. A longer period of utilization depends on developing safe breeder reactors or fusion converters, both of which are uncertain or unproven. Only for renewable energy sources are the required technologies in place, but so far the market penetration has been limited, due to costs being (in some cases slightly, in other cases substantially) higher than the prevailing energy prices, when disregarding externalities (Sørensen, 2005b).

In the light of history one might fear that the situation will lead to energy wars and increasing disparity between population groups and nations. The more benign avenue would be that societies get used to higher energy prices, responding by increasing the lifetime of products and by using energy more efficiently to obtain the services desired, perhaps with more consideration of which services are really necessary.

The present high density of population in many parts of the world has led to increased focus on fundamental questions such as that of sustainability. Sustainability will have to be required not just for the use of finite resources, but also for the economic activities (Barbier, 1993). However, the emergence of sustainability as a discussion issue has not significantly changed the course of political decision-making, which is still based on 18th century liberalistic paradigms and actions taken under strong influence from industry lobbying groups, largely ignoring the on-going sustainability debate. Perhaps the liberal paradigm could have worked, if prices reflected the total cost of things, including all types of externalities, by incorporating indirect costs to society or environment. But they do not, precisely due to pressure from those who benefit from the liberal type of economic order (Sørensen, 2011).

*Figure 12.10. In earlier times, it was common for people to give their houses names. This name of this one translates as something like "enough is enough" (photo by author, 1982).*

Sustainability is related to issues of growth and renewability. Renewable energy sources are fundamentally more sustainable than depletable ones. Whether growth is compatible with sustainability requires a deeper debate. Unlimited material growth is clearly difficult to make compatible with indefinite continuation of the processes involved, whether for reasons of the mineral resources employed or of the impacts from the manufacturing processes, which are never zero, even in the case of extensive recycling. On the other hand, no limitations have to be placed on non-material growth, as linked to cultural and other human activities based on mind and intellect rather than on equipment.

Advocates of attaining fulfilling human conditions where "enough is enough" have been around for a long time (Fig. 12.10), including in the last half-century the advocates of curbed materialism (Meadows *et al.*, 1971; unfortunately linking their arguments to oversimplified computer models) and more recently Gemynthe (1998) and O'Neill *et al.* (2010). Moving towards an achievable alternative may be to follow the idea of the basic income movement, accepting that societies consist of both people who, after having reached a certain state of goal fulfilment ("welfare"), are able to say enough is enough, and those who want more and are willing to put in extra work to get it. The challenge is for societies to set rules that ensure that the latter group do not reach their desired growth by using methods that jeopardize the overall sustainability. Current technology can in principle create adequate welfare through 1–2 hours of daily work by members of the healthy adult population.

The ability to adapt to strange external conditions is a key human strength. However, sometimes we may be adapting too quickly to conditions that we ought to have fought more strongly against. Danes seem particularly good at adapting. The Cimbrians (Chapter 5) had to adapt to their Roman capturers; the Vikings, and particularly the farmers accompanying them, adapted so well to conditions in France and England that after a few generations they were no longer seen as a separate group. This is in contrast to, for example, the Jewish people migrating to several European countries. For millennia they retained their identity, avoided intermarriage, and were repeatedly mistrusted or hated by the inhabitants of the countries to which they immigrated. So one may ask to what extent adaptation is a good thing. The answer is probably to stick to the middle road and not adapt immediately to any insulting condition, but also trying not to keep separate and feel forever superior. This balanced recipe may hold also for the attitude we should take towards the present changing environment as regards liberalism contra solidarity, an environment where debates are manipulated by commercial, party-political or other advertisements. In contrast to such external influences, we have the desire to control as much as possible of our own lives. Acceptance of absolute rulers (whether media operators or despots, hereditary or elected) contrasts with our insisting on a democracy entailing not only occasional voting, but also the right to be properly informed about the issues and candidates voted upon. The democratic requirement of tolerance further implies, among other things, a respect for minorities, so that 51% of the population is not ignoring or marginalizing a 49% minority.

When political parties first emerged, they were expected to formulate party programmes covering all important aspects of life in society. The grassroots organizations, on the other hand, typically started as single-issue campaigners. As in the nuclear debate case (Chapter 11), this could be a narrow, technical field, but the interesting point is that in that case it took the participants in the debate from the special issue to a range of general questions: should the development of a society necessarily involve unlimited growth in materialistic activities, or was there a plateau of life satisfaction where you could say "enough is enough"? Should the organization of society be centralized or decentralized? Are some citizens disqualified from voting (such as the children, the senile or criminals)? These issues, touched upon during the discussions earlier in this chapter, are part of the many threads that need to be followed, even if, as in the nuclear case, the starting point was limited, precisely because the use of such a single issue, being more easy to appreciate, may have caused people who otherwise would not have engaged themselves to actively embark on the further thoughts and active manifestations of real participation in democracy.

The present world is still dominated by people convinced that an invisible god is watching them, and will possibly intervene to ameliorate their lives and help them eradicate their perceived enemies. Many are not concerned about the severe pollution and possible climate change caused by their actions (a modest example is the burning of fossil fuel and of wood and other biomass; see Fig. 12.11). Everywhere, such people seem to form a majority, maybe well-intended but not too well-informed. An environmental concern does exist, but it is not strong enough to make serious changes in the lifestyles and the attitudes that most people have towards industrial products, and far from loud enough when protesting against uneducated media entertainment increasingly replacing informed debate and active commitment.

I have argued that features such as political corruption, nationalism, engagement in unnecessary violence and war, and pawning national assets are not just features of the rulers of a distant past, but have prevailed right to the present day. Most politicians and most voters react to the energy problem by burying their heads in the sand. Will we ever be

*Figure 12.11a, b. Perhaps a small example of our sticking to a mindset that should have been abandoned as a consequence of present knowledge… Midsummer Night is still being celebrated by bonfires: combustion of wood in a manner causing maximum air pollution (Danish Environment Agency, 2007; DMU, 2010), rivalled only by the fireworks of New Year's Eve and its content of toxic metals (photos by author, 2009).*

able to rid ourselves of the mindset associated with such behaviour? Here lies the fundamental gap between optimism and pessimism, and I do not feel that I have the polished answer. We have come a long way since the first Neanderthals set foot on our soil, but the question is whether we are moving in the right direction. The positive outlook on the situation of humanity is that we still have a healthy social debate between two sides of human positioning. One is characterized by cautious people, who want to obtain an understanding of possible impacts from a proposed new technology before agreeing to introduce it. The other is carried by the audacious people, who are willing to take a risk by stepping into uncharted territory, and who may even claim that should something unforeseen happen, then perhaps the time lapsed until the problem arises may have brought new remedies online. The latter group may in the past have helped technological progress, while the former may have prevented more disasters than the ones that actually happened. I hope that we will continue to allow ourselves to listen to both sides, and I am sure that the presence of both orientations in substantial amounts is providing a clear advantage to humanity.

## 12.5 Epilogue

Two threads of exposition have characterized this book. One is the scientific analysis of events and artefacts, occasionally revealing features not previously identified. The other is the interpretation of energy matters and the general social setting into which they are embedded. I have not refrained from criticizing policy as executed in the past or in recent times, nor from making suggestion for better social management in the future. There have always been divided opinions on how far in this direction a scholarly text may allow itself to go. However, my view is that omitting the author's interpretations, such as those that may be described as normatively based, will not remove any subjectivity from the text; it will just make it more difficult to identify the nature of the interpretations that in this way have been smuggled in. Instead, by explicitly including such interpretations in a way I hope lets the separation between objective analysis and subjective reflections remain clear, I have furnished the reader with adequate tools for opposing, changing or expanding the thoughts and suggestions I have ventured to make. It is my deep conviction that there is no single "correct" interpretation of history, just as there is no single "correct" path to follow in the future social and political development of human settlements on Earth. Therefore the most useful thing to do is to present views and solutions to identified problems that may be subjective, but also to attempt to establish their consistency and possible implementation. Subjectivity should not approach fundamentalism, but not approach fundamentalism, but must contain argued reasoning and a personal willingness to listen to other views, and perhaps change direction if these are deemed convincing. I find this kind of approach much more amenable when the subjective interpretations are made explicit rather than having to be guessed, and hope my readers will agree.

***References:***

Australian Government Commission, 1975. *Poverty in Australia.* Ch. 6, Commission of Inquiry into Poverty, Canberra. Web version at https://qutvirtual.qut.edu.au, HHP011.

Axel, G., 1987. *Babette's feast.* Danish Oscar-awarded film. Nordic Film Co.

Balslev, J., 2010. Iraq-krigen sluttede i højesteret. *Fyens Stiftstidende* 18th March, p. 12.

Barbier, E. (ed.), 1993. *Economics and Ecology.* Chapman & Hall, London.

Basic Income Earth Network, 2008. *History of Basic Income.* Web page at http://www.basicincome.org/bien/

Citizen's Income Trust, 2011. Citizen's income – an introduction. Web page at http://www.citizensincome.org, 2011.

Danish Academy of Technical Sciences, 1947. Betænkning II fra Energiudvalget vedr. Danmarks samlede energiforbrug 1938-1944. *Beretning* No. 8, ATV, Gad Publ., Copenhagen.

Danish Energy Agency, 2006. *Energifortider.* An occasionally updated webpage on main events related to energy, http://www.ens.dk

Danish Energy Agency, 2010a. *Energistatistik 2009* (with some data back to 1980). Copenhagen.

Danish Energy Agency, 2010b. *Energy statistics in tabular form 1972–2009.* Spreadsheet available at http://www.ens.dk

Danish Environmental Agency, 2007. *Brændeovne og små kedler – partikelemissioner og reduktionstiltag.* Project Report # 1164, Danish Ministry of Environment.

Danish Ministry of Food, Agriculture and Fisheries, 2010. *Statistics on organic farms 2009.* Plantedirektoratet, Copenhagen. http://pdir.fvm.dk

Danmarks Statistik, 1959. Danmarks Energiforsyning 1900–1958. *Statistiske Undersøgelser* No. 2, Danish Department of Statistics, Copenhagen.

Danmarks Statistik, 1967. Danmarks Energiforsyning 1948–1965. *Statistiske Undersøgelser* No. 20, Danish Department of Statistics, Copenhagen.

Danmarks Statistik, 2007. Energibalancer 2006, Miljø og Energi. *Statistiske Efter-*

*retninger* 2007:8, Danish Department of Statistics, Copenhagen.

Danmarks Statistik, 2010. *Statistikbank.* Energi og miljø. Energi. Energiforbrug efter branche. http://www.statistikbanken.dk, Danish Department of Statistics, Copenhagen.

DK/DEFU, 1974. Danmarks Energibalance 1972. *Elektroteknikeren,* 7th November (plus seven annual follow-up balances).

DMU, 2010. *Brændefyrings bidrag til luftforurening.* Report 779, Danmarks Miljøundersøgelser, Aarhus University.

Drescher, J., Samus, Y., Storozhuk, O., 2010. *Lectures on basic income.* Project Jovialismus, Kiev. http://basicincome.iovialis.org

Estes, R., 1984. *The social progress of nations.* Praeger Publ., New York.

Evald, A., 2006, 2009. *Brændeforbrug i Danmark.* Reports from Force Technology, Danish Energy Agency, Copenhagen.

Gemynthe, F., 1998. *Rigdom uden arbejde.* Christian Ejler's Forlag, Copenhagen.

IMO, 2000. *Prevention of air pollution from ships* (R. Henningsen, K. Skjølsvik, A. Andersen, J. Corbett, J. Skjelvik). Report 45/8 from the International Maritime Organization, Geneva.

Kaarsholm, L., 2007. Mærsk tjener milliarder på Iraqkrigen. *Information* 16–17. June, p. 10.

Kuiper, J., 1976. Arbeid en inkomen: twee plichten en twee rechten. *Social Maandblat Arbeid,* pp. 501–512 (Dutch language).

Legatum Institute, 2011. *2010 Legatum Prosperity Index Report.* 325 pp, Legatum Institute, London. The report is available at http://www.prosperity.com

Meadows, D., Meadows, D., Randers, J., Behrens III W., 1972. *The Limits to Growth.* Universe Books, New York.

Meyer, N., Petersen, K., Sørensen, V., 1978. *Oprør fra midten.* Gyldendal, Copenhagen (in Danish). English translation by C. Hauch, Marion Boyars, London, 1981.

Milner, D., Milner, M., 1918. *Scheme for a State bonus.* Pamphlet. Followed by D. Milner: *Higher Production by a Bonus on National Output. A Proposal for a Minimum Income for All, Varying with National Productivity.* Allen & Unwin, London, 1920. Web version at https://qutvirtual.qut.edu.au, HHB213.

More, T., 1516. *Utopia.* Louvain (in Latin). English translation by P. Turner, 1963, Penguin Classics, Hammondsworth. http://www.gutenberg.org E-text # 2130.

Munkers, C., 2010. *Dansk, danskere, danskhedstest...*Master's Thesis, Copenhagen Business School, Copenhagen. 77 pp. http://www.kommunikationog sprog.dk/file_upload/1374.pdf

O'Neill, D., Dietz, R., Jones, N. (eds.), 2010. *Enough is Enough: Ideas for a Sustainable Economy in a World of Finite Resources.* Centre for the Advancement of the Steady State Economy and Economic Justice for All, Leeds.

Phillimore, A., 1989. University research performance indicators in practice: The University Grants Committee's evaluation of British universities 1985–86. *Research Policy* **18**, 255–271.

Rasmussen, A., 1993. *Fra socialstat til minimalstat.* Samleren Publ., Copenhagen.

Russell, B., 1918. *Proposed Roads to Freedom.* Cornwall Press, New York & Henry Holt, London. Web version at http://www.gutenberg.org, E-text # 690.

Schipper, L., Marie-Lilliu, C., 1999. Carbon Dioxide Emissions From Travel and Freight in IEA Countries: The Recent Past and Long-Term Future. In *Proceedings of a Conference on Policies for Fostering Sustainable Transportation Technologies*, pp. 83–118, US Transportation Research Board, Asilomar, CA (as quoted in IMO. 2000).

Smidt, P., 2007. EU, reformtraktaten og den grinagtige grundlov. *Information* 13th August, p. 3.

Sørensen, B., 2005a. *Another constitution?* Discussion paper at "downloads", http://energy. ruc.dk

Sørensen, B., 2005b. *Hydrogen and Fuel Cells.* Elsevier/Academic Press, Burlington (2nd edition: end of 2011).

Sørensen, B., 2007. Assessing current vehicle performance and simulating the performance of hydrogen and hybrid cars. *Int. Journal of Hydrogen Energy* **32**, 1597–1604.

Sørensen, B., 2010. *Renewable Energy. Physics, Engineering, Environmental Impacts, Economics & Planning.* 4th edition. Academic Press/Elsevier, Burlington.

Sørensen, B., 2011. *Life-Cycle Analysis of Energy Systems: From Methodology to Applications.* RSC (Royal Society of Chemistry) Publishing, Cambridge.

Tomlinson, J., 2001. *Income Insecurity: The Basic Income Alternative,* Ch. 10. E-book at http://pandora.nla.au/tep/42199

United Nations (1948). *The Universal Declaration of Human Rights.* May be found in different languages at http://www.un.org/en/documents/udhr/

US Department of Commerce, 1975. *Historical Statistics of the United States Colonial Times to 1970,* Part 2. US Bureau of the Census and Department of Commerce, Washington DC.

US DoE, 2005. *Transportation Energy Data Book.* 24th edition, Oak Ridge National Laboratory & US Department of Energy, Washington DC.

Vangkilde, J., Seidelin, M., 2010. Farvel til Rønns tipskupon. *Politiken* 15th June, p. 4.

Wikipedia, 2010. *Programme for International Student Assessment (PISA)*. Results and caveats explained, http://en.wikipedia.org

Wikipedia, 2011. *Reinheitsgebot.* http://en.wikipedia.org

Ytzen, F., 2011. I faraonernes dybe kviksand. *Politiken* 18th February, p. 6.

Zakaria, F., 2011. Are America's best days behind us? *Time*, 14th March, 16–21.

# Index